Quantum Communication and Quantum Internet Applications

Quantum Communication and Quantum Internet Applications

Daniel Minoli and Benedict Occhiogrosso

CRC Press
Taylor & Francis Group

Boca Raton and London

Designed cover image: ©shutterstock

First published 2025
by CRC Press
2385 NW Executive Center Drive, Suite 320, Boca Raton FL 33431

and by CRC Press
4 Park Square, Milton Park, Abingdon, Oxon, OX14 4RN

CRC Press is an imprint of Taylor & Francis Group, LLC

Library of Congress Cataloguing-in-Publication Data
A catalogue record for this book is available from the Library of Congress

ISBN: 9781032775548 (hbk)
ISBN: 9781032779669 (pbk)
ISBN: 9781003485612 (ebk)

DOI: 10.1201/9781003485612

Typeset in A Garamond Pro
by DerryField Publishing Services

Trademarks Used in This Book

Contents

List of Figures

List of Tables

Preface

Proponents refer to the emerging quantum internet (Quaint) as being "highly anticipated", and while large-scale/commercial quantum networks do not exist at this juncture, they are expected to enjoy some embryonic deployment by the end of the current decade. Quantum computing, particularly in conjunction with a high level of interconnectivity facilitated by quantum connectivity, can facilitate a host of new, previously unattainable applications—these include blind quantum computation, (information) teleporting, clock synchronization, quantum key distribution, secure access to remote quantum computers, superdense coding, secure identification, distributed sensors, and other scientific and commercial applications.

This book aims at a basic introduction for practitioners to emerging quantum communications foundations and applications—it is not targeted to researchers but telecommunications professionals. Basic quantum mechanics principles are discussed, within a minimalistic, mostly self-contained mathematical framework. A handful of other text sources emerged in the recent past, but our goal is to make the science accessible to the "person of ordinary skill in the art"—a typical telecom practitioner—and to provide an up-to-date description of the discipline. Naturally, a reader should not expect to get a fully inclusive learning of the field, especially the underlying quantum mechanics, from a single text. For the interested reader, this book can serve as a point of departure, coupled with a set of other references that focus on specific aspects of the quantum computing and quantum mechanics discipline.

After an introductory overview chapter, Chapter 2 covers basic concepts in quantum physics by describing some mathematics needed for the discussion, while trying to keep the machinery to a minimum. Chapter 3 continues the discussion by addressing some more advanced mathematics, but with an eye for clarity. Chapter 4 covers basic concepts in quantum physics by describing some physical technologies, while also keeping the science to a sufficient minimum. Chapter 5 provides an overview of quantum computers, being that these nodes constitute the endpoints to be supported by the quantum internet over the horizon. Chapter 6 focuses on quantum communications, including logical architectures and the Quaint. Next, in Chapter 7, we endeavor to provide an accessible discussion of quantum-based security. Chapter 8 offers information on software tools that can be employed to develop networked quantum applications. The glossaries not only define key terms in each of the chapters, but also elaborate on various concepts introduced in the respective chapters and should be read through as part of the instructive process.

This text is not intended as review of current research activities or a complete tutorial on this matter. Furthermore, the authors have not developed any of the science or mathematics of quantum mechanics. Therefore, the background material included in a number of chapters originates from the industry at large, with liberal acknowledgements to the sources of the information. The goal of this text is to synthetize the background material and highlight the applicability to the newly emerging field of quantum communication and quantum internet, in an approachable manner, so that managers and other non-practitioners can get a basic understanding of the discipline as subsetted and applied to the field of remote connectivity and networking.

January 14, 2025
Daniel Minoli
Benedict Occhiogrosso

How to read this textbook

Following are the levels of understanding readers will reach through careful perusal of the various chapters.

Key:

Level	
Basic	▢
Intermediate	▢
Advanced	▢
Implementation	▢

Reader's Goal	Material	Level
Basic, not-too-technical overview of Quantum Communication and the Quantum Internet	Chapter 1	Basic
Intermediate, not-too-technical overview of Quantum Communication and the Quantum Internet	Chapters 1 and 6	Intermediate
Overview of Quantum Communication and the Quantum Internet with an understanding of some of the basic mathematical machinery	Chapters 1, 2, 3, and 6	Implementation
Overview of Quantum Communication and the Quantum Internet with an understanding of some of the basic physics involved	Chapters 1, 2, 4, and 6	Implementation
Basic overview of Quantum Computing	Chapters 1 and 5	Basic
Intermediate overview of Quantum Computing	Chapters 1, 4, and 5	Intermediate
More inclusive overview of Quantum Computing	Chapters 1, 2, 3, 4, and 5	Advanced
Basic understanding of quantum-based security	Chapter 1 and 7	Basic
More inclusive overview of quantum-based security	Chapters 1, 2, 3, 6, and 7	Intermediate
Application of Quantum Communications and Quantum Computing	Chapters 1, 6, and 7	Implementation
Basic development of application of Quantum Communications and Quantum Computing	Chapters 1 and 8	Implementation
More inclusive overview of development of application of Quantum Communications and Quantum Computing	Chapters 1, 2, 3, 6, and 8	Intermediate
A more all-inclusive overview of Quantum Communication and the Quantum Internet	Chapters 1 through 8, along with the Super Glossary	Advanced

About the Authors

Daniel Minoli, Principal Consultant, DVI Communications, New York, graduate of New York University, has published 60 technical books and 350 papers and made 90 conference presentations. Over 8,000 academic researchers cite his work in their own peer-reviewed publications, according to Google Scholar, and over 250 U.S. patents and 40 U.S. patent applications cite his published work. He has many years of technical and managerial experience in planning, designing, deploying, and operating secure IP/IPv6-, VoIP-, telecom-, wireless-, satellite- and video networks for carriers and financial companies. He has published and lectured in the area of M2M/IoT, network security, satellite systems, wireless networks, IP/IPv6/Metro Ethernet and has taught as adjunct for many years at New York University, Stevens Institute of Technology, Rutgers University, and Fairleigh Dickinson University. He also has served as a testifying expert witness in over two dozen cases.

This 2025 book caps 50 years of fundamental, leading-edge telecom-related technology work by Mr. Minoli, including research, design, analysis, engineering, deployment, forecasting, economic evaluation, and patent assessment and development.

Benedict Occhiogrosso is a Co-Founder of DVI Communications, New York. He is a graduate of New York University Polytechnic School of Engineering. For over 25 years, he has served as the CEO of a multi-disciplinary consulting firm and advised both technology producers and consumers on adoption and deployment of technologies. He is also a technology expert in various aspects of telecommunications, security, and information technology with concentration in speech recognition, video surveillance and more recently IoT. He has supported high-tech litigation encompassing Intellectual Property (patent and trade secrets) as a testifying expert witness and advised inventors and acquirers on patent valuation. He has also served as both a technology and business advisor to several start-up and operating companies with respect to product planning, company organization and capital formation.

Chapter 1

Introduction

1.1 Basic Concepts and Terminology

There is an expectation that quantum-based networks/networking (Quanets) will become a reality in the near future. Just as the classical Internet fostered a high-speed hyperconnected ecosystem that enhanced innumerable aspects of human life and global scientific research, a quantum-based ecosystem will greatly enhance the capabilities to address and solve complex problems of all types and will provide an enhanced panoply of tools for researchers in practically every scientific field. Network connectivity refers to the ability to go beyond point-to-point communication and support exchange data among multiple end nodes. Thus, basic desiderata for Quanets entail the support of adequate functionality, multi-endpoint connectivity, and the ability of bridging meaningful geographic distances. Quanets are seen as a disruptive technology, although this technology is still in its infancy. Currently Quanets are perceived as being complementary to classical networks and are not intended to replace them.

Quantum technologies, including networking, make use of quantum mechanics (QM) phenomena and mechanisms such as superposition, interference, entanglement (quantum correlation, coordinated behavior), and measurement—QM is the (current) scientific *model* of the atomic and subatomic environment. QM proceeds by partitioning the environment into two components: (i) the system being studied or modeled, and (ii) the observer. An interaction between the observer and the system under consideration is a *measurement*. Properties of the constituent elements of the system that are measurable (position, momentum, angular momentum, energy, and so on) are known as *observables*. The state of the system is often represented in terms of a wave function. QM principles, when properly applied and exploited, enable one to design quantum-based computing systems that can effectively analyze or simulate multi-dimensional physical phenomena and/or can undertake complex algorithmic computations, enable one to develop sensors that achieve very precise measurements, and enable one to deploy networks that can securely interconnect such quantum devices.

In fact, in recent years the field has seen quantifiable progress in various quantum subsystems such as subsystems for quantum computation, communication, and sensing; the next step, in the evolution toward practical Quanets is the effective integration of such subsystems to create a unified, symbiotic ecosystem. Proponents claim that Quantum-based Communication (Quacom) and Quantum Internet (Quaint) are areas of research and development (R&D) that are gaining increasing attention due to their "potential to revolutionize the way we communicate and exchange information" [IEE24]; some proponents advance the claim that "are currently experiencing a second quantum revolution" [YAM24]. Quacom and Quaint have current applicability in several fields, such as secure communication, distributed computing, and simulation of complex systems, with additional areas of applicability being forthcoming. Commercially deploying and scaling Quanets is a major interdisciplinary effort requiring new, continuous, and practical advancements in physics, computer science, engineering, protocol design, and software development. Nonetheless, while large scale/commercial Quanets do not exist at this juncture, there is an expectation for some embryonic deployment by the end of this decade. There is already quite an extensive set of research literature related to this topic. Short(est)-term Quanet applications include efficient, secure communication and distributed measurements enabled by interconnected quantum sensors.[1]

Quanets enable the transmission quantum bits (qubits). A qubit[2] is a quantum computing (Quacomp) analog of a classical computer system's bit. However, qubits are intrinsically different from classical bits because, while classical bits can take only two values, 0 or 1, qubits can simultaneously be in a superposition mode of being $|0\rangle$ and $|1\rangle$—this duality is in fact called *superposition*. It is not the case that the qubit can have an intermediate value, such as 0.83: when the state of the qubit is measured, the result is either $|0\rangle$, (0), or $|1\rangle$, (1); but in the process of a computation, a qubit can act as if it were a mixture of states—for example: 83 percent in $|0\rangle$ and 17 percent in $|1\rangle$. Qubits thus exist as a simultaneous superposition of $|0\rangle$, $|1\rangle$ (or other comparable) states. Superposition is the QM-intrinsic concept that any linear combination of superposed states is also itself a quantum state. A single qubit is a state represented by a vector in a two-dimensional Hilbert space. A qubit is a fundamental data-storage unit in a Quacomp environment, providing inherent parallelism in support of algorithmic computations.

Consider an analogy: in classical computing bits are either 1 or 0, say the state of a coin that has been flipped and ultimately results in a head (= 0) or tail (= 1), or a computer gate's electrical signal having a high value (switched on, or 1) or no value (switched off, or 0). In this analogy qubits can be pedagogically perceived as a "spinning coin", which carries the possibility of being

---

[1] As noted in the Preface, these authors did not invent any of aspects of quantum mechanics, quantum computing, or quantum communications science; thus, our discussion builds on the exact definitions, terminology, experimentations, results, theorems, and standards developed and presented by the more than 600 references we cite throughout the text, with ample acknowledgments. The goal of this text is to synthetize the background material and highlight the applicability to the newly emerging field of quantum communication and quantum internet, in an approachable manner, to the extent possible.

[2] As noted in [KOZ23], "qubit" can have two related meanings. In the first meaning, "qubit" refers to a physical quantum *system* whose quantum state can be expressed as a superposition of basis states; here, "qubit" refers to a *physical implementation*, a circuit, typically in a cryogenic enclosure/environment. In the *second* meaning, "qubit" refers to the abstract quantum *state* of a quantum system with such basis states; in this case, the meaning of "qubit" is comparable to the logical value of a bit, from classical computing—i.e., "logical 0" or "logical 1". The two concepts are related, because a physical "qubit" (*first* meaning) can be used to store the abstract "qubit" (*second* meaning). Both meanings are used interchangeably in the literature, and the specific meaning is generally clear from the context.

either or both heads and tails; once the qubits are measured, they acquire a 1 or a 0 state. A qubit is a two-state QM system. With each additional qubit, the number of states that can be represented by the qubits doubles.

In addition, qubits can be entangled with each other, fostering correlations over distances that are stronger than are achievable with classical information [SCH95]. Namely, the measurement outcome of one part of the system predicts a measurement outcome of the other part of the system. Entanglement is a form of correlation between quantum particles: the state of a qubit can influence the state of another qubit. Thus, entanglement describes a "non-classical" relationship between quantum systems: when one element of an entangled quantum system is measured, the impact propagates instantaneously amongst the other elements of that system; for example, when two electrons are, or become, entangled, they no longer exhibit independent behaviors, and if one of the entangled electrons is measured and is found to have a clockwise spin, the second electron will have an anti-clockwise spin. One can envision the concept of entanglement contemplating a (simple, naïve) illustration of flipping two coins, in physical proximity or at any distance, where both coins together always land on "heads" or both land on "tails", as if they were correlated by some invisible coordinating, matching, or joining phenomenon. Entanglement is the phenomenon of quantum correlation between two qubits.

Entanglement makes possible the exploitation of non-classical correlations for the support of new types of application methods that are not achievable with classical communication. Quantum entanglement is an intrinsic process that is fundamentally required to support Quacom, Quanets, and Quacomp: in Quacom remote parties are connected by sharing long-lived entangled states, and any quantum-based applications (e.g., superdense coding, teleportation, quantum metrology) are based on entanglement correlations. The creation, modification, control, and practical application of entanglement—quantum correlation—not only have become burgeoning research fields in the past decades but also have become areas of engineering development and implementation, particularly when it comes to the ability to distribute entanglement between geographically separated endsystems.

A qubit can be perceived as a quantum system with an associated quantum state that may be used to encode information. In QM a "state" is synonymous with a state vector (or a function). The quantum state of an atom or particle is a description of its physical and electromagnetic properties such as its energy, location, motion, magnetic field, orientation (spin), and possibly other properties. Another way of describing a quantum state is that it is a mathematical representation of a QM system that allows one to establish certain properties of the system that can be observed—for example its position, energy, or momentum; however, due to the uncertainty principle, these properties cannot be determined with definitive precision but may be described probabilistically with probabilistic tools.

QM allows for phenomena that are not observed at the macroscopic level and are perceived by the novice to be obscure, abstract, or unintuitive in various instances, at least at a first pass. Qubits can be prepared in linear superpositions of the values 0 and 1, and multiple quantum bits can be entangled. Quantum entanglement (which Einstein characterized as "spooky action at a distance") is a phenomenon in which the quantum state of two or more quantum systems cannot be described separately, even when they are separated by a large distance; the systems are "paired together", they are "correlated".

Quantum entanglement has numerous applications in computing, sensing, and communications applications. Entanglement allows two qubits q_1 and q_2 to become firmly interlinked, regardless of how much space exists between them: one can say that entangled qubits "mirror" one another, each qubit being correlated with the measurement of the other qubit, and if one qubit, say q_1, is switched to a 0, its entangled partner qubit q_2 will also be switched, but to the opposite state. This phenomenon is utilized to forward quantum information securely—a process known as *teleportation*. Teleportation is a transfer of "quantum states" between separate atoms (or particles)—transferring key properties of one atom (or particle) to another atom (or particle). Quantum entanglement has been observed between stable particles, such as photons or electrons; more recently, physicists have demonstrated quantum entanglement in top quarks and their antimatter partners (observing spin correlation between the particles) [AND24]. Quantum teleportation employs quantum entanglement.

In a QM system, observation is synonymous with measurement; however, as discussed later on, the measurement of a quantum resource collapses the associated quantum state or states in an irreversible manner. Upon measurement, a qubit loses its superposition status and collapses into one of the two basis states, $|0\rangle$ or $|1\rangle$; in which of the two states the qubit ends up can be established from the readout of the measurement: the measurement result is a classical bit, 0 or 1, corresponding to $|0\rangle$ and $|1\rangle$, respectively. A priori, the result is not deterministic: the probability of measuring the state in the $|0\rangle$ state is $|\alpha|^2$, and the probability of measuring the state in the $|1\rangle$ state is $|\beta|^2$.

Quantum algorithms instantiate a sequence of interference patterns that yield the correct answer to a problem under study. Interference relates to manipulating qubits in such a manner that their states combine constructively during computations in order to amplify correct solutions and combine destructively to attenuate or suppress incorrect solutions. Constructive interference occurs when the peaks of two waves combine to yield a higher peak; destructive interference occurs when a wave peak and a wave trough combine and cancel each other out.

This chapter starts the discussion of this text by defining some quantum-related terminology and gives a preview of some key technology concepts that are discussed in greater detail in the chapters that follow. Also see Table 1.1. The International Telecommunication Union (ITU) uses the following terminology [ITU22]: (i) Quantum Information Technology (QIT): a class of emerging technologies that improves information processing capability by harnessing principles of QM; (ii) Quantum Information Network (QIN): any network that incorporates quantum communication technologies for the purpose of transporting quantum states—this text follows the nomenclature of Table 1.1.

Table 1.1   Basic Nomenclature

Term	Description	Term	Description
QC	Quantum computer	Quaintco	Quantum internet computing
QM	Quantum mechanics	Quanet	Quantum-based network/networking
Quaco	Quantum control	Quarep (aka QR)	Quantum repeater
Quacom	Quantum-based communication	Qubits	Quantum bits
Quacomp	Quantum computing	QuICs	Quantum interconnects
Quaint	Quantum internet	QIST	Quantum Information Science and Technology

1.2 Introductory Overview

Advances in quantum theory and QM over the years have led to development of technological devices such transistors, lasers, superconducting materials, and nanotechnology materials. Quantum information science (QIS) is a field that combines the principles of QM with ICT (information and communication technologies) science—a "marriage" of the two—to assess and enhance the processing, analysis, and remote transmission of information; some researchers call this area Quantum Information Science and Technology (QIST). Quantum information science (or "theory") has emerged in the past quarter century, aiming to exploit QM for communication and computation algorithms; thus, QM has become a tool for extracting information from matter [BEN98]. QIST encompasses three subdisciplines: Quacomp, Quacom, and Quantum Control (Quaco); in turn, each of these entails a number of subdisciplines. Ultimately, many QIST applications require communications between quantum nodes, and Quanets support this functionality.

Quecomp utilizes the principles of the science of QM to undertake computing functions. Quantum computers (QCs) are seen as offering improved solutions to many complex problems in the sciences as well as in artificial intelligence. Applications include but are not limited to logistics, drug development, scheduling, financial modeling, and materials sciences. QCs can solve certain computational problems much faster than can conventional computers. A QC is a device that takes advantage of QM effects to perform certain computations faster than a classical (Turing-type) computer is able to; they typically comprise a group of qubits that can perform quantum operations on data [ABE23], [BAU20], [DIA22], [HUA22], [LIU21], [MIN24], [NIE10].

Quanets utilize Quantum Interconnects (QuICs). QuICs are devices or processes that allow the transfer of quantum states between two specified physical degrees of freedom (e.g., material, electromagnetic [EM]), or, more broadly, connect a quantum system with a classical one. QuICs are fundamental elements of all QIST systems, including quantum computing, quantum sensing, and quantum communication [AWS21]. A Quanet uses photons—the smallest possible quantum of light—to transfer information: they utilize the quantum properties of photons to encode information, thus allowing the quantum state of photons to carry information from sender to receiver through a communications infrastructure. Other atomic particles (e.g., bosons, electrons) can also be used. A generic view of a (future) Quanet is shown in Figure 1.1.

A key fact is that qubits cannot be copied, and any efforts to do so can be detected; these conditions complicate the transmission of qubits and require new technological mechanisms and implementation. Quantum repeaters (Quareps) are needed to deploy the Quanets given the fact, as just noted, that qubits cannot be copied[1] (cloned) or amplified; thus, these devices are utilized to transmit quantum information over non-trivial distances—traditional repetition or signal amplification cannot be utilized in a Quanet setting for transmission of qubits [BRI98], [PIR17]. A Quarep is a device that can relay an entangled state from one set of qubits to a distant set without physically sending an entangled qubit over the entire distance. The quantum state of each entangled photon is correlated with that of its entangled partners, regardless of their distance apart. Entanglement implies that changing the spin or other state of one photon also changes the paired photon, regardless of how far apart they are; this linkage is what makes Quacom possible.

Quanets will be separate, but co-existent, networks alongside classical networks such as the Internet; they may use photons carrying quantum information being transmitted in classical

---

[1] Some refer to this predicament as the "no-cloning theorem".

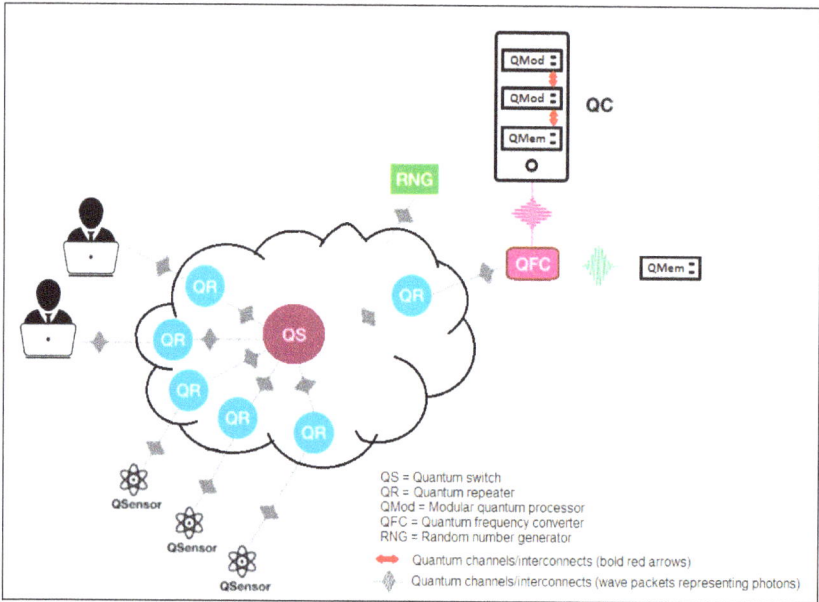

QMod
QMod
QMem
O
QC

RNG

QFC

QMem

QS = Quantum switch
QR = Quantum repeater
QMod = Modular quantum processor
QFC = Quantum frequency converter
RNG = Random number generator

Quantum channels/interconnects (bold red arrows)
Quantum channels/interconnects (wave packets representing photons)

QSensor

Figure 1.1 Generic view of a (future) Quanet (inspired by [AWS21])

fiberoptic links (classical networks) at telecom wavelengths. Current-generation Quanets are based on the use of Bell pairs and entanglement swapping in the subtending Quareps. Bell pairs express the quantum states of two qubits that represent the simplest examples of entanglement, as elaborated in Chapters 2 and 3. Some see a progression of Quacom development over the foreseeable time horizon along the following lines [LAC23]:

- **Stage 1:** Point-to-point. A network built with direct point-to-point connections, where entanglement can be established between two separate quantum devices—comparable logically to the physical layer of the Open Systems Interconnection Reference Model (OSIRM).
- **Stage 2:** Point-to-multipoint. At this stage, a quantum device can support and manage connections with many sites, delivering entanglement to any two as required—comparable logically to the data link layer of the OSIRM.
- **Stage 3:** Multipoint-to-multipoint. Reliable long-distance quantum communication through a complex network, which relies on resilient quantum hardware at the sites—comparable logically to the network layer of the OSIRM.

Quacom and Quanets, which as noted enable one to transmit qubits, can be leveraged for a variety of ICT purposes, including cryptography management and support of large-scale distributed computing. Small-scale Quanets have been used to demonstrate network protocols such as non-local quantum gates, entanglement purification (also known as distillation), and entanglement swapping—these networks are currently limited to a few nodes, distances of up to one mile or so, and entanglement generation rates in the Hz to kHz range, but enhancements are on the horizon [RUF21].

At present three classes of Quacom applications are within usability scope, while others are expected to emerge in the near future: (1) quantum cryptography, where being able to exchange quantum information provides novel methods for trusted communication between remote parties while monitoring a communication channel for eavesdropping; (2) distributed generic environment monitoring utilizing shared quantum resources, where quantum sensors measure data simultaneously in dispersed physical locations; and (3) interconnecting QCs into clusters where such clusters synergistically enhance the aggregated computing power. As implied in the list just cited, currently the best-developed application of Quacom is Quantum Key Distribution (QKD), which enables two remote nodes to establish an encryption key where security relies only on the laws of QM; it is a secure communication method that implements a cryptographic QM protocol for distributing cryptographic keys, enabling the two parties to generate a shared random secret key or cryptographic key that is by design known only to them; the key is then used to encrypt and decrypt messages in communications between them (or more precisely, their endpoint devices). Long-distance QKD has already been realized over hundreds of miles in free space transmission and over fiberoptic links.

QM is more than a century old. In 1900, Planck established that one can explain the spectrum of blackbody radiation by introducing a new physical constant with the dimensions of angular momentum—the Planck constant; he also proposed that energy is quantized, not contiguous. The incremental development of wave mechanics spearheaded by Bohr in the next few years saw rapid progress that led to De Broglie's work asserting that particles were also waves. Schrödinger introduced the concept of probabilities into the energy states of particles, electrons in particular. By 1927, QM theory was mostly in place and was fortified by Dirac's seminal publications in 1930, leading to a comprehensive theory of atoms' behavior and their electromagnetic interactions. Dirac also introduced what is now a well-accepted notation for QM states. The principle of wave-particle duality is intrinsic to the modern understanding of the EM field. Fundamental to the particle side of the duality is the concept of *quantization,* which postulates that the EM field is composed of discrete packets of energy (photons). Additional work has continued to establish new results, perspectives, refinements, and revisions in the intervening decades.

QM is a mathematical construct—a model, a theory—for predicting the behaviors of microscopic particles. QM describes a set of facts about the way the microscopic world influences the macroscopic one, how it affects measuring instruments, described in the language of classical mechanics. In classical mechanics, a particle has an exact momentum and an exact position at all times. In QM, observables such as momentum and position are not actual numbers but operators; it follows that there are uncertainty relations—e.g., $\Delta x \Delta p \geq \hbar, \hbar$, the reduced Planck constant—which implies that as some observables become more precisely (accurately and exactly) defined, other observables become more uncertain[1]. In particular, the state of a quantum system is given by a wave function $\psi(x)$. The wave function is complex and gives the amplitude for finding the particle at position x. Predictions in QM are *probabilistic*. The probability density is the square modulus of the amplitude; thus in one dimension the probability to find the particle located between x and $x + dx$ is $P(x)dx = |\psi(x)|^2 dx$. A way to grasp this is as follow: if one starts out with a group of identical copies of the system, all with the same wave function $\psi(x)$, and in each copy of the system one measures the position of the particle, then one will obtain different results from different members of the group [ALF08]. The probability of getting particular answers is $P(x)dx = |\psi(x)|^2 dx$. Naturally,

---

[1] The formal inequality relating the standard deviation of position σ_x and the standard deviation of momentum σ_p is $\sigma_x \sigma_p \geq \hbar/2$ where $\hbar$ is the reduced Planck's constant $\hbar = h/2\pi$; this is known as the *uncertainty principle,* also known as *Heisenberg's indeterminacy principle.*

$$\int |\psi(x)|^2 dx = 1$$

Each observable corresponds to a linear operator. A (linear) operator is a transformation/mapping that acts on a state and generates another state—that is, it changes one function into another.

An interpretation of the theory provides an account of what the world is like according to QM, intrinsically and from the bottom up. QM is anchored on the Hilbert spaces that represent the state-spaces of quantum systems. A quantum state ψ of a particle such as a photon is a vector in a vector in complex-valued vector space known as Hilbert space[1] and can be expressed or perceived as a kind of projection (or factorization) onto a set of basis vectors (or functions) that span that space. QM is the modern theory of one particle (or, perhaps, very few particles); Quantum Field Theory (QFT) as an extension of QM for the analysis of systems with many particles—and therefore with a large number of degrees of freedom [ISM21], [STA06]. Although the mathematics of QM can be overwhelming, this self-contained text aims at presenting sufficient basics in the simplest possible form to provide an introductory description of the field, particularly in the context of applications and applicability.

In a Quacom system, information is transferred between a transmitter and a receiver by encoded single quanta, such as single photons: each photon carries one bit of information that can be encoded upon a property of the photon, such as its polarization[2] [WOO23]. Early work on Quacom was undertaken by researchers such as, but not limited to, Wiesner, Bennett, and Brassard (e.g., see [WIE83]). Quacom deals with quantum-based communication between two (or more) endpoints while Quanet deals with a more comprehensive any-to-any quantum-based communication. Quacom entails the (long-distance) exchange of quantum information or the sharing of entangled quantum state between two or more parties (entanglement relates to the concept that the quantum state of each particle in the group cannot be described independently of the state of the other particles). Quantum teleportation is the foundation of quantum networking. Among other phenomena, quantum entanglement allows the generation of shared multiparty secrets, distributed Quecomp, and advanced sensing.

1.3 A First View of a Qubit

The analogue of the classical bit is the qubit, cited earlier, which embodies quantum information. In the past 75 years bits became the universal language of information, transcending physical modality; now, the world is at the inflection point of the next information revolution, as quantum bits—qubits—have become the universal units fueling a new generation of "quantum information technologies". A qubit is implemented in hardware by nanoscale physical components with QM characteristics that are intrinsically associated with such nanoscale subsystems.

---

[1]   Hilbert space is a complete, complex vector space with basis, equipped with an inner product operation which allows distances and angles to be defined. The concepts are elaborated in Chapters 2 and 3.

[2]   There is a qubit associated with photon, captured by its polarization: a photon traveling along a trajectory such as the z-axis in 3-D has an associated electric field in the x-y plane, and the 2-D quantity specifying this field represents the polarization of the photon.

An example of physical qubits is depicted in Figure 1.2, which includes four physical qubits. These physical components can have a state $|\psi\rangle$ that can take on a large (even infinite) number of states instantiations. When the state of a qubit is physically measured, the measurement produces one of two different basis[1] states—for example, basis states $|0\rangle$ and $|1\rangle$, being that the QM state of a qubit is a state vector captured by a superposition of the two states $|0\rangle$ and $|1\rangle$ as $|\psi\rangle = \alpha|0\rangle + \beta|1\rangle$.

Thus, a qubit has two states and may be perceived as a weighted superposition of these two states; this enables certain functions to be evaluated for both values simultaneously. A qubit represents information in a (true) two-level system, such as the spin of an electron or the polarization of a photon or the direction of current in a superconducting circuit; it can also represent information in a pseudo-two-level system, such as two energy levels of an atom. A qubit serves the same

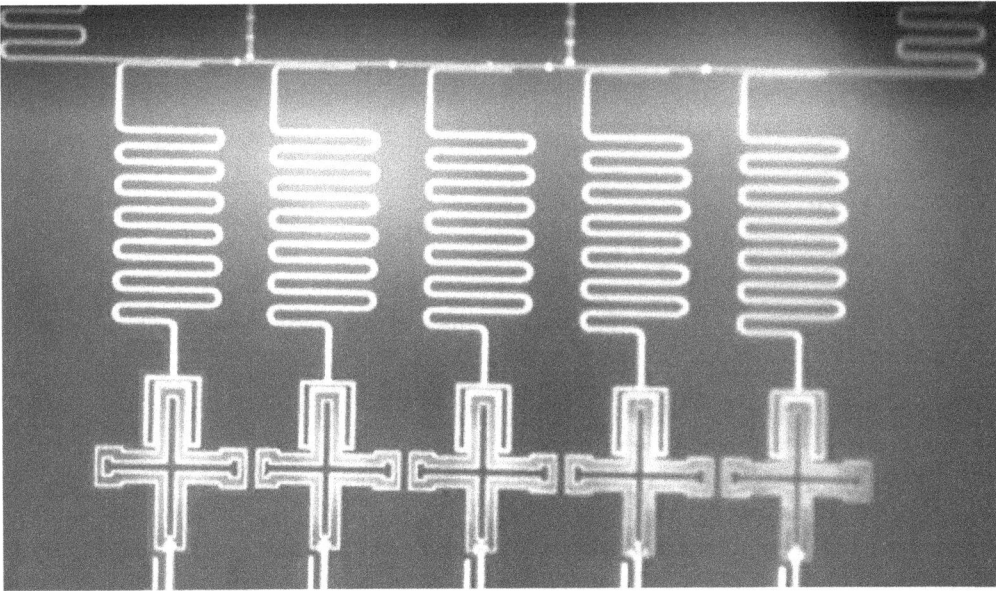

Figure 1.2   Example of a physical qubit implementation in hardware (Courtesy Google Quantum AI)

---

[1] A basis is a set of linearly independent (orthogonal) vectors that can be employed to represent any vector in a given vector space. The basis of a vector space provides a coordinate system that enables one to represent vectors utilizing numerical coordinates. A vector space is a mathematical structure of a set of constituent elements in that space, constituent vectors where the operation of addition and scalar multiplication is supported. A set of vectors and operations that are defined on those vectors constitute the mathematical structure called a vector space. In a vector space, if a (small) set of vectors can be used to express every vector in the space as a unique linear combination of those vectors, and those vectors are linearly independent (meaning that none of them can be expressed as a linear combination of the others), then they act as basis vectors for that vector space. Example: in the familiar two-dimensional plane, say $b_1 = [1,0]$ and $b_2 = [0,1]$, then any vectors (points) in two dimensions can be written as $a \times b_1 + b \times b_2$; for example, the vector $v = 3x + 2y$ is expressible as $v = 3 \times b_1 + 2 \times b_2$.

purpose as a traditional bit in classical computing. The superposition principle of QM implies that it only requires a linearly growing number of qubits to encode information, while it is exponentially costly to encode the equivalent information on conventional computer or memory (when multiple qubits are combined in a single system, the number of *possible QM states* that coexist in the superposition grows exponentially with the linear number of qubits) [RUF21], [NIE10], [SIL08]. Qubits and qudits may be implemented in a variety of physical quantum systems.

- As noted, a qubit embodies the superposition of the two states. The state of a qubit is a vector in a two-dimensional complex vector space. Using $|0\rangle$ and $|1\rangle$ as *computational basis states*— namely, an orthonormal basis for this vector space[1], the *state* $|\psi\rangle$ of a qubit can be written as a linear combination:

$$|\psi\rangle = \alpha|0\rangle + \beta|1\rangle$$

where α and β are complex numbers with $|\alpha|^2 + |\beta|^2 = 1$, and

$$|0\rangle = \begin{bmatrix} 1 \\ 0 \end{bmatrix} \text{ and } |1\rangle = \begin{bmatrix} 0 \\ 1 \end{bmatrix}$$

so that

$$|\psi\rangle = \begin{bmatrix} \alpha \\ \beta \end{bmatrix} = \alpha|0\rangle + \beta|1\rangle = \alpha\begin{bmatrix} 1 \\ 0 \end{bmatrix} + \beta\begin{bmatrix} 0 \\ 1 \end{bmatrix}.$$

$|0\rangle$ and $|1\rangle$ are orthonormal representations of eigenstates of a physical system. A more specific representation is

$$|\psi\rangle = \{\cos(\theta/2)\}|0\rangle + \{e^{i\phi}\sin(\theta/2)\}|1\rangle$$

where numbers θ and ϕ are angular displacements; $|\psi\rangle$ then represents a point on a three-dimensional unit sphere (this sphere being known as Bloch Sphere, which utilizes polar coordinates—see Figure 1.3).

Thus, the state of a qubit can be perceived as a two-dimensional vector with basis $|0\rangle$ and $|1\rangle$ and value $|\psi\rangle = \cos(\theta/2)|0\rangle + e^{i\phi}\sin(\theta/2)|1\rangle$. A related set of two or more qubits is called a *quantum register*. At the practical level, mathematically-defined qubits are realizable in physical hardware qubits that can be implemented utilizing a variety of physical implementations, including, but not limited to, trapped ions, optical cavities, elementary particles, molecules, or groups of molecules[2].

An inner product space is a vector space with an associated operation, a scalar product, which is the inner product; inner product spaces generalize Euclidean vector spaces; the inner product supports formal definitions of basic geometric concepts, such as lengths, angles, and orthogonality. In the Dirac notation, a state or vector $|\psi\rangle$ is referred to as the "ket" notation for vectors of an inner product vector space; the inner product vector space also has a *dual vector,* represented by the Dirac "bra" notation $\langle\psi|$ (kets are represented by columns vectors and bras by row vectors). See

---

[1]   Mathematical concepts are introduced and discussed in more detail in Chapter 2 and Chapter 3.

[2]   For example, qubits in nuclear magnetic resonance (NMR) computing utilize the spin states of nuclei within molecules; NMR uses an ensemble of systems (molecules) instead of a single pure state; quantum states are measured by exploiting nuclear magnetic resonances (e.g., implemented as a variation of nuclear magnetic resonance spectroscopy).

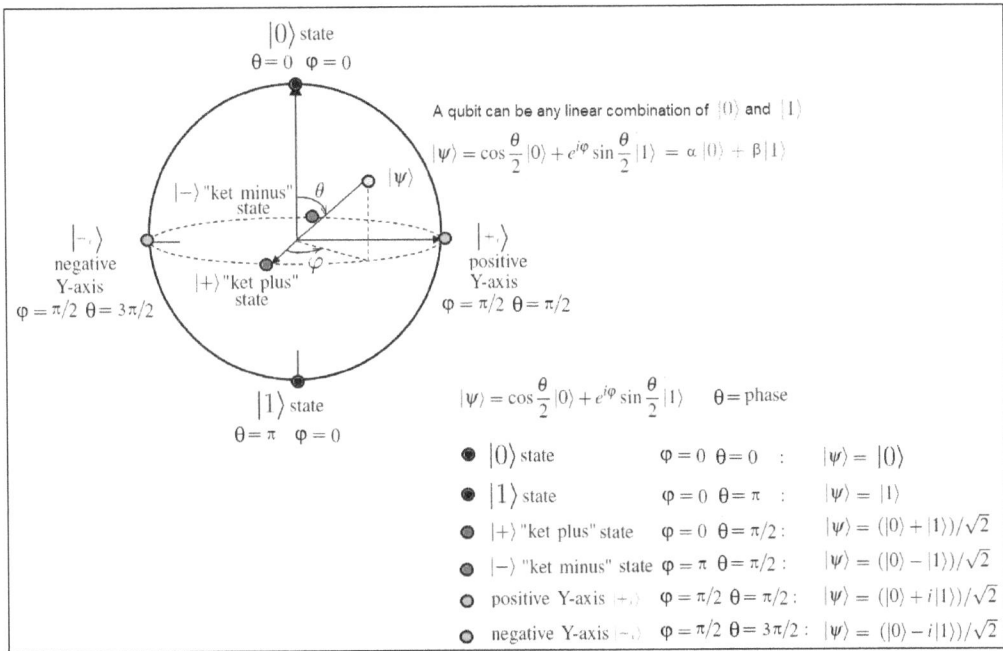

Figure 1.3   Qubit representation via the Bloch Sphere, where any point on the surface of the unit sphere represents a possible qubit state (vector) $|\psi\rangle$

Figure 1.4 (on next page) for concepts and notation. The magnitude (norm) of a qubit state vector is calculated as the square root of the inner product of the state vector and its corresponding dual state vector. State vectors describing qubit states are normalized to a magnitude of 1. A complex row vector of size 2×1 multiplied by a complex column vector of size 2×1 (e.g., the inner product of the state vector $|\psi\rangle$) produces a real number (a 1×1 number/result), while a column vector of size 2×1 multiplied by a row vector of size 2×1 produces a complex-valued 2×2 matrix/result.

In QM terms, one is only able to acquire restricted information about the quantum state via a measurement or sampling operation. The QM state $|\psi\rangle$ of a qubit is not directly observable by measurement, but physical measurement operations are performed on a qubit to produce a measured state in a computational basis, such as the basis $|0\rangle$ or $|1\rangle$. More specifically, there is a probability distribution capturing the value of the states, such that when one endeavors to "measure" a qubit, one obtains the result value 0, with probability $|\alpha|^2$, and/or the result 1, with probability $|\beta|^2$. The measurement operations can be represented as 2×2 matrices M_0 and M_1, which cause projections of $|\psi\rangle$ onto its basis $|0\rangle$ or $|1\rangle$:

$$M_0 = |0\rangle\langle 0| = \begin{bmatrix} 1 & 0 \\ 0 & 0 \end{bmatrix},$$

$$M_1 = |1\rangle\langle 1| = \begin{bmatrix} 0 & 0 \\ 0 & 1 \end{bmatrix}.$$

The squares of the magnitudes of the coefficients α and β in the expression $|\psi\rangle = \alpha|0\rangle + \beta|1\rangle$ correspond to the probabilities that a measurement of a qubit in state $|\psi\rangle$ will produce states $|0\rangle$ and $|1\rangle$, respectively: the probability that a measurement of a qubit in state $|\psi\rangle$ will return a state $|0\rangle$ or $|1\rangle$ is given by the inner products:

State vector, "ket"

$$|\psi\rangle = \begin{bmatrix} \alpha \\ \beta \end{bmatrix} = \alpha|0\rangle + \beta|1\rangle = \alpha\begin{bmatrix} 1 \\ 0 \end{bmatrix} + \beta\begin{bmatrix} 0 \\ 1 \end{bmatrix}$$

Dual vectors, "bra":

$$\langle\psi| = [\alpha^*, \beta^*] \quad \alpha^*, \beta^* \text{ being the complex conjugates of } \alpha, \beta$$

$$\langle 0| = [1, 0]$$

$$\langle 1| = [0, 1]$$

Magnitude (norm) of a vector is the square root of the inner product of the vector with its dual vector

$$\||\psi\rangle\| = \|\langle\psi|\| = \sqrt{\langle|\psi\rangle\,|\,\langle\psi|\rangle} = \sqrt{\langle\langle\psi|\,|\,|\psi\rangle\rangle}$$

$$= \sqrt{\langle\,\psi\,|\,\psi\,\rangle} \quad \text{(simplified notation)}$$

$$= \sqrt{\alpha\alpha^* + \beta\beta^*}$$

$$= 1 \quad \text{(the norm is 1)}$$

The norms of $|0\rangle$ and $|1\rangle$ are also 1 and the basis is called orthonormal

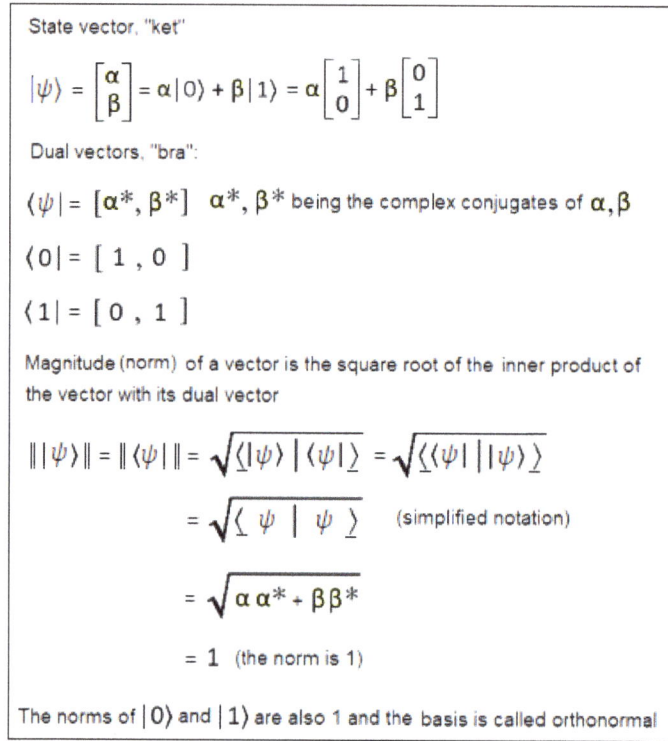

Figure 1.4   Basic bra/ket notation for states

$$p_0 = \langle\psi\,|\,M_0^\dagger M_0\,|\psi\rangle = [\alpha^*, \beta^*]\begin{bmatrix} 1 & 0 \\ 0 & 0 \end{bmatrix}\begin{bmatrix} \alpha \\ \beta \end{bmatrix} = \alpha\alpha^* = |\alpha|^2.$$

$$p_1 = \langle\psi\,|\,M_1^\dagger M_1\,|\psi\rangle = [\alpha^*, \beta^*]\begin{bmatrix} 0 & 0 \\ 0 & 1 \end{bmatrix}\begin{bmatrix} \alpha \\ \beta \end{bmatrix} = \beta\beta^* = |\beta|^2.$$

($M_0^\dagger$ being the conjugate transpose of M_0 and $M_1^\dagger$ being the conjugate transpose of M_1).

To further highlight the nomenclature and the concepts, if one lets $\alpha = v_0$ and $\beta = v_1$, then the vector representation of a single qubit is $|a\rangle = v_0|0\rangle + v_1|1\rangle$, also representable as the vector

$$\begin{bmatrix} v_0 \\ v_1 \end{bmatrix}$$

where v_0 and v_1 are the complex probability amplitudes of the qubit and establish the probability of measuring a 0 or a 1 when measuring the state of the qubit. When $v_0 = 1$ and $v_1 = 0$, one obtains the state $|0\rangle$, and clearly (consistent with the earlier definition)

$$|0\rangle = \begin{bmatrix} 1 \\ 0 \end{bmatrix}.$$

When $v_0 = 0$ and $v_1 = 1$, one obtains the state $|1\rangle$ and clearly (consistent with the earlier definition)

$$|1\rangle = \begin{bmatrix} 0 \\ 1 \end{bmatrix}.$$

Single qubit states that are not entangled and lack global phase can be represented as points on the surface of the Bloch sphere.

The tensor product,[1] represented as $\otimes$, is used to combine quantum states. The combined state for a qubit register is the tensor product of the constituent qubits. The vector representation of two qubits $|ab\rangle$ is (this tensor product being a simple equation in this instance)

$$|ab\rangle = |a\rangle \otimes |b\rangle = v_{00}|00\rangle + v_{01}|01\rangle + v_{10}|10\rangle + v_{11}|11\rangle,$$

also representable as the vector

$$\begin{bmatrix} v_{00} \\ v_{01} \\ v_{10} \\ v_{11} \end{bmatrix}$$

Each state represents a possible outcome of a measurement of the two-qubit system; for example, $|10\rangle$ denotes a state in which the first qubit is in the state $|1\rangle$ and the second is in the state $|0\rangle$.

The action of the gate (operator) U on a specific quantum state is found by multiplying the vector $|\psi_1\rangle$ by the matrix representing the gate; the result is a new quantum state $|\psi_2\rangle$ where $|\psi_1\rangle = U |\psi_1\rangle$. *Quacomp* and *Quacom* applications are realized by applying a series of quantum gates to one or several qubits and then performing a measurement of the qubits to read out information in the quantum states [VAR22]. A set of qubits obeys the principles of *QM*, similar to the way an electronic wave function does. Because a qubit exists in a superposition of its basis states, elementary quantum gates act on all states of the superposition simultaneously. For example, the action of a NOT gate is:

$$\text{NOT} (\alpha|0\rangle + \beta|1\rangle) \rightarrow \alpha|1\rangle + \beta|0\rangle$$

In QM, quantum entanglement arises when a group of particles is generated, interacts, or shares spatial proximity in such a manner that the quantum state of each particle in the group cannot be described independently of the state of the other particles even when the particles are at a distance. The entanglement phenomenon is unique to QM and is a foreign concept in classical mechanics. Distributing entanglement over meaningful, long distances is a fundamental prerequisite for the application of quantum protocols to real-world practical communication problems.

Engineering challenges to widespread adoption of Quacomp and Quacom include the fact that physical qubits (for example atoms, ions, or photons), are intrinsically unstable and prone to collapsing out of their quantum states, giving rise to errors. Quantum Error Correction (QEC) algorithms aim at mitigating the effects of noise by encoding single-qubit information

---

[1] The tensor product is an operation (but not the usual matrix multiplication) on two matrices that results in a block matrix. The tensor product is the mathematical mechanism for combining quantum states together. These concepts are discussed in more detail in Chapter 2.

into a larger quantum system—these error-correcting algorithms are needed in Quacomp/Quacom, but they remain complex to implement. Quantum Bit Error Rate (QBER) measures are of interest (e.g., see [AHN23] for an example).

1.4 Quantum Elements in Quanets

Typical quantum elements in a Quanet, which can be characterized as *interconnects*, or even as Quantum Network Elements, include the following [AWS21]:

- *Communication channels* between two quantum systems (including but not limited to optical channels, acoustic channels, free space channels); the systems can be on the same chip or separated by a significant distance. Examples include an optical cavity, a waveguide, a fiber connecting two quantum emitters, or a cold microwave waveguide connecting two superconducting-qubit processors.
- *Quantum repeaters* capable of relaying (but not amplifying) quantum information signals.
- *Quantum memories* and the associated interface to the communication channel; a memory can, for example, be a color center in diamond, trapped ion node, all-photonic cluster-state-based.
- *Quantum transducers* used to connect qubits of different kinds (for example, acousto-optical, spin-photon, spin-phonon), or of the same kind but at different energy (microwave-optical photon, visible-telecom photon).
- *Converters* between different qubit-encoding schemes or degrees of freedom (for example, polarization, temporal, spectral encodings of photons).
- *Small-scale and application-specific quantum computers* (for example, a quantum repeater), to extend the reach of quantum communication channels.
- *Entanglement sources*, these being physical processes that create quantum-entangled states.

At a coarser level of granularity, a Quanet consists of three high-level hardware elements implementing quantum principles:

- End nodes such as QCs and other quantum processors. End nodes may either be simple nodes that can only prepare and measure single qubits or be high-end QCs.
- Quantum channel: a physical medium, including EM spectrum, that enables the transmission of qubits. Traditional optical fibers can be used by transmitting light pulses (photons). An engineering fact is that quantum channels, like other telecommunications links, are lossy.
- Intermediate nodes, Quareps, are necessary to extend these short distances. Such a repeater is placed at appropriate intervals along the optical fiber connection, in theory allowing qubits to be transmitted over desired long distances. (In the future, repeaters may also function as long-distance routers in a Quanet; some end nodes may also act as Quareps).

It is well recognized by practitioners [ALS21] that useful Quanets must consist of spatially separated nodes with independent, heterogeneous quantum resources (stationary qubits, detectors, photon sources, and so on) synchronized to a common clock; the network architecture must be compatible with other networking topologies to enable the establishment of larger networks. A Quanet also needs classical network capabilities in the form of a control plane for management and a parallel data plane for classical communications between the network nodes.

A technical and deployment evolution is expected in the next few years in reference to the Quanet: initially, simple nodes are interconnected utilizing basic quantum repeaters to enable a certain, but limited, distance. Further along a more comprehensive network can be deployed. Recent technical advancements (including, but certainly not limited to, [OUR23], [DIB18],

[AWS18]) support the hypothesis that small-scale implementations of Quanets may emerge by the end of this decade.

A basic requirement in both Quacomp and quantum cryptography is the need to transmit information from one point to another in a reliable manner. As already implied, Quacom is the methodology employed to transmit quantum signals (or even classical signals) across significant distances utilizing the principles of QM; it is obvious that a fundamental step to allow QM to play a direct role in Quacom is the ability to connect quantum devices over long distances. Quantum channels are used to transmit or store quantum information. Quantum channels may intuitively be perceived as a "pipe" with which one transmits, say, a spin-half particle, thus conveying a single qubit. However, even small magnetic fields inside the channel may perturb the information carrier, producing noise; or, in an optical fiber, a single photon "carries" one qubit in via its polarization, but inhomogeneities in the fiber may disturb the polarization—or the photon may be consumed and disappear through some absorption process; both of these situations introduce undesired noise. For example, photons at microwave frequency levels, such as photons that can be produced in a QC while in the very cold environment of the cryostat chamber, typically decohere rapidly when placed in a room-temperature environment, and thermal or background noise associated with the transmission of these photons at such microwave frequency levels can be undesirably high.

For classical channels there are well-understood ways of overcoming noise in a channel through error correction; in principle, this is also possible in quantum channels, although it is more difficult [GRI12]. As is typically the case in traditional communications, a basic engineering challenge of Quacom systems is the transmission of quantum information at high rates over practical distances in the presence of losses and noise; to address this challenge, one needs to deploy Quareps. While classical data signals can be amplified across any pertinent distance, quantum signals cannot. The transmission of quantum signals entails the creation of an entangled state (or the use of a previously established entangled state) and the transfer of the (or a) quantum state (whether it is entangled or not).

Quacom allows the endpoints to generate long-distance quantum entanglement, which serves a number of information conveyance tasks. Quantum signals must be "repeated" in intervals—that is, processed and passed on by specialized repeaters [LYO23], [OUR23], [ABR13], [ABR13a]. To develop Quareps, the designer needs to balance local computation and networked functions (entanglement generation). Quarep technologies, in particular the ones built using quantum memories, has been a subject of major R&D work in recent years; however, an operational demonstration of a quantum repeater has proven challenging. Some repeater protocols without quantum memory have been proposed of late; however, their implementation is also still challenging [TAK14].

Light clearly plays an important role in many of the QM phenomena, due to its high speed and relatively inert interactions with its immediate environment. For example, light can be utilized to transmit quantum information when such information is encoded in the quantum state of the light. Light at its most basic level is composed of photons—quantized wave-packets of electromagnetic (EM) energy. There are several quantum states of light: a single photon is one such example of a quantum state of light, being one quantum of energy occupying a single mode of the EM field; a single photon is a Fock state of the EM field with photon number $n = 1$. Nonclassical light is used to transmit qubits and/or entanglement over long distances; this entails generating nonclassical states in the context of a quantized EM field and QM (nonclassical light cannot be described with the classical EM theory). Typical forms of nonclassical light used in quantum information are Fock states[1] (particularly single-photon states) and squeezed states. The just-cited Fock state (named after Vladimir Fock, a Soviet physicist) is a quantum state that describes a specified integer number of non-interacting identical particles; for example, in quantum optics, a single-photon state is a Fock state of one photon. Fock states have applications to Quareps.

The short-term method of linking remote nodes is to employ fiber or free-space photonic communication channels to distribute entanglement. While all photon-based schemes are associated with losses that scale with distance (thus necessitating the use of Quareps), positioning entanglement generation on successful photon transmission events manages these losses into reduced entanglement generation rates without lowering entanglement fidelities. Optically mediated remote entanglement of individually controllable qubits has been generated for several materials platforms [RUF21], including quantum dots, trapped ions, neutral atoms, and Nitrogen-Vacancy (NV) centers in diamond.

Information transfer may be realized by direct transmission in fiber, by being relayed utilizing Quareps, or by teleportation using preshared entanglement. However, quantum signals are rather delicate and ephemeral and cannot be amplified or copied. Thus, critical issues in the world of Quacom/Quanets relate to addressing noise and loss utilizing quantum error correction techniques. Because unknown arbitrary qubits cannot be perfectly duplicated, quantum information cannot be amplified when traveling over long distances. Therefore, one needs to utilize technologies that deal with minimizing and mitigating communication loss. Basic required technologies and procedures include the following, among others[2]:

- Sources of nonclassical light
- Single-photon detectors
- Quantum memory and repeaters
- Transducers

---

[1] The quantization of the electromagnetic field leads to the existence of QM states, called Fock states, with an exact integer number of photons. This quantization is expressed by the Fock states $|n\rangle$ ($n = 0, 1, 2, \ldots$), which are eigenstates of the photon number operator $a^\dagger a$, where a is the annihilation operator of a quantized mode of the electromagnetic field. Fock states are the most elementary quantum states of light. Being photon number eigenstates, Fock states have an exactly defined integer photon number, with zero uncertainty. Fock states are not subject to the so-called shot noise. Large Fock states are considered a type of "holy grail" for quantum science and technology. These highly nonclassical states have long been considered for precision measurements (in the field of quantum metrology) because they have no uncertainty in their photon number (or, equivalently, their intensity); they are also considered valuable for the fields of quantum simulation and quantum information processing (quantum computation) [RIV23]. These concepts are revisited in following chapters.

[2] For a material platform to be suitable for a node in a Quanet where entanglement is mediated by photons, it must meet the following key requirements [RUF21]:

i.   The capability to interface at least one qubit efficiently with optical photons (at telecommunication wavelengths for fiber-based systems) to establish remote entanglement at high rates.

ii. The ability to store quantum states during entanglement generation; in particular, this requires qubit coherence times under full network activity to be longer than the time it takes to generate entanglement between nodes.

iii. The capability to store several entangled states per node with a capability for high-fidelity operations between them to enable multi-qubit protocols such as error correction.

These are discussed in more detail in Chapter 4.

Quareps endeavor to relay an entangled quantum state across a distance; they may interface to telecom fibers or utilize other transducers for non-fiber channels (these convert, say, a spin state or superconducting flux into a transmittable signal). In basic terms, a Quarep operates as follows: the Quarep initially generates entanglement between itself and each of the tow remote end nodes, node 1 and node 2, on the assumption that the distance of each end point to the repeater is small enough to allow direct entanglement generation by transmitting photons over telecom-grade fiber. Thereafter, in a procedure known as *entanglement swapping*, the repeater teleports one of the qubits entangled with node 1 onto node 2, allowing the establishment of entanglement over long distances where direct transmission is not achievable. After establishing long-distance entanglement, a data qubit is sent utilizing quantum teleportation. Figure 1.5, based on [LEE20], depicts the components of a Quarep network; the Quanet comprises client nodes, which end users directly access, and repeater nodes, which connect clients by propagating entanglement through the network.

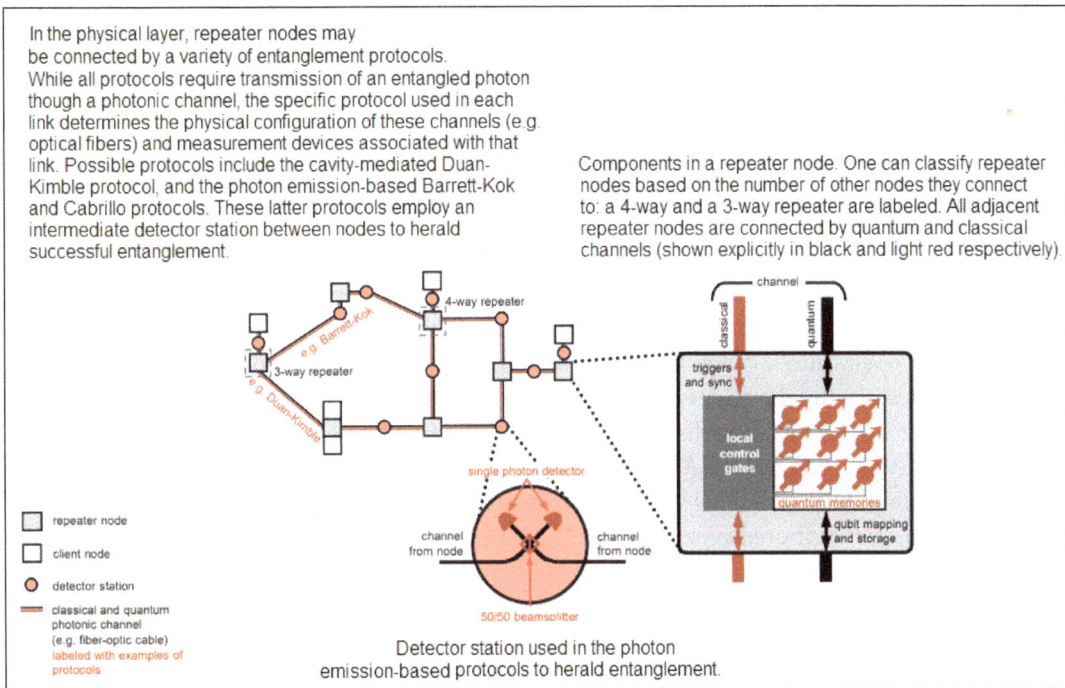

In the physical layer, repeater nodes may be connected by a variety of entanglement protocols. While all protocols require transmission of an entangled photon though a photonic channel, the specific protocol used in each link determines the physical configuration of these channels (e.g. optical fibers) and measurement devices associated with that link. Possible protocols include the cavity-mediated Duan-Kimble protocol, and the photon emission-based Barrett-Kok and Cabrillo protocols. These latter protocols employ an intermediate detector station between nodes to herald successful entanglement.

Components in a repeater node. One can classify repeater nodes based on the number of other nodes they connect to: a 4-way and a 3-way repeater are labeled. All adjacent repeater nodes are connected by quantum and classical channels (shown explicitly in black and light red respectively).

Detector station used in the photon emission-based protocols to herald entanglement.

Figure 1.5   Components of a Quarep-based network (based on [LEE20])

Consistent with the earlier OSIRM-like description, the topological characterization of generic networks, as well as current or near-term implementations of Quanets, can be classified as being (i) point-to-point, (ii) trusted-node, (iii) point-to-multipoint, and (iv) fully connected [ALS21]:

- A simple *point-to-point* quantum link can be instantiated between two remote parties, such as is found in QKD implementations;
- A *trusted-node* quantum network consists of multiple point-to-point links in a partial mesh arrangement, in which optical links terminate in trusted nodes supporting a given endpoint

(Alice-Bob) pair; communications between distant nodes are enabled by intermediate nodes in a hop-by-hop model (e.g., [EVA21], [WAN14], [MAO18], among others). End-to-end security requires that all intermediary nodes be trusted.

- In a simple *point-to-multipoint* quantum network, a passive beam splitter may be utilized to enable a node to communicate with some of the available remote nodes. In another approach, dedicated entanglement can be established between various pairs of users by assigning frequency-correlated wavelength channels.

- Fully connected Quanets may be realized utilizing nested Dense Wavelength Division Multiplexers (DWDMs); reconfigurable quantum links can be obtained by combining such DWDMs with transparent space division switches or Wavelength-Selective Switches (WSSs).

Quantum Local Area Networks (QLANs) that share information among systems in separate buildings using entangled photons have already been developed in labs. In one documented case involving three buildings, a protocol called *remote state preparation* was used, in which a successful measurement of one half of an entangled photon pair converts the other photon to the preferred state (the conversion was achieved across all the paired links in the QLAN: the QLAN connected a laboratory containing the photon source and the first node in the network to the second and third nodes using a fiber-optic network) [DOE22], [ALS21].

1.5 Large-Scale Quanets: The Quantum Internet

In addition to communication and cryptography applications, Quacom/Quanet are perceived by researchers as being an intrinsic element for large-scale distributed Quecomp, where QCs are interconnected with each other via an appropriate networking apparatus. This paradigm conceives an open, public Quaint, where, at a point in the future, one can expect a large number of devices (in the thousands or millions) being connected over a global "quantum network of networks" enabling distributed quantum internet computing (Quaintco) [KIM08], [DIA21], [DIA22], [MON14], [ABE23], [CAS18], [SIM21].

The recently-published RFC 9340, *Architectural Principles for a Quantum Internet*, describes how Quanets are distributed systems of quantum devices that utilize QM phenomena such as superposition, entanglement, and quantum measurement to achieve capabilities beyond what is currently possible with non-quantum (classical) networks (RFC 9340 describes the framework of a Quanet and introduces some basic architectural principles for a Quaint). Depending on the evolutionary stage of implementation of a Quanet, such devices range from simple photonic devices capable of preparing and measuring only one quantum bit (qubit) at a time, all the way to large-scale QCs. Using Quacom, in a few years a Quaint may supplement the Internet in place today or even, perhaps further along, supplant it; at this time, however, Quanets do not aim at replacing classical networks but support hybrid classical-quantum networks capturing new functional capabilities that are unachievable by the classical networks [KOZ23], [KIM08], [WEH18], [VAN2014]. Proponents refer to Quaint as being "highly anticipated"; however, current technology is still in very early stages [ALS21]. Figure 1.6 [RUF21] provides a simplified view of a Quaint.

For Quaint in general and Quacom in particular, quantum interconnects are used to convert quantum states from one physical system to states of another physical system in a reversible

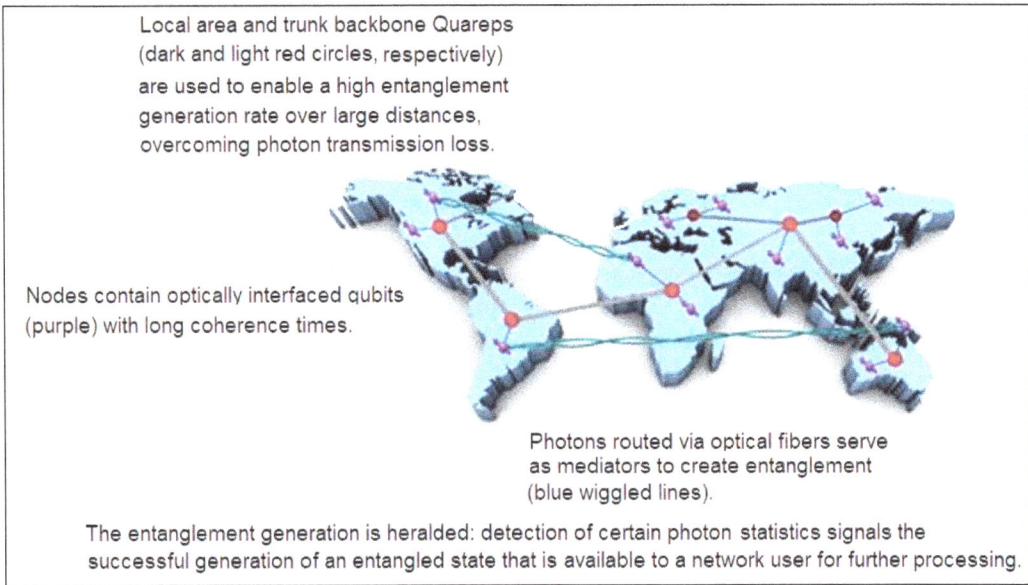

Local area and trunk backbone Quareps (dark and light red circles, respectively) are used to enable a high entanglement generation rate over large distances, overcoming photon transmission loss.

Nodes contain optically interfaced qubits (purple) with long coherence times.

Photons routed via optical fibers serve as mediators to create entanglement (blue wiggled lines).

The entanglement generation is heralded: detection of certain photon statistics signals the successful generation of an entangled state that is available to a network user for further processing.

Figure 1.6   A view of a Quaint (modeled after [RUF21])

manner; quantum connectivity in Quanets can be achieved by the optical interactions of single photons and atoms, facilitating the distribution of entanglement over the network and the teleportation of quantum states between nodes [KIM08]. Quanets can be perceived as private or institutional quantum-based networks, while the Quaint can be perceived as an open, public, global, wide-reaching quantum network—a global network interconnecting QCs and smaller heterogeneous Quanets, enabling unimpeded global transmission of quantum information (that is, qubits) and generate and distribute entangled states.

The goal of Quaintco is interconnecting discrete quantum computers over a global Wide Area Network (WAN), similar to how traditional computers achieved their full computing potential after the emergence of the traditional Internet [ROH21], [FUR20], [DOW20], [LOK23], [VAN14]. Quaint and Quaintco aim at connecting quantum information processors (that is, QCs) in order to achieve enhanced capabilities that are impractical or impossible by only utilizing classical information.

Quaint aims at significantly scaling up the number of qubits for wide area communication of quantum as well as classical information. The long-term successful denouement of Quanets in general and Quaint in particular depends on technological advances that reliably link quantum devices together at a large scale. The National Institute of Standards and Technology (NIST) [NIS21] notes that at present, based on proof-of-principle experiments, many material platforms capable of processing quantum information have been developed, with no clear "champion technology": some physical qubits appear best for storing quantum information, while others are best for manipulating and exchanging quantum information. Modular quantum systems in which different physical qubits are used for different purposes is postulated to be necessary to develop a scalable Quaint.

Further to the preceding observations, the full development of Quanets/Quaint requires the establishment of mechanisms and protocols that allow the transparent interconnection of the underlying hardware and allow for the rapid generation of entanglement in order to mitigate evanescent qubit lifetimes. As is the broad case in the classical OSIRM, a series of layered protocols are needed to provide an abstraction that permits applications to exchange data between two end nodes without having to know the specific details on how this connection is actually achieved (see Figure 1.7).

Potential architectures for large-scale Quacom/Quaint internetworking are under development at this time. However, Quanet architectures have distinct variations compared with classical networks and classical network architectures. In particular, the no-cloning predicament precludes making a perfect copy of an arbitrary unknown quantum state, thereby preventing the use of traditional detection, correction, and retransmission methodologies that are common in classical optical networking.

A challenge for the near-term future is to transition from the current proof-of-principle experiments to large-scale Quanets that can be utilized in areas such as distributed quantum computation, enhanced quantum sensing, and secure quantum communication.

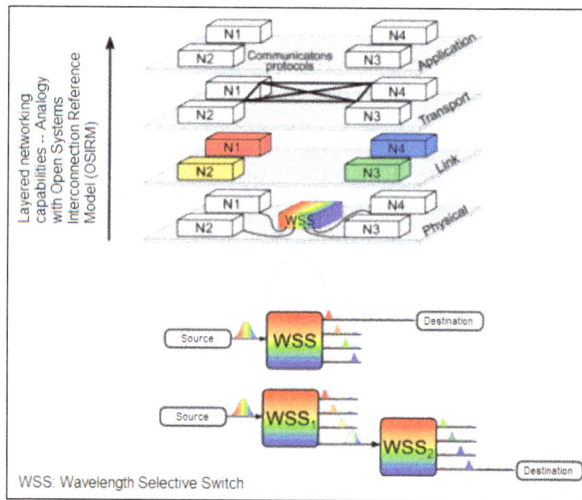

The physical layer includes the optical components where the photons travel and are manipulated.
The link layer slices the spectrum and routes each slice to a particular user.
The transport layer is the quantum-correlated network where a pair of users shares entanglement.
The application layer uses the entangled pairs to perform a service.

Figure 1.7   (Desirable) layered architecture of Quanets (inspired by [ALS21])

1.6 A First View of Quantum Computers

Quanets are ultimately designed to interconnect QCs and other quantum-based QIST elements. Although still in its infancy, Quacomp is expected to support a broad array of industrial sectors, including financial services, healthcare, scientific and industrial simulation, and artificial intelligence, to name a few. As noted, a qubit is a quantum system that may exist in one of two orthogonal states—$|0\rangle$ and $|1\rangle$—or in a superposition of the two states. QCs operate by storing information in the form of qubits and, in the circuit model of Quacomp, processing these qubits using quantum gates; the operations performed by the quantum gates are defined by linear algebra operations over a (Hilbert) vector space, and the circuit behavior is governed by QM. The mathematical model of qubits and the operations on them enable QCs to solve a class of problems much faster than is the case with classical computers or to solve problems that are computationally intractable for classical computers. Unlike a classical bit, a qubit cannot be stored as a single voltage or signal value in a medium. Instead, a qubit is physically realized using and operating on a two-level $\begin{bmatrix} \alpha \\ \beta \end{bmatrix}$ state of the QM system.

A QC is a (fully) programmable quantum processor that can implement or approximate any unitary dynamics defined within its full Hilbert space [ISO23]. By operating on a system (an ensemble) of qubits, a QC is able to rapidly undertake certain categories of computations that would require large or even impractical amounts of time in a classical computer[1]. A number of physical implementations of qubits have been developed in recent years; some examples of qubits implementations include superconducting circuits, spin photonic qubits, and trapped ions. Early theoretical prototypes of QCs were conceived in the 1980s.

QCs rely on the dynamics of quantum objects such as photons, electrons, atoms, ions, molecules, and nanostructures, which operate under the rules of QM. In QM, the quantum state of a quantum object is described by a set of physical properties; the complete set is referred to as a *mode*. For example, in the case where the quantum object is a photon, modes may be defined by the position in space of the photon (e.g., which waveguide or superposition of waveguides the photon is propagating within), frequency of the photon, the associated direction of propagation (e.g., the k-vector for a photon in free space), the polarization state of the photon (e.g., the direction—horizontal or vertical—of the photon's electric and/or magnetic fields), a time window in which the photon is propagating, the orbital angular momentum state of the photon, and so on [LIT23].

As another example, Trapped Ion Quantum Computing (TIQC) is a system that utilizes ions (an atom or molecule with a net electrical charge) as qubits for computation, with the excitation state of an electron indicating a logical value or logic state—qubits are stored in the electronic states of each ion, and information can be transferred through the aggregate quantized motion of the ions in a trap [FUR23]. Trapped atoms may be utilized to implement QIST systems and/or Quacomp: atomic-based qubits can be used for quantum memories, QC quantum gates, and nodes for Quanets. Figures 1.8 and 1.9 are high-level system diagrams illustrating generic QC systems.

---

[1] For example, among a plethora of scientific and commercial/financial applications, quantum computers could very easily solve many mathematical problems, including, but not certainly limited to, finding a very large set of hyperperfect numbers (e.g., [MIN75], [MIN80], [MIN80a], [MIN02], [RIE84], [CRA00], [NAS02]), finding a very large set of inductive formulas for general sum operations (e.g., [MIN80b]), finding a large set of closed-form expression for the generalized constrained average of functions (e.g., [MIN83], [MIN20]), or computing a complexity index for large molecules, networks, or graphs (e.g., [MIN75a], [JAN18], [BON03], [EMM12]).

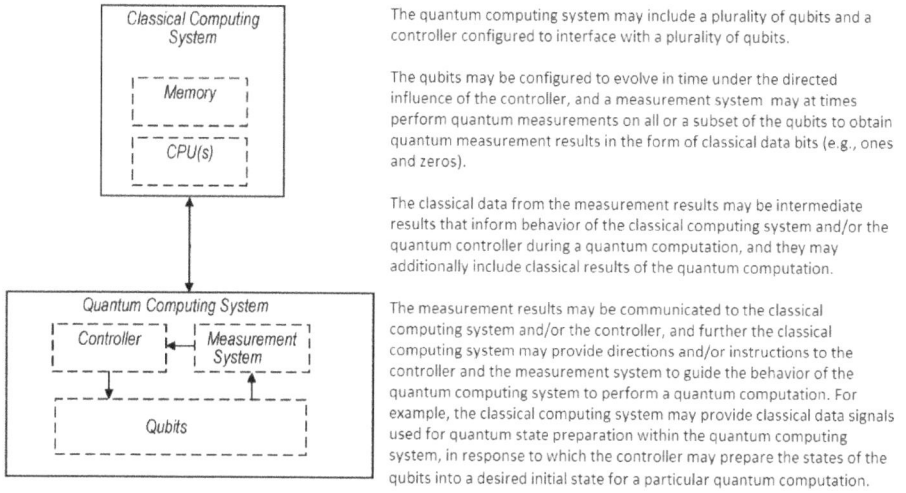

The quantum computing system may include a plurality of qubits and a controller configured to interface with a plurality of qubits.

The qubits may be configured to evolve in time under the directed influence of the controller, and a measurement system may at times perform quantum measurements on all or a subset of the qubits to obtain quantum measurement results in the form of classical data bits (e.g., ones and zeros).

The classical data from the measurement results may be intermediate results that inform behavior of the classical computing system and/or the quantum controller during a quantum computation, and they may additionally include classical results of the quantum computation.

The measurement results may be communicated to the classical computing system and/or the controller, and further the classical computing system may provide directions and/or instructions to the controller and the measurement system to guide the behavior of the quantum computing system to perform a quantum computation. For example, the classical computing system may provide classical data signals used for quantum state preparation within the quantum computing system, in response to which the controller may prepare the states of the qubits into a desired initial state for a particular quantum computation.

Figure 1.8   High-level system diagram illustrating a quantum computing system (adapted from [LIT23])

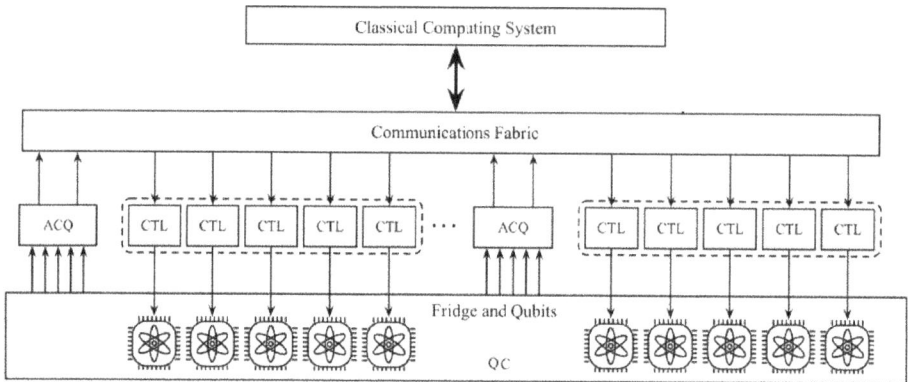

ACQ = Acquisition nodes CTL = control nodes

Control/ACQ nodes are components that control individual qubits. Control nodes are also network endpoints that receive the messages or signals sent by the classical computing system. Each control node is a computing device that may execute codes or run parts of an algorithm or program. Each control node may also make local decisions based on qubit values or measurements.

Acquire/ACQ nodes are components that receive measurements from qubits. Each acquire node may be a shared receiver that gathers measurements from multiple qubits. The acquire nodes can also be network endpoints that relay the measurements back to the classical computing system.

Figure 1.9   High-level system diagram illustrating a quantum computing system (adapted from [HAV23])

IBM®, Google®, Amazon®, and others have already developed lab-oriented QCs[1,2]. At the time of this writing, some refer to QCs as being "third-generation" superconducting QCs.

---

[1]   For example, among other endeavors, IBM announced the completed installation of a 127-qubit quantum computing system at the University of Tokyo in the Fall of 2023, featuring the company's Eagle processor.

[2]   Major vendors involved in Quacomp at the time of this writing include, but are not limited to, IBM, Google Quantum AI, Amazon, Microsoft®, Intel®, D-Wave®, Quantinuum®, Rigetti®, Xanadu®, IonQ®, and Atos Quantum. See Chapter 5.

Applications include scientific computations, quantum simulation, and quantum machine learning—quantum simulation entails using a QC to simulate, among other physical systems, molecules, and physical materials. However, while QCs have already shown superiority in some computational arenas and classes of problems, as of the time of this writing, QCs had not yet outperformed a supercomputer across the board on a full range of real-world applications; thus, in the short term, QCs will operate in a complementary manner to classical computers, outperforming them for some classes of problems, but not for all classes of problems [DAR24]. Quacomp was a $1 billion market in 2023 and was expected to grow to $6.5 billion by 2030; the U.S. government spent about $3 billion on Quacomp between 2019 and 2022 [GRE23], [SMI23a]. See Figure 1.10 for an example of a quantum chip.

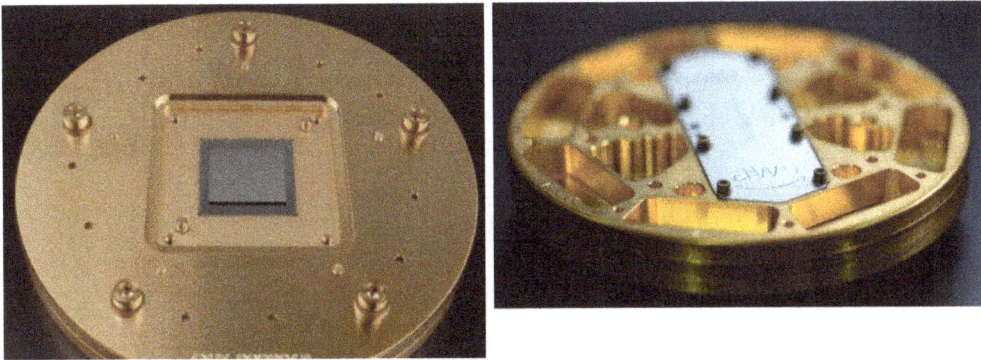

Figure 1.10   QC chips (Courtesy AWS Quantum Computing)

Classical computers utilize (large assemblies of) transistors. Transistors are able to turn "on" or "off" to symbolize data as ones and zeros. In turn, transistors are connected to construct the building blocks of digital circuits—namely, logic gates. Logic gates implement logical operations such as AND, OR, and NOT.

QCs, on the other hand, rely on qubits, which as noted have a quantum state known as superposition, thus they are expressed as being both 1 and 0 at the same time; superposition allows each qubit to perform two calculations at once. Specifically, QCs process information in qubits that can represent a coherent superposition of both binary information states at the same time; two or more qubits may be entangled, hence their physical properties are correlated even when separated by considerable distances. QC may simultaneously perform a large number of operations on these entangled qubits; this massive parallelism allows QCs to perform complex calculations at speeds exceeding the most powerful supercomputers on the market for some classes of problems [MEI23]. The more qubits that a QC links together, the more (basic) computations known as *quantum gates* it can perform, thus the greater its computational power. Each additional qubit doubles the computer's power. QCs operate by running quantum algorithms that describe sequences of elementary operations called quantum logic gates applied to a set of qubits. QCs and quantum algorithms are poised to provide computational improvements over existing systems. The description of the state of an n-particle quantum system grows exponentially with n; however, this information capacity cannot be easily accessed because any measurement of the system yields only *n* elements of information. Thus, the basic challenge of a quantum algorithm is to practically manipulate the exponential amount of information in the quantum state of the system and thereupon extract some desired elements utilizing a final measurement [WHA09]. QCs can currently outperform classical computers in a specialized set of computational problems; by and large, QCs have demonstrated superiority in solving optimization problems.

Applications of QCs, in the quantum simulation arena, span science and industry, ranging from quantum chemistry and many-body physics to optimization, finance, bioinformatics, materials science, and machine learning. Documented areas of current QC applications include [DAL23] condensed matter physics (simplified models designed to capture the universal physics of material systems), quantum chemistry, nuclear and particle physics, combinatorial optimization, continuous optimization, cryptanalysis, solving differential equations, finance, and machine learning with classical data (quantum machine learning via quantum linear algebra, quantum machine learning via energy-based models). Networking will greatly enhance the scope, range, applicability, and accessibility of these systems.

Since the early 2000s physicists and engineers have demonstrated that Quacomp is achievable by manipulating quantum particles (atoms, ions, or photons) to realize physical qubits. But in order to advance to the next stage of (commercial) utilization, QCs must run reliably on logical qubits. Logical qubits are sets of redundant, error-corrected physical qubits, which can store information for use in a quantum algorithm; in logical qubits information is encoded across a set of physical qubits for redundancy [MAN23]. Eliminating or minimizing errors is fundamental to being able to achieve practical large-scale Quacomp. The overhead in the realization of error-corrected logical qubits remains a challenge and an ongoing area of research.

Progress has been made recently to demonstrate ever more complex QC systems. For example, in 2021, IBM unveiled Eagle, the first QC with more than 100 qubits. In 2022 it brought out Osprey, which operates with 433 qubit [CHO22],[1] tripling the number of qubits on a chip in just one year. At the time of this writing, IBM introduced Condor, the world's first universal quantum computer with more than 1,100 qubits.[2,3] All the qubits must operate in unison, a process called *(quantum) coherence*. Quantum coherence is the ability of a quantum system to retain a well-defined state over time without being impacted by nearby disturbances. When

---

[1] In 2016 IBM placed the first QC on the cloud; it had 5 qubits, each a superconducting circuit cooled to near-absolute-zero temperatures (−273° C). In 2019, the company introduced the 27-qubit Falcon; in 2020, the 65-qubit Hummingbird; and in 2021, the 127-qubit Eagle.

[2] In the fall of 2023 IBM unveiled a QC, called Condor, which has 1,121 superconducting qubits arranged in a honeycomb pattern; IBM also unveiled a chip called Heron, which has 133 qubits, but with a record low error rate—three times lower than that of its previous quantum processor [CAS23]. Generally state-of-the-art error-correction techniques require more than 1,000 physical qubits for each logical qubit. An alternative error-correction scheme called "quantum Low-Density Parity Check" (qLDPC) could reduce that number by a factor of 10 or more [BRA23]. Reportedly, the short-term goal is building chips designed to hold a few qLDPC-corrected qubits in just 400 or so physical qubits, and then networking those chips together. The challenge, however, is that the qLDPC approach requires each qubit to be directly connected to at least six other qubits—in typical superconducting chips, each qubit is connected only to two or three neighboring qubits.

[3] Proponents observe that ". . . a quantum computer with three hundred qubits could possess the processing power equivalent to the number of atoms in the known universe" [MUK23].

coherence breaks down, errors ensue, and mitigating those errors and extending coherence time while scaling up to larger machines are some of the current QC/Quacomp challenges.

Recently one saw the introduction of the first *programmable* QC capable of both encoding up to 48 logical qubits and executing hundreds of logical gate operations [BLU24], [MAN23], [RYA22]. This QC utilizes rubidium atoms as the physical qubits allowing the atoms to move and be connected into entangled pairs mid-computation.

To be implemented, quantum algorithms must be compiled on specific quantum hardware to decompose complex operations into the naturally available elementary gates. Considering the resource constraints imposed by the Noisy Intermediate-Scale Quantum (NISQ) technology that is foreseeable and/or available in the next few years, it is critical to optimize the use of every qubit and every gate cycle to facilitate the deployment of near-term applications. A concatenation of several global or multi-qubit operations are circuit components that are required for building quantum algorithms: the best-known examples are the quantum arithmetic circuits used in Shor's factoring algorithm[1] and the multiply controlled gates used in Grover's search algorithm.[2] As just stated, implementing quantum algorithms on physical devices necessitates translating high-level global operations into sequences of hardware-native logic gates in a circuit; this process is known as *quantum compiling*. A quantum circuit is a model for computation in which the computation is achieved with a sequence of quantum gates. Physical constraints and/or limitations (e.g., constraints in gates and connectivity) typically result in high implementation costs; thus, to enable practical near-term applications, it is important to optimize compilation by making effective use of the capabilities of constituent gates. To that end, a quantum version of AND logic that can, for example, decrease the compilation overhead, enabling the execution of key quantum circuits, is desirable. Such a gate has recently been demonstrated—see Figure 1.11 [CHU23]. Also, various types of quantum information storage devices that store and then allow the reading of a quantum bit output from a quantum computer have been proposed and/or emerged.

Quacomp still faces significant hurdles in hardware and software development. QCs are extremely sensitive to any unintentional interactions with their environments. Relatively trivial changes in their environment, such as vibrations or heat, cause noise that can alter the superposition state of qubits. This results in the phenomenon of *decoherence,* in which qubits rapidly degrade to the 0 or 1 states of classical bits. As already noted, a key factor in building reliable QCs is controlling these errors; building large-scale Quacomp systems capable of delivering on the promise of quantum speed-ups requires overcoming decoherence [LID23].

The high error rates of qubits preclude the use of QCs for solving complex problems that might entail billions of error-free operations. Qubits can experience errors in two areas: *bit flips*

---

[1]   Shor's algorithm (developed in 1994 by Peter Shor) is a quantum algorithm for finding prime factors. It provides super-polynomial speedup compared to best classical algorithms. It is currently one of the few known quantum algorithms for practical applications. However, factoring very large numbers requires far more qubits than are available in the near term; another consideration is that noise in quantum circuits may degrade results and may require additional qubits for quantum error correction.

[2]   Grover's algorithm (aka *quantum search algorithm*) (developed by Lov Grover in 1996) is a quantum algorithm for unstructured search that identifies with high probability the specific input to a black box function that produces a particular output value, using just order-of-the-square-root-of-N evaluations of the function (N is the size of the function's domain); classical methods require order-of-N evaluations. This algorithm can be employed to speed up a number of algorithms for NP-complete problems where exhaustive searches are often undertaken.

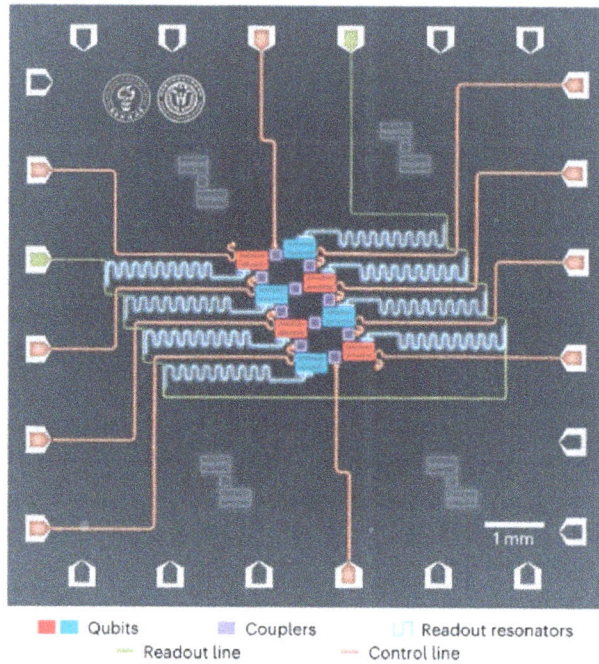

Red and blue indicate the lower and higher fixed-frequency transmon qubits, respectively

Figure 1.11   Example of a QuAND gate (false-color micrograph) on a high-scalability superconducting quantum processor; eight physical qubits are shown. (*Source:* J. Chu, X. He, et al. [CHU23])

and *phase flips.* To mitigate the effects of phase flip errors, Quantum Error Correction (QEC) techniques are employed by encoding information in multiple qubits in a way that allows detection and correction of the phase flip errors. During the past decade, progress has been achieved in reducing qubit error rates from one error every 10 quantum operations to one error every 1,000 operations. However, despite this significant improvement, qubits are still too noisy to be practically useful for solving highly complex problems [SMI23]. The practical design target for QCs is to develop effective methods of suppressing and correcting quantum errors. Engineering and research work continues in earnest to address these issues. The technologies needed to support scalable Quacomp and Quacom/Quanets are closely interrelated.

1.7 Quanet Applications

Quacomp, Quacom, and Quanets, specifically, are receiving considerable research attention of late in the context of bleeding-edge emerging technologies for ICT applications. As already discussed, Quacomp and Quacom use quantum information in lieu of classical information; in particular, Quacom and Quanets support quantum-secure communication and entanglement-based communication [PIR20], [BEN92]. Recent Quacomp breakthroughs are driven by advances in quantum device hardware, as highlighted in the previous section; at the same time, new quantum algorithms and extensions of already known methods (for example, Grover's search) have emerged during the last few years, including for graph problems and image processing [ENG23]. Another field of growing interest is Quantum Machine Learning, with algorithms to accelerate classical machine-learning algorithms. The QC/Quacom market is forecast by some to be worth $93 billion by 2040 [FAR24].

At this juncture, three broad classes of Quacom applications are emerging: (i) quantum cryptography entailing new mechanisms for trusted communication between distant parties on a network while monitoring the communication channel for unwanted eavesdropping; (ii) distributed sensing utilizing distributed quantum sensors to measure data simultaneously in different physical locations (quantum sensors enjoy higher measurement sensitivity than classical sensors); and (iii) interconnecting QCs into clusters to exponentially enhance the aggregate computing power. More specifically, some of the currently envisioned Quacomp/Quacom applications are shown in Figure 1.12 and Table 1.2.

In the context of Quacom, there have been several major Quaint projects worldwide in the recent past. For example, but not to the exclusion of other initiatives, the European Commission has funded the Quantum Internet Alliance (QIA), a consortium of research institutions and firms that aims to develop a Quaint infrastructure and demonstrate its applications in a number of fields. QIA is part of a broader effort, the Quantum Flagship, a research initiative funded by the European Union (EU). Other efforts include the Quantum Internet and Networked Computing (QuInC) project, which focuses on the development of quantum algorithms and protocols for Quanets [IEE24]. However, a Quaint is not intended to replace classical communication but rather to supplement it with quantum communication; often one assumes that all nodes can communicate classically (e.g., over the IPv4 Internet) in order to exchange (quantum) control information [WEH18].

The majority of the currently-conceived Quaint applications utilize one of two features or aspects of quantum entanglement, *consistent measurement* and *unsharability*:

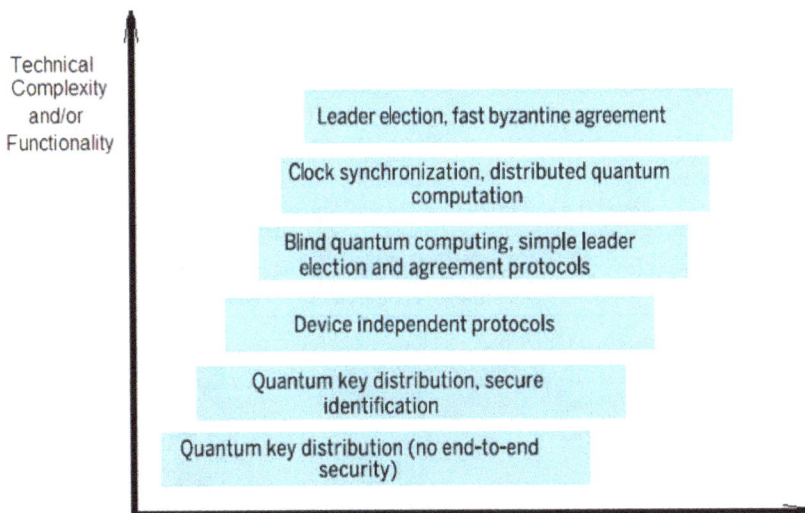

Figure 1.12   Near-term Quacomp/Quacom applications as a function of technical complexity (inspired by [WEH18])

Text continues on page 34

Table 1.2 Some Currently-Envisioned Quacom or Quacom-Related Applications

Application	Description	Key References
(Information) Teleporting/ Teleportation	A mechanism for transmitting quantum information across a geographic distance. Entanglement is a particular relationship in which two qubits share a state correlation even if these two qubits are located or stored at distant network nodes; once two qubits are confirmed to be entangled, a datum can be sent deterministically. Teleporting is a process for transmitting quantum information, in which, given that Alice and Bob have a pre-shared Bell pair, a "transfer" of one qubit from Alice to Bob is accomplished by transmitting two classical bits between the parties. An unknown quantum state $\lvert\phi\rangle$ can be disassembled into, then later reconstructed from, purely classical information and purely nonclassical Einstein-Podolsky-Rosen (EPR) correlations. To do so, the sender, Alice, and the receiver, Bob, must prearrange the sharing of an EPR-correlated pair of particles. Alice makes a joint measurement on her EPR particle and the unknown quantum system, then sends Bob the classical result of this measurement (teleportation is undertaken with the assistance of classical communication). Knowing this result, Bob can convert the state of his EPR particle into an exact replica of the unknown state $\lvert\phi\rangle$, which Alice destroyed (consumed). Teleportation requires users to share entanglement and classical communications. In practical terms, teleportation is the process of using a preshared entangled state between two endpoints using (i) preshared, and (ii) classical communication to transport a qubit from one party to another.	[BEN93], [HER22], among others
Blind Quantum Computation (BQC)	Mechanisms that provide the capability for a user to execute *Quacomp* using one or more remote quantum servers supporting distributed quantum computing, while keeping the structure of the computation hidden. In this secure delegated *Quacomp* the client hides its computation from the server by sending qubits to the server and then requesting (in a classical message) that the server measure them in an encoded basis; the client then decodes the results it receives from the server to obtain the result of the computation [BAR12]. BQC protocols allow one party to outsource the computation of a quantum algorithm to another party without revealing the input, output, or even the algorithm that was run. The goal of BQC protocols is to ensure only the privacy of the computation; many BQC protocols also allow for verification of the computation being performed, by embedding hidden tests within the computation.	[ARR06], [BAR12], [BRO10], [FIT17] among others
Clock synchronization	A quantum, cooperative protocol for the operation of a network consisting of geographically remote optical atomic clocks, combining precision metrology and quantum networks utilizing nonlocal entangled states.	[KOM13], [KOM14], [ZAD17], among others

Application	Description	Key References
Clock synchronization *(cont.)*	Proponents state that such a network can be operated near the fundamental limit set by quantum theory yielding an ultra-precise clock signal. In addition, the internal structure of the network along with basic techniques from quantum communication enhances internal and external security from inimical agents. Realization of such a global quantum network of clocks may allow construction of a real-time single international time scale (world clock) with great stability and accuracy.	
Distributed sensors	Quantum sensors can be used to measure data simultaneously in different physical locations using shared quantum resources, likely improving the performance of traditional Wireless Sensor Networks [MIN07]. Quantum sensing enables the detection and measurement of numerous physical properties such as pressure, temperature, electromagnetic energy with improved sensitivity and precision than possible non-quantum instruments.	[GUO19], [MAZ08], [QIA21], [QIA19], among others
Distributed quantum computing	*Quacomp* over distributed quantum systems	
Hybrid quantum machine learning	Methods of improved training of Machine-Learning (ML) models by using approaches that combine the best features of classical and *QCs*. Quantum Neural Networks (QNNs) are hybrid classical-quantum ML models that are composed of parameterized quantum circuits (Ansatz) and classical feedforward neural networks that are trained iteratively, typically by minimizing a cost function [ROZ23]. QC-based ML techniques have been advanced for data fitting, pattern recognition, generative machine learning, and other Artificial Intelligence (AI) applications. Classical computers and QCs can both be used to train machine-learning models to solve equations in various application-specific high-dimensional spaces. Hybrid classical/QC models contain both curved and jagged regions in such mathematical spaces: classical computers can solve the equations that govern the jagged areas, but they are able to approximate the curved areas only by using stepwise functions; QCs operate by wavefunctions and are well suited for solving curves in multidimensional space, but are inefficient at solving the areas with jagged edges [XU23], [KOR23]. Parallel hybrid networks exploit the neural-network technique's ability to combine results into a single model. Proponents and advocates see quantum AI represents a transformative advancement in technology [DAW23]. Typical target applications include speech recognition and computer vision [MIN24] and may entail some kind of (classical or evolving *Quacom*) communication. Telecommunications-related applications include solution to wireless resource allocation problems, cognitive radio spectrum sensing, network traffic forecasting, unmanned aerial vehicle trajectory planning, imaging, WiFi sensing, and human pose recognition tasks.	[KOI22], [KER19], [KOR23], [LIU16], [NAR21], [NAR22], [SHA22], [WIE15], [WIE16], among others

Table 1.2 Some Currently-Envisioned Quacom or Quacom-Related Applications (*cont.*)

Application	Description	Key References
Post-Quantum Cryptography (PQC)	Encryption algorithms that can resist attacks by QCs. NIST is set to standardize encryption algorithms that can resist attack by quantum computers. Three new algorithms are expected to be ready for use in 2024; others will follow (more discussion in Chapter 7). Although at this time, there are no functional QCs outside of a laboratory environment, and common public-key cryptography schemes may soon be at risk. Commonly used algorithms such as Rivest-Shamir-Adleman (RSA) and Diffie-Hellman (DH) schemes rely on the inability of classical computers to complete certain types of complex mathematical calculations, such as integer factorization and discrete logarithm computation within a reasonable amount of time. On the other hand, a QC implementing Shor's algorithm can complete these complex calculations in a relatively short time and so determine the private keys used for current public-key systems from the corresponding public keys [MEI23]. Although QCs capable of impacting classical encryption algorithms are still a few years away, the threat of a "harvest now and decrypt later attack" is being given consideration, even if the threat will not be actionable until a sufficiently robust QC is developed in the future. The "harvest now and decrypt later attack" is a long-game attack where an inimical agent collects, records and stores encrypted data waiting for the time when QCs can decrypt it.	[NIS23a]
Quantum blockchains	Blockchains support distributed data storage, point-to-point transmission, consensus mechanics, and encryption algorithms [MIN19]. It is a "tamper-proof" database that contains records, such as a history of financial transactions or other transactions, which does not require a centralized entity to maintain its accuracy. However, QCs may (soon) successfully break the cryptographic protocols that are used to secure blockchains. Quantum blockchains have been proposed. A quantum blockchain is a decentralized, encrypted and distributed database based on quantum computation and quantum information theory; once the data are recorded in the quantum blockchain, they will not be maliciously tampered with. One quantum blockchain proposal is a quantum blockchain using entanglement in time (as opposed to an entanglement in space). As noted in [LI19] and [RAJ19], the notion of the quantum blockchain is captured by the inseparability (entanglement) of quantum systems such as photons; the blockchain is encoded as the Greenberger-Horne-Zeilinger state of the photons that have never coexisted. In the conceptual design of this quantum blockchain, the data represented in the classical block are simplified into a string of two bits. The encoding procedure converts the record of each block, say r_1, r_2, into a temporal Bell state generated at a specific time such as t=0. In particular, the first photon of a block is absorbed immediately. When records are generated, the system encodes them into a temporal Bell state. These photons are then created and absorbed at their respective times.	[LI19], [RAJ19], among others

Application	Description	Key References
Quantum dynamic network server load management	Leveraging the capabilities of quantum computing to analyze formatted load management information and determine an optimized load leveling process for distributing traffic across IT infrastructure devices in real time or near real time. Quantum approaches have been proposed that provide a system and method for quantum dynamic network server load management. Such a system receives a large amount of load management information from its plurality of devices within the information technology infrastructure. Given that the system will need to quickly determine how to redirect traffic across the devices to prevent bottlenecks, slow-downs, or outages for the devices, the system can identify at least a set of the load management information that needs to be processed by a quantum optimizer to meet the time and accuracy requirements of a load management system. In such instances, load management or load balancing systems that run on conventional computer apparatuses are not adequate to make the determinations in a fast enough time (with enough accuracy) to resolve load leveling issues on a large (e.g., global) scale.	[MUK23]
Quantum Key Distribution (QKD)	As early as 2010 [CHE10] could state "It has been nearly 3 decades since the proposal of quantum key distribution". QKD allows two parties to share a secret key which can be used for symmetric-key encryption: Alice and Bob run a key generation protocol aiming to generate the secret key at a rate close to the secret key agreement capacity of that channel (the practical challenge, however, is that it is hard in general to calculate this capacity precisely, even more so to come up with protocols that can attain the key rates close to that capacity) [TAK14]. Thus, QKD provides a low-level quantum capability of eavesdropping detection and enables one to create shared, secret random numbers. See Figures 1.13, 1.14 [WOO23], and 1.15 [LUK23]. In 1984, Bennett and Brassard published the BB84 QKD protocol based on Heisenberg's Uncertainty Principle (HUP). The protocol is named for the authors and the year it was published. It is one of the most well-known quantum protocols; other protocols based on HUP are basically variants of BB84. Another protocol, the E91 protocol proposed by Ekert in 1991, uses Bell states emitted by a common source and distributed between Alice and Bob; Alice and Bob then use randomly chosen polarization bases. The protocol uses a central device to distribute entangled pairs to the end users. The advantage of using quantum states instead of classical bits is that they permit the detection of a possible eavesdropper who tries to acquire information about the secret key. Therefore, QKD has attracted significant attention in recent years. After the initial developments and experimental realizations. there are now very complex protocols and quantum devices to realize such protocols. QKD is limited in distance to a few hundred kilometers in optical fiber, possibly longer distances using satellites (quantum key distribution between a low-Earth-orbit satellite	[BEN84], [CHE22], [DUE06], [GIS02], [GIS07], [KRA05], [SHO00], among others

(continues)

Table 1.2 Some Currently-Envisioned Quacom or Quacom-Related Applications (*cont.*)

Application	Description	Key References
Quantum Key Distribution (QKD) *(cont.)*	and multiple ground stations has already been documented [LIA18]). However, it would be desirable to enable quantum key distribution over large distances—that is, continental and intercontinental distances; the rate of secret key generation decays exponentially with distance. To achieve longer distances, quantum repeaters have been proposed. These are protocols that exploit entanglement and quantum information primitives as entanglement swapping and entanglement purification (distillation). The final aim is to create a long-distance entangled pair. This pair may be then used for quantum key distribution [ABR13a].	
Quantum metrology	Metrology is the science of measurement and precision; quantum phenomena are utilized to accurately estimate physical parameters with a precision exceeding the best possible classical methods. It implies the deployment of entangled sensor networks: the service utilizes the phenomenon measuring half of an entangled system instantaneously affects the other half, regardless of the distance between them, to enable precise timing and position verification. The goal is to push the standard quantum limit of detection to the Heisenberg Limit, thus improving the sensitivity beyond the shot noise limit of interferometers and/or improving the resolution of imaging beyond the Rayleigh diffraction limit [LAC23], [LAI23]. Also known as Quantum Metrology & Sensing and quantum enhanced Imaging (QMSI).	[DEG17], [GUO19], [MAZ08], [OVA22], [PAR09], among others
Quantum-supported Natural Language Processing (NLP)	A chatbot is a device that simulates human conversation using natural language input, such as a text or a speech input. Chatbots use Natural Language Processing (NLP) in conjunction with Artificial Intelligence (AI) and Machine Learning (ML) techniques and may employ stochastic techniques. NLP entails receiving a sample comprising natural language; processing the sample, wherein processing the sample comprises generating a plurality of response hypotheses and generating a plurality of confidence values, wherein each response hypothesis is associated with a corresponding confidence value; and selecting a response, comprising selecting the response randomly among the plurality of response hypotheses based at least in part on the corresponding confidence value by means of a quantum random number generator. Quantum objects are known to constitute the only carriers of true randomness, and quantum random number generators may demonstrate truly unpredictable behavior. Employing a quantum random number generator in selecting the response among the plurality of response hypotheses will enhance the capabilities of the natural language processing, and in particular may permit a chatbot to output responses that avoid repetitive patterns and emulate the response behavior and free will of human dialogue partners much better than what can be achieved with conventional NLP techniques [LES23]. A quantum random number generator is (typically) a hardware element that is capable of generating random numbers from a quantum physical process (these phenomena may comprise nuclear decay, shot noise, or photons traveling through a semi-transparent mirror or other quantum optical processes).	[LES23]

Application	Description	Key References
Scientific and commercial applications	Applications expected to emerge include chemical simulation, optimization, banking/finance (e.g., derivative pricing, credit valuation adjustments, value at risk, portfolio optimization, currency transaction monitoring, post-quantum cryptography, quantum-secure networking and quantum key distribution), medicine, secure navigation, and machine learning (the science-oriented applications are typically known as 'quantum simulation'). To undertake advanced computing to support complex scientific and business applications, quantum nodes, such as QCs, that are located across various geographies and locations throughout the world need to be connected; extant technologies cannot support this goal, and *Quanets* are needed. See Figure 1.16 (on page 37) for an example of financial application.	[AWS21], [BAK23], [BOU20], [BRA21], [CHO19], [DAL23], [EGG20], [LAD10], [NIC14], [ORU19], [REB18], [STA20], [WEH18], [WOE19], etc.
Secure access to remote quantum computers	Secure access to remote QCs in the cloud (secure cloud *Quacomp*). Here a quantum terminal that can generate and measure individual qubits can access a remote QC in such a way that the remote processor is unable to detect that activity and which computation the quantum terminal has performed. Quantum cloud services typically are accessed via Application Programming Interfaces (APIs), supporting the execution of quantum circuits on real Quantum Processing Units (QPUs) or on quantum simulators running on classical hardware.	[BHA20], [DEN08], [KAI19], [KIM08], [MAR16], [VIE21], etc.
Secure identification	Here user U proves to server S that he knows an agreed (possibly low-entropy) password *w*, while giving away as little information on *w* as possible—namely, the adversary can exclude at most one possible password for each execution of the scheme. Solutions in the bounded-quantum-storage model have been proposed, where U and S may exchange qubits, and a dishonest party is assumed to have limited quantum memory. An improved version deals with a man-in-the-middle attack, but requires U and S to additionally share a high-entropy key *k*; however, security is still guaranteed if one party loses *k* to the attacker but notices the loss. In both versions of the scheme, the honest participants need no quantum memory, and noise and imperfect quantum sources can be tolerated.	[DAM14], etc.
Superdense Coding (SD)	Superdense coding is a fundamental *Quacom* protocol. It enables one to transmit two classical bits of information by transmitting only one quantum bit from sender to receiver. Superdense coding employs quantum entanglement. Superdense coding enables one to transmit several classical bits of information simply by transmitting a (much) smaller number of qubits; the protocol operates under the assumption that the sender and the receiver have pre-shared an entangled resource. For example, when looking at a simple link with Alice and Bob at respective endpoints, the protocol involves having the two parties share a pair of maximally entangled qubits, thus allowing Alice to transmit a semi-nibble (two bits—i.e., one of 00, 01, 10, or 11) to Bob by sending only one qubit.	[SRI21], among others *(continues)*

Table 1.2 Some Currently-Envisioned Quacom or Quacom-Related Applications (*cont.*)

Application	Description	Key References
Superdense Coding (SD), (*cont.*)	This protocol was developed in the early 1970s and validated experimentally in the mid-1990s. Superdense coding can be perceived as being the opposite of quantum teleportation; in the latter, given that Alice and Bob have a pre-shared Bell pair, a transfer of one qubit from Alice to Bob is accomplished by transmitting two classical bits between the parties.	
Voting using a *Quanet*	Proposals have been made for conducting a vote using a quantum network. A tally quantum node generates a first plurality of ballot quantum states encoding a first bit string, each of the first plurality of ballot quantum states comprising a plurality of qubits. Each of the first plurality of ballot quantum states are distributed to exactly one of a plurality of voter quantum nodes via a quantum network. At least one of the plurality of voter quantum nodes performs a projective measurement of its one of the first plurality of ballot quantum states, thereby determining a parity of a random pair of bits of the first bit string; reads a vote; computes a first encoded vote based on the parity and vote; and broadcasts the first encoded vote and an identifier of the pair of bits to each other of the plurality of voter quantum nodes. The first bit string is provided to decode the first encoded vote. These methods allow casting a vote using a quantum network. A ballot quantum state is received via a quantum network, the ballot quantum state encoding a first bit string in a plurality of qubits. A projective measurement of its one of the first plurality of ballot quantum states is performed, thereby determining a parity of a random pair of bits of the first bit string. A vote is read. A first encoded vote is computed based on the parity and vote. The first encoded vote and an identifier of the pair of bits is broadcast to each other of the plurality of voter quantum nodes.	[KHA23]

- If two qubits at different network nodes are entangled with each other, then any measurement on qubit 1 at node A and the *same* measurement on qubit 2 at node B instantaneously result in the same answer. This feature makes entanglement well suited for applications that require coordination (for example, but not limited to, clock synchronization, or achieving consensus about data in a cloud-based environment).

- Quantum entanglement cannot be shared—namely, if two qubits are entangled with each other, then it is impossible for a third qubit to be entangled with either of them. This makes entanglement private, supporting security-oriented applications such as generating encryption keys or secure identification.

To support distributed quantum applications, a Quanet node must (i) enable local quantum computation—that is, the execution of quantum gates and measurements at each end node in the network on which applications are run; and (ii) enable the generation of quantum entanglement between any two nodes in such a network [VAR22]. A typical Quanet application consists of both local quantum computations and the generation of entanglement. Some applications have more of a need for local quantum processing (being computation-heavy), while other applications have more of a need for entanglement generation (being network-heavy). An example of the former is secure delegated quantum computation; an example of the latter is QKD, which is characterized by entanglement generation—the only local operations are measurements.

In practical terms, at the time of this writing, quantum applications were typically hybrid, consisting of both classical and quantum components; the latter require QCs for execution, which were often offered as cloud services. It follows that to implement quantum applications, developers need the integration of quantum and classical components of the application [VIE21]. Most applications need long-lived superpositions and entanglement; the fact is that currently, relatively short-lived superpositions and entanglement are achievable; short-lived superpositions are used in demonstrations of devices known as intermediate-scale quantum processors [NIS23].

As the world becomes more interconnected—particularly in an Internet of Things (IoT) setting [MIN07], [MIN13]—both the threats and the consequences of cyberattacks are becoming worrisome. These threats are particularly pernicious in energy delivery systems, in which cyberattacks not only have an economic impact, but also endanger safety. Ensuring a more secure, resilient grid for the future is of utmost importance, requiring countermeasures at all levels of the security stack, from the physical layer to the human users themselves. QCs could break public key cryptosystems based on number theory such as RSA (Rivest, Shamir, Adleman) and Diffie-Hellman (DH) key exchange schemes, which are widely in use today; thus quantum-based methods are of interest. Because QKD relies on the laws of physics, and not the hardness of a mathematical problem, QKD represents a viable tool for future-proofing the security of various communication networks, including the security of networks supporting energy-delivery systems (e.g., smart grids [LUK23]), IoT sensor networks, and 5G/6G networks.

The fact that qubits cannot be copied, and that any efforts to do so can be detected, makes quantum techniques well suited for security applications. QKD is a technique that facilitates the sharing of cryptographic keys between two parties (between a transmitter often referred to as "Alice"; and a receiver often referred to as "Bob"). See Figures 1.13, 1.14 and 1.15. The salient advantage of this technique is that it provides a test of whether any part of the key can be known to an unauthorized eavesdropper, often referred to as "Eve". In a typical QKD setting, Alice and Bob use two or more non-orthogonal bases in which to encode the bit values [WOO23]. The laws of QM dictate that measurement of the photons by Eve without prior knowledge of the encoding basis of each causes an intrinsic modification to the state of some of the photons [LO99], [MAY01], [REN08], [WOO82]. These state changes of the photons will give rise to errors in the bit values sent between Alice and Bob, and thus, by comparing a part of their common bit string, Alice and Bob can determine if Eve has gained information: one attributes all channel impairments (such as loss or noise) measured by Alice and Bob during a channel-estimation step of the protocol to inimical actions by Eve. While QKD's potential is demonstrable, it faces some practical challenges that are retarding widespread adoption, including speed of communication and lack of direct compatibility with existing fiber-optic infrastructure.

Teleportation allows Alice to "send" a quantum state to Bob without "really sending anything". It entails the transfer of "quantum states" between separate atoms (or particles), transferring specified properties of one atom (or particle) to another atom (or particle)—the mechanism by which quantum information is transmitted from one endpoint to another endpoint, with the support of classical communication and previously shared quantum entanglement between the two endpoints. Quantum teleportation is the basic process for reliably transmitting qubits across (lossy) links, and it is an important primitive function for Quanet protocols and applications. Although experimental demonstrations of qubit teleportation have been performed and documented on several quantum network platforms in recent years, progressing beyond directly connected nodes has been frustrated by the onerous requirements on the pre-shared remote entanglement, joint qubit readout, and coherence times [HER22].

[Text continues on page 39]

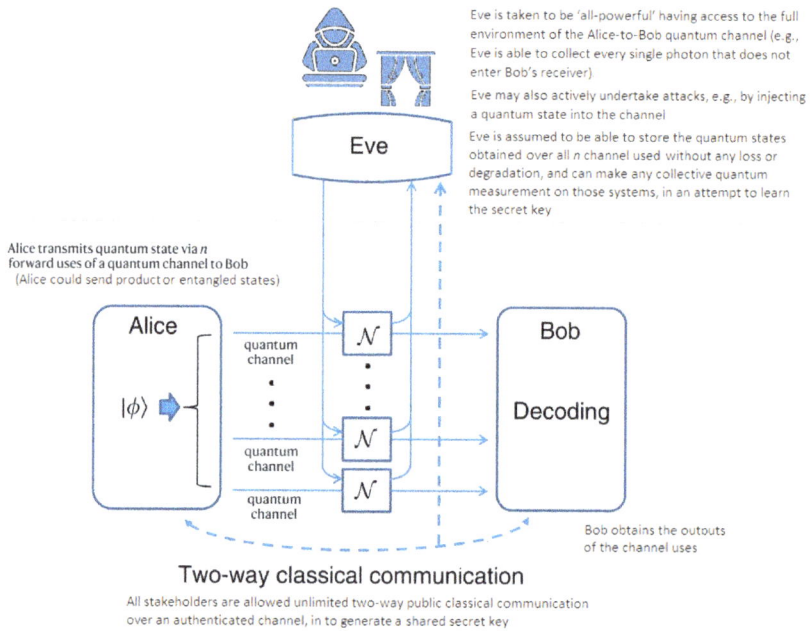

Eve is taken to be 'all-powerful' having access to the full environment of the Alice-to-Bob quantum channel (e.g., Eve is able to collect every single photon that does not enter Bob's receiver)

Eve may also actively undertake attacks, e.g., by injecting a quantum state into the channel

Eve is assumed to be able to store the quantum states obtained over all n channel used without any loss or degradation, and can make any collective quantum measurement on those systems, in an attempt to learn the secret key

Eve

Alice transmits quantum state via n forward uses of a quantum channel to Bob
(Alice could send product or entangled states)

Alice

$|\phi\rangle$

$\mathcal{N}$
quantum channel

$\mathcal{N}$
quantum channel

$\mathcal{N}$
quantum channel

Bob

Decoding

Bob obtains the outouts of the channel uses

Two-way classical communication

All stakeholders are allowed unlimited two-way public classical communication over an authenticated channel, in to generate a shared secret key

Figure 1.13   Generic point-to-point QKD protocol

Schematic of a point to point QKD system

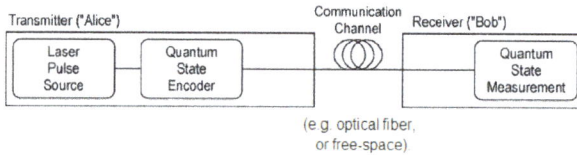

(e.g. optical fiber, or free-space)

Schematic of a point to point QKD system with control electronics

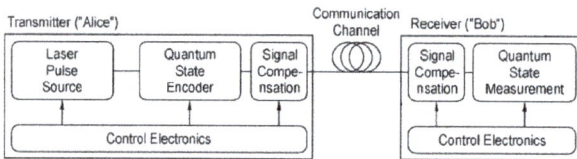

The transmitter (often referred to as "Alice") comprises a source of pulsed radiation and a state encoder. The transmitter generates quantum states, which are coherent states formed by the pulsed laser emission of the source of pulsed radiation and the state encoder (which modulates the light with a bit value (0 or 1) and in a random basis (e.g., X or Y basis)). Various different encoding schemes exist (e.g., polarization encoding, time-bin encoding etc.); all QKD protocols rely on high-speed high-quality state generation (i.e., negligible error between expected encoding value and actual encoded value).

The encoded light is transmitted along the communication channel, where it may experience optical phenomena that change some of the pulse properties—for example, dispersion may broaden the pulse, polarization fluctuations may alter the polarization state, channel timing delays may change the expected pulse arrival time, etc.

At the receiver ("Bob"), the quantum state is measured, which involves randomly choosing a basis for measurement (e.g., by applying some unitary operation on the quantum state) using a demodulator (not shown) and then detecting the signal using single-photon detectors (not shown) to form a quantum state measurement. Although various detection approaches exist, they usually rely on the ability to accurately determine the encoded bit value while rejecting noise sources such as detector noise or other noise introduced by the channel.

Following the generation, transmission and measurement of quantum states, a post-processing stage is performed involving authenticated classical communication between Alice and Bob, where they reveal a subset of their random choices to perform sifting, information reconciliation, error correction and privacy amplification. This results in an information theoretically secure quantum key distributed on both the remote nodes, which can then be delivered to other equipment for e.g. data encryption.

A basic quantum communication protocol such as polarization is used.

Figure 1.14   QKD system (Adapted from [WOO23])

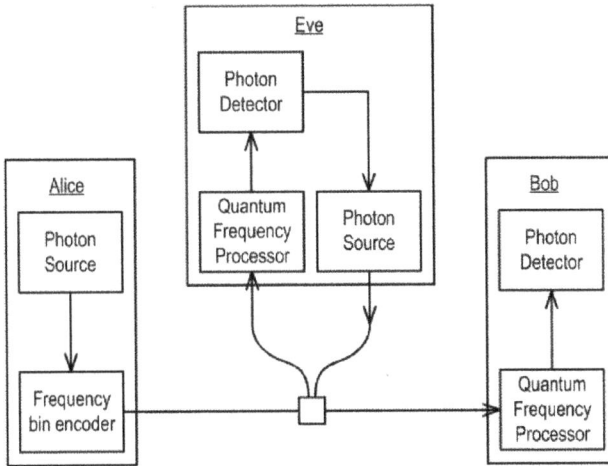

Figure 1.15   Schematic of a quantum key distribution network ([LUK23])

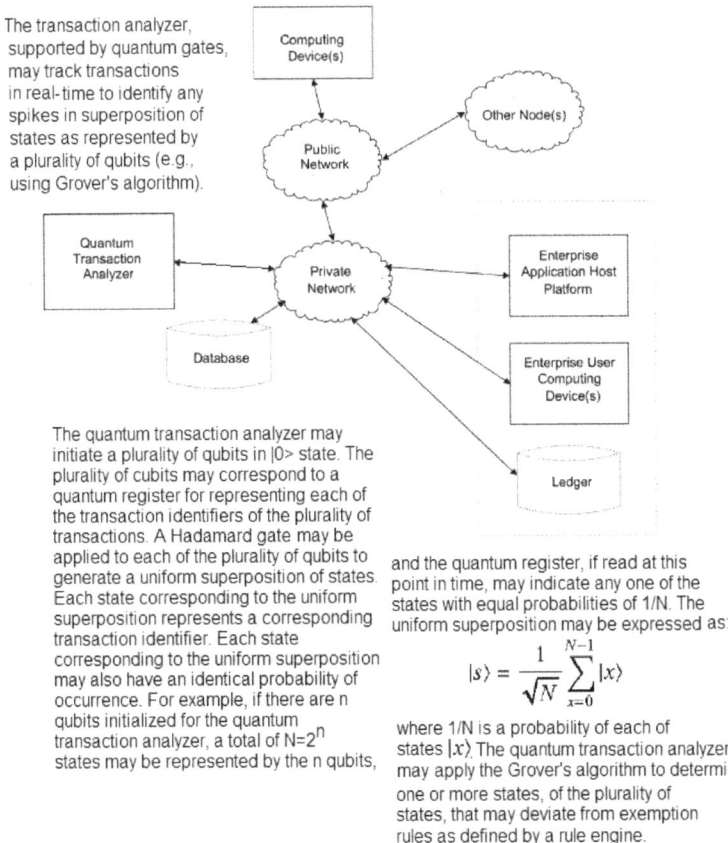

The transaction analyzer, supported by quantum gates, may track transactions in real-time to identify any spikes in superposition of states as represented by a plurality of qubits (e.g., using Grover's algorithm).

The quantum transaction analyzer may initiate a plurality of qubits in |0> state. The plurality of cubits may correspond to a quantum register for representing each of the transaction identifiers of the plurality of transactions. A Hadamard gate may be applied to each of the plurality of qubits to generate a uniform superposition of states. Each state corresponding to the uniform superposition represents a corresponding transaction identifier. Each state corresponding to the uniform superposition may also have an identical probability of occurrence. For example, if there are n qubits initialized for the quantum transaction analyzer, a total of $N=2^n$ states may be represented by the n qubits, and the quantum register, if read at this point in time, may indicate any one of the states with equal probabilities of 1/N. The uniform superposition may be expressed as:

$$|s\rangle = \frac{1}{\sqrt{N}} \sum_{x=0}^{N-1} |x\rangle$$

where 1/N is a probability of each of states $|x\rangle$. The quantum transaction analyzer may apply the Grover's algorithm to determine one or more states, of the plurality of states, that may deviate from exemption rules as defined by a rule engine.

Figure 1.16   Example of Quacomp applied to financial transactions (synthetized from [BAK23])

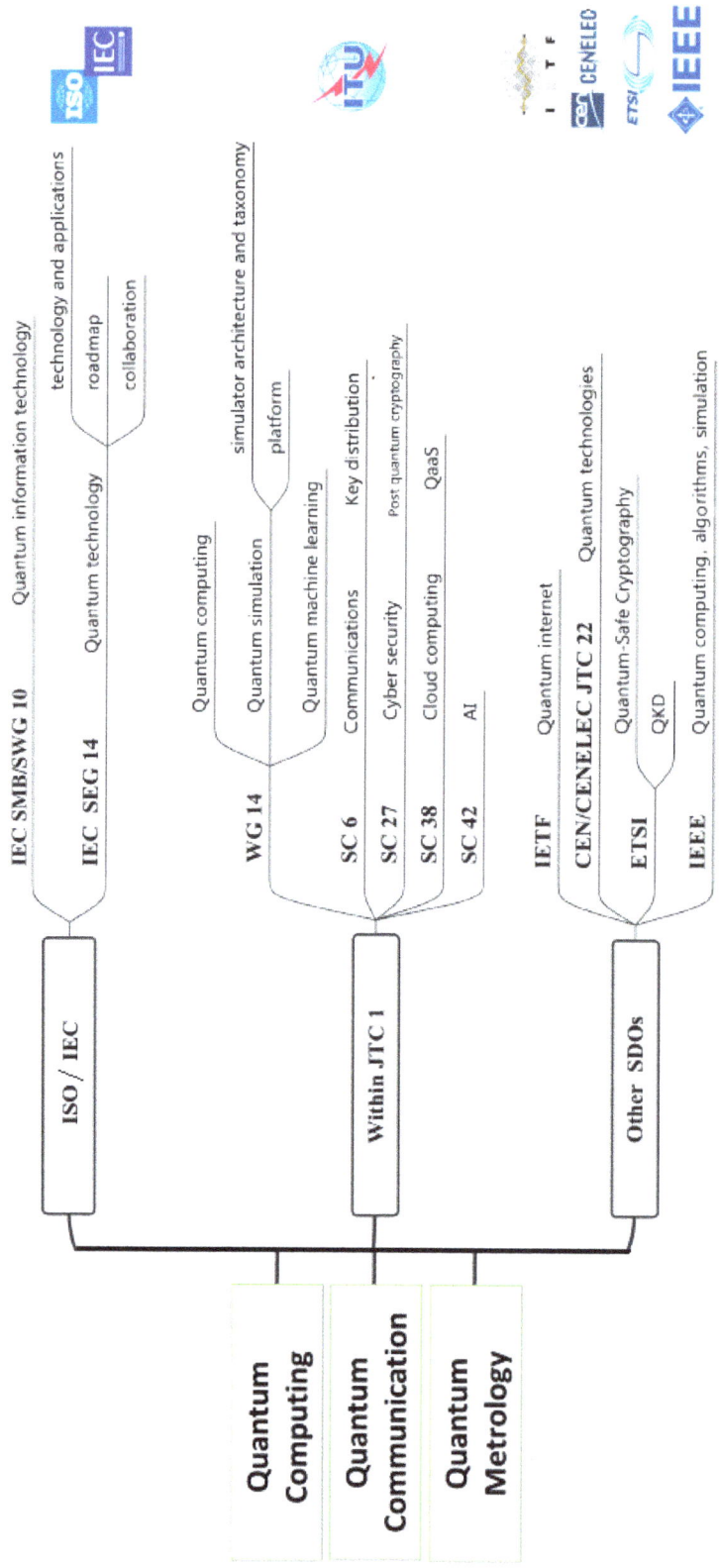

Figure 1.17 Standardization efforts related to Quacom/Quanets (*Source: ISO/IEC JTC 1/WG 14 [YAN23]*)

Table 1.3 Key Areas of Applied Research and Development Focus for QIST

Area of QIST focus	Description of activities related to *QIST* area
Quacomp	Gate-based *QCs*, non-gate-based *QCs*, simulators, hybrid computing, error correction technologies
Quacom/Quanets	Quantum (entanglement) networks; quantum/classical hybrid network; *Quanet* network elements (e.g., memories, quantum repeaters, routers, switches, transducers); standardization
Quantum application software/algorithm	Domain-specific software and algorithms; Quantum Key Distribution (QKD); quantum sensing (e.g., inertial, electric/magnetic fields, imaging, precision time, mass, temperature)
Quantum benchmarks	Application-oriented benchmarks; system-oriented benchmarks; quantum energy metrics
Quantum-enabling technologies and research	Quantum materials; 2D and 3D structure materials; photonics; single photon sources; single electron sources; photon counter/coincidence detectors; control electronics; cryogenics; electromagnetic compatibility (EMC); optical fibers

Many Quacom applications are significantly more complex than the QKD application cited above, and networks capable of supporting them are just nascent at this juncture; but significant overall progress is anticipated in the next five years—that is, by 2030.

1.8 Quanet Standardization Initiatives

There are several standardization efforts related to Quacom/Quanets, as depicted in Figure 1.17 from [YAN23]. Standardization facilitates the commercialization of the technology. The topic is revisited in Chapter 6.

1.9 Scope of This Book

Table 1.3 (partially based on ISO/[ALL23]) and Figure 1.18 (CEN-CENELEC/[CEN24]) summarize some key areas of applied research and development focus for QIST that will promote and expedite commercialization of the technology in the next few years before or at the turn of the decade. Many of these topics are addressed in the chapters that follow.

There is a large body of academic information on the topic of quantum communications and quantum computing, but either it is not organized in an organic, pedagogical format or it is highly mathematically focused. Even extensive surveys of the Quacomp field (e.g., [DAL23] among others) fail to address the Quacom/Quanets/Quaint aspect. A handful of other book text sources emerged in the recent past, but the goal of this book is to make the science accessible to the "person of ordinary skill in the art"—a typical telecom practitioner—and to provide an up-to-date description of the discipline.[1]

The senior author's 2006 book *Nanotechnology Applications to Telecommunications and Networking* [MIN06] contained several fairly extensive supplementary appendices addressing substantive QM and Quacomp concepts, as follows:

---

[1] Apparently (just) the following at the time of this writing, in reverse chronological order: [LOK23], [ROH21], [FUR20], [DOW20], and [VAN14].

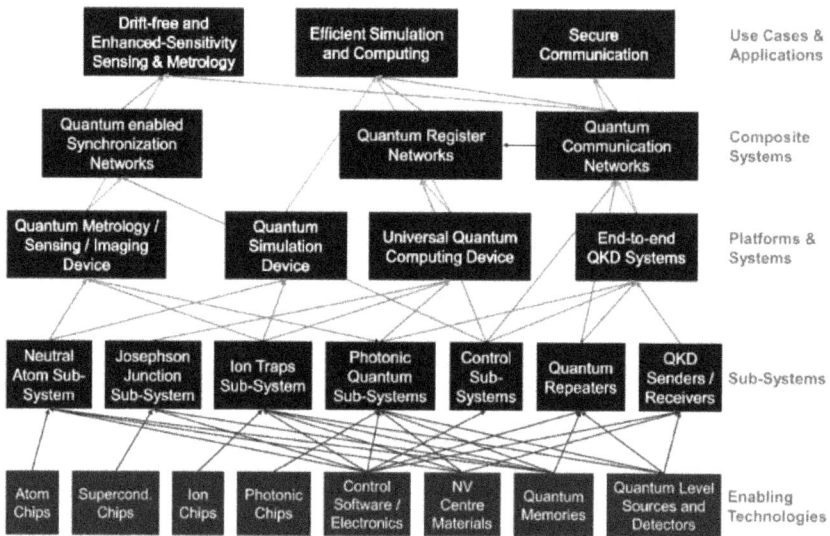

Figure 1.18   A taxonomy of key quantum elements (*Source:* CEN-CENELEC FGQT [CEN24])

- Appendix B: Introduction to Hilbert Spaces
- Appendix D: Basic Nanotechnology Science—Quantum Physics
- Appendix E: Mechanical Molecular Models and Quantum Aspects of Chemistry
- Appendix G: Quantum Computing

Two decades hence, this text revisits the topic at a more pragmatic level, documenting significant practical progress made in the implementation of the QM theory into near-commercial Quacomp and Quacom solutions.

After this introductory overview chapter, Chapter 2 covers basic concepts in quantum physics by describing some mathematics needed for the discussion, while trying to keep the machinery to a minimum. Chapter 3 continues the discussion by addressing some more advanced mathematics, but with an eye for clarity. Mathematical foundations of quantum information, QIST in particular, deal with defining quantum states, qubits, multiple qubits and quantum entanglement, mixed states, vector and matrix representation, quantum operations (e.g. Pauli Operators), quantum measurements, collapse of the wave function, distance measures, and fidelity, among other concepts.

Chapter 4 covers basic concepts in quantum physics by describing some physical technologies, while also keeping the science to a sufficient minimum. Chapter 5 provides a basic overview of quantum computers, being that these nodes constitute the endpoints to be supported by the quantum Internet over the horizon. Chapter 6 focuses on quantum communications, including logical architectures and the Quaint. Next, in Chapter 7, we endeavor to provide an accessible discussion of Quantum-based security. Chapter 8 offers information on software tools and languages that can be employed to develop networked quantum applications.

Although it is desirable to work through Chapter 2 and 3 (which, parenthetically, only cover a subset of the QM-supportive mathematics), some readers may opt to either skip or do a superficial reading of these two chapters and endeavor to absorb the material in the rest of the text without the formal mathematics.

One of the challenges in QM for the novice is that the nomenclature has "a life of its own": one ends up trusting the formality of the nomenclature and "hope" that the concatenated sequence of (linear algebra) transforms "carry the day" (which obviously it does), but where the intuitive, granular understanding is sublimated to the symbology; or, one can endeavor to struggle through the actual algebraic (matrix) operations and validate to one's satisfaction that the end result is actually correct (provable) with the detailed mathematical steps. The initiated may be comfortable with the formal nomenclature, while the novice may desire to work through (some of) the stated relationships.

References

[ABE23] A. J. G. Abelém, G. Vardoyan, and D. Towsley, "Quantum Internet: The Future of Internetworking". April 2023, arXiv:2305.00598v1.

[ABR13] S. Abruzzo, S. Bratzik, et al., "Quantum Repeaters and Quantum Key Distribution: Analysis of Secret-Key Rates". Physical Review A, 2013, 87(5):052315.

[ABR13a] S. Abruzzo, "Long Distance Quantum Key Distribution with Quantum Repeaters". Inaugural-Dissertation zur Erlangung des Doktorgrades der Mathematisch-Naturwissenschaftlichen Fakultat der Heinrich-Heine-Universit at Dusseldorf, Dusseldorf, November 2013.

[AHN23] B. Ahn, S. Lee, et al., Method and Device for Estimating Quantum Bit Error Rate on Basis of Maximum Bit Group and Two-Dimensional Parity. U.S. Patent Application 20230299950, 2023-09-21. Uncopyrighted material.

[ALB18] T. Albash, D. A. Lidar, "Adiabatic Quantum Computing". Rev. Mod. Phys. 90, 015002, 2018. 2 Feb 2018. https://doi.org/10.48550/arXiv.1611.04471

[ALF08] M. Alford, "The Essentials of Quantum Mechanics". Department of Physics, Washington University, St. Louis, Missuri. 2008-Oct-22. Available online on 11/29/2023 at https://web.physics.wustl.edu/alford/physics/essentials.pdf.

[ALL23] C. Allocca, "IEC/SEG14 Quantum Technologies". ISO/IEC JTC 1 Workshop on Quantum Information Technology Standardization. 03 April 2023. National Institute of Standards and Technology (NIST).

[ALS21] M. Alshowkan, B. P. Williams, et al., "Reconfigurable Quantum Local Area Network Over Deployed Fiber". PRX Quantum 2, 040304; 2021. doi: 10.1103/PRXQuantum.2.040304.

[AND24] D. Andreatta, "Faster Than the Speed of Light: Information Transfer Through "Spooky Action at a Distance" at the Large Hadron Collider". The SciTechDaily. June 24, 2024. Available online on June 24, 2024 at https://scitechdaily.com/faster-than-the-speed-of-light-information-transfer-through-spooky-action-at-a-distance-at-the-large-hadron-collider/

[ARR06] P. Arrighi, L. Salvail, "Blind Quantum Computation". Int. J. Quant. Inf. 4, 883; 2006.

[AWS18] D. D. Awschalom, R. Hanson, et al., "Quantum Technologies with Optically Interfaced Solid-State Spins". Nat. Photon. 12, 516–527; 2018.

[AWS21] D. Awschalom, K.K. Berggren, et al., "Development of Quantum Interconnects (QuICs) for Next-Generation Information Technologies". PRX Quantum 2, 017002— Published 24 February 2021. Available online on 11/12/2023 at https://journals.aps.org/prxquantum/abstract/10.1103/PRXQuantum.2.017002.

[BAK23] S. Bakshi, S. K. Paini, et al., Dynamic Quantum Enabled Method for Large Currency Transaction Exemption Using Distributed Hash Chain. U.S. Patent Application 20230334489, 2023-10-19. Uncopyrighted material.

[BAR12] S. Barz, E. Kashefi, et al., "Demonstration of Blind Quantum Computing". Science

335, 303; 2012.

[BAU20] B. Bauer, S. Bravyi, et al., "Quantum Algorithms for Quantum Chemistry and Quantum Materials Science". Chem. Rev. 120, 12685; 2020.

[BAY23] M. J. Bayerbach, S. E. D'aurelio, et al., "Bell-State Measurement Exceeding 50% Success Probability with Linear Optics". Science Advances, 9 Aug 2023, Vol 9, Issue 32. doi: 10.1126/sciadv.adf4080.

[BEL04] J. S. Bell, *Speakable and unspeakable in quantum mechanics: Collected papers on quantum philosophy*. Cambridge University Press, 2004.

[BEL64] J. S. Bell, "On the Einstein Podolsky Rosen Paradox". Physics Physique Физика. 1 (3): 195–200. November 1964. doi:10.1103/PhysicsPhysiqueFizika.1.195.

[BEN84] C. H. Bennett, G. Brassard, "Quantum Cryptography: Public-Key Distribution and Coin Tossing". In Proceedings of IEEE International Conference on Computers, Systems and Signal Processing, Bangalore, India, 1984, pp. 175-179. Also, Theor. Comput. Sci. 2014;560:7. doi: 10.1016/j.tcs.2014.05.025.

[BEN92] C. H. Bennett, S. J. Wiesner, "Communication via One and Two-Particle Operators on Einstein-Podolsky-Rosen States". Phys. Rev. Lett. 69, 2881; 1992.

[BEN93] C. H. Bennett, G. Brassard, et al., "Teleporting an Unknown Quantum State Via Dual Classical and Einstein-Podolsky-Rosen Channels". Physical Review Letters, 70(13):1895, 1993. https://doi.org/10.1103/PhysRevLett.70.1895.

[BEN98] C.H. Bennett, P.W. Shor, "Quantum Information Theory". IEEE Transactions on Information Theory, Volume: 44, Issue: 6, October 1998. 44(6):2724–2742, 1998.

[BHA20] M. K. Bhaskar, R. Riedinger, et al., "Experimental Demonstration of Memory-Enhanced Quantum Communication". Nature 580, 60–64, 2020.

[BLU24] D. Bluvstein, S.J. Evered, et al., "Logical Quantum Processor Based on Reconfigurable Atom Arrays". Nature 626, 58–65, 2024. https://doi.org/10.1038/s41586-023-06927-3.

[BON03] D. G. Bonchev, D.H. Rouvray, *Complexity: Introduction and Fundamentals*, CRC Press, 2003. ISBN-13: 978-0415287913.

[BOU20] A. Bouland, W. van Dam, et al., "Prospects and Challenges of Quantum Finance". arXiv:2011.06492, 2020.

[BRA21] M. C. Braun, T. Decker, et al., "A Quantum Algorithm for the Sensitivity Analysis of Business Risks". arXiv:2103.05475, 2021.

[BRA23] S. Bravyi, A. W. Cross, et al., "High-Threshold and Low-Overhead Fault-Tolerant Quantum Memory". August 15, 2023, https://doi.org/10.48550/arXiv.2308.07915.

[BRI98] H. J. Briegel, W. Dúr, et al., "Quantum Repeaters: The Role of Imperfect Local Operations in Quantum Communication". Phys. Rev. Lett. 81, 5932–5935; 1998. https://doi.org/10.1103/PhysRevLett.81.5932.

[BRO10] A. Broadbent, J. Fitzsimons, E. Kashefi, "Measurement-Based and Universal Blind Quantum Computation". In: A. Aldini, M. Bernardo, et al (eds), *Formal Methods for Quantitative Aspects of Programming Languages*. SFM 2010. Lecture Notes in Computer Science, vol 6154. Springer, Berlin, Heidelberg. https://doi.org/10.1007/978-3-642-13678-8_2.

[CAS18] D. Castelvecchi, "The Quantum Internet Has Arrived (And It Hasn't)". Nature 554, 289–292; 22 February 2018. doi: https://doi.org/10.1038/d41586-018-01835-3

[CAS23] D. Castelvecchi, "IBM Releases First-Ever 1,000-Qubit Quantum Chip". 04 December 2023, Nature. Available online on December 5, 2023 at https://www.nature.com/articles/d41586-023-03854-1.

[CEN24] CEN-CENELEC FGQT, FGQT Q03: Towards Standardization for Quantum Technologies. Available online on April 18, 2024 at https://www.cencenelec.eu/areas-of-work/cen-cenelec-topics/quantum-technologies/

[CHE10] T.-Y. Chen, J. Wang, et al., "Metropolitan All-Pass and Inter-City Quantum Communication Network". Opt. Express 18, 27217; 2010. https://doi.org/10.1364/OE.18.027217.

[CHE22] X. Chen, L. Chen, Y. Yan, "Detecting a Photon-Number Splitting Attack in Decoy-State Measurement-Device-Independent Quantum Key Distribution via Statistical Hypothesis Testing". Entropy (Basel). 2022 Sep 2;24(9):1232. doi:10.3390/e24091232.

[CHE23] Y. Chen, S. Zhang, Controlling a Quantum Computing System Using Imaginary-Time Evolution. U.S. Patent Application 2023/0289639, 2023-09-14. Uncopyrighted material.

[CHI23] A. M. Childs, H. Fu, et al., "Streaming Quantum State Purification". arXiv: 2309.16387v1. 28 Sep 2023. https://doi.org/10.48550/arXiv.2309.16387.

[CHO19] H. Choi, M. Pant, et al., "Percolation-Based Architecture for Cluster State Creation Using Photon-Mediated Entanglement Between Atomic Memories". npj Q. Inform. 5, 104; 2019. https://doi.org/10.1038/s41534-019-0215-2.

[CHO22] C. Q. Choi, "IBM Unveils 433-Qubit Osprey Chip - Next Year Entanglement Hits the Kilo-Scale with Big Blue's 1,121-qubit Condor". IEEE Spectrum, 09 November 2022.

[CHU23] J. Chu, X. He, et al., "Scalable Algorithm Simplification Using Quantum AND logic". Nat. Phys. 19, 126–131; 2023. https://doi.org/10.1038/s41567-022-01813-7.

[CRA00] J. S. McCranie, "A Study of Hyperperfect Numbers". Journal of Integer Sequences, 3: 13, 2000. #P00.1.3.

[DAL23] A. M. Dalzell, S. McArdle, et al., "Quantum Algorithms: A Survey of Applications and End-to-End Complexities". arXiv:2310.03011v1; 4 Oct 2023. https://arxiv.org/pdf/2310.03011.pdf.

[DAM14] I. Damgård, S. Fehr, et al., "Secure Identification and QKD in the Bounded-Quantum-Storage Model". Theor. Comput. Sci. 560, 12; 2014. Lecture Notes in Computer Science book series (LNSC, Volume 4622). doi: 10.1016/j.tcs.2014.09.014

[DAW23] R. Daws, "Quantum AI Represents a 'Transformative Advancement'". AI News, November 14, 2023. Available online on 11/18/2023 at https://www.artificialintelligence-news.com/2023/11/14/quantum-ai-represents-transformative-advancement/

[DEC24] M. DeCross, A. Ellinor, et al., "Quantum Tunneling". Available online on January 12, 2024 at https://brilliant.org/wiki/quantum-tunneling/.

[DEG17] C. L. Degen, F. Reinhard, P. Cappellaro, "Quantum Sensing". Reviews of Modern Physics, vol. 89, no. 3, p. 035 002, 2017.

[DEN08] V. S. Denchev, G. Pandurangan, "Distributed Quantum Computing". ACM SIGACT News 39, 77; 2008. doi: 10.1145/1412700.1412718.

[DIA21] S. DiAdamo, M. Ghibaudi, and J. Cruise, "Distributed Quantum Computing and Network Control for Accelerated VQE". IEEE Trans. Quantum Eng. 2, 1; 2021.

[DIA22] S. DiAdamo, B. Qi, et al., "Packet Switching in Quantum Networks: A Path to the Quantum Internet". Physical Review Research 4, 043064; 2022.

[DIB18] A. M. Dibos, M. Raha, et al., "Atomic Source of Single Photons in the Telecom Band". Phys. Rev. Lett. 120, 243601; 2018.

[DOE22] U.S. DOE, "New Quantum Network Shares Information at a Scale Practical for Future Real-World Applications". April 21, 2022. Available on November 14, 2023 at https://www.energy.gov/science/ascr/articles/new-quantum-network-shares-informa

tion-scale-practical-future-real-world.

[DOW20] J. P. Dowling, *Schrödinger's Web: race to build the quantum Internet*. CRC Press, 08/25/2020, ISBN-10:0367322315, ISBN-13: 9780367322311.

[DUE06] M. Dušek, N. Lütkenhaus, M. Hendrych, "Quantum Cryptography". Prog. Opt. 2006; 49:381.

[EGG20] D. J. Egger, C. Gambella, et al., "Quantum Computing for Finance: State-of-the-Art and Future Prospects". IEEE Trans. Quantum Eng. 1, 1, 2020.

[EMM12] F. Emmert-Streib, M. Dehmer, "Exploring Statistical and Population Aspects of Network Complexity". PLOS ONE. May 8, 2012. https://doi.org/10.1371/journal.pone.0034523.

[ENG23] A. Engelsberger, T. Villmann, et al "Quantum Computing Approaches for Vector Quantization—Current Perspectives and Developments". Entropy (Basel). 2023 Mar; 25(3): 540. Published online 2023 Mar 21. doi: 10.3390/e25030540.

[EVA21] P. G. Evans, D. Earl, et al., "Trusted Node QKD at an Electrical Utility". IEEE Access 9, 105220; 2021. doi: 10.1109/ACCESS.2021.3070222.

[FAR24] F. Farooque, "Rags to Riches: 3 Quantum Computing Stocks That Could Make Early Investors Rich". InvestorPlace, June 2024. Available online on June 21, 2024 at https://investorplace.com/2024/06/rag-to-riches-3-quantum-computing-stocks-that-could-make-early-investors-rich/

[FIT17] J.F. Fitzsimons, "Private Quantum Computation: An Introduction to Blind Quantum Computing and Related Protocols". NPJ Quantum Inf 3, 23; 2017. https://doi.org/10.1038/s41534-017-0025-3.

[FIT23] R. Fitzpatrick, "Eigenstates and Eigenvalues". Available online on 10/25/2023 at https://phys.libretexts.org/Bookshelves/Quantum_Mechanics/Introductory_Quantum_Mechanics_(Fitzpatrick)/03%3A_Fundamentals_of_Quantum_Mechanics/3.08%3A_Eigenstates_and_Eigenvalues.

[FUR20] G. Fürnkranz, A. Aglibut (Translator), R. Ursin (Foreword by), *The Quantum Internet: Ultrafast and Safe from Hackers*, 04/07/2020, Springer International Publishing, ISBN-13: 9783030426644.

[FUR23] W. Furtner, Multi Dimensional Electrode Controller for Quantum Computing. U.S. Patent Application 20230325698, 2023-10-12. Uncopyrighted material.

[GIS02] N. Gisin, G. Ribordy, et al., "Quantum Cryptography". Rev. Mod. Phys. 2002;74:145. doi:10.1103/RevModPhys.74.145.

[GIS07] N. Gisin, R. Thew, "Quantum Communication". Nat. Photon. 2007;1:165. doi: 10.1038/nphoton.2007.22.

[GRE23] T. Greene, "IBM Brings 'Utility-Scale' Quantum Computing to Japan as China and Europe Struggle to Compete". Cointelegraph, November 27, 2023. Available online on 11/29/2023 at https://cointelegraph.com/news/ibm-brings-utility-scale-quantum-computing-to-japan-china-europe-struggle-compete.

[GRI12] R. B. Griffiths, "Quantum Channels, Kraus Operators, POVMs". 22 March 2012. Available online on December 20, 2023 at https://quantum.phys.cmu.edu/QCQI/qitd412.pdf.

[GUO19] X. Guo, C. R. Breum, et al., "Distributed Quantum Sensing in a Continuous-Variable Entangled Network". Nat. Phys. 16, 281–284, 2019.

[HAV23] F. Haverkamp, J. Saalmueller, et al., Multitasking Scheme for Quantum Computers. U.S. Patent Application 20230342652. 2023-10-26. Uncopyrighted material.

[HER22] S. N. L. Hermans, M. Pompili, et al., "Qubit Teleportation Between Non-Neighbouring

Nodes in a Quantum Network". Nature. 2022; 605(7911):663-668. doi: 10.1038/s41586-022-04697-y. Epub 2022 May 25. PMID: 35614248; PMCID: PMC9132773. https://www.ncbi.nlm.nih.gov/pmc/articles/PMC9132773/.

[HUA22] H.-Y. Huang, M. Broughton, et al., "Quantum Advantage in Learning from Experiments". Science 376, 1182, 2022.

[IEE24] IEEE, "Quantum Communications, Quantum Internet, and Applications". Special Issue Communications Magazine, IEEE Communications Society, February 2024. Available online on December 13, 2023 at https://www.comsoc.org/publications/magazines/ieee-wireless-communications/cfp/quantum-communications-quantum-internet-and

[INS23] Institute for Quantum Computing, "Quantum 101 Glossary". University of Waterloo, Waterloo, ON, Canada. Available online on December 19, 2023 at https://uwaterloo.ca/institute-for-quantum-computing/quantum-101/quantum-101-glossary#quantumsimulation

[ISM21] J. Ismael, "Quantum Mechanics". Stanford Encyclopedia of Philosophy, Edward N. Zalta (Editor). Fall 2021 Edition); Available online (10/24/2023) at https://plato.stanford.edu/entries/qm/ or https://plato.stanford.edu/archives/fall2021/entries/qm/

[ISO23] ISO/IEC DIS 4879, *Information Technology - Quantum Computing Terminology and Vocabulary*, Main Editor: Gen Lei. April 2023.

[ITU22] International Telecommunication Union—Study Period 2022-2024. "LS on Work Progress on Quantum Key Distribution (QKD) Network in SG13 (as of November 2022)". ITU-T Working Party 3/13.

[JAN18] D. Janežič, A. Miličević, et al., "Topological Complexity of Molecules. In: Meyers, R. (eds) Encyclopedia of Complexity and Systems Science". Springer, Berlin, Heidelberg. 2018. https://doi.org/10.1007/978-3-642-27737-5_554-3.

[KAI19] M. Kaiiali, S. Sezer and A. Khalid, "Cloud Computing in the Quantum Era". 2019 IEEE Conference on Communications and Network Security (CNS), Washington, DC, USA, 2019, pp. 1-4, doi: 10.1109/CNS44998.2019.8952589.

[KER19] I. Kerenidis, J. Landman, et al., "qmeans: A Quantum Algorithm for Unsupervised Machine Learning". Adv. Neural Inf. Process. Syst. 32, 2019.

[KHA23] E. T. Khabiboulline, M. Lukin, et al., Efficient Quantum Voting with Information-Theoretic Security. U.S. Patent Application 20230327862, 2023-10-12. Uncopyrighted material.

[KIM08] H.J. Kimble, "The Quantum Internet". Nature Vol. 453, Iss. 7198, pp. 1023-1030; 2008. June 2008. https://www.nature.com/articles/nature07127. doi: 10.1038/nature07127.

[KOI22] T. Koike-Akino, P. Wang, Y. Wang, "Quantum Transfer Learning for Wi-Fi Sensing". In: ICC 2022-IEEE International Conference on Communications, pp. 654–659. 2022. doi: 10.1109/ICC45855.2022.9839011.

[KOM13] P. Komar, E. M, Kessler, et al., "A Quantum Network of Clocks". October 22, 2013, arXiv:1310.6045, https://doi.org/10.48550/arXiv.1310.6045

[KOM14] P. Kómár, E. M. Kessler, et al., "A Quantum Network of Clocks". Nature Phys 10, 582–587, 2014. https://doi.org/10.1038/nphys3000.

[KOR23] M. Kordzanganeh, D. Kosichkina, A. Melnikov, "Parallel Hybrid Networks: An Interplay Between Quantum and Classical Neural Networks". Intelligent Computing, 9 Oct 2023, Vol. 2, Article ID: 0028. doi: 10.34133/icomputing.0028.

[KOZ23] W. Kozlowski, S. Wehner, et al., RFC 9340 - Architectural Principles for a Quantum Internet, March 2023, https://doi.org/10.17487/RFC9340.

[KRA05] B. Kraus, N. Gisin, R. Renner, "Lower and Upper Bounds on the Secret-Key Rate for

Quantum Key Distribution Protocols Using One-Way Classical Communication". Phys. Rev. Lett. 2005;95:080501. doi: 10.1103/PhysRevLett.95.080501.

[LAC23] B. Lackey, "Quantum Networking: A Roadmap to a Quantum Internet". November 1, 2023. Available online on December 20 at https://cloudblogs.microsoft.com/quantum/2023/11/01/quantum-networking-a-roadmap-to-a-quantum-internet/

[LAD10] T. D. Ladd, F. Jelezko, et al., "Quantum Computers". Nature 464, 45–53, 2010.

[LAI23] J-Y. Lai, C. Le Touze, M-H. Chou, "Cavity-enhanced Frequency Mixer for Classical and Quantum Applications". U.S Patent 11,762,262; 2023-09-19. Uncopyrighted material.

[LEE20] Y. Lee, E. Bersin, et al., "A Quantum Router Architecture for High-Fidelity Entanglement Flows in Multi-User Quantum Networks". 2020/05/04, https://doi.org/10.48550/arXiv.2005.01852.

[LES23] G. Lesovik, A. Marchenko, et al., Natural Language Processing by Means of a Quantum Random Number Generator. U.S. Patent Application 20230325152. 2023-10-12. Uncopyrighted material.

[LI19] C. Li, Y. Xu, et al., "Quantum Blockchain: A Decentralized, Encrypted and Distributed Database Based on Quantum Mechanics". Journal of Quantum Computing 2019, 1(2), 49-63. https://doi.org/10.32604/jqc.2019.06715.

[LIA18] S-K Liao, W-Q Cai, et al., "Satellite-Relayed Intercontinental Quantum Network". Phys. Rev. Lett. 120, 030501—Published 19 January 2018. https://doi.org/10.1103/PhysRevLett.120.030501.

[LID23] D. Lidar, "A Scientist Explains an Approaching Milestone Marking the Arrival of Quantum Computers". phys.org. November 20, 2023. Available online on 11/25/2023 at https://phys.org/news/2023-11-scientist-approaching-milestone-quantum.html

[LIT23] D. Litinski, First-Quantization Block Encoding for Quantum Emulation. U.S. Patent Application 20230297866, 2023-09-21. Uncopyrighted material.

[LIU16] K. Liu, "Optimization Algorithm of Cognitive Radio Spectrum Sensing Based on Quantum Neural Network". Automat. Control Comput. Sci. 50(5), 324–331, 2016. https://doi.org/10.3103/s0146411616050072.

[LIU21] Y. Liu, S. Arunachalam, and K. Temme, "A Rigorous and Robust Quantum Speed-Up in Supervised Machine Learning". Nat. Phys. 17, 1013, 2021.

[LO99] H. K. Lo, H.F. Chau, "Unconditional Security of Quantum Key Distribution Over Arbitrarily Long Distances". Science. 1999;283:2050. doi:10.1126/science.283.5410.2050.

[LOK23] S. W. Loke, *From Distributed Quantum Computing to Quantum Internet Computing: An Introduction*, 12/12/2023, Wiley, ISBN-13:9781394185511.

[LUK23] J. Lukens, P. Lougovski, Quantum Frequency Processor for Provable Cybersecurity. U.S. Patent Application 20230353351, 2023-11-02. Uncopyrighted material.

[LYO23] S. Lyon, "A New Route to a Quantum Internet". Princeton University, August 30, 2023. Available online at https://engineering.princeton.edu/news/2023/08/30/simpler-way-connect-quantum-computers#:~:text=Researchers%20have%20a%20new%20way,an%20ocean%2C%20quantum%20signals%20cannot

[MAN20] A. Manzalini, "Quantum Communications in Future Networks and Services". Quantum Rep. 2020, 2(1), 221-232; 11 March 2020. https://doi.org/10.3390/quantum 2010014

[MAN23] A. J. Manning, "Researchers Create First Logical Quantum Processor". The Harvard Gazette. December 8, 2023. Available online on December 11, 2023 at https://news.harvard.edu/gazette/story/2023/12/researchers-create-first-logical-quantum-processor/

[MAO18] Y. Mao, B.-X. Wang, et al., "Integrating Quantum Key Distribution with Classical Communications in Backbone Fiber Network". Opt. Express 26, 6010; 2018. https://doi.org/10.1364/OE.26.006010.

[MAR16] S. Muralidharan, L. Li, J. Kim, et al., "Optimal Architectures for Long Distance Quantum Communication". Sci. Rep. 6, 20463, 2016.

[MAY01] U. Mayers, "Unconditional Security in Quantum Cryptography". J. ACM. 2001; 48:351. doi: 10.1145/382780.382781.

[MAZ08] J. R. Maze, P. L. Stanwix, et al., "Nanoscale Magnetic Sensing with an Individual Electronic Spin in Diamond". Nature 455, 644–647, 2008.

[MAZ23] G. Mazzola, P. Ollitrault, I. Tavernelli, Procedure to Speed-up Variational Quantum Eigensolver Calculations in Quantum Computers. U.S. Patent 11803611, 2023-10-31. Uncopyrighted material.

[MEI23] M. E. Meinholz, P. Bordow, et al., Quantum Phenomenon-Based Obfuscation of Memory. U.S. Patent Application 20230306142. 2023-09-28. Uncopyrighted material.

[MIN02] D. Minoli, *Voice over MPLS*, McGraw-Hill, New York, NY, 2002, ISBN 0-07-140615-8 (p. 114-134).

[MIN06] D. Minoli, *Nanotechnology Applications to Telecommunications and Networking*, Wiley 2006, New York, NY. ISBN: 978-0-471-71639-6.

[MIN07] D. Minoli, K. Sohraby and T. Znati, *Wireless Sensor Networks*, Wiley 2007, New York, NY.

[MIN13] D. Minoli, *Building the Internet of Things with IPv6 and MIPv6,* Wiley, 2013.

[MIN19] D. Minoli, Editor-in-Chief, Special Issue on Blockchain Applications in IoT Environments, Overview, Elsevier IoT Journal, November 2019, https://www.sciencedirect.com/science/article/pii/S2542660519302537.

[MIN20] D. Minoli, B. Occhiogrosso, "Constrained Average Design Method for QoS-based Traffic Engineering at the Edge/Gateway Boundary in VANETs and Cyber-Physical Environments". Chapter in book *Managing Resources for Futuristic Wireless Networks*, IGI Global, 2020. Mamata Rath, Editor. https://www.igi-global.com/chapter/constrained-average-design-method-for-qos-based-traffic-engineering-at-the-edgegateway-boundary-in-vanets-and-cyber-physical-environments/262549

[MIN24] D. Minoli, B. Occhiogrosso, *AI Applications to Communications and Information Technologies: The Role of Ultra Deep Neural Networks*, December 2023, Wiley-IEEE Press, ISBN: 978-1-394-18999-1.

[MIN75] D. Minoli, R. Bear, "Hyperperfect Numbers". Pi Mu Epsilon Journal, Fall 1975, 6 (3): 153–157.

[MIN75a] D. Minoli, "Combinatorial Graph Complexity". Att. Accad. Naz. Linc, Rendiconti, Volume LIX, 2nd Semester, Fasc. 6, December 1975, pp. 651-661.

[MIN80] D. Minoli, "Issues in Non-Linear Hyperperfect numbers". Mathematics of Computation, April 1980, 34 (150): 639–645, doi:10.2307/2006107, JSTOR 2006107.

[MIN80a] D. Minoli, W. Nakamine, "Mersenne Numbers Rooted on 3 for Number Theoretic Transforms". ICASSP '80. IEEE International Conference on Acoustics, Speech, and Signal Processing. 1980. Vol. 5. pp. 243–247. doi:10.1109/ICASSP.1980.1170906.

[MIN80b] D. Minoli "Inductive Formulae for General Sum Operations". Mathematics of Computation. Vol. 34, No. 150, April 1980, pp. 543-545.

[MIN81] D. Minoli, "Structural Issues for Hyperperfect Numbers". Fibonacci Quarterly, Feb 1981, 19 (1): 6–14.

[MIN83] D. Minoli, "A New Design Criterion for Store-and-Forward Networks". Computer Networks, 7(1983), 9-15.

[MON14] C. Monroe, R. Raussendorf, et al., "Large-Scale Modular Quantum Computer Architecture with Atomic Memory and Photonic Interconnects". Phys. Rev. A 89, 022317 (2014).

[MUK23] M. Mukherjee, Quantum Dynamic Network Sever Load Management System. U.S. Patent Application 20230306296, 2023-09-28. Uncopyrighted material.

[NAR21] B. Narottama, S. Y. Shin, "Quantum Neural Networks for Resource Allocation in Wireless Communications". IEEE Trans. Wireless Commun. 21(2), 1103–1116, 2021. https://doi.org/10.1109/twc.2021.3102139.

[NAR22] B. Narottama, T. Q. Duong, "Quantum Neural Networks for Optimal Resource Allocation in Cell-Free MIMO systems". In: Proceedings of the IEEE Global Communications Conference, GLOBECOM 2022, pp. 2444–2449. 2022. ISBN: 9781665435413.

[NAS02] J. Nash, "Hyperperfect Numbers". Periodica Mathematica Hungarica. Volume/Issue: Volume 45: Issue 1-2. Sep 2002. doi: https://doi.org/10.1023/a:1022306315474.

[NIC14] N. H. Nickerson, J. F. Fitzsimons, and S. C. Benjamin, "Freely Scalable Quantum Technologies Using Cells of 5-to-50 Qubits with Very Lossy and Noisy Photonic Links". Phys. Rev. X 4, 041041; 2014. https://doi.org/10.1103/PhysRevX.4.041041

[NIE10] M. Nielsen, L. I. Chuang, *Quantum computation and quantum information: 10th Anniversary Edition.* 2010. Cambridge University Press, New York, NY, USA. ISBN: 9781107002173. http://dx.doi.org/10.1017/cbo9780511976667.016.

[NIS23] NIST, "Quantum Networks at NIST". Available on Nov. 11, 2023 at https://www.nist.gov/pml/productsservices/quantum-networks-nist.

[NIS23a] NIST, Post-Quantum Cryptography, Computer Security Resource Center, *August 24, 2023, available online on December 17, 2023 at* https://csrc.nist.gov/projects/post-quantum-cryptography.

[ORU19] R. Orús, S. Mugel, E. Lizaso, "Quantum Computing for Finance: Overview and Prospects". Rev. Phys. 4, 100028, 2019.

[OUR23] S. Ourari, L. Dusanowski, et al., "Indistinguishable Telecom Band Photons from a Single Er Ion in the Solid State". Nature 620, 977–981; 2023. https://doi.org/10.1038/s41586-023-06281-4.

[OVA22] O. van Deventer, N. Spethmann, et al., "Towards European Standards for Quantum Technologies". 3 Oct 2022, arXiv:2203.01622. https://doi.org/10.48550/arXiv.2203.01622. EPJ Quantum Technol. 9 (2022) 33. https://doi.org/10.1140/epjqt/s40507-022-00150-1.

[PAR09] M. G. Paris, "Quantum Estimation for Quantum Technology". International Journal of Quantum Information, vol. 7, no. supp01, pp. 125–137, 2009.

[PIR17] S. Pirandola, R. Laurenza, et al., "Fundamental Limits of Repeaterless Quantum Communications". Nat. Commun. 8, 1–15; 2017. https://doi.org/10.1038/ncomms15043.

[PIR20] S. Pirandola, U. L. Andersen, et al., "Advances in Quantum Cryptography". Adv. Opt. Photon. 12, 1012; 2020.

[QIA19] K. Qian, Z. Eldredge, et al., "Heisenberg-Scaling Measurement Protocol for Analytic Functions with Quantum Sensor Networks". Physical Review A, 100, 042304; 2019. https://doi.org/10.1103/PhysRevA.100.042304.

[QIA21] T. Qian, J. Bringewatt, et al., "Optimal Measurement of Field Properties with Quantum Sensor Networks". Physical Review A, 103, L030601 (2021), https://doi.org/10.1103/PhysRevA.103.L030601.

[RAJ19] D. Rajan, M. Visser, "Quantum Blockchain Using Entanglement in Time". 18 Apr 2019; arXiv:1804.05979; https://doi.org/10.48550/arXiv.1804.05979. Also, Quantum Rep. 2019, *1*(1), 3-11; 2019. https://doi.org/10.3390/quantum1010002.

[REB18] P. Rebentrost, B. Gupt, T. R. Bromley, "Quantum Computational Finance: Monte Carlo Pricing of Financial Derivatives". Phys. Rev. A 98, 022321, 2018.

[REN08] R. Renner, "Security of Quantum Key Distribution". Int. J. Quantum Inf. 2008;6:1. doi: 10.1142/S0219749908003256.

[RIE84] H. J. J te Riele, "Rules for Constructing Hyperperfect Numbers". Fibonacci Q., 1984, 22: 50–60, Zbl 0531.10005.

[RIV23] N. Rivera, J. Sloan, et al., "Creating Large Fock States and Massively Squeezed States in Optics Using Systems with Nonlinear Bound States in the Continuum". PNAS 120 (No. 9) e2219208120, February 24, 2023. https://doi.org/10.1073/pnas.2219208120.

[ROH21] P. P. Rohde, *The Quantum Internet: The Second Quantum Revolution*, 09/30/2021, Cambridge University Press, ISBN-13:9781108491457.

[ROZ23] G. G. Rozenman, N. K. Kundu, et al., "The Quantum Internet: A Synergy of Quantum Information Technologies and 6G Networks". IET Quantum Communication, Volume 4, Issue4, December 2023. Pages 147-166. This article also appeared in: Quantum Industry: Applications in Quantum Communication (Quantum. Tech Europe 2022). https://doi.org/10.1049/qtc2.12069.

[RUF21] M. Ruf, N. H. Wan, et al., "Quantum Networks Based on Color Centers in Diamond". J. Appl. Phys. 130, 070901, 2021. https://doi.org/10.1063/5.0056534.

[RYA22] C. Ryan-Anderson, N. C. Brown, et al., "Implementing Fault-tolerant Entangling Gates on the Five-qubit Code and the Color Code". arXiv:2208.01863v1, 3 Aug 2022, https://arxiv.org/pdf/2208.01863.pdf

[SCH95] B. Schumacher, "Quantum Coding". Phys. Rev. A, 51:2738–2747, Apr 1995. https://doi.org/10.1103/PhysRevA.51.2738

[SHA22] K. Sharma, M. Cerezo, et al., "Reformulation of the No-Free-Lunch Theorem for Entangled Datasets". Phys. Rev. Lett. 128, 070501—Published 18 February 2022. https://doi.org/10.1103/PhysRevLett.128.070501.

[SHO00] P.W. Shor, J. Preskill, "Simple Proof of Security of the BB84 Quantum Key Distribution Protocol". Phys. Rev. Lett. 2000;85:441. doi:10.1103/PhysRevLett.85.441.

[SIL08] M. Silverman, *Quantum Superposition: Counterintuitive Consequences of Coherence, Entanglement, and Interference*. Berlin: Springer; 2008. ISBN 3540718834.

[SIM21] C. Simon, "Towards a Global Quantum Network". Nat. Photonics 11(11), 678–680, 2017. https://doi.org/10.1038/s41566-017-0032-0. [SMI23] C. S. Smith, "Amazon Introduces New Quantum Chip to Reduce Errors". Forbes. Available online on 11/30/2023 at https://www.forbes.com/sites/craigsmith/2023/11/28/amazon-introduces-new-quantum-chip-to-reduce-errors/?sh=18fa665d1f58

[SMI23a] C. S. Smith, "Top 10 Quantum Computing Companies Making Change". Forbes, Dec 11, 2023. Available online on Decembre 13, 2023 at https://www.forbes.com/sites/technology/article/top-quantum-computing-companies/?sh=57b3a4253a94

[SRI21] G. Srinivasan, S. K. Borah and S. Bitragunta, "Superdense Coding Through Repeaterless Hybrid Network of Depolarizing Quantum Communication Channels". 2021 IEEE International Conference on Electronics, Computing and Communication Technologies (CONECCT), Bangalore, India, 2021, pp. 1-6, doi: 10.1109/CONECCT52877.2021.9622362.

[STA06] Stanford Encyclopedia of Philosophy, "Quantum Field Theory". June 22, 2006; available online (on 10/25/2023) at https://plato.stanford.edu/entries/quantum-field-theory/

[STA20] N. Stamatopoulos, D. J. Egger, et al., "Option Pricing Using Quantum Computers". Quantum 4, 291, 2020.

[TAK14] M. Takeoka, S. Guha, S. and M. Wilde, "Fundamental Rate-Loss Tradeoff for Optical Quantum Key Distribution". *Nat Commun* **5**, 5235 (2014). https://doi.org/10.1038/ncomms6235.

[UTK24] The University of Tennessee, Department of Physics and Astronomy, "Quantum Mechanics - Elements of Physics II". Available online on January 4, 2024 at https://labs.phys.utk.edu/mbreinig/phys222core/modules/m10/quantum_mechanics.html

[VAN14] R. Van Meter, *Quantum Networking*. ISTE Ltd/John Wiley and Sons. Inc., ISBN 978-1-84821-537-5. April 2014. https://onlinelibrary.wiley.com/doi/book/10.1002/9781118648919. doi 10.1002/9781118648919.

[VAR22] G. Vardoyan, M. Skrzypczyk, S. Wehner, "On the Quantum Performance Evaluation of Two Distributed Quantum Architectures". Elsevier, Performance Evaluation 153, 2022. https://doi.org/10.1016/j.peva.2021.102242. Open Access: *"You are free to: Share—copy and redistribute the material in any medium or format for any purpose, even commercially. Adapt—remix, transform, and build upon the material for any purpose, even commercially. The licensor cannot revoke these freedoms as long as you follow the license terms. Under the following terms: Attribution—You must give appropriate credit, provide a link to the license, and indicate if changes were made. You may do so in any reasonable manner, but not in any way that suggests the licensor endorses you or your use. No additional restrictions—You may not apply legal terms or technological measures that legally restrict others from doing anything the license permits"*.

[VIE21] D. Vietz, J. Barzen, et al., "An Exploratory Study on the Challenges of Engineering Quantum Applications in the Cloud". 2nd Quantum Software Engineering and Technology Workshop, co-located with IEEE International Conference on Quantum Computing and Engineering (QCE21) (IEEE Quantum Week 2021), October 18–22, 2021.

[WAN14] S. Wang, "Field and Long-Term Demonstration of a Wide Area Quantum Key Distribution Network". Opt. Express 22, 21739; 2014). https://doi.org/10.1364/OE.22.021739.

[WEH18] S. Wehner, D. Elkouss, R. Hanson, "Quantum Internet: A Vision for the Road Ahead". Science 362, 303, 19 October 2018. Available online at https://www.science.org on November 01, 2023.

[WHA09] B. Whaley, D. Gorman, "C/CS/Phys191: Qubits, Quantum Mechanics and Computers - Fall 2009". berkeley.edu, Available online on 11/27/2023 at https://inst.eecs.berkeley.edu/~cs191/fa09/ (e.g., https://inst.eecs.berkeley.edu/~cs191/fa09/lectures/lecture1_fa09.pdf, https://inst.eecs.berkeley.edu/~cs191/fa09/lectures/lecture2_fa09.pdf, and so on).

[WIE15] N. Wiebe, A. Kapoor, K. M. Svore, "Quantum Algorithms for Nearest-Neighbor Methods for Supervised and Unsupervised Learning". Quantum Info. Comput. 15, 316, 2015.

[WIE16] N. Wiebe, A. Kapoor, K. M. Svore, "Quantum Deep Learning". Quantum Info. Comput. 16, 541, 2016.

[WIE83] S. Wiesner, "Conjugate Coding". ACM SIGACT News, 15, 78-88. 1983. https://doi.org/10.1145/1008908.1008920

[WOE19] S. Woerner, D. J. Egger, "Quantum Risk Analysis". npj Quantum Inf. 5, 15, 2019.

[WOO23] R. I. Woodward, Y. S. Lo, A. J, Shields, Transmitter for a Quantum Communication System, a Receiver for a Quantum Communication System and a Method of

Controlling a Quantum Communication System. U.S. Patent Application 20230283370, 2023-09-07. Uncopyrighted Material.

[WOO82] W. K. Wootters, W. H. Zurek, "A Single Quantum Cannot Be Cloned". Nature. 1982;299:802. doi:10.1038/299802a0.

[YAM24] A. Yamauchi, K. Tanaka, et al., "Room-Temperature Quantum Coherence of Entangled Multiexcitons in a Metal-Organic Framework". Science Advances, 3 Jan 2024, Vol 10, Issue 1. doi: 10.1126/sciadv.adi3147.

[YAN23] H. Yang, "Quantum Information Technology standardization updates". ISO/IEC JTC 1 Workshop on Quantum Information Technology Standardization, ISO/IEC JTC 1/WG 14 Convenor. April 2023.

[ZAD17] J. M. Zadrozny, A. T. Gallagher, et al., "A Porous Array of Clock Qubits". J. Am. Chem. Soc. 139, 7089–7094, 2017.

[ZUB05] S. Zubairy, "Quantum Squeezing". Journal of Optics B: Quantum and Semiclassical Optics, Volume 7, Number 5, 156. doi 10.1088/1464-4266/7/5/B01.

Basic Glossary of Key Concepts in Chapter 1

This Glossary is based on various industry sources, including [ABE23], [ALB18], [AND24], [AWS21], [BAY23], [CHE23], [CHI23], [DAL23], [DEC24], [HER22], [INS23], [KOZ23], [LIT23], [MAN20], [MAZ23], [MEI23], [MUK23], [NIST23], [UTK24], [VAR22], [YAM24], and [ZUB05]. This Glossary should be used not only as a repository of terms and terminology, but as a tool that further elaborates, explains, and amplifies on various fundamental quantum concepts.

Term/Concept	Description
Basis	A set of linearly independent (orthogonal) vectors that can be employed to represent any vector in a given vector space. The basis of a vector space provides a coordinate system that enables one to represent vectors utilizing numerical coordinates. In a vector space, if a (small) set of vectors can be used to express every vector in the space as a unique linear combination of those vectors, and those vectors are linearly independent (meaning that none of them can be expressed as a linear combination of the others), then they constitute as basis vectors for that vector space.
Bell State Measurement (BSM)	State measurements that function as basic primitives for several other quantum protocols. For example, in Quacom, BSMs facilitate entanglement swapping and, thus, support the implementation of Quareps; they can be used to link distant QCs via optical channels, thus creating the foundation for the/a Quaint. In Quacomp, BSMs play a role in photonic quantum computing, particularly in measurement- and fusion-based approaches, for example for fusing small-scale units to large resource states for the realization of quantum error correction [BAY23].
Bell states	Quantum states of two qubits, a qubit pair, with maximal entanglement where when one of the two qubits is measured it acquires a specific value, and the second qubit is forced to also take on a specific value, being that the entangled state collapses. $$\mid \Psi^{\pm} \rangle = \frac{1}{\sqrt{2}}(\mid 01 \rangle \pm \mid 10 \rangle)$$ $$\mid \Phi^{\pm} \rangle = \frac{1}{\sqrt{2}}(\mid 00 \rangle \pm \mid 11 \rangle)$$ They are named after John Stewart Bell [BEL64], [BELL04]. Bell (28 July 1928-1 October 1990) was a physicist and the originator of Bell's theorem dealing with hidden-variable theories.
Circuit	See *Quantum Circuit*.
Coherence	A description of the stability of a quantum system. The ability of a quantum system to retain a well-defined state over time without being impacted by nearby disturbances. The coherence time of a qubit measures how long coherence can be maintained before decoherence takes over.
Decoherence	The process by which a quantum state breaks down, often due to unwanted and uncontrolled interactions. Decoherence occurs when the interaction with the environment around a quantum system causes it to collapse and lose all information. Decoherence causes the state to collapse as if it was being measured. Decoherence is the progressive decay of the state of a system, typically due to interactions of the QM system under discussion with its environment: quantum states in various physical implementations of a QM system are fragile; for example, excited atoms lose energy, thus decaying; spins of electrons and atomic nuclei may flip, thus introducing

(Continues)

Term/Concept	Description
Decoherence (cont.)	errors; these interactions with the environment result in the leaking of information about the state outward into the environment, such that the original intended state cannot be recovered. In the case of decoherence, measurements of the system likely will not produce the desired results, possibly (or likely) resulting in the failure of quantum algorithm that relied on related measurements.
Demand for resources: "exponentially many resources"	Term means that a demand for resources increases exponentially with the increase of the quantity of qubits. The exponentially many resources may mean that a total quantity of quantum circuits to be measured is exponentially many, that is, correspondingly, exponentially more computing time is required.
Eigenstate	If O is an operator such that $$O(|\psi_e\rangle) = e \times |\psi_e\rangle$$ where e is a complex number, then $|\psi_e\rangle$ *is called an eigenstate of operator O* corresponding to the eigenvalue e; this equation implies that if $|\psi_e\rangle$ is the eigenstate with eigenvalue e, then acting on $|\psi_e\rangle$ with the operator O just returns the same state back, but multiplied by the eigenvalue e; a state described by an eigenfunction is called an eigenstate of the operator [FIT23]. Also see *Eigenvalue*. In QM, eigenstates that are possible numerical values of a mechanical quantity are all eigenvalues of its operator. A state described by an eigenfunction is called an eigenstate of this operator. In its own eigenstate, the mechanical quantity takes a certain definite value, that is, an eigenvalue to which the eigenstate belongs. For a Hamiltonian matrix H, solutions satisfying the equation $H|\psi\rangle = E|\psi\rangle$ are called eigenstates $|\psi\rangle$ of H, having eigenenergy E. A ground state corresponds to an eigenstate with the lowest energy of a quantum system. The quantum state after a measurement becomes the eigenstate corresponding to that measurement.
Eigenvalue	If an operator O is an operator such that $$O(|\psi_e\rangle) = e \times |\psi_e\rangle$$ where e is a complex number, then $|\psi_e\rangle$ *is called an eigenstate of operator O* corresponding to the eigenvalue e; this equation implies that if $|\psi_e\rangle$ is the eigenstate with eigenvalue e. Application: a measurement of an observable yields a value, called an eigenvalue of the observable. Also see *Eigenstate*.
Entanglement	(Also known as *quantum entanglement*) A phenomenon where the quantum state of two or more quantum systems (particles) cannot be described separately, even when they are separated by a large distance. It relates to the coordinated behavior of particles, qubits, that have interacted but then are separated apart. In addition, in an entangled system, a measurement on one part can affect the system as a whole, so that a measurement outcome of one part of the system can predict a measurement outcome of the other. Hence, the result of a measurement on one system can instantaneously influence the result of a measurement on the second system; that is, measuring properties such as position or momentum or spin of one of the separated pair of particles instantaneously changes the properties of the other particle, regardless of how far the second particle has moved away from its twin; the state of one entangled particle is inseparable from the state of the other.
Entropy	A measure of uncertainty regarding information.
Fidelity	A metric characterizing similarity between quantum states. In particular, fidelity (ranging from 0 to 1.0) is used to compare a measured quantum state to a desired quantum state. It is used to assess the quality of a state; essentially, it is the probability that a qubit is actually in the state it ought to be in, or how 'imperfect' the state is in compared to the ideal (desired) state.

Basic Glossary of Key Concepts in Chapter 1 *(cont.)*

Term/Concept	Description
Fock state	A quantum state that describes a definite number of non-interacting identical particles. Example: in quantum optics, a single-photon state is a Fock state of one photon.
Gate	See *Quantum Gate*
Graph state	A way of representing certain types of entangled states; a special type of multi-qubit state that can be represented by a graph G = (V, E) with V the set of vertices and E the set of edges E. Each qubit is represented by a vertex of the graph; and, there is an edge between every interacting pair of qubits. In particular, graph states are a convenient way of representing certain types of entangled states. Graph states find applications in quantum error-correcting codes, entanglement measurement and Quacomp.
Ground state of a QM system	The system's stationary state, this being the state at the lowest energy (such lowest energy known as the zero-point energy of the system). A ground state corresponds to an eigenstate with the lowest energy of a quantum system. An excited state is any state with energy greater than the ground state energy.
Grover's algorithm	(Aka quantum search algorithm) A quantum algorithm for unstructured search that identifies with high probability the specific input to a black box function that produces a particular output value, using just order-of-the-square-root-of-N evaluations of the function (N is the size of the function's domain).
Hamiltonian	A Hamiltonian is an operator that mathematically describes a physical system in terms of its energies; a mathematical *operator* describing the total energy of a system, usually denoted by . A Hermitian conjugate matrix describing the total energy of a quantum system.
	It is an operator that fully defines a quantum system. The lowest eigenstate of such operators is called a ground state that can be the target of quantum calculations. "Energy" denotes an expectation value of a Hamiltonian on a given normalized quantum state. Energy can be minimized when the state is the ground state.
	"Stoquastic" Hamiltonians are Hamiltonian operators having only non-positive off-diagonal elements in the computational basis. Adiabatic quantum algorithms used for optimization problems typically use "stoquastic" Hamiltonians.
	(Sir William Rowan Hamilton was a 19th century Irish mathematician, astronomer, and physicist.)
Heisenberg's Uncertainty Principle (HUP)	In a quantum system, only one property of a pair of conjugate properties (such as position and momentum) can be known with certainty; for example, a measurement of a particle's position will disturb its speed. As an application, quantum cryptography relies on HUP by utilizing the polarization of photons, exchanged over fiber optic links, on different bases as being the conjugate properties.
Ions	An atom or molecule with a net electrical charge
Measurement, Quantum Measurement	The act of observing or interacting with a quantum system: an interaction between the observer and the system under consideration is a measurement. A single n-qubit read-out (that is, a projection of each qubit on either the 0 or 1 state) that can be obtained after a single quantum circuit execution. A quantum measurement causes the quantum state to collapse. Although quantum measurements results are random, the probabilities can be predicted using QM.
	For example, before a photon is measured, it exists in a superposition of all its possible quantum states, each with an associated probability, and a measurement selects one among these states; in addition, the photon's quantum state cannot be measured without causing a disturbance that discloses the attempt (when used in applications, this behavior can be used to support inherent security).

Term/Concept	Description
No cloning theorem	QM-related 'law' that it is impossible to create a duplicate of an arbitrary, unknown quantum state. It can be used to detect eavesdroppers.
Nonclassical (quantum) light	Light that can only be described by quantum mechanics. Examples: squeezed vacuum and single photons.
Non-unitary operators/ matrices	Non-unitary operators/matrices are matrices where $U^{\dagger}U{\neq}I$, $U^{\dagger}$ being the conjugate transpose of U.
	All evolution processes directly allowed by quantum mechanics can be described by the unitary matrix. Non-unitary matrices tend to have stronger expressive capability and a faster ground-state projection effect.
Obtaining a ground state	(That is, obtaining a lowest eigenstate) of a quantum system represents obtaining the most stable state of the quantum system; this has important applications in the study of basic properties of quantum physics and quantum chemistry systems, solutions of combinatorial optimization issues, and pharmaceutical research.
Polarization	When an electromagnetic wave is polarized it oscillates in a single plane. There are two types of generic electromagnetic polarization: linear and elliptical. Linear polarization has two states: rectilinear (horizontal and vertical) and diagonal (diagonal and anti-diagonal). Photon polarization is a two-level system: one level can be associated with $|0\rangle$ and another can be associated with $|1\rangle$. The two states of rectilinear polarization, horizontal and vertical, are represented as $|H\rangle$ and $|V\rangle$; the two states of diagonal polarization, diagonal and anti-diagonal, are represented as $|D\rangle$ and $|A\rangle$. Thus, one can consider $|H\rangle$ and $|V\rangle$ as $|0\rangle$ and $|1\rangle$ on the Z-axis and $|D\rangle$ and $|A\rangle$ as $|+\rangle$ and $|-\rangle$ on the X-axis.
Post-Quantum Cryptography (PQC)	Encryption algorithms that can resist attacks by QCs. Although at this time, there are no functional QCs outside of a laboratory environment, the expectation is that QCs will disrupt current security protocols that protect governmental institutions and global financial markets, including the threat of a "harvest now and decrypt later attack", even if the threat is not actionable until a sufficiently robust QC is developed in the future.
Quantum algorithm	An algorithm that is implemented on a QC. A step-by-step procedure, where each of the steps is performed on a QC. An algorithm is a step-by-step procedure, instructions, to perform a calculation, or a sequence of instructions to solve a problem. Examples include but are not limited to: Deutsch–Jozsa's algorithm, Bernstein–Vazirani's algorithm, Simon's algorithm, Shor's large number factoring algorithm, Grover's search algorithm, Harrow–Hassidim–Lloyd (HHL) algorithm for solving a system of linear equations. In particular, the 1994 Shor's algorithm for factoring integers and computing discrete logarithms (with impacts to cryptography) is still the best example of a substantial quantum speedup that targets a "real-world" problem, although many additional quantum algorithms and subroutines have been discovered, optimized, and generalized in recent years.
Quantum circuit	A sequence of gates that can be implemented on a quantum computing device to formally realize a unitary operator that acts on a given initial quantum state and produces a final quantum state. Circuits can be parametrized via gate parameters such as one or more angles of a single qubit rotation gate.
	Basic quantum gates act on a fixed number of qubits. Quantum circuits identify the steps in a quantum algorithm using a series of symbols. A representation of a quantum general-purpose computer, which represents a hardware implementation of a corresponding quantum algorithm/program in a quantum gate model. If the quantum circuit contains adjustable parameters controlling the quantum gate, it is called a Parameterized Quantum Circuit (PQC) or a Variational Quantum Circuit (VQC), both of which are the same concept.

Basic Glossary of Key Concepts in Chapter 1 *(cont.)*

Term/Concept	Description
Quantum communications (Quacom)	The exchange of information between two or more systems that cannot be accomplished with classical information exchange; the exchange of 'information carriers' (qubits) that can support superpositions and entanglement and cannot be operationally accomplished with classical information exchange. Special technologies are needed for exchanging qubits.
Quantum computation	A computation method based on quantum logic, a basic unit for storing data being a qubit.
Quantum computer (QC)	Any computer that utilizes the principles of QM to perform computational operations. Several variations of QC design are available, including photonic quantum computing, superconducting quantum computing, nuclear magnetic resonance quantum computing, and/or ion-trap quantum computing. Irrespective of the particular type of QC implementation, all QCs encode data onto qubits. QCs encode data by placing a qubit into one of two identifiable quantum states; qubits exhibit quantum behavior, allowing the QC to process a vast number of calculations simultaneously.
	With n qubits the QC deals with (operates on) an aggregate of particles that can be in a superposition of up to 2^n states simultaneously; in a classical computer bit strings of n bits can only be in one of the 2^n states at a single time. That is to say: A 'classical' bit encodes one of two mutually exclusive states at any given time; a qubit can exist in a superposition of the two states: it can be simultaneously zero and one; thus, while n classical bits encode only one of 2^n possible states at a given time, n qubits can simultaneously encode all the 2^n possible states at the same time. This supports an exponential speed-up in computation.
Quantum computing (Quacomp)	A computing method based on quantum logic, and a basic unit for storing data is a qubit.
Quantum Error Correction (QEC)	The assessment, measurement, mitigation and correction of quantum errors to improve the performance, size, and reliability of a QC. QEC algorithms aim at mitigating the effects of noise by encoding single-qubit information into a larger quantum system.
Quantum gate	(Also known as quantum logic gate) a basic quantum circuit operating on a small quantity of qubits. A mechanism or operator that transforms a sequence of quantum bits. An operation on a quantum system that transforms a quantum state. Quantum gates are mathematically represented by matrices.
	Known quantum gates include the following: I Identity, H Hadamard, X Pauli-X, Y Pauli-Y, Z Pauli-Z, Rx Arbitrary x-rotation rx, Ry Arbitrary y-rotation ry, Rz Arbitrary z-rotation rz, X90, Y90, mX90, mY90, S Phase, Sdag Phase dagger, T, Tdag T dagger, CNOT, Toffoli, CZ CPHASE, SWAP, CRK Controlled Phase Shift ($\pi/2^k$), CR Controlled Phase Shift (arbitrary angle), c-X Binary-Controlled X, c-Z Binary-Controlled Z.
Quantum information	Information that can be associated with or identified by the QM state of atomic, electronic or photonic particles. Represented as qubits.
Quantum Information Network (QIN)	(International Telecommunication Union term): a network that incorporates quantum communication technologies for the purpose of transporting quantum states. A network that incorporates Quacom technologies such as quantum teleportation and quantum repeating, for the purpose of transporting or storing of quantum states, which is to connect quantum information processing nodes, including QKD nodes, QCs and quantum sensors. Quanet is the term used in this text.
Quantum Information Technology (QIT)	(International Telecommunication Union term): A class of emerging technology that improves information processing capability by harnessing principles of QM. QIST is the term used in this text.

Term/Concept	Description
Quantum Interconnects (QuICs)	Devices or processes that allow the transfer of quantum states between two specified physical degrees of freedom (material, electromagnetic, and so on), or, more broadly, connect a quantum system with a classical one.
Quantum Internet (Quaint)	A network infrastructure to support the exchange of quantum information. A global network system that makes use of the laws of QM, such as entanglement and superposition, to send, receive, and store information among groups of quantum devices. A Quaint enables the distribution of entanglement, which is used to perform quantum encryption, quantum sensing, and other communication and application tasks.
Quantum Key Distribution (QKD)	A family of protocols that generate secret keys for two or more parties in a secure manner where the security of the key distribution is guaranteed by the laws of QM.
Quantum memory	(Aka quantum storage) Memory that can store one or more quantum states. Storing quantum information at a given point is a nontrivial task, since quantum information tends to decay or degrade relatively quickly.
Quantum network (Quanet)	A collection of quantum nodes connected via Quacom channels and supplementary classical channels for stabilization, timing, and/or routing. Network nodes are linked by quantum channels that enable the transmission of quantum state and entanglement from one location to another with high fidelity. The network supports the transmission of information in the form of qubits.
Quantum neural network	(Not a Quanet, but) the use a QC as a machine learning model.
Quantum operations	Transformation, mappings, measurements of qubits undertaken using 'unitaries' (that is, unitary operators) U, also known as "gates". To physically implement these gates, one needs to implement the underlying Hamiltonian operators associated with U. Rotations and reflections are unitary; furthermore, the composition of two unitary transformations is also unitary
Quantum operator	Operators are used to describe transformations (mappings) to quantum states. A quantum operator Λ is a linear map from an initial Hilbert space H_1 to a target Hilbert space H_2: $\Lambda : H_1 \rightarrow H_2$. Λ is typically expected to be a Completely Positive Trace-Preserving Map (CPTPM), namely, having the following properties: (i) trace-preserving: $\Lambda(\rho)$ to have unit-trace; (ii) $\Lambda(\rho)$ to be positive semi-definite; (iii) $\Lambda(\rho)$ to be Completely Positive (CP) mapping, namely, if a partial trace is taken then the remaining subsystem is also positive semi-definite. A CPTP map can be written in the form $\Lambda(\rho) = \sum_j A_j \rho A_j^\dagger$ where $\{A_j\}$ is such that $\sum_j A_j A_j^\dagger = 1$ (topic is further covered in Chapter 3).
Quantum Purification	Quantum states are known to be susceptible to the effects of noise-induced decoherence. Quantum state purification is the process of recovering a nearly pure copy of an unknown pure quantum state using multiple noisy copies of the state. This process has obvious applications to Quacom over noisy channels and quantum computation with imperfect devices. A basic challenge in QIST is thus to identify and utilize practical mechanisms for protecting quantum systems from noise, and/or for removing noise that has already occurred in order to (partially) reverse the effect of decoherence and produce less noisy states out of mixed ones. These processes comprise the task of state purification [CHI23].

Basic Glossary of Key Concepts in Chapter 1 *(cont.)*

Term/Concept	Description
Quantum repeater (Quarep, aka QR)	A device that enables quantum state transmission over long distances. Given that quantum states decay exponentially with transmission distance due to loss, a quantum repeater is used to mitigate that effect. Quareps perform entanglement swapping to reliably extend the distance between which two devices can become entangled Quareps allow one to create entangled state between the end points of the network by first segmenting the network into segments, creating entanglement across the segments, and then, connecting those entanglements to create the long-range entanglement (thus, in order to distribute entanglement across a long link, entanglement is generated via shorter links). Quareps utilize both local quantum operations (e.g., to perform entanglement purification), as well as entanglement generation with neighboring repeater nodes.
Quantum simulation	The process of simulating large quantum systems using another quantum system as a simulator that is easier to control and study. Quantum simulation entails using of a QC to simulate, among other physical systems, molecules and physical materials. A quantum simulator operates on the basis of QM.
Quantum simulator	A quantum device that can be used to study quantum systems
Quantum software	Application software developed with quantum programming languages and compilers to support quantum information processing.
Quantum state	An abstract object that provides a complete description of the system at any particular moment t_0; when combined with the rules of the system's evolution in time it provides a complete description of the system at all times t. In the Quacomp/Quacom context, the classical equivalent of a quantum state is a string or stream of logical bit values. A QM system can consist of a single particle, e.g., an atom or a photon of light, or it can be more complex. Then, a state is characterizable as a description of the atomic energy levels, the photon polarization, an electron spin, and so on.
	In QM a quantum state is a microscopic state determined by a set of quantum numbers. Quantum state can be thought of as a mathematical element of a Hilbert vector space H, formally one writes $|\psi\rangle \in$ H.
	For a qubit, H is a two-dimensional space and $\{|0\rangle, |1\rangle\}$ is an orthonormal basis. Other basis are, for example, $\{|+\rangle, |-\rangle\}$ and $\{|+_i\rangle, |-_i\rangle\}$, where
	$$|+\rangle = (1/\sqrt{2})(|0\rangle + |1\rangle) \text{ and } |-\rangle = (1/\sqrt{2})(|0\rangle - |1\rangle);$$ $$|+_i\rangle = (1/\sqrt{2})(|0\rangle + i|1\rangle) \text{ and } |-_i\rangle = (1/\sqrt{2})(|0\rangle - i|1\rangle).$$
	For any orthonormal basis $\{|x\rangle, |y\rangle\}$, the inner product of two quantum states (two vectors)
	$$|\psi\rangle = \alpha|x\rangle + \beta|y\rangle \text{ and } |\phi\rangle = \gamma|x\rangle + \delta|y\rangle$$ is defined (here) as $$\langle\psi|\phi\rangle = \langle\phi|\psi\rangle^\dagger = \alpha\ \gamma^* + \beta\ \delta^*$$ where the asterisk denotes the complex conjugate.
Quantum Switch (QS)	A device that can route optical signals between different channels while maintaining quantum coherence and entanglement.
Quantum transducer	A device that converts a physical quantum state from one physical system to another. Typically, a transducer maps the *flying qubit*'s state to a *matter qubit*, a qubit suitable for storage and/or information processing.

(continues)

Term/Concept	Description
Quantum transducer *(cont.)*	An example of use can be seen in the context of connecting two ion-based quantum nodes over a large distance. In this scenario, an ion excitation is converted to a photon using a quantum transducer, so that its carrier frequency is compatible with a telecom fiberoptic band; typically, this is done with nonlinear frequency conversion. After propagation, the photon is converted back to an ion excitation, again using an appropriate quantum transducer.
Quantum tunneling	A wave-like phenomenon that allows quantum particles to randomly appear on the other side of a barrier, even if it does not have the energy to go through or over it. A QM phenomenon when a particle is able to penetrate through a potential energy barrier that is higher in energy than the particle's kinetic energy. A particle "tunnels" through what would appear to be an insurmountable barrier. Quantum tunneling refers to the nonzero probability that a particle in QM can be measured to be in a state that is forbidden in classical mechanics; quantum tunneling occurs because there is a nontrivial solution to the Schrödinger equation in a classically-forbidden region, which corresponds to the exponential decay of the magnitude of the wave function.
Qubit (qbit, quantum bit)	A basic unit of quantum computing. The binary unit of quantum information. A quantum bit can simultaneously store a coherent superposition of both binary values. qubits store or represent information as both the binary states and superpositions of the binary states. One way to notate the value that a qubit holds, is $\lvert qubit\rangle = \alpha\,\lvert 0\rangle + \beta\,\lvert 1\rangle$. A "qubit" can be perceived as a quantum system with an associated quantum state that may be utilized to encode information. A quantum state may be used to encode one bit of information where the quantum state space is modeled as a (complex) two-dimensional vector space, with one dimension in the vector space being mapped to logical value 0 and the other to logical value 1. In contrast to classical bits, a qubit may have a state that is a superposition of logical values 0 and 1. Thus, qubits can be manipulated to exist in a quantum state known as superposition, where they are both 1 and 0 at the same time, as well as being in all the possible states in between: a qubit can be in one of its basis states, $\lvert 0\rangle$, $\lvert 1\rangle$, or in a state that is a linear combination of both "0" and "1": $\lvert\psi\rangle = \alpha\lvert 0\rangle + \beta\lvert 1\rangle$, where α and β are complex numbers $(\alpha, \beta \in \mathbf{C})$ and $\lvert\alpha\rvert^2 + \lvert\beta\rvert^2 = 1$, being complex probability amplitudes of the system at 0 and 1; their modulo squares $\lvert\alpha\rvert^2$ and $\lvert\beta\rvert^2$ respectively represent probabilities at 0 and 1. This equates to a two-state quantum system. The implication of the linear superposition is that the qubit is in the state (0) with probability *amplitude* $\alpha \in \mathbf{C}$ and in the excited state (1) with probability *amplitude* $\beta \in \mathbf{C}$. $\lvert 0\rangle$, $\lvert 1\rangle$ are the basis for the quantum state. Qubits may be realized in a variety of quantum systems. Examples of qubits include: polarization states of photons; presence of photons in waveguides; or energy states of molecules, atoms, ions, nuclei, or photons. Other examples include flux qubits, phase qubits, or charge qubits (e.g., formed from a superconducting Josephson junction); topological qubits (e.g., Majorana fermions); or, spin qubits formed from vacancy centers (e.g., nitrogen vacancies in diamond). (There are some ongoing research efforts to achieve quantum coherence at room temperature—photonic Quacomp can, in fact, operate at room temperature, but there is also interest in qubits of other types having this property).

Basic Glossary of Key Concepts in Chapter 1 *(cont.)*

Term/Concept	Description
Qudit	Any quantum system having a quantum state space that may be modeled as a (complex) d-dimensional vector space (for any integer *d*), which may be used to encode *d* bits of information. Quantum states belong to general d-dimensional Hilbert space and are representable as $$\sum_{k=0}^{d-1} \alpha_k |k\rangle$$ where $\Sigma_k |\alpha_k|^2 = 1$ and $\{|0\rangle, |1\rangle, \dots, |d\text{-}1\rangle\}$ are a basis for that space.
Shor's algorithm	A quantum algorithm for finding the prime factors. It offers super-polynomial speedup compared to best classical algorithms.
Single photon	A single excitation of a mode of electromagnetic field. Single photons are quantum states; single photons cannot be prepared by classical light sources.
Squeezed state	A quantum state is squeezed when the noise in one variable is reduced below the symmetric limit at the expense of the increased noise in the conjugate variable, such that the Heisenberg uncertainty relation is not violated. Squeezed states exhibit a squeezed uncertainty of electric field strength; for example, a beam of squeezed light has a lower quantum uncertainty than a beam without photons. Squeezed light enables better measurement accuracy and facilitates secure communication. Squeezing properties of light are manifest in nonlinear systems such as parametric oscillators (for example, efficient sources of squeezing have been developed based on whispering gallery mode resonators—see Chapter 4). Squeezed states have applications in communications and measurement being that there are advantages of using nonclassical light over classical light in these settings [ZUB05].
Superdense coding (SD)	Superdense coding is a fundamental Quacom protocol in quantum communication. SD enables one to transmit two classical bits of information by transmitting only one quantum bit from sender to receiver. The protocol operates under the assumption that the sender and the receiver have pre-shared an entangled resource, sharing a pair of maximally entangled qubits. Here Alice can transmit a semi-nibble (two bits, i.e., one of 00, 01, 10 or 11) to Bob by sending only one qubit.
	Superdense coding is the underlying mechanism of secure Quacom: the requirement of having both qubits in order to decode the information being transferred precludes eavesdroppers from being able to intercept and interpret the transmissions, because in order for Bob to find out which classical bits Alice sent he performs a CNOT unitary operation, with A as a control qubit (previously received during the entanglement preparation stage) and B as target (received) qubit.
Superposition	Any quantum state that may be represented as a sum of several other quantum states. Namely, any linear combination of valid quantum states is also a valid quantum state.
Teleportation	A transfer of "quantum states" between separate atoms (or particles); transferring key properties of one atom (or particle) to another atom (or particle) without using any physical link. Here given that Alice and Bob have a pre-shared Bell pair, a transfer one qubit from Alice to Bob is accomplished by transmitting two classical bits between the parties.
	Making use of a teleporter in the form of a pre-shared entangled state, the quantum information is transferred by undertaking a joint Bell State Measurement (BSM) on the sender's part of the entangled state and the qubit state to be teleported. The state is recovered on the receiving node by a gate operation conditioned on the BSM outcome. A deterministic BSM combined with real-time feed-forward enables teleportation where state transfer is achieved each time a qubit state is inserted into the teleporter [HER22], [BEN93].

Term/Concept	Description
Three dimensional (3D) networks	6G networks comprised of terrestrial fiber and wireless networks, geosynchronous and low-earth satellite networks, drone networks, and also possibly underwater communication networks.
Unitary matrix/ operator U	Unitary matrices are matrices that satisfy $U^\dagger U = I$, where $U^\dagger$ is a conjugate transposition of U; time evolution processes in QM may be described by a unitary matrix.
Variational Quantum Eigensolver (VQE)	A (near-term quantum) algorithm that can find a ground state of a Hamiltonian on a quantum computing device. An estimation of the ground state energy of a specific quantum system implemented through a variational circuit is a typical quantum-classical hybrid computing environment (which is widely applied in the field of quantum chemistry).
Walltime	The time to run (or execute) a quantum algorithm. Walltime can be directly proportional to a number of measurements (hence, circuit repetitions) involved in executing a quantum algorithm. Computational gain denotes a ratio between a walltime associated with running a quantum algorithm by implementing some new procedure compared with a walltime associated with running the algorithm using some other/earlier procedure.
Wave function	A mathematical model of the quantum state of a quantum system. It describes the characteristics of a particle, such as its position, momentum, time and/or spin. The wave function has no direct physical meaning: it is simply one way of capturing the information about the system that can be available to the observer.
	To make predictions about the outcome of all measurements, at any time, one has to "do" something to the wave function to extract the information: specifically, one has to perform some mathematical operation on it, such a squaring it, multiplying it by a constant, differentiating it, and so on. To achieve this, one has to operate on the wave function with some specific operator; the operator is a set of one or more instructions. For example, to make predictions about the energy (or momentum) of a particle, one has to operate on its wave function with the energy (or momentum) operator. Every observable is associated with its own operator [UTK24].
Wave-particle duality	The concept that quantum matter and light cannot be specifically described as a "wave" or a "particle", but are perceived to have some properties in common with waves and others in common with particles.

Chapter 2

Basic Concepts in Quantum Physics—Mathematical Tools

This chapter presents some of the basic mathematical background and nomenclature needed to gain an introductory understanding of Quantum Mechanics (QM). The more direct application of the math to QM per se is covered in chapters that follow.

This chapter focuses on vector spaces, qubit representation as being vector(s), the joint state of multiple individual qubits, the tensor (outer) product of two vector spaces, the concept of a density matrix, entanglement, a first view of state measurements, and the purpose of gates.

The nomenclature and mathematical notation in QM can easily become overwhelming, not only because of the complexity of the underlying physical (scientific) concepts, but also because three notations are employed across the literature to represent the "same thing": the nomenclature/symbolics used by mathematicians, the nomenclature/symbolics used by physicists, and the Dirac nomenclature/symbolics. We endeavor to provide a basic, intuitive, and self-contained formulation.

Although the mathematics may at first look overwhelming, *relatively basic concepts from linear algebra (broadly speaking, matrix manipulation/multiplications and vector operations) are used* in QM; the standard notation commonly used in quantum computing is based on vector spaces, but the nomenclature is somewhat pedantic. There is some redundancy in our discussion below, but it is intended for the purpose of solidifying the understanding of the non-specialist reader. After basic familiarization with basic vector algebra and nomenclature is accomplished, the mathematics should become somewhat pedestrian. It should also always be remembered that a mathematical model (and the associated machinery) is just that—a model of reality (with intrinsic simplifications), and not reality itself.

This treatment is intended only to provide the most basic understanding of QM for the non-scientist, non-researcher practitioner. We seek to introduce a skeletal formal machinery so that the concepts and scope of quantum-based networks/networking (Quanets) can be understood and appreciated. For the audacious reader seeking a most authoritative discussion reference for QM, [DIR1958] is the canonical text. The Appendix also provides an extensive tabulation of other sources, compiled from [ISM21].

2.1 Basic Mathematical Background of a Vector Space

At a broad level, QM describes the behavior of nanoscale, atomic, and sub-atomic entities (particles—for example, photons [MIN05]). (Advanced) mathematical models (mathematical abstractions) are used to study their behavior. Classical computer science deals with a deterministic *two-state system* or entity: the bit: a bit has two well-defined values or states: "0" or "1", say /**0**/ and /**1**/, electrical signal "on" or "off", light pulse either "on" or "off", and so on. In the quantum version of the two-state system, the qubit (quantum bit) plays a fundamental role, where states $|0\rangle$ and $|1\rangle$ play a bit-comparable (but certainly not an identical) role as information containers. A qubit is a quantum system with an associated quantum state that is used to encode information. A quantum state may be used to encode one bit of information if the quantum state space can be modeled as a (complex) two-dimensional vector space, with one dimension (one basis vector) in the vector space mapped to logical value 0 and the other dimension (other basis vector) mapped to logical value 1. More generally, a "qudit" describes any quantum system having a quantum state space that may be modeled as a (complex) n-dimensional vector space (for any integer n), which may be used to encode n bits of information [LIT23].

The *state-space* of a system is the space formed by the set of its possible states—namely, the physically possible ways of combining the values of quantities that characterize the system internally [ISM21]. In QM, the state-space can be perceived as coordinates in an appropriate (mathematical) vector space. The state of a single qubit is a vector in a two-dimensional complex[1] vector space. A *vector space* is a set of vectors closed under the operation of vector addition and vector multiplication by constants (scalars). A set of vectors and operations that are defined on those vectors constitute the mathematical structure of a vector space. In a vector space, if a (small) set of vectors can be used to express every vector in the space as a unique linear combination of those "basic" vectors, and those vectors are linearly independent (meaning that *none* of them can be expressed as a linear combination of the others), then they act as *basis vectors* for that vector space; a basis is, thus, a set of linearly independent (orthogonal) vectors that can be employed to represent any vector in a given vector space. The basis of a vector space provides a coordinate system that enables one to represent vectors in that space utilizing numerical coordinates. Table 2.1 provides some of the key terminology in use, which should be studied by the reader. Also see Figure 2.1 (on page 65).

Paul Dirac[2] introduced the basic formal notation for QM: the *bra* and *ket* notation. The Dirac notation $|v\rangle$ is used to represent a column vector, such as v, and is referred to as a *ket*, while the conjugate transpose of that vector, a row vector $\langle v|$, is referred to as a *bra*. The *conjugate transpose* is written as $\langle v| = |v\rangle^\dagger$ and also written as $(|v\rangle^*)^T$, so that $\langle v| = |v\rangle^\dagger = (|v\rangle^*)^T$. The dagger symbol represents the transpose of a vector also, along with the conjugation of the numerical elements; the asterisk symbol by itself represents the complex conjugate of a number.

---

[1] *Complex* refers to complex numbers or elements of **C**, the field of complex numbers. The field of complex numbers, also shown as (**C**,+,×), is the set of complex numbers under the two operations of addition and multiplication. A complex vector space is a vector space where all elements are specifiable as linear combinations of defined basis vectors, utilizing complex numbers.

[2] Paul Dirac was Professor of Mathematics at the University of Cambridge (1932-1968) and worked at the University of Miami in Coral Gables, Florida, and Florida State University in Tallahassee, Florida, in the 1970s and 1980s (e.g., see https://en.wikipedia.org/wiki/Paul_Dirac).

Table 2.1 Basic Nomenclature

Concept	Description
General terminology	Complex conjugate of the complex number z z^* Vector, a ket $\lvert\psi\rangle$ Vector complex conjugate to $\lvert\psi\rangle$, a bra $\langle\psi\rvert$ Inner product between the vectors $\lvert\varphi\rangle$ and $\lvert\psi\rangle$ $\langle\varphi\mid\psi\rangle$ Hermitian transpose or adjoint of the A matrix $A^\dagger$ Inner product between $\lvert\varphi\rangle$ and $A\lvert\psi\rangle$ $\langle\varphi\mid A\mid\psi\rangle$ Trace of a matrix $tr(A)$ Density operator or density matrix ρ
Dirac nomenclature	The Dirac notation $\lvert v\rangle$ refers to a column vector—called a "ket". In k space, a vector is represented as a column (or row) of numbers: $$\text{vector } \mathbf{v}: \lvert v\rangle = \begin{bmatrix} \alpha_0 \\ \alpha_1 \\ \cdot \\ \cdot \\ \alpha_{k-1} \end{bmatrix}$$ The Dirac notation $\langle v\rvert$ refers to a row vector—called a "bra"—specifically, it denotes the row vector, the conjugate-transpose of $\lvert v\rangle$, or $\lvert v\rangle^\dagger$. $$\text{vector } \langle v\rvert = \lvert v\rangle^\dagger = \begin{bmatrix} \alpha_0^* & \alpha_1^* & \cdots & \alpha_{k-1}^* \end{bmatrix}$$ One considers **states as being vectors** (and as being spanned by a basis). The **basis is also considered to be a (set of) states**: $$\lvert\psi\rangle = \sum_{j=0}^{k-1} \alpha_j \lvert j\rangle$$ where $\lvert 0\rangle = [1, 0, ..., 0]^T$, $\lvert 1\rangle = [0,1,...,0]^T$, . . . , $\lvert k\text{-}1\rangle = [0, 0, ...,1]^T$; the set $\{\lvert 0\rangle,\lvert 1\rangle,..,\lvert k\text{-}1\rangle\}$ is the basis of the vector space. For $k=2$ $$\lvert\psi\rangle = \alpha\lvert 0\rangle + \beta\lvert 1\rangle = \begin{bmatrix} \alpha \\ \beta \end{bmatrix}$$ $$\langle\psi\rvert = \begin{bmatrix} \alpha \\ \beta \end{bmatrix}^\dagger = \begin{bmatrix} \alpha^* & \beta^* \end{bmatrix} = \alpha^*\langle 0\rvert + \beta^*\langle 1\rvert$$ $$\lvert 0\rangle = \begin{bmatrix} 1 \\ 0 \end{bmatrix} \quad \langle 0\rvert = \begin{bmatrix} 1 & 0 \end{bmatrix}$$ $$\lvert 1\rangle = \begin{bmatrix} 0 \\ 1 \end{bmatrix} \quad \langle 1\rvert = \begin{bmatrix} 0 & 1 \end{bmatrix}$$ The inner product of two vectors $\lvert\psi\rangle = \sum_i \alpha_i \lvert i\rangle$ and $\lvert\phi\rangle = \sum_i \beta_i \lvert i\rangle$ is $\sum_i \alpha_i^* \beta_i$. $\lvert A\rangle = a_1\lvert A_1\rangle + a_2\lvert A_2\rangle$ and $\lvert B\rangle = b_1\lvert A_1\rangle + b_2\lvert A_2\rangle$ one has $\langle A\mid B\rangle = (a_1^*)(b_1)+(a_2^*)(b_2)$ Regarding the conjugates, note: $[a+bi]^* = a-bi$; $[a-bi]^* = a+bi$; $[a]^* = a$ when $a \in \mathbf{R}$.

A quantum system may have k distinguishable states (for example a system that can be in one of k distinct quantized energy states); the state of such a system can be represented as a *(column)* vector in a k dimensional (complex) (Hilbert) vector space C^k. k distinguishable states of unit length are postulated to exist that form an orthogonal basis for the vector space and constitute the set $\{|0\rangle, |2\rangle, ... |k\text{-}1\rangle\}$; sometimes the set $\{|1\rangle, |2\rangle, ... |k\rangle\}$ is used. A generic vector (that is, state) in C^k may be $|v\rangle = \sum \alpha_i |i\rangle$; when $|v\rangle$ is transposed it becomes a *row* vector $|v\rangle' = \langle v|$; in particular, $|v\rangle' = \langle v| = \sum \alpha_i^* |i\rangle'$. $\langle v|$ is called a bra and $|v\rangle$ is called a ket.

2-dimensional space example $|v\rangle = \alpha_1|0\rangle + \alpha_2|1\rangle$

$$|v\rangle = \begin{bmatrix} \alpha_1 \\ \alpha_2 \end{bmatrix} = \alpha_1|0\rangle + \alpha_2|1\rangle$$

$$|0\rangle = \begin{bmatrix} 1 \\ 0 \end{bmatrix} \quad |1\rangle = \begin{bmatrix} 0 \\ 1 \end{bmatrix}$$

$$\langle v| = |v\rangle' = \begin{bmatrix} \alpha_1^* & \alpha_2^* \end{bmatrix} = \alpha_1^* \begin{bmatrix} 1 & 0 \end{bmatrix} + \alpha_2^* \begin{bmatrix} 0 & 1 \end{bmatrix} = \alpha_1^*|0\rangle' + \alpha_2^*|1\rangle' = \alpha_1^*\langle 0| + \alpha_2^*\langle 1|$$

$$\langle 0| = |0\rangle' = \begin{bmatrix} 1 \\ 0 \end{bmatrix}' = \begin{bmatrix} 1 & 0 \end{bmatrix} \quad \langle 1| = |1\rangle' = \begin{bmatrix} 0 \\ 1 \end{bmatrix}' = \begin{bmatrix} 0 & 1 \end{bmatrix}$$

Figure 2.1 Bra and ket nomenclature of vectors/states

The quantum state of the system can be written as the linear superposition (column) vector $\begin{bmatrix} \alpha \\ \beta \end{bmatrix} \in \mathbf{C}^2$. Using $|0\rangle$ and $|1\rangle$ as *computational basis states*—namely, an orthonormal basis for this vector space—the *state* $|\psi\rangle$ of a qubit can be *written* as:

$$|\psi\rangle = \alpha|0\rangle + \beta|1\rangle$$

where α and β are complex numbers ($\alpha, \beta \in \mathbf{C}$) with $|\alpha|^2 + |\beta|^2 = 1$ and

$$|0\rangle = \begin{bmatrix} 1 \\ 0 \end{bmatrix} \quad |1\rangle = \begin{bmatrix} 0 \\ 1 \end{bmatrix}.$$

More generally, the quantum superposition principle posits that a quantum system can be any one of number of different states, the state of the system then being expressed as a linear superposition of these states with complex coefficients. Hence, the quantum state of a k-state system is described by a sequence of k complex numbers $\alpha_0, \ldots, \alpha_{k-1} \in \mathbf{C}$ with normalized amplitudes $\sum_j |\alpha_j|^2 = 1$. α_j is the (complex) amplitude with which the system exists in state j. In the generic setting one can describe the state of the system as a k dimensional vector in a k dimensional space:

$$\begin{bmatrix} \alpha_0 \\ \alpha_1 \\ \cdot \\ \cdot \\ \cdot \\ \alpha_{k-1} \end{bmatrix}$$

QM utilizes the Dirac's *ket* notation to denote the vector:

$$|\psi\rangle = \sum_{j=0}^{k-1} \alpha_j |j\rangle$$

where $|0\rangle = [1, 0,..., 0]^T$, $|1\rangle = [0,1,...,0]^T$,..., $|k\text{-}1\rangle = [0, 0,...,1]^T$; the set $\{|0\rangle,|1\rangle,...,|k\text{-}1\rangle\}$ is the basis of the vector space; the state of the qubit is a vector, and that it encapsulates data (0 or 1) to be processed.

The following basic concepts are fundamental for the models and/or descriptions used in QM:

- A *Hilbert vector space* is as follows: a Hilbert space (named after mathematician David Hilbert) is a real or complex *inner product space* that is also a *complete metric space* with respect to the distance function produced by the inner product. A Hilbert vector space is effectively synonymous to an inner product space.
- The *inner product* of two vectors in the inner product space is a scalar, normally notated with angle brackets, e.g., $\langle a,b \rangle$; *here we use the following to enhance clarity* $\langle a,b \rangle$. Inner products support formal definitions of basic geometric concepts, such as lengths, angles, and orthogonality of vectors (orthogonality equates to zero inner product).
- An *inner product space* is a real (or complex) vector space with an associated operation that is the inner product. Inner product spaces generalize Euclidean vector spaces, where the inner product is the dot product (scalar product) of values of the Cartesian coordinates.
- A vector space is *complete* (is a complete metric space, a Cauchy space) when any Cauchy sequence of vectors in the space converges to a vector in the space. A space S is called complete if every *Cauchy sequence* of points in S has a limit that is also in S.
- A sequence $s_1, s_2, s_3, \ldots$ in a metric space (S, d) is called a *sequence Cauchy* if for every positive real number $r > 0$ there is a positive integer N such that for all positive integers $m, n > N$, $d(s_m, s_n) < r$.

In classical mechanics each possible state corresponds to a point in the vector space, and each point in the vector space represents a possible state of such a system. In QM the *state-spaces* are a particular type of vector spaces: *Hilbert spaces*.[1] Hilbert vector spaces are vector spaces (i) on which an *inner product* is defined and (ii) which are *complete*.

Recapitulating, in the context of QM, the finite dimensional complex vector space(s) that underpin the theory is the Hilbert space—namely, a vector space V with an inner product on V. All the information about the internal relations among states and quantities is embodied in the mathematical relations among the vectors and operators that represent them. All the physically relevant features of the behaviors of QM systems are derivative of the mathematical properties of those relations.

Tables 2.A1, 2.A2, and 2.A3 in Appendix A (starting on page 88) to this chapter provide some of the basic mathematical concepts for understating and/or describing QM. Table 2.A1 lists the more basic concepts, Table 2.A2 lists more advanced concepts, and Table 2.A3 lists yet more advanced concepts (which can be skipped on first reading). Readers should familiarize themselves with these underlying concepts by working through the tables, here focusing in particular on Table 2.A1 and Table 2.A2. Some of the key concepts and nomenclature follow in the next sections.

---

[1] However, in that environment there exists mathematically describable ways of combining the values of the quantities that do not represent physically possible states.

2.2 Qubit Representation as Vector(s)

Expanding the discussion above, quantum information in Quantum Information Science and Technology (QIST) is encoded using qubits: beyond holding information as discrete values such as 0 or 1, qubits can hold quantum states that are linear combinations of these values.

Consider the vector space $\mathbf{C}^n$, the space of all n-tuples of complex numbers, $[z_1; ::::; z_n]$. The elements of a vector space are obviously called vectors. With the *ket* $|\psi\rangle$ notation a *vector is represented* as a column as:

$$|\psi\rangle = |\psi_A\rangle = \begin{bmatrix} a_0 \\ a_1 \\ \vdots \\ a_{N-1} \end{bmatrix}$$

where $a_i \in \mathbf{C}$; $i = 0;\ 1; :::: ;\ N\text{-}1$.

The *bra* $\langle\psi|$ is the *adjoint (conjugate transpose) row vector* of column vector $|\psi\rangle$ and is defined as:

$$\langle\psi| = \langle\psi_A| = |\psi^*\rangle^T = [a_0^*, a_1^*, \ldots, a_{N-1}^*]$$

where a^* is the complex conjugate of a; namely, if $a = x + iy$, then $a^* = x - iy$; and if $a = x - iy$, then $a^* = x + iy$.

As stated earlier, an inner product is a function that takes as input two vectors, $|A\rangle$ and $|B\rangle$, from a vector space and generates a complex number. It is depicted as $(|A\rangle,|B\rangle)$ or as $\langle A\,|\,B\rangle$ (the former is commonly utilized in linear algebra, while the latter is typically used in QM); here, as noted, for enhanced clarity *we use* $\langle A\,|\,B\rangle$. This inner product can be expressed as

$$\langle\,|\psi_A\rangle,|\psi_B\rangle\rangle = \langle\psi_A|\psi_B\rangle = \sum_{i=0}^{N-1} a_i^* b_i$$

We noted in Chapter 1 and above that the state of a qubit can be perceived as a two-dimensional vector with basis $|0\rangle$ and $|1\rangle$ and value $|\psi\rangle = \cos(\theta/2)|0\rangle + e^{i\phi}\sin(\theta/2)|1\rangle$ or $|\psi\rangle = \alpha|0\rangle + \beta|1\rangle$ where $\alpha = \cos(\theta/2)$ and $\beta = e^{i\phi}\sin(\theta/2)$. Namely, a qubit (with state) $|\psi\rangle$ can be in a generic *superposition* of the basis states $|0\rangle$ and $|1\rangle$; states $|0\rangle$ and $|1\rangle$ are vectors forming a basis of

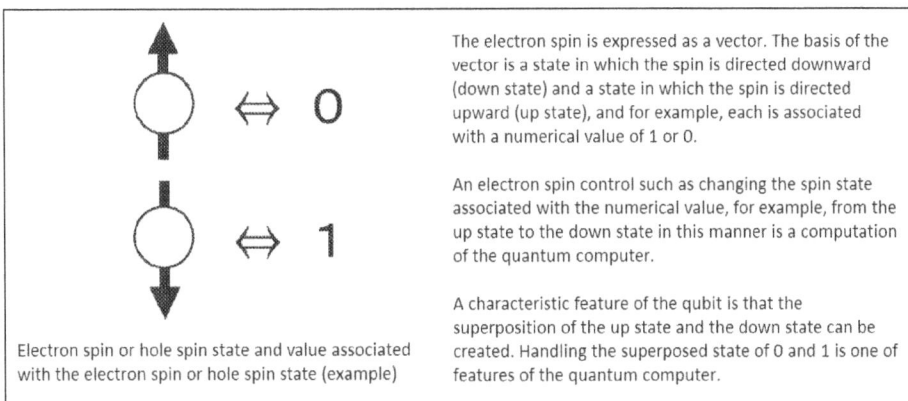

	The electron spin is expressed as a vector. The basis of the vector is a state in which the spin is directed downward (down state) and a state in which the spin is directed upward (up state), and for example, each is associated with a numerical value of 1 or 0.
	An electron spin control such as changing the spin state associated with the numerical value, for example, from the up state to the down state in this manner is a computation of the quantum computer.
Electron spin or hole spin state and value associated with the electron spin or hole spin state (example)	A characteristic feature of the qubit is that the superposition of the up state and the down state can be created. Handling the superposed state of 0 and 1 is one of features of the quantum computer.

Figure 2.2 Example of spin-based qubits (loosely based on [SHI23]) *(Figure described on next page)*

a two-dimensional Hilbert space *H*. The complex coefficients α and β are independent of each other, and the only requirement is that $|\alpha|^2 + |\beta|^2 = 1$; this physically corresponds to the requirement that the probability to find the state $|\psi\rangle$ in precisely *one or the other* of the basis states $|0\rangle$ or $|1\rangle$ is 1. Examples of qubits include the polarization states of a photon or the spin of electrons[1,2]; e.g., see Figure 2.2 (above).

More generally, a (pure) quantum state can be expressed as a *vector* of length 1, $|\psi\rangle \in \mathbf{C}^d$, where *d* is typically a finite value (a finite dimensional vector) and **C** is the field of complex numbers. Let $\rho = |\psi\rangle \times \langle\psi| = |\psi\rangle \langle\psi|$; quantum states where tr(ρ^2) = 1 are called *pure states,* otherwise they are called *mixed states* (tr(ρ^2) <1). When the system is in a definite quantum state defined as $|\psi\rangle = \alpha|0\rangle + \beta|1\rangle$, it is said to be in a pure state—even though a measurement has a probabilistic outcome.

Note: Using this machinery and given states /1/ and /0/, the representation of a traditional, classical bit (state) would be $|\psi_{\text{Trad}}\rangle = \alpha$ /0/ + (1− α)/1/, with $|\alpha|^2 + |1-\alpha|^2 = 1$. This leads to $|\alpha|^2 + (1 - 2|\alpha| + |\alpha|^2) = 1$, which leads to $2|\alpha|^2 = 2|\alpha|$ or the only two solutions: $\alpha = 0$ or $\alpha = 1$. Notice that with the classical bit there is only one state-defining parameter—namely, α. When $\alpha = 0$, $|\psi_{\text{Trad}}\rangle$ = /1/ with probability 1 and when $\alpha = 1$, $|\psi_{\text{Trad}}\rangle$ = /0/, also with probability 1. α cannot have any other value. In the qubit environment complex coefficients α and β are independent and establish the state $|\psi\rangle$ as a weighted superimposed states of $|0\rangle$ and $|1\rangle$.

Although one can represent the state of the qubit as a vector in a two-dimensional plane, another representation of all the possible states of single-qubit is the *Bloch Sphere* already mentioned in Chapter 1, and based on spherical coordinates θ and ϕ with $0 \le \theta \le \pi, 0 < \phi \le 2\pi$; here,

$$|\psi\rangle = \{\cos(\theta/2)\}|0\rangle + \{e^{i\phi}\sin(\theta/2)\}|1\rangle$$

where numbers θ and ϕ are angular displacements; $|\psi\rangle$ then represents a point on a three-dimensional unit sphere, and every point on the sphere represents a possible qubit. All possible qubits can be perceived as vectors on this unit sphere.

---

[1] Elementary particles (such as electrons and protons) have a spin, based on the intrinsic angular momentum called *spin*: the angular momentum vector that can be said to point up $|\uparrow\rangle$ or point down $|\downarrow\rangle$. The quantum mechanical spin state of the particle is then $|\psi\rangle = \alpha|\uparrow\rangle + \beta|\downarrow\rangle$--- $|\uparrow\rangle$ and $|\downarrow\rangle$ are analogous/equivalent to $|0\rangle$ and $|1\rangle$. The intrinsic angular momentum of a particle is the "spin" = $\hat{S}$. *In* QM the intrinsic magnetic moment $\mu = - (g \times e /2m) \hat{S}$; *g* is a unitless QM "correction" and *m* the mass of the particle (for electrons, $g \approx 2$ and for protons, $g \approx 5.6$). To characterize the behavior of the electron's intrinsic magnetic moment μ (which is an observable one is able to measure) one must characterize the behavior of its intrinsic angular momentum—namely, the spin.

[2] There are some research efforts to achieve quantum coherence at room temperature. While most qubits require cryogenic temperatures, some molecular qubits can be initialized by photoexcitation, possibly enabling Quacomp at room temperature and quantum sensing of living systems. One such effort at the time of this writing entailed a dye molecule that absorbs light and emits color, chromophore, in a metal-organic framework (MOF)—coherence was observed only for nanoseconds. The ultimate goal would be to develop materials that can generate multiple qubits at room temperatures [YAM24]. Photonic Quacomp can, in fact, operate at room temperature, but there is also interest in qubits of other types having this property [FUR23], [TOK23].

A complex number z is represented as $z = a + ib$, where a and b are real numbers and $i = \sqrt{(-1)}$. On a reference circle one has $a = r \times \cos\theta$ and $b = r \times \sin\theta$, thus the complex number z can be alternatively expressed in polar form as $z = a + ib = r\,(\cos\theta + i\sin\theta) = re^{i\theta}$ (or $e^{i\theta}$ when $\|r\| = 1$). This relationship enables one to express the state vector $|\psi\rangle$ with two angles θ and ϕ on the sphere. Using the polar form for the complex coefficients α and β, state vector $|\psi\rangle$ can be expressed as $|\psi\rangle = r_\alpha \times e^{i\phi_\alpha}|0\rangle + r_\beta \times e^{i\phi_\beta}|1\rangle$, from which, with some mathematical manipulation, one obtains $|\psi\rangle = \cos(\theta/2)|0\rangle + e^{i\phi}\sin(\theta/2)|1\rangle$, as derived in [BOC23]:

$$e^{-i\phi_\alpha}|\psi\rangle = e^{-i\phi_\alpha}r_\alpha e^{i\phi_\alpha}|0\rangle + e^{i\phi_\alpha}r_\beta e^{i\phi_\beta}|1\rangle,$$

$$= r_\alpha|0\rangle + r_\beta e^{i(\phi_\beta - \phi_\alpha)}|1\rangle$$

$$= r_\alpha|0\rangle + r_\beta e^{i\phi}|1\rangle,$$

$$|\psi\rangle = r_\alpha|0\rangle + r_\beta e^{i\phi}|1\rangle,$$

$$x = r\sin\theta'\cos\phi,\ y = r\sin\theta'\sin\phi,\ z = r\cos\theta',$$

$$r = 1,\ r_\alpha \to z,$$

$$|\psi\rangle = \cos\theta'|0\rangle + e^{i\phi}\sin\theta'|1\rangle,$$

$$\theta = 2\theta',$$

$$|\psi\rangle = \cos\!\left(\frac{\theta}{2}\right)|0\rangle + e^{i\phi}\sin\!\left(\frac{\theta}{2}\right)|1\rangle.$$

The state vector $|\psi\rangle$ can be multiplied by an arbitrary complex number with unit modulus, $e^{i\gamma}$, without altering the magnitudes of the squares of the coefficients α and β. Therefore, there is a global phase-shift degree of freedom in the expression of a state vector that does not impact the measurable properties associated with the state.

Before moving on, some additional formalism for the Dirac notation follows. Say $|v_1\rangle = a_1|0\rangle + b_1|1\rangle$ and $|v_2\rangle = a_2|0\rangle + b_2|1\rangle$; then the expression $\langle v_1\|v_2\rangle$ is the matrix product of the 1×2 matrix $\langle v_1|$ and the 2×1 matrix $|v_2\rangle$; this turns out to be a scalar. The scalar is typically represented with the symbol $\langle v_1|v_2\rangle$, where

$$\langle v_1|v_2\rangle = \begin{bmatrix} a_1^* & b_1^* \end{bmatrix}\begin{bmatrix} a_2 \\ b_2 \end{bmatrix} = a_1^* a_2 + b_1^* b_2$$

and it is the inner product—**here we prefer the symbol $\langle v_1|v_2\rangle$ as noted**. A few results are as follows:

- $\langle v_1|v_2\rangle = (a_1^*\langle 0| + b_1^*\langle 1|)(a_2|0\rangle + b_2|1\rangle) = a_1^* a_2 + b_1^* b_2$
- $\langle v_1|v_2\rangle = \langle v_2|v_1\rangle$;
- $\langle 0|0\rangle = 1$
- $\langle 1|1\rangle = 1$
- $\langle 0|1\rangle = 0$
- $\langle 1|0\rangle^* = 0$

In terms of inner products with $|\psi\rangle = \alpha|0\rangle + \beta|1\rangle$ one obtains

- $\alpha = \langle 0\||\psi\rangle = \langle 0|\psi\rangle = \langle 0|\psi\rangle$ in our notation
- $\beta = \langle 1\||\psi\rangle = \langle 1|\psi\rangle = \langle 1|\psi\rangle$ in our notation
- $\langle\psi|\psi\rangle = 1,\ \langle\psi|\psi\rangle = 1$ in our notation

These relationships follow given the following observations:

$$|0\rangle = \begin{bmatrix} 1 \\ 0 \end{bmatrix} \quad \langle 0| = [1\ 0] \ \text{and} \ |0\rangle\langle 0| = \begin{bmatrix} 1 & 0 \\ 0 & 0 \end{bmatrix}$$

$$|1\rangle = \begin{bmatrix} 0 \\ 1 \end{bmatrix} \quad |1\rangle = [0\ 1] \ \text{and} \ |1\rangle\langle 1| = \begin{bmatrix} 0 & 0 \\ 0 & 1 \end{bmatrix}$$

$$|0\rangle\langle 0| + |1\rangle\langle 1| \ \text{is the } 2 \times 2 \text{ identity matrix } \begin{bmatrix} 1 & 0 \\ 0 & 1 \end{bmatrix}$$

$$|\alpha|^2 + |\beta|^2 = 1$$

The concept of projections is covered next. Define P as the matrix (the operator)

$$P = |v\rangle \times \langle v| = |v\rangle\langle v|$$

(the last term is by notational definition). For example, in two dimensions, if

$$|\psi\rangle = \begin{bmatrix} \alpha \\ \beta \end{bmatrix}$$

$$\langle \psi| = \begin{bmatrix} \alpha \\ \beta \end{bmatrix}^\dagger = [\alpha^* \ \beta^*]$$

$$|\psi\rangle \times \langle \psi| = |\psi\rangle\langle \psi| = \begin{bmatrix} \alpha \\ \beta \end{bmatrix} \times [\alpha^* \ \beta^*] = \begin{bmatrix} \alpha\alpha^* & \alpha\beta^* \\ \beta\alpha^* & \beta\beta^* \end{bmatrix}$$

$$\quad\quad\quad\quad\quad\quad\quad 2\times1 \quad\quad 1\times2 \quad\quad\quad\quad 2\times2$$

then for any vector $|w\rangle$, the following relation holds:

$$P \times |w\rangle = |v\rangle \times \langle v|w\rangle$$

which can be interpreted as a projection (via projection operator P) of vector $|w\rangle$ onto another vector $|v\rangle$, which is expressible as a scalar multiplication of the original vector via the inner product between the two vectors. Operator $P = |v\rangle\langle v|$ projects an arbitrary state $|\psi\rangle$ onto the state $|v\rangle$; thus, P acts as the projection operator onto $|v\rangle$. See Figure 2.3 for a generic example of vector projection.

In particular, $|v\rangle$ could be a vector in the vector basis set. $P^2 = |v\rangle \times \langle v|v\rangle \times \langle v| = P$, when/if $|v\rangle$ has norm 1. That is, for an orthonormal basis $\{|j\rangle\}$ one can define the set of projection operators $P_j = |j\rangle\langle j|$, which obey the "completeness relation" $\Sigma_j P_j = \Sigma_j |j\rangle\langle j| = 1$.

An example follows:

$$|v\rangle = \alpha|0\rangle + \beta|1\rangle = \begin{bmatrix} \alpha \\ \beta \end{bmatrix}$$

$$|w\rangle = a|0\rangle + b|1\rangle = \begin{bmatrix} a \\ b \end{bmatrix}$$

$$\boxed{P \times |w\rangle} = \begin{bmatrix} \alpha\alpha^* & \alpha\beta^* \\ \beta\alpha^* & \beta\beta^* \end{bmatrix} \begin{bmatrix} a \\ b \end{bmatrix} = \begin{bmatrix} \alpha\alpha^*a + \alpha\beta^*b \\ \beta\alpha^*a + \beta\beta^*b \end{bmatrix}$$ (Note: P was computed above for this example)

$$\quad\quad\quad 2\times2 \quad 2\times1 \quad 2\times1$$

The inner product of $|A\rangle$ and $|B\rangle$ for complex spaces is defined in terms of the *conjugates* of complex coefficients as follows, using basis (unit) vectors $|A1\rangle$ and $|A2\rangle$; then if $|A\rangle = a_1|A_1\rangle + a_2|A_2\rangle$ and $|B\rangle = b_1|A_1\rangle + b_2|A_2\rangle$ one has

$$\langle A | B\rangle = (a^*_1)(b_1) + (a^*_2)(b_2)$$

$$\langle v|w\rangle = \alpha^*a + \beta^*b$$

$$\boxed{|v\rangle \times \langle v|w\rangle} = \begin{bmatrix} \alpha \\ \beta \end{bmatrix} [\alpha^*a + \beta^*b] = \begin{bmatrix} \alpha\alpha^*a + \alpha\beta^*b \\ \beta\alpha^*a + \beta\beta^*b \end{bmatrix}$$

Thus, indeed, $P \times |w\rangle = |v\rangle \times \langle v|w\rangle$ in this illustrative example.

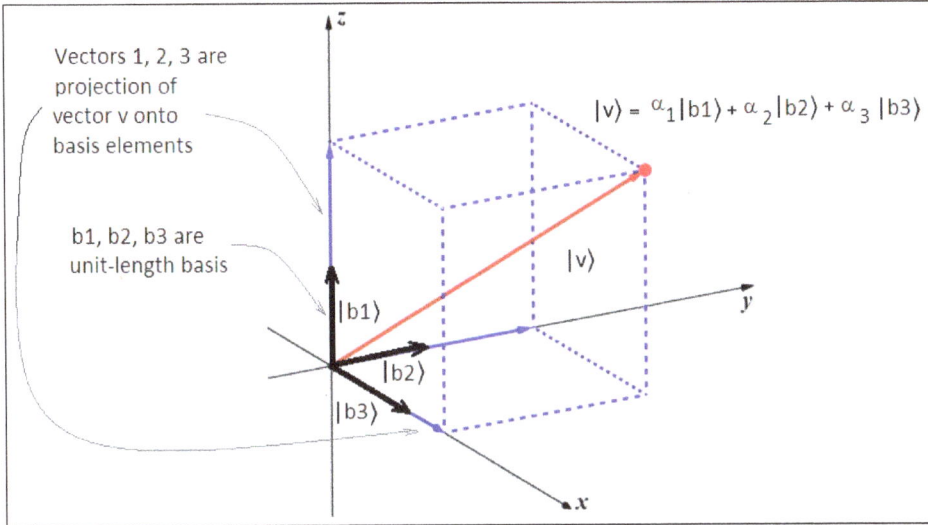

Vectors 1, 2, 3 are projection of vector v onto basis elements

$|v\rangle = \alpha_1|b1\rangle + \alpha_2|b2\rangle + \alpha_3|b3\rangle$

b1, b2, b3 are unit-length basis

Figure 2.3   Generic example of basis vectors and projection vectors

An example of the projection in a two-dimensional setting is as follows:

$$|v\rangle = \alpha|0\rangle + \beta|1\rangle = \begin{bmatrix} \alpha \\ \beta \end{bmatrix} \qquad |0\rangle = \begin{bmatrix} 1 \\ 0 \end{bmatrix} \qquad |1\rangle = \begin{bmatrix} 0 \\ 1 \end{bmatrix}$$

The scalar projection of a ket onto a basis vector is given by the inner product

$$\langle 0|v\rangle = \begin{bmatrix} 1 & 0 \end{bmatrix}\begin{bmatrix} \alpha \\ \beta \end{bmatrix} = \alpha \qquad \langle 1|v\rangle = \begin{bmatrix} 0 & 1 \end{bmatrix}\begin{bmatrix} \alpha \\ \beta \end{bmatrix} = \beta.$$

Thus, one can write

$$|v\rangle = \begin{bmatrix} \alpha \\ \beta \end{bmatrix} = \langle 0|v\rangle|0\rangle + \langle 1|v\rangle|1\rangle$$

or

$$|v\rangle = \alpha|0\rangle + \beta|1\rangle,$$

where $|0\rangle$ and $|1\rangle$ are the basis.

2.3 Joint State of *n* Individual Qubits

When two QM systems interact, say systems V and U, a new state space for the combined state W is generated. The composite state defines uniquely the states of its components.

If H_V and H_U are Hilbert spaces for V and U, then $H_W = H_V \otimes H_U$ is also a Hilbert space. The state-space of a composite system W is obtained from the "*tensor product*", also referred to as "*outer product*", of the individual state-spaces, H_V and H_U: $H_W = H_V \otimes H_U$.

If $\{v_i\}$ is an orthonormal basis for H_V and $\{u_j\}$ is an orthonormal basis for H_U, then the set of pairs (v_i, u_j) can be taken to form an orthonormal basis for the tensor product space $H_V \otimes H_U$. Notation-wise $v_i \otimes u_j$ is used to represent the *pair set* $\{(v_i, u_j)\}$.

The state *set* of an *n*-tuple of qubits, say a *n*-qubit register, has 2^n values, say $d = 2^n$, then the basis elements of C^d are represented by strings $x = x_1, \ldots, x_n \in \{0, 1\}^n$. The joint state of *n* individual qubits $|\psi_1\rangle, \ldots, |\psi_n\rangle$ is obtained by the *tensor (outer) product* of the individual vectors—namely, $|\psi\rangle = |\psi_1\rangle \otimes \cdots \otimes |\psi_n\rangle$. The tensor product is an operation combining two smaller vector spaces into one larger vector space; elements of this larger space are called *tensors*. For a system of *n* qubits, the basis is of the form $|x_1 x_2 x_3, \ldots, x_n\rangle$, and a quantum state of such a system is specified by 2^n amplitudes.

Here,

$$|\psi\rangle = \Sigma_{ij} \alpha_{ij} x_{ij}\rangle \text{ with } \Sigma_{ij} |\alpha_{ij}|^2 = 1$$

For example, for $n=2$ one has four values for the state: say, $v_1 = a|0\rangle + b|0\rangle$ and $v_2 = c|0\rangle + d|1\rangle$; then the convolution of two qubits $v_1 \otimes v_2$ is

$$|\psi\rangle = \alpha_{00}|00\rangle + \alpha_{01}|01\rangle + \alpha_{10}|10\rangle + \alpha_{11}|11\rangle$$

where $|00\rangle = [1,0,0,0]^T$, $|01\rangle = [0,1,0,0]^T$, $|10\rangle = [0,0,1,0]^T$, $|11\rangle = [0,0,0,1]^T$, with $\alpha_{00} = ac$, $\alpha_{01} = ad$, $\alpha_{10} = bc$, $\alpha_{11} = bd$. Note that clearly $|\psi\rangle \in C^4$

$$|\psi\rangle = ac\begin{bmatrix}1\\0\\0\\0\end{bmatrix} + ad\begin{bmatrix}0\\1\\0\\0\end{bmatrix} + bc\begin{bmatrix}0\\0\\1\\0\end{bmatrix} + bd\begin{bmatrix}0\\0\\0\\1\end{bmatrix} = \begin{bmatrix}ac\\ad\\bc\\bd\end{bmatrix}$$

The α's are complex amplitudes in C^4 and parametrize the basis "vectors" (here $|00\rangle$, $|01\rangle$, $|10\rangle$ and $|11\rangle$); these elements can be seen as states or a basis). $|\alpha_{00}|^2 + |\alpha_{01}|^2 + |\alpha_{10}|^2 + |\alpha_{11}|^2 = 1$.

From another perspective, the qubits can also be in weighted superpositions of these four discrete states $|00\rangle$, $|01\rangle$, $|10\rangle$ and $|11\rangle$. See Figure 2.4 for the set of states and Figure 2.5 for the bra/ket notations. An example is the state of an electron pair (e_1, e_2) in two closely located hydrogen atoms: given that each electron can be in either of the ground state (g) or the excited (e) state, classically the two electrons are in one of four states, *gg, ge, eg,* or *ee*, and represent two bits of classical information. In QM terms, they are in a superposition of those four states; one can store two qubits of information in the quantum state of the two electrons in the superposition just cited.

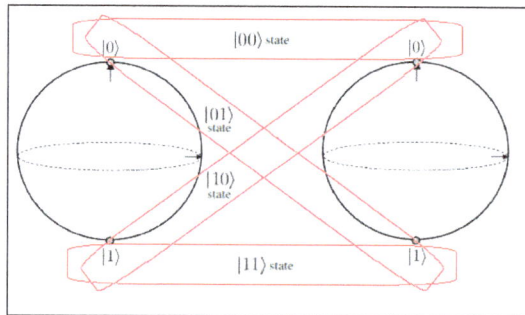

State $|00\rangle$ encapsulates qubit 1 being in state $|0\rangle$ and qubit 1 being in state $|0\rangle$,
State $|01\rangle$ encapsulates qubit 1 being in state $|0\rangle$ and qubit 1 being in state $|1\rangle$, And so on . . .

Figure 2.4   The state of an *n*-tuple of qubits (*n=2*)

State of one qubit

$$|\psi\rangle = \alpha\,|0\rangle + \beta\,|1\rangle = \alpha\begin{bmatrix}1\\0\end{bmatrix} + \beta\begin{bmatrix}0\\1\end{bmatrix} = \begin{bmatrix}\alpha\\0\end{bmatrix} + \begin{bmatrix}0\\\beta\end{bmatrix} = \begin{bmatrix}\alpha\\\beta\end{bmatrix} \quad \text{with basis} \quad |0\rangle = \begin{bmatrix}1\\0\end{bmatrix} \quad |1\rangle = \begin{bmatrix}0\\1\end{bmatrix}$$

State of two qubits: composite system

$$\mathbf{u} = \begin{bmatrix}\alpha_1\\\beta_1\end{bmatrix} \quad \mathbf{v} = \begin{bmatrix}\alpha_2\\\beta_2\end{bmatrix} \quad \begin{array}{c}\mathbf{u}\otimes\mathbf{v}=\mathbf{A}=\\ \text{(outer product)}\\ \text{(tensor product)}\end{array} \begin{bmatrix}\alpha_1\alpha_2 & \alpha_1\beta_2\\ \beta_1\alpha_2 & \beta_1\beta_2\end{bmatrix} = \begin{bmatrix}\alpha_{00} & \alpha_{01}\\ \alpha_{10} & \alpha_{11}\end{bmatrix}$$

defining the basis vectors as

$$|00\rangle = \begin{bmatrix}1 & 0\\0 & 0\end{bmatrix} \quad |01\rangle = \begin{bmatrix}0 & 1\\0 & 0\end{bmatrix} \quad |10\rangle = \begin{bmatrix}0 & 0\\1 & 0\end{bmatrix} \quad |11\rangle = \begin{bmatrix}0 & 0\\0 & 1\end{bmatrix}$$

one gets

$$|\psi\rangle = \alpha_{00}\,|00\rangle + \alpha_{01}\,|01\rangle + \alpha_{10}\,|10\rangle + \alpha_{11}\,|11\rangle$$

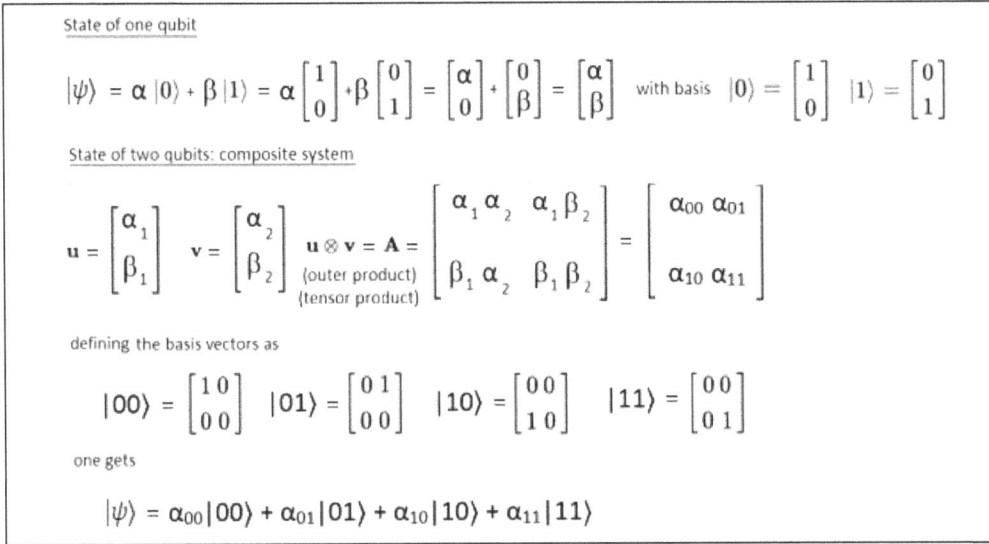

Figure 2.5   Basis representation of the state of an *n*-tuple of qubits (*n* = 2).

Suppose the first qubit is in the state $|\psi_1\rangle = \alpha_1|0\rangle_1 + \beta_1|1\rangle_1$ and the second qubit is in the state $|\psi_2\rangle = \alpha_2|0\rangle_2 + \beta_2|1\rangle_2$. The joint state of the two qubits is given by the outer (tensor) product:

$$\psi = |\psi_1\rangle \otimes |\psi_2\rangle = \alpha_1\alpha_2\,|00\rangle + \alpha_1\beta_2|01\rangle + \beta_1\alpha_2|10\rangle + \beta_1\beta_2|11\rangle$$

Effectively (by definition of the outer product), one simply multiplies the amplitudes of the corresponding individual basis elements $|0\rangle$, $|1\rangle$ to determine the amplitude of the new basis elements $|00\rangle$, $|01\rangle$ and so on (for example, the amplitude of the state $|01\rangle$ is the product of the amplitude of $|0\rangle_1$, $|1\rangle_2$). The two qubits are not entangled, and individual measurements of/on the two qubits can be obtained independently. However, having knowledge (having a measurement) of the state of the combined entity does not provide information of the state of the original discrete qubits. Say that one knew that $\alpha_1\alpha_2 = \frac{1}{16}$; one cannot recover α_1 and/or α_2; one could have $\alpha_1 = \frac{1}{8}$ and $\alpha_2 = \frac{1}{2}$ or $\alpha_1 = \frac{1}{4}$ and $\alpha_2 = \frac{1}{4}$, and so on; thus, the original sates cannot be uniquely determined. The combined state is entangled; it cannot be decomposed into the state of the individual two qubits.

2.4 Tensor (Outer) Product

Continuing the discussion of the previous section, additional definitions and examples of tensor product follow next, herewith.

- As seen, the entries of the matrix are obtained by taking the *outer product*, the operator $\otimes$, of the *ket* and the *bra* of the state $|\psi\rangle$. In general, the outer product is equivalent to a *matrix multiplication* uv^*, (or uv^T if vectors are real) when u is represented as a column $m\times1$ vector, and v is represented as a column $n\times1$ vector (whereby v^T is row vector). The outer product of two tensors (multidimensional arrays of numbers) is also a tensor. So, the outer product of two coordinate vectors u and v is the matrix whose entries are all products of an element in the first vector u with an element in the second vector v.

- More specifically, when two vectors have dimensions n and m, then their *outer product* is an $n \times m$ matrix.[1] Given two vectors u and v of size $m \times 1$ and $n \times 1$, respectively, their outer product, denoted as $u \otimes v$, is a $n \times m$ matrix A defined as follows (see Figure 2.6 for an example):

$$\mathbf{u} = \begin{bmatrix} u_1 \\ u_2 \\ \vdots \\ u_m \end{bmatrix}, \quad \mathbf{v} = \begin{bmatrix} v_1 \\ v_2 \\ \vdots \\ v_n \end{bmatrix} \quad \mathbf{u} \otimes \mathbf{v} = \mathbf{A} = \begin{bmatrix} u_1 v_1 & u_1 v_2 & \cdots & u_1 v_n \\ u_2 v_1 & u_2 v_2 & \cdots & u_2 v_n \\ \vdots & \vdots & \ddots & \vdots \\ u_m v_1 & u_m v_2 & \cdots & u_m v_n \end{bmatrix}$$

(say $m = 2$ and $n = 2$, then the resulting matrix is 2×2; if $m = 3$ and $n = 2$, then the resulting matrix is 2×3).

Note: One could (more intuitively) represent the matrix obtained by the outer product multiplication of the vectors representing the states $|\psi\rangle$ and $\langle\psi|$—namely, $|\psi\rangle \otimes \langle\psi|$, but the normal convention is to use $|\psi\rangle\langle\psi|$.

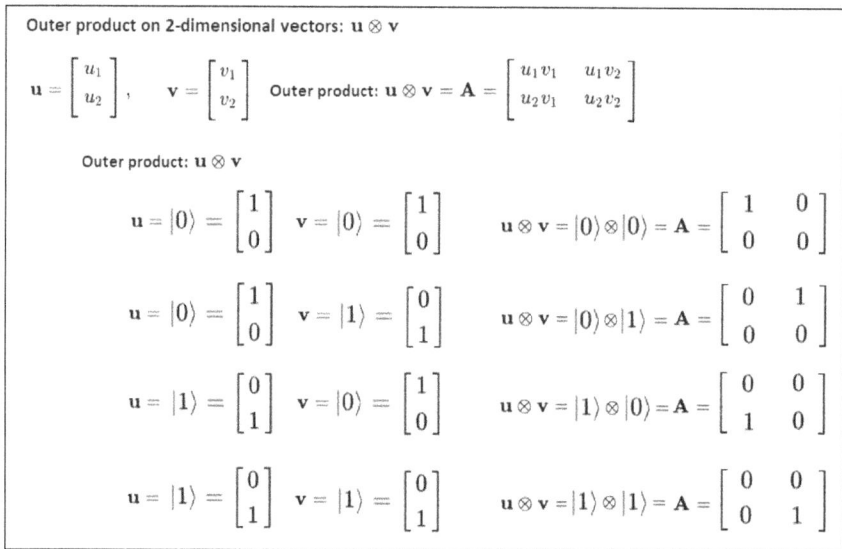

Figure 2.6   Examples of outer product

The outer product and the *Kronecker product* (which is also used), *are related and the same symbol is commonly used to denote both operations*. See Figure 2.7 for an example of the two products.

More *formally* and in summary, one has the following two definitions:

- *Tensor product of two vector spaces, say $V \otimes U$:* Say V and U are two finite dimensional spaces over the same field of scalars $\mathbb{F}$. Let $\alpha = \{v_1, v_2, \ldots, v_n\}$ and $\beta = \{u_1, u_2, \ldots, u_m\}$ be their respective bases, then the tensor product $V \otimes U$ of vector spaces V and U is spanned on the basis $\{v_i \otimes u_j \text{ with } i = 1, 2, \ldots, n; j = 1, 2, \ldots, m\}$. Elements of $V \otimes U$, the tensors, are linear combinations of nm of basis vectors

---

[1] By contrast the dot product maps a pair of vectors to a scalar; the dot product is a special case of the "inner product" operation. On the other hand, the *Kronecker product* maps a pair of matrices to produce a block matrix, and the *standard matrix multiplication* produces a new simple matrix.

The outer product and Kronecker product are related and the same symbol is commonly used to denote both operations

The Kronecker product can be viewed as a form of vectorization (or flattening) of the outer product

Example: $\mathbf{u} = \begin{bmatrix} 1 & 2 & 3 \end{bmatrix}^{\mathsf{T}}$, $\mathbf{v} = \begin{bmatrix} a & b \end{bmatrix}^{\mathsf{T}}$

$$\mathbf{u} \otimes_{\mathrm{Kron}} \mathbf{v} = \begin{bmatrix} a \\ b \\ 2a \\ 2b \\ 3a \\ 3b \end{bmatrix}, \quad \mathbf{u} \otimes_{\mathrm{outer}} \mathbf{v} = \begin{bmatrix} a & b \\ 2a & 2b \\ 3a & 3b \end{bmatrix}$$

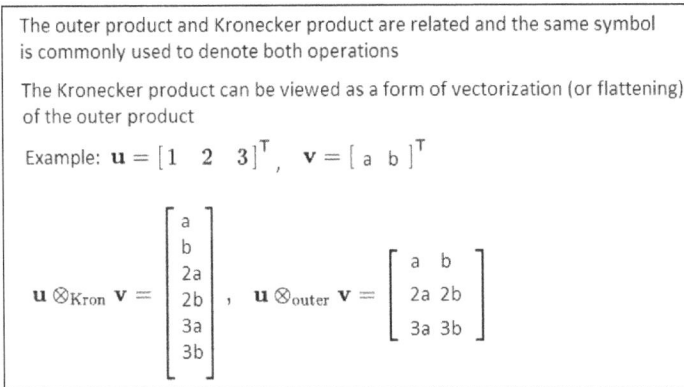

Figure 2.7   Comparisons with the Kronecker product

$$\sum_{i,j} \tau^{i,j} v_i \otimes u_j, \qquad \tau^{i,j} \in \mathbb{F},$$

satisfying the following two conditions:

$$c(v \otimes u) = (cv) \otimes u = v \otimes (cu), \ c \in \mathbb{F} \text{ (scalar multiplication)},$$

and

$$a \otimes u + b \otimes u = (a+b) \otimes u,$$
$$v \otimes u + v \otimes z = v \otimes (u+z)$$

Another description is as follows: $V \otimes U$, *the tensor product space of U and V, is the (Hilbert) (vector) space spanned by the basis elements* $|v\rangle \otimes |u\rangle$, *where* $|v\rangle \in V$ *and* $|u\rangle \in U$, such that the following relations hold for any $|v\rangle$, $|v'\rangle \in V$ and $|u\rangle$, $|u'\rangle \in U$:

$$(|v\rangle + |v'\rangle) \otimes |u\rangle = |v\rangle \otimes |u\rangle + |v'\rangle \otimes |u\rangle$$
$$|v\rangle \otimes (|u\rangle + |u'\rangle) = |v\rangle \otimes |u\rangle + |v\rangle \otimes |u'\rangle$$

- The *Kronecker Product* also plays a role. The Kronecker Product is an operation (not the usual matrix multiplication) on two matrices that results in a block matrix. It is a particularization of the generic tensor product (and is shown by the same symbol). If A is an $m \times n$ matrix and B is a $p \times q$ matrix, then the Kronecker product $A \otimes B$ is the $pm \times qn$ block matrix:

$$\mathbf{A} \otimes \mathbf{B} = \begin{bmatrix} a_{11}\mathbf{B} & \cdots & a_{1n}\mathbf{B} \\ \vdots & \ddots & \vdots \\ a_{m1}\mathbf{B} & \cdots & a_{mn}\mathbf{B} \end{bmatrix}$$

Properties include: $\mathbf{A} \otimes (\mathbf{B} + \mathbf{C}) = \mathbf{A} \otimes \mathbf{B} + \mathbf{A} \otimes \mathbf{C}$ and $(\mathbf{A} \otimes \mathbf{B})(\mathbf{C} \otimes \mathbf{D}) = (\mathbf{AC}) \otimes (\mathbf{BD})$. A simple numerical example of product and block matrix follows:

$$\begin{bmatrix} 1 & 2 \\ 3 & 4 \end{bmatrix} \otimes \begin{bmatrix} a & b \\ c & d \end{bmatrix} = \begin{bmatrix} 1\begin{bmatrix} a & b \\ c & d \end{bmatrix} & 2\begin{bmatrix} a & b \\ c & d \end{bmatrix} \\ 3\begin{bmatrix} a & b \\ c & d \end{bmatrix} & 4\begin{bmatrix} a & b \\ c & d \end{bmatrix} \end{bmatrix} = \begin{bmatrix} 1 \times a & 1 \times b & 2 \times a & 2 \times b \\ 1 \times c & 1 \times d & 2 \times c & 2 \times d \\ 3 \times a & 3 \times b & 4 \times a & 4 \times b \\ 3 \times c & 3 \times d & 4 \times c & 4 \times d \end{bmatrix}$$

2.5 Density Matrix

So far we have dealt with *pure* quantum states where $|\psi\rangle$ is expressed by a deterministic formulation:

$$|\psi\rangle = \sum_x \alpha_x |x\rangle.$$

A *mixed state environment* is characterized as a collection of pure states $|\psi_i\rangle$, each with associated weight probability p_i, $\{p_i, |\psi_i\rangle\}$, with the conditions $0 \le p_i \le 1$ and $\sum p_i = 1$. One needs to consider mixed states because quantum states are difficult to isolate and are often entangled with the environment.

Consider the mixture of quantum states $|\psi_i\rangle$ with probability p_i. *Each* $|\psi_i\rangle$ can be represented by a vector in $\mathbf{C}^d$ ($d = 2^n$), and thus we can associate the outer product $|\psi_i\rangle\langle\psi_i| = \psi_i \psi_i^*$, which is a $2^n \times 2^n$ matrix

$$|\psi_i\rangle\langle\psi_i| = \begin{bmatrix} a_{1i} \\ a_{2i} \\ \vdots \\ a_{Ni} \end{bmatrix} \begin{bmatrix} a_{1i}^* & a_{2i}^* & \cdots & a_{Ni}^* \end{bmatrix} = \begin{bmatrix} a_{1i}a_{1i}^* & a_{1i}a_{2i}^* & \cdots & a_{1i}a_{Ni}^* \\ a_{2i}a_{2i}^* & a_{1i}a_{2i}^* & \cdots & a_{2i}a_{Ni}^* \\ \vdots & & & \vdots \\ a_{Ni}a_{1i}^* & a_N a_{2i}^* & \cdots & a_{Ni}a_{Ni}^* \end{bmatrix}$$

The *density matrix* of the mixture (of states) $\{p_i, |\psi_i\rangle\}$ is defined as

$$\rho = \sum_i p_i |\psi_i\rangle\langle\psi_i|.$$

(the matrix elements $\rho_{ij} = \delta_{ij} \times p_i$). In fact, ρ is an operator. States in which ($\mathrm{tr}(\rho^2) = 1$) are called pure states; otherwise they are called mixed states ($\mathrm{tr}(\rho^2) < 1$). A diagonal density matrix with identical matrix elements along the diagonal is referred to as "the completely mixed state".

Figure 2.8 provides an example of some density matrices. Notice that the two density matrices are identical, although the mixed state with which one started out was different. It is possible for two different mixed states to have the same density matrix; this implies that one cannot distinguish the two states (if one measured the mixed state, one cannot obtain any information about the different possible mixed states that give rise to a specific density matrix).

The expected value of the operator over the mixed states captured by the density matrix is

$$\langle A \rangle = \sum_i p_i \langle\psi_i|A|\psi_i\rangle = \mathrm{tr}[\rho A]$$

Environment with mixed states
$|0\rangle$ with $p_0 = 0.5$
$|1\rangle$ with $p_1 = 0.5$

$$|0\rangle\langle0| = \begin{bmatrix} 1 \\ 0 \end{bmatrix} \begin{bmatrix} 1 & 0 \end{bmatrix} = \begin{bmatrix} 1 & 0 \\ 0 & 0 \end{bmatrix}$$

$$|1\rangle\langle1| = \begin{bmatrix} 0 \\ 1 \end{bmatrix} \begin{bmatrix} 0 & 1 \end{bmatrix} = \begin{bmatrix} 0 & 0 \\ 0 & 1 \end{bmatrix}$$

$$\rho = \frac{1}{2}|0\rangle\langle0| + \frac{1}{2}|1\rangle\langle1| = \begin{bmatrix} 1/2 & 0 \\ 0 & 1/2 \end{bmatrix}$$

$$\rho^2 = \begin{bmatrix} 1/4 & 0 \\ 0 & 1/4 \end{bmatrix}$$

$\mathrm{tr}\,(\rho^2) = 1/2$: mixed states

Environment with fixed states
$|0\rangle$ with $p_0 = 1$
$|1\rangle$ with $p_1 = 0$

$$|0\rangle\langle0| = \begin{bmatrix} 1 \\ 0 \end{bmatrix} \begin{bmatrix} 1 & 0 \end{bmatrix} = \begin{bmatrix} 1 & 0 \\ 0 & 0 \end{bmatrix}$$

$$|1\rangle\langle1| = \begin{bmatrix} 0 \\ 1 \end{bmatrix} \begin{bmatrix} 0 & 1 \end{bmatrix} = \begin{bmatrix} 0 & 0 \\ 0 & 1 \end{bmatrix}$$

$$\rho = 1|0\rangle\langle0| + 0|1\rangle\langle1| = \begin{bmatrix} 1 & 0 \\ 0 & 0 \end{bmatrix}$$

$$\rho^2 = \begin{bmatrix} 1 & 0 \\ 0 & 0 \end{bmatrix}$$

$\mathrm{tr}\,(\rho^2) = 1$: pure (fixed) states

Environment with mixed states
$|+\rangle = \frac{1}{\sqrt{2}}(|0\rangle + |1\rangle)$ with probability $1/2$
$|-\rangle = \frac{1}{\sqrt{2}}(|0\rangle - |1\rangle)$ with probability $1/2$

$$|+\rangle\langle+| = (1/2)\begin{bmatrix} 1 \\ 1 \end{bmatrix} \begin{bmatrix} 1 & 1 \end{bmatrix} = \frac{1}{2}\begin{bmatrix} 1 & 1 \\ 1 & 1 \end{bmatrix},$$

$$|-\rangle\langle-| = (1/2)\begin{bmatrix} 1 \\ -1 \end{bmatrix} \begin{bmatrix} 1 & -1 \end{bmatrix} = \frac{1}{2}\begin{bmatrix} 1 & -1 \\ -1 & 1 \end{bmatrix}$$

$$\rho = \frac{1}{2}|+\rangle\langle+| + \frac{1}{2}|-\rangle\langle-| = \begin{bmatrix} 1/2 & 0 \\ 0 & 1/2 \end{bmatrix}$$

Environment with fixed states
$|0\rangle$ with $p_0 = 0$
$|1\rangle$ with $p_1 = 1$

$$|0\rangle\langle0| = \begin{bmatrix} 1 \\ 0 \end{bmatrix} \begin{bmatrix} 1 & 0 \end{bmatrix} = \begin{bmatrix} 1 & 0 \\ 0 & 0 \end{bmatrix}$$

$$|1\rangle\langle1| = \begin{bmatrix} 0 \\ 1 \end{bmatrix} \begin{bmatrix} 0 & 1 \end{bmatrix} = \begin{bmatrix} 0 & 0 \\ 0 & 1 \end{bmatrix}$$

$$\rho = 0|0\rangle\langle0| + 1|1\rangle\langle1| = \begin{bmatrix} 0 & 0 \\ 0 & 1 \end{bmatrix}$$

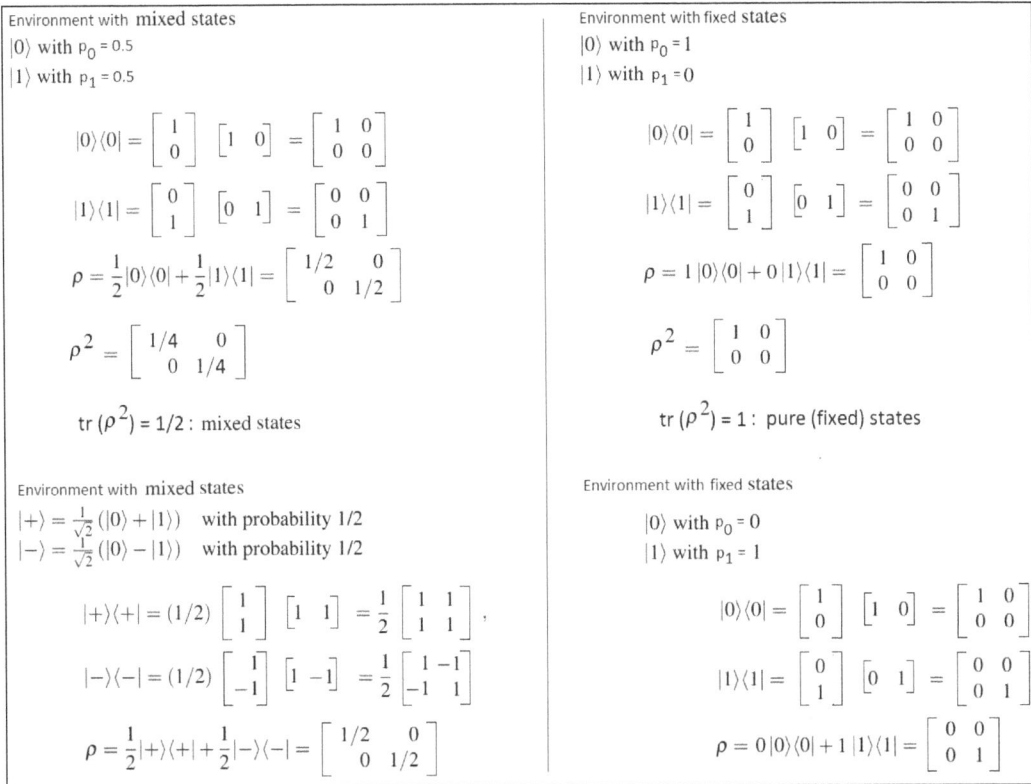

Figure 2.8   Examples of density matrix (Left: mixed states; right: pure states)

In summary, make note that a quantum state $|\psi\rangle$ can (also) be represented (for convenience) as a *density matrix*:

$$\rho = |\psi\rangle\langle\psi|.$$

Density matrices are Hermitian ($\rho = \rho^\dagger$), positive semi-definite ($\rho > 0$) and with trace equal to one: $\mathrm{tr}(\rho) = 1$. The Hermitian conditions arise from the fact that ρ is a sum of Hermitian outer products $(\psi\psi^*)^* = \psi\psi^*$.

In terms of the density matrix, quantum states for which the density operator ρ is also a projector are called *pure states*—namely, $\mathrm{tr}(\rho^2) = 1$; otherwise, if $\mathrm{tr}(\rho^2) < 1$, they are called *mixed states*.

Say $|\psi\rangle = \alpha|0\rangle + \beta|1\rangle$ and $\langle\psi| = \alpha^*\langle0| + \beta^*\langle1|$, where α and β are complex numbers with $|\alpha| + |\beta|^2 = 1$; and $|\phi\rangle = \gamma|0\rangle + \delta|1\rangle$ and $\langle\psi| = \gamma^*\langle0| + \delta^*\langle1|$, where γ and δ are complex numbers with $|\gamma|^2 + |\delta|^2 = 1$. Then the mixed state $\{(p_1, \alpha|0\rangle + \beta|1\rangle), (p_2, \gamma|0\rangle + \delta|1\rangle)$, is represented by the matrix

$$p_1 \begin{bmatrix} \alpha \\ \beta \end{bmatrix} \begin{bmatrix} \alpha^* & \beta^* \end{bmatrix} + p_2 \begin{bmatrix} \gamma \\ \delta \end{bmatrix} \begin{bmatrix} \gamma^* & \delta^* \end{bmatrix} = \begin{bmatrix} p_1|\alpha|^2 + p_2|\gamma|^2 & p_1\alpha\beta^* + p_2\gamma\delta^* \\ p_1\alpha^*\beta + p_2\gamma^*\delta & p_1|\beta|^2 + p_2|\delta|^2 \end{bmatrix}$$

$\mathrm{tr}\,(\rho) = p_1|\alpha|^2 + p_2|\gamma|^2 + p_1|\beta|^2 + p_2|\delta|^2 = p_1 + p_2 = 1.$ $\mathrm{tr}\,(\rho^2)$ can be calculated from the above matrix.

2.6 Composite Systems/Entanglement

Most quantum applications rely on an entanglement that qubits can share entangled states, forming the basic building block of the majority of Quanet applications. Quantum entanglement is a QM phenomenon in which the quantum states of two (or more) objects need to be described with reference to each other, even though the individual objects may be spatially separated. Figure 2.9 from descriptions in RFC 9340 [KOS23] provides one explanation of entanglement. More generally, consider two quantum systems, A and B, that interact with each other. Let H_A and H_B be the Hilbert spaces as the state spaces of these systems, respectively, and let $|x\rangle_A$ and $|y\rangle_B$ be the respective basis

$$|\phi\rangle_A = \sum \alpha_A |x\rangle_A = \sum c_x |x\rangle_A$$

$$|\zeta\rangle_B = \sum \alpha_B |y\rangle_B = \sum d_y |y\rangle_B$$

for appropriate c_x and d_y. Then the state space of the combined interacting system is described by the outer (tensor) product discussed above, $H_{AB} = H_A \otimes H_B$:

$$|\psi\rangle_{AB} = \sum_{x,y} q_{x,y} |x\rangle_A \otimes |y\rangle_B$$

where $q_{x,y}$ are the appropriate parameters from the outer (tensor) product. The composite state $|\psi\rangle_{AB}$ is said to be *separable* if $q_{x,y} := c_x d_y$; then $|\psi\rangle_{AB} = |\phi\rangle_A \otimes |\zeta\rangle_B$. If the composite state is not separable, then it is *entangled*.

Consider a two-qubit register with

- First qubit is in the superposed state $1/\sqrt{2}|0\rangle + 1/\sqrt{2}|1\rangle$
- Second qubit is in the state $|0\rangle$

Combined state is $(1/\sqrt{2}|0\rangle + 1/\sqrt{2}|1\rangle) \otimes |0\rangle = (1/\sqrt{2}|00\rangle + 1/\sqrt{2}|10\rangle)$

Consider applying the CNOT gate on the combined state with the first qubit being the control

- CNOT $(1/\sqrt{2}|00\rangle + 1/\sqrt{2}|10\rangle) \rightarrow (1/\sqrt{2}|00\rangle + 1/\sqrt{2}|11\rangle)$

This final state after the CNOT is applied is *entangled* because there is no possible way of representing that quantum state $1/\sqrt{2}|00\rangle + 1/\sqrt{2}|11\rangle$ as a product of two individual qubits; they are no longer independent, it is not possible to describe the quantum state of either of the individual qubits in a way that is independent of the other qubit; only the quantum state of the system that consists of both qubits provides a physically complete description of the two-qubit system. The states of the two individual qubits are now correlated: neither qubit is in a definite $|0\rangle$ or $|1\rangle$ state, but if one performed a measurement on either one, the outcome of the partner qubit will *always* yield the exact same outcome.

After the measurement, the final state, whether it is $|00\rangle$ or $|11\rangle$, is fundamentally random, but the states of the two qubits following a measurement will always be identical. Once a measurement is performed, the two qubits are once again independent: the final state is either $|00\rangle$ or $|11\rangle$, and both of these states can be decomposed into a product of two individual qubits: the entanglement has been consumed, and the entangled state must be prepared again.

Figure 2.9   Description of entanglement as narrated in RFC 9340 [KOS23]

One can also look at *entanglement from a density matrix perspective* as follows. Define $expA = d_A \times d_A$ and $expB = d_B \times d_B$. A quantum state $\rho_{AB} \in C^{expA} \otimes C^{expB}$ of a combined quantum system of qubits A and B is said to be *separable* when ρ_{AB} can be written as a convex combination of tensor (outer) products of single-node states[1]; separable states have only classical correlations between A and B. *Any state ρ_{AB} that is not separable is called entangled.* Using the density matrix as the representation of the mixed states, a state is separable if

$$\rho_{AB} := \sum_i p_i \phi_A^{(i)} \otimes \zeta_B^{(i)}.$$

where p_i are positive numbers such that $\Sigma_i \, p_i = 1$ and $\phi_A^{(i)}$ and $\zeta_B^{(i)}$ are density matrices of quantum systems A and B.

For example, Bell's states (named after John Bell), also known as Einstein, Podolsky, and Rosen (EPR) pairs (further discussed in Chapter 3):

$$|\phi^+\rangle_{AB} := \frac{1}{\sqrt{2}}(|00\rangle_{AB} + |11\rangle_{AB})$$

$$|\phi^-\rangle_{AB} := \frac{1}{\sqrt{2}}(|00\rangle_{AB} - |11\rangle_{AB})$$

$$|\psi^+\rangle_{AB} := \frac{1}{\sqrt{2}}(|01\rangle_{AB} + |10\rangle_{AB})$$

$$|\psi^-\rangle_{AB} := \frac{1}{\sqrt{2}}(|01\rangle_{AB} - |10\rangle_{AB})$$

They express quantum states of two qubits that represent the simplest examples of entanglement (entangled states).[2] In particular, the $|\phi^+\rangle$, $|\phi^-\rangle$ EPR pair are in states $(|00\rangle + |11\rangle)/\sqrt{2}$ and $(|00\rangle-|11\rangle)/\sqrt{2}$: here the measurement outcomes are entangled—interrelated—namely, a measurement of one qubit results in both qubits being zero or both qubits being one, with identical probability; specifically, upon measuring the first qubit, one obtains two possible results:

- 0 with probability 0.5, leaving the post-measurement state to be $|00\rangle$ and
- 1 with probability 0.5, leaving the post-measurement state to be $|11\rangle$.

Consequently, a measurement of the second qubit invariably gives the same result as the measurement of the first qubit—namely, measuring one qubit has determined the state of the other (consider the example where Alice holds one qubit in a Bell pair, and Bob, at some distance, holds the other one; when Alice measures her qubit and finds a "one", she can be assured that when Bob measures his qubit, it will be a "one"; and, if she measures "zero", Bob will measure "zero"; see Figure 2.10). Figure 2.11 provides additional information on the Bell states.

(c = control, t = target)

---

[1] Namely, $\rho_{AB} = \Sigma_j \, p_j \, \sigma^j_A \otimes \tau^j_B$ for some distribution $\{p_j\}_j$, and states $\{\sigma^j_A\}_j$ on A and states $\{\tau^j_B\}_j$ on B [VAR22].

[2] There are other multi-party entangled states—for example, the Greenberger-Horne-Zeilinger (GHZ) state involving three or more subsystems (particle states, or qubits); GHZ states are utilized in a number of Quacom protocols and in cryptography.

There are also depolarized Bell states. These are states of the form

$$\rho(p) := p|\phi^+\rangle\langle\phi^+| + \frac{1-p}{4}\mathbb{I}$$

with $0 < p \leq 1$. The density matrix is [ABR13a]:

$$\rho(p) := p|\phi^+\rangle\langle\phi^+| + \frac{1-p}{4}\mathbb{I} = \frac{1+3p}{4}|\phi^+\rangle\langle\phi^+| + \frac{1-p}{4}\left(|\phi^-\rangle\langle\phi^-| + |\psi^-\rangle\langle\psi^-| + |\psi^+\rangle\langle\psi^+|\right)$$

and

$$\mathbb{I} = \begin{bmatrix} 1 & \cdots & 0 \\ \vdots & \ddots & \vdots \\ 0 & \cdots & 1 \end{bmatrix}$$

defined earlier. It is interesting that it turns out that for $p > \frac{1}{3}$ the state above is entangled, otherwise it is separable.[1]

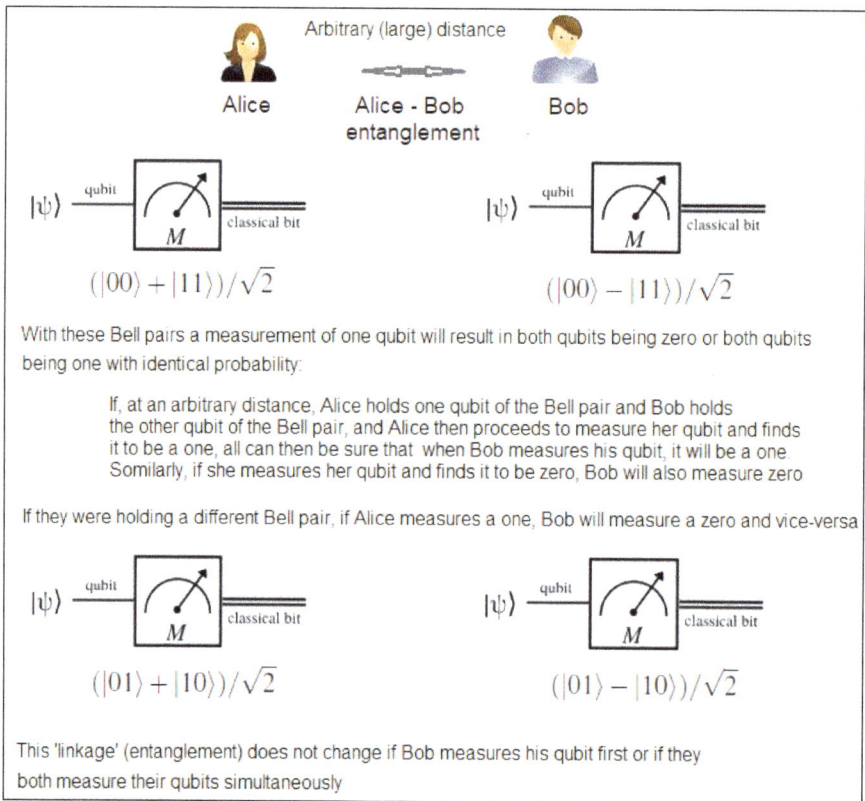

Figure 2.10 Entanglement of Bell states

---

[1] The coefficient $F = (1+3p)/4$ defines the fidelity of $\rho(p)$ with respect to the quantum state $|\phi^+\rangle\langle\phi^+|$, and it represents the overlap between these two states—that is, $F = \mathrm{tr}(\rho(p)|\phi^+\rangle\langle\phi^+|)$. Thus, fidelity is a metric characterizing similarity between quantum states; in particular, fidelity is used to compare a measured quantum state to a desired quantum state. This topic is further discussed in Chapter 3.

	Input		Output	
	Control	Target	Control	Target
=	|0⟩	|0⟩	|0⟩	|0⟩
	|0⟩	|1⟩	|0⟩	|1⟩
	|1⟩	|0⟩	|1⟩	|1⟩
	|1⟩	|1⟩	|1⟩	|0⟩

$$H = \begin{bmatrix} \frac{1}{\sqrt{2}} & \frac{1}{\sqrt{2}} \\ \frac{1}{\sqrt{2}} & -\frac{1}{\sqrt{2}} \end{bmatrix} \qquad X = \begin{bmatrix} 0 & 1 \\ 1 & 0 \end{bmatrix}$$

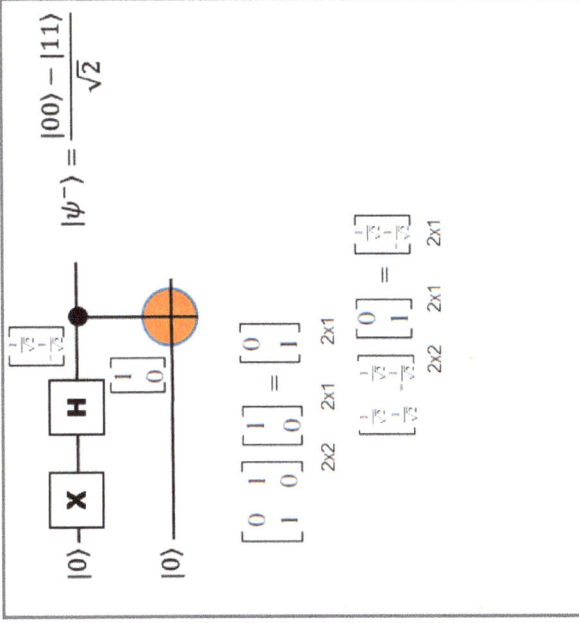

CNOT

$$|\psi^+⟩ = \frac{|00⟩ + |11⟩}{\sqrt{2}}$$

$$\begin{bmatrix} \frac{1}{\sqrt{2}} & \frac{1}{\sqrt{2}} \\ \frac{1}{\sqrt{2}} & -\frac{1}{\sqrt{2}} \end{bmatrix} \begin{bmatrix} 1 \\ 0 \end{bmatrix} = \begin{bmatrix} \frac{1}{\sqrt{2}} \\ \frac{1}{\sqrt{2}} \end{bmatrix}$$

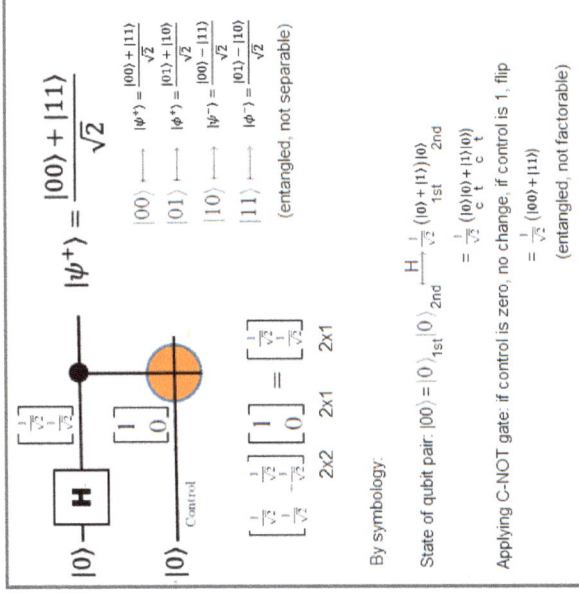

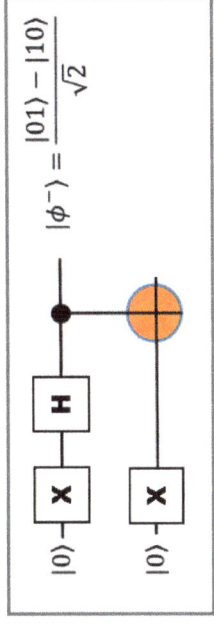

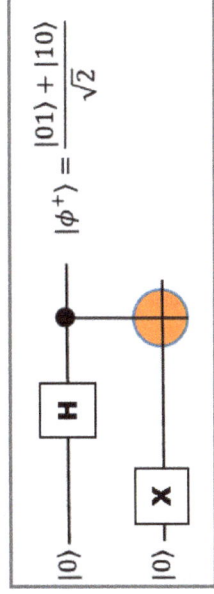

2x2   2x1   2x1

By symbology:

$$|\psi^+⟩ = \frac{|00⟩ + |11⟩}{\sqrt{2}}$$

$$|\phi^+⟩ = \frac{|01⟩ + |10⟩}{\sqrt{2}}$$

$$|\psi^-⟩ = \frac{|00⟩ - |11⟩}{\sqrt{2}}$$

$$|\phi^-⟩ = \frac{|01⟩ - |10⟩}{\sqrt{2}}$$

(entangled, not separable)

$$|00⟩ \longrightarrow |\psi^+⟩ = \frac{|00⟩ + |11⟩}{\sqrt{2}}$$

$$|01⟩ \longrightarrow |\phi^+⟩ = \frac{|01⟩ + |10⟩}{\sqrt{2}}$$

$$|10⟩ \longrightarrow |\psi^-⟩ = \frac{|00⟩ - |11⟩}{\sqrt{2}}$$

$$|11⟩ \longrightarrow |\phi^-⟩ = \frac{|01⟩ - |10⟩}{\sqrt{2}}$$

(entangled, not separable)

State of qubit pair $|00⟩ = |0⟩_{1st}|0⟩_{2nd} \xrightarrow{H} \frac{1}{\sqrt{2}}(|0⟩ + |1⟩)|0⟩$

$= \frac{1}{\sqrt{2}}(|0⟩|0⟩ + |1⟩|0⟩)$

Applying C-NOT gate: if control is zero, no change, if control is 1, flip

$= \frac{1}{\sqrt{2}}(|00⟩ + |11⟩)$

(entangled, not factorable)

$$|\psi^-⟩ = \frac{|00⟩ - |11⟩}{\sqrt{2}}$$

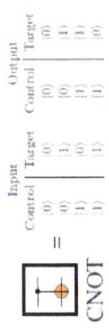

$$\begin{bmatrix} 0 & 1 \\ 1 & 0 \end{bmatrix} \begin{bmatrix} 1 \\ 0 \end{bmatrix} = \begin{bmatrix} 0 \\ 1 \end{bmatrix}$$

2x2   2x1   2x1

$$\begin{bmatrix} \frac{1}{\sqrt{2}} & \frac{1}{\sqrt{2}} \\ \frac{1}{\sqrt{2}} & -\frac{1}{\sqrt{2}} \end{bmatrix} \begin{bmatrix} 0 \\ 1 \end{bmatrix} = \begin{bmatrix} \frac{1}{\sqrt{2}} \\ -\frac{1}{\sqrt{2}} \end{bmatrix}$$

2x2   2x1   2x1

$$|\phi^+⟩ = \frac{|01⟩ + |10⟩}{\sqrt{2}}$$

$$|\phi^-⟩ = \frac{|01⟩ - |10⟩}{\sqrt{2}}$$

Figure 2.11   Quantum circuits for entangled Bell states preparation

2.7 Time Evolution of Quantum States

In general, systems, including quantum systems, are dynamic—namely, they evolve in time. Time evolution is the change of state brought about by the passage of time. QM predicts how the state of the system evolves over time and, therefore, how the information the observer has about the system changes during the time progression, after a given initial sate at t_0. The evolution of the state is deterministically formulated, and the Schrödinger equation describes it for any follow-on time t passed t_0. In general, the wave function of the system changes after a measurement; said measurements at a follow-on time t provide new state information, but note that the state of the system (typically) will change after a (follow-on) measurement.

There is a need for a formulation by which the dynamic properties of the system can be described; that is, given the state $|\psi(0)\rangle$ of a quantum system, one is interested in having a mathematical expression that defines what the state will be at some other time t; this is known as the *quantum law of evolution*. The *Hamiltonian operator* $\hat{H}$, or simply H, *generates the time evolution of quantum states*: if $|\psi(t)\rangle$ is the state of the system at time t, then the evolution is described by the Schrödinger equation:

$$H\,|\psi(t)\rangle = i\hbar\frac{\partial}{\partial t}\,|\psi(t)\rangle$$

If H is independent of t, then given the state at some initial time $t = 0$, then the unitary time evolution operator $U(t)$ is the exponential operator

$$|\psi(t)\rangle = U(t)\,|\psi(0)\rangle = e^{-iHt/\hbar}\,|\psi(0)\rangle$$

Let $|\psi(0)\rangle$ be the initial state of the system, and $|\psi(t)\rangle$ be the state at some other time t. As noted elsewhere, the state vector at time t representing the system can differ by is a *multiplicative factor*, $u(t)$—that is, $|\psi(t)\rangle = u(t)|\psi(0)\rangle$. Since one would expect that if the initial state $|\psi(0)\rangle$ is normalized to unity, then, similarly, would the state $|\psi(t)\rangle$; thus one can write $\langle\psi(t)|\psi(t)\rangle = |u(t)|^2\langle\psi(0)|\psi(0)\rangle$, and hence $|u(t)|^2 = 1$; then one must conclude that $u(t) = e^{-i\phi(t)}$, where $\phi(t)$ is a time-dependent phase. Thus, one would write $|\psi(t)\rangle = e^{-i\phi(t)}\,|\psi(0)\rangle$.

$\phi(t)$ can be determined as follows when the system is assumed to be isolated. An isolated system has no interaction with any other system—the dynamics of the system are determined only by its own endogenous interactions, not by exogenous interactions; the state of the system after the t time slots is the same, no matter what the initial t_0; what matters is how long the system evolves for, not what the starting time was ([CRE11] notes that if the system were open—that is, not isolated—then there is the prospect of time-dependent external influences acting on the system—for example, an externally produced laser pulse fired at an atom; thus, the evolution of the state of the atom would differ depending on the starting time. In such a situation, the system might not have any stationary states at all, considering that it is potentially always being forced to change state because of these external influences). For an isolated system, one can choose the initial time t_0 be any arbitrary time, and the evolution of $|\psi(t_0)\rangle$ up to the time t is given by $|\psi(t)\rangle = e^{-i\phi(t-t_0)}\,|\psi(t_0)\rangle$.

Consider the evolution of the system in a stationary state over a time interval (t, t'), then one has $|\psi(t')\rangle = e^{-i\phi(t'-t)}\,|\psi(t)\rangle = e^{-i\phi(t'-t)}\,e^{-i\phi(t-t_0)}\,|\psi(t_0)\rangle$, but over the interval (t_0, t') one has $|\psi(t')\rangle = e^{-i\phi(t'-t_0)}\,|\psi(0)\rangle$. Thus, by comparing the last two equations, one finds that $e^{-i\phi(t'-t_0)} = e^{-i\phi(t'-t)}\,e^{-i\phi(t-t_0)}$ or, $\phi(t-t_0) = \phi(t'-t) + \phi(t-t_0)$, this being an equation for ϕ, with the solution $\phi(t) = \omega t$, where ω is a constant [CRE11].

Therefore, one can conclude that
$$|\psi(t)\rangle = e^{-i\omega t}|\psi(0)\rangle,$$

which is the basic expression for the evolution in time of a stationary state of an isolated system. Note that as $|\psi(t)\rangle$ evolves, other states decay faster, and only the ground state is left.

2.8 Measurements in QM—A First View

In order to physically determine the state of a particle, one needs to perform a measurement. As noted in Chapter 1, while in classical mechanics a particle has an exact position and an exact momentum at all times, in QM position and momentum are not numbers but operators; the ultimate consequence is that as some observables become more precisely (accurately and exactly) defined, other observables become more uncertain. A measurement is the testing or access to a physical system in order to obtain a numerical result about that system. The result is expressed in terms of probabilities. As discussed, a qubit exists in a continuum superposition of states between $|0\rangle$ and $|1\rangle$. When a qubit is measured, the result is a 0 or 1 with an attached probability. Furthermore, the measurement changes the state of a qubit, collapsing it from its superposition of $|0\rangle$ and $|1\rangle$ to the specific state consistent with the measurement result. Say the qubit was in state $|1\rangle$; if measurement gives out a 0 (with a calculated or statistically derived probability), then the post-measurement state of the qubit will be $|0\rangle$. See Figure 2.12.

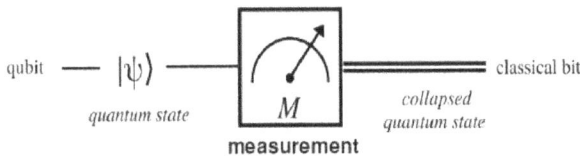

Measurement is the testing or access to a physical system in order to obtain a numerical result
The results have an attached probability
The measurement changes the state of a qubit, collapsing it from its (say) basis state to the other basis state

Figure 2.12   Measurement process

The same concepts apply to the measurement of a quantum system of multiple qubits with respect to a given orthonormal basis. One can measure the entire system or only part. Again, measuring a single qubit will alter the state of the system. For example, consider the state vector of two qubit in superpositions: $|\psi\rangle = \alpha_{00}|00\rangle + \alpha_{01}|01\rangle + \alpha_{10}|10\rangle + \alpha_{11}|11\rangle$; the measurement result, one of $x = 00$, $x = 01$, $x = 10$, or $x = 11$, occurs with probability $\|\alpha_x\|^2$ respectively, with the state of the qubits after the measurement being that state $|x\rangle$ (with x being, as applicable, $x = 00$; $x = 01$; $x = 10$; or $x = 11$).

Thus, a measurement of an observable yields a value which is called an *eigenvalue* of the observable. Immediately after the measurement, the state of the system is an *eigenstate* of the observable. In a QM environment, the measurement process alters the state of a quantum system. The effect of the measurement is that the new state is exactly the outcome of the measurement process: if the outcome of the measurement is j, then following the measurement, the qubit ends up in state $|j\rangle$, and one cannot collect any additional information about the amplitudes α_j by repeating the measurement. Measuring $|\psi\rangle$ in the standard basis yields j with probability $|\alpha_j|^2$. More specifically, a measurement may be perceived as a probabilistic mechanism for projecting the vector representing the state onto one of the vectors of the orthonormal basis [WHA09]. It is important to make note that measurements always alter quantum states—for example, causing superpositions to collapse. Considering an orthonormal basis of the k-dimensional vector space, the complex vector space Hilbert space

$$|\psi\rangle = \sum_{j=0}^{k-1} \alpha_j |j\rangle$$

where $|0\rangle = [1, 0, ..., 0]^T$, $|1\rangle = [0,1, ...,0]^T$, ...,$|k-1\rangle = [0, 0, ...,1]^T$ then the outcome of the measurement is j with probability equal to the square of the length of the projection of the state vector ψ onto the j–th basis vector; in addition, if the outcome is j, then the new state is the j–th basis vector, $|j\rangle$. This topic is further discussed in the chapters that follow.

2.9 Gates—A First View

Quantum computing (Quacomp) employs physical properties of quantum states to perform calculations utilizing quantum algorithms. In Quantum Computers (QCs) one undertakes quantum operations using "gates": in order to manipulate a qubit $|\psi\rangle = a_1|0\rangle + b_1|1\rangle$), one must manipulate its state by applying gates—unitary operators—to undertake the transformation $U|\psi\rangle = a_2|0\rangle + b_2|1\rangle$, where U is an appropriate 2×2 matrix (more generally the matrix can be $2^n \times 2^n$). Gates are particular unitaries (operators) acting on the space of qubits. (Figure 2.11 cited earlier also depicted some examples of gates).

A gate is a basic quantum circuit that operates on a small number of qubits. Quantum gates form the building blocks of quantum circuits, similar to classical logic gates in classical digital circuits. A quantum computation is achieved by performing an appropriate operation on each qubit, or on each of a two-qubits pair. An arbitrary operation can *typically* be realized by a certain combination of some one-quantum bit computation (one-quantum gate operation) and some two-quantum bit computation (two-quantum gate operation).

To physically realize these gates, one needs to implement the corresponding Hamiltonian operators [WHA09]. The Hamiltonian operator $\hat{H}$ (Hermitian conjugate) describes the total energy of a system, usually denoted by $\hat{H}$ or just H. An operator U is unitary if $U^\dagger U = UU^\dagger = \mathbb{I}$ (that is, an operator is unitary if and only if each of its matrix representations is unitary). A unitary operator U satisfies the requirement $U^\dagger = U^{-1}$ (that is, the adjoint operator is equal to its inverse).

$$U = \begin{bmatrix} a & b \\ c & d \end{bmatrix} \Rightarrow U^\dagger = \begin{bmatrix} \bar{a} & \bar{c} \\ \bar{b} & \bar{d} \end{bmatrix} = \begin{bmatrix} a^* & c^* \\ b^* & d^* \end{bmatrix}$$

Certain values of a, b, c, d yield a unitary transformation U (other values do not).

In summary, QCs process data by applying a universal set of certain quantum gates (e.g., Pauli gates, Hadamard gates) that can emulate any rotation of the quantum state vector. A set of quantum gates may be considered to be "universal" when any unitary transformation can be approximately written as a product of gates from this set—namely, there exist gates G_1, G_2, ... G_N from the gate set such that $G_N \times G_{N-1} \times ... \times G_2 \times G_1 \approx U$ [BOC23] (the convention for matrix multiplication is to multiply from right to left, so that G_N is actually the last gate applied to the input quantum state vector). The idea is to decompose a (more complex) unitary transform into a series of basic gates (or operations) that are easier to implement.

The development of quantum algorithms to solve problems on QCs is an ongoing R&D industry effort. Until recently these algorithms were studied primarily from a theoretical standpoint. Efficient and accurate implementation of quantum algorithms allows the algorithms to be tested and improved upon. *Implementing* quantum algorithms, however, can be challenging because quantum algorithms are often presented in scientific journals and are intended to be analyzed from a theoretical perspective; it is advantageous that quantum algorithms can also be represented by graphical models called *quantum circuits*. Quantum circuits present the steps in a quantum algorithm using a series of symbols.

Quantum circuits can be implicit or explicit. *Implicit* quantum circuits can represent subroutines intended to cause the Quacomp system to evolve from one state to a subsequent state as abstract "black boxes". *Explicit* quantum circuits, which contain all of the programming instructions necessary to implement the quantum algorithm on a Quacomp system, have several advantages: because quantum circuits represent quantum algorithms symbolically in a diagram rather than in writing using mathematical formalisms, explicit quantum circuits are typically able to represent necessary information compactly and concisely; furthermore, programmers need not have expertise in QM to be taught how to read and understand quantum circuits [CLA23]. Figure 2.13 depicts one illustrative example of such a circuit.

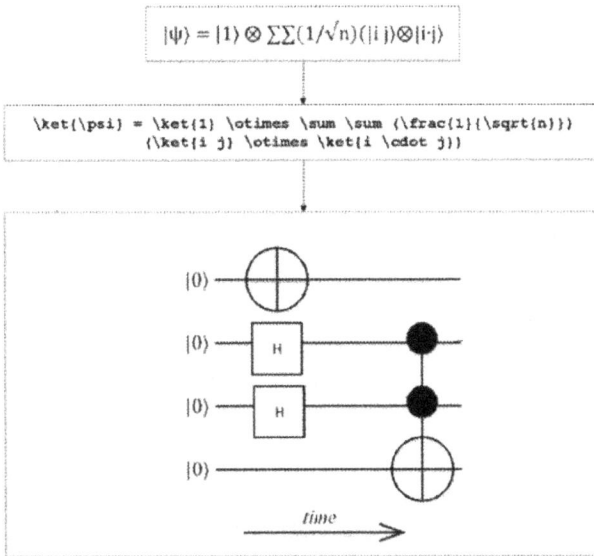

Figure 2.13 Example of a quantum algorithm implemented in a circuit ([CLA23])

This topic is further discussed in the chapters that follow.

References

[ABR13a] S. Abruzzo, "Long Distance Quantum Key Distribution with Quantum Repeaters". Inaugural-Dissertation zur Erlangung des Doktorgrades der Mathematisch-Naturwissenschaftlichen Fakultat der Heinrich-Heine-Universit at Dusseldorf, Dusseldorf, November 2013.

[ALB18] T. Albash, D. A. Lidar, "Adiabatic Quantum Computing". Rev. Mod. Phys. 90, 015002, 2018. 2 Feb 2018. https://doi.org/10.48550/arXiv.1611.04471

[ARE22] Z. Areej, "The Hamiltonian Operator". August 15th, 2022. Available online on November 27, 2023 at https://psiberg.com/hamiltonian-operator/

[BOC23] A. V. Bocharov, E. Kemp, *et al.*, Quantum Bit Prediction. U.S. Patent 11797872, 2023-10-24. Uncopyrighted material.

[CHE23] Y. Chen, S. Zhang, Controlling A Quantum Computing System Using Imaginary-Time Evolution. U.S. Patent Application 2023/0289639, 2023-09-14. Uncopyrighted material.

[CLA23] J. Clapis, R. H. Preston, K. J. Bergollo Lorenzo, Quantum State Vector Implementation Toolkit. U.S. Patent Application 20230325704, 2023-10-12. Uncopyrighted material.

[CRE11] J. D. Cresser, *Quantum Physics Notes*. 2011, Department of Physics Macquarie University. Available online on December 29, 2023 at http://physics.mq.edu.au/~jcresser/Phys304/Handouts/QuantumPhysicsNotes.pdf.

[DAR24] J. Dargan, "Google Quantum AI Team Demystifies Quantum Computing & Its Future Potential". The Quantum Insider, June 7, 2024. Available online on June 7, 2024 at https://thequantuminsider.com/2024/06/07/google-quantum-ai-team-demystifies-quantum-computing-its-future-potential/

[DIR1958] P.A.M. Dirac, *The Principles of Quantum Mechanics*, Oxford: Clarendon Press, 1930; 4th edition, revised, 1958.

[FIT23] R. Fitzpatrick, "Eigenstates and Eigenvalues". Available online (on 10/25/2023) at https://phys.libretexts.org/Bookshelves/Quantum_Mechanics/Introductory_Quantum_Mechanics_(Fitzpatrick)/03%3A_Fundamentals_of_Quantum_Mechanics/3.08%3A_Eigenstates_and_Eigenvalues.

[FUR23] W. Furtner, Multi-Dimensional Electrode Controller for Quantum Computing. U.S. Patent Application 20230325698, 2023-10-12. Uncopyrighted material.

[ISM21] J. Ismael, "Quantum Mechanics". *Stanford Encyclopedia of Philosophy*, Edward N. Zalta (Editor). Fall 2021 Edition); Available online (10/24/2023) at https://plato.stanford.edu/entries/qm/ or https://plato.stanford.edu/archives/fall2021/entries/qm/

[JAC23] J. D. Jackson, *A Course in Quantum Mechanics*, August 2023, Wiley, 978-1-119-88041-7.

[KOZ23] W. Kozlowski, S. Wehner, *et al.*, RFC 9340 - Architectural Principles for a Quantum Internet, March 2023, https://doi.org/10.17487/RFC9340.

[MIN05] D. Minoli, *Nanotechnology Applications to Telecommunications and Networking*, New York, Wiley, 2005.

[NIE10] M. Nielsen, L. I. Chuang, *Quantum computation and quantum information: 10th Anniversary Edition*. 2010. Cambridge University Press, New York, NY, USA. http://dx.doi.org/10.1017/cbo9780511976667.016.

[REU19] D. Reutter, J. Vicary, *Categorical Quantum Mechanics an Introduction. Lectures and Notes*, Department of Computer Science, University of Oxford, Hilary Term 2019. Available online on January 8, 2024 at https://www.cs.ox.ac.uk/files/10510/notes.pdf

[SHI23] G. Shinkai, R. Tsuchiya, Y. Kanno, Control Method of Quantum Bit and Quantum Computer. U.S. Patent Application 20230297873, 2023-09-21. Uncopyrighted material.

[STA23] Staff, "Hamilton Operator". Encyclopedia of Mathematics, available online on November 27, 2023 at https://encyclopediaofmath.org/wiki/Hamilton_operator

[TOK23] M. Tokunari, R.D. Schilling, Electro-Optomechanical Quantum Transduction. U.S. Patent Application 20230204861, 2023-06-29. Uncopyrighted material.

[VAR22] G. Vardoyan, M. Skrzypczyk, S. Wehner, "On the Quantum Performance Evaluation of Two Distributed Quantum Architectures". Performance Evaluation 153, 2022. https://doi.org/10.1016/j.peva.2021.102242.

[WHA09] B. Whaley, D. Gorman, "C/CS/Phys191: Qubits, Quantum Mechanics and Computers - Fall 2009". berkeley.edu, Available online on 11/27/2023 at https://inst.eecs.berkeley.edu/~cs191/fa09/ (e.g., https://inst.eecs.berkeley.edu/~cs191/fa09/lectures/lecture1_fa09.pdf, https://inst.eecs.berkeley.edu/~cs191/fa09/lectures/lecture2_fa09.pdf, and so on).

[YAM24] A. Yamauchi, K. Tanaka, *et al.*, "Room-Temperature Quantum Coherence of Entangled Multiexcitons in A Metal-Organic Framework". Science Advances, 3 Jan 2024, Vol 10, Issue 1. doi: 10.1126/sciadv.adi3147.

Appendix A

Mathematical Building Blocks

Tables 2.A1, 2.A2, and 2.A3[1] in this appendix provide some of the basic mathematical concepts for understating and/or describing QM.

Table 2.A1 lists the more basic concepts, Table 2.A2 lists more advanced concepts, and Table 2.A3 lists yet more advanced concepts (which can possibly be skipped on first reading).

Readers should familiarize themselves with these underlying concepts by working through the tables, here focusing in particular on Table 2.A1 and Table 2.A2.

These tables can also be used retrospectively when reading chapters that follow and mathematical concepts are used, introduced, or re-introduced.

Table 2.A1 Basic Mathematical Concepts Applicable to Quacom

Concept	Description
Basis	See *Orthogonal Vector Basis* further below
Bra (Dirac)	A "bra" $\langle v|$ is "dual" to a vector $|v\rangle$; the bra is an adjoint of the vector $\langle v| = (|v\rangle)^\dagger$. This means that, with a ket, the bra generates a complex number, $\langle v|v\rangle \in \mathbf{C}$ when the dot product is taken—at times also shown as $\langle v|\,|v\rangle$, which for further clarity could be shown (here) as $\langle\langle v|\,|v\rangle\rangle$ or abbreviated as $\langle v|v\rangle$.
Complete vector space	A vector space such that any Cauchy sequence of vectors in that space converges to a vector in that space. A metric space S is called *complete* (also known as a *Cauchy space*) if every Cauchy sequence of points in S has a limit that is also in S. In this formulation, a sequence $s_1, s_2, s_3, \ldots$ in a metric space (S, d) is called Cauchy if for every positive real number $r > 0$ there is a positive integer N, such that for all positive integers $m, n > N$, $d(s_m, s_n) < r$.
	The space $\mathbf{R}$ of real numbers as well as the space $\mathbf{C}$ of complex numbers (with the metric given by the absolute difference) are complete; the Euclidean space $\mathbf{R}^n$, with the usual distance metric, is also complete.
Complex conjugate c^* of complex number $c = a \pm bi$	The complex conjugate c^* of $c = a \pm bi$ is defined as: $$[a+bi]^* = c^* = a - bi$$ $$[a-bi]^* = c^* = a + bi$$ Note that for all complex numbers c, $[c^*]^* = c$; and when c is real (namely, $c = a$) then $c^* c = a$.
Complex vector space	A complex vector space is a vector space where all elements are specifiable as linear combinations of defined basis vectors, utilizing complex numbers.
Dimension of an inner product space	The maximum number of nonzero, mutually orthogonal vectors it contains; namely, the number of vectors in the orthogonal basis needed to describe it.
Hilbert space	A vector space on which an inner product is defined, and which is complete. Complete means that any Cauchy sequence of vectors in the space converges to a vector in the space. By definition, any manner of adding vectors in a Hilbert space, or multiplying them by scalars, results in a vector that is also in the same space. Hilbert spaces represent the state-spaces of quantum mechanical systems. *(continues)*

---

[1] Note: These tables are based on various industry sources, including [ABR13a], [ARE22], [BOC23], [FIT23], [ISM21], [JAC23], [NIE10], [REU19], [STA23], [VAR22], and [WAH09].

Concept	Description
Hilbert space *(cont.)*	A Hilbert space is an abstract multidimensional vector space with an inner product that allows lengths and angles to be measured. Hilbert spaces enjoy certain "geometric intuition"—e.g., analogs of the parallelogram law and Pythagorean theorem hold in such a space. An element (a point, a member) of a Hilbert space can be specified uniquely by its coordinates with respect to an orthonormal basis, similar to the concept of Cartesian coordinates in classical geometry. Hilbert spaces have applications in partial differential equations, quantum mechanics, Fourier analysis, and numerous other fields. A Hilbert space can be a finite or an infinite dimensional space. A well-known example of a Hilbert space is the Euclidean vector space R^3 consisting of three-dimensional vectors and equipped with the dot product.
Identity operator	An operator $I_{op} = I$ mapping a vector space A onto itself, defined by the equation $I_{op}(|A\rangle) = |A\rangle$ for all vectors $|A\rangle$.
Inner product of tensor space	The inner product on $H_A \otimes H_B$, $\langle v_A i \otimes u_B m | v_A j \otimes u_B n\rangle$ is defined as the product of the individual inner products, specifically, $$\langle v_A i \otimes u_B m | v_A j \otimes u_B n\rangle = \langle v_A i | v_A j\rangle \times \langle u_B m | u_B n\rangle$$
Inner product of two vectors—abstract formulation	The notion of inner product generalizes the notion of dot product of vectors in $\mathbf{R}^n$. Definition. Let V be a vector space. A function $\delta: V \times V \to \mathbf{C}$, usually denoted $\delta(x, y) = \langle x, y\rangle$ (here we prefer to use $\langle x|y\rangle$) is called an inner product on V if it is positive, symmetric, and bilinear. That is, if (i) $\langle x|x\rangle \geq 0$, $\langle x|x\rangle = 0$ only for $x = 0$ (positivity) (ii) $\langle x|y\rangle = \langle y|x\rangle$, (symmetry) (iii) $\langle rx|y\rangle = r\langle x|y\rangle$, (homogeneity) (iv) $\langle x+y|z\rangle = \langle x|z\rangle + \langle y|z\rangle$, (distributive law) An inner product space is a vector space endowed with an inner product. A specific inner product of interest is the scalar equal to the product of their lengths times the cosine of the angle, θ, between them—namely, $$\langle A \mid B\rangle = |A||B|\cos\theta.$$ **Note**: the inner product between two vectors, say $|v\rangle$ and $|w\rangle$, should be formally shown as (here) $\langle \, |v\rangle||w\rangle \, \rangle$; for example, if $|v\rangle = |\psi\rangle$ and $|w\rangle = |0\rangle$, then $\langle \, |v\rangle||w\rangle \, \rangle = \langle \, |\psi\rangle||0\rangle \, \rangle$. For expediency, the abbreviated form is customarily used: $\langle \psi|0\rangle$ (or $\langle\psi|0\rangle$ by others); the reader needs to keep this notation simplification in mind when working through various discussions and mathematical formulation.
Inner product of two vectors in complex spaces $\langle A \mid B\rangle$	The inner product of $|A\rangle$ and $|B\rangle$ for complex spaces is defined in terms of the *conjugates of complex coefficients* as follows, using basis (unit) vectors $|A1\rangle$ and $|A2\rangle$; then if $|A\rangle = a_1|A_1\rangle + a_2|A_2\rangle$ and $|B\rangle = b_1|A_1\rangle + b_2|A_2\rangle$ one has $$\langle A \mid B\rangle = (a^*_1)(b_1) + (a^*_2)(b_2)$$ Regarding the conjugates, note: $[a+bi]^* = a-bi$; $[a-bi]^* = a+bi$; $[a]^* = a$ when $a \in \mathbf{R}$. More generally, for $|\psi\rangle = \Sigma_k a_k|k\rangle$ and $|\phi\rangle = \Sigma_k b_k|k\rangle$ $$\langle \psi|\phi\rangle = \Sigma_k a_k^* \times b_k$$ Refer to next entry below for a clarification on inner products in $\mathbf{C}^n$ versus inner products in $\mathbf{R}^n$.

Table 2.A1 Basic Mathematical Concepts Applicable to Quacom (cont.)

Concept	Description
Inner product of two vectors in complex spaces $(A \mid B)$, comparison with traditional inner product	For vectors $\mathbf{x} = (x_1, x_2, \ldots, x_n)$ and $\mathbf{y} = (y_1, y_2, \ldots, y_n)$ in $\mathbf{R}^n$ the inner product $\langle \mathbf{x}, \mathbf{y} \rangle$ for the (real) n-dimensional Euclidian space $\mathbf{R}^n$ is defined to be $\langle \mathbf{x}, \mathbf{y} \rangle = x_1 y_1 + x_2 y_2 + \cdots + x_n y_n$. and the norm (or the magnitude) $\|\mathbf{x}\|$ is given by $$\|\mathbf{x}\| = \sqrt{\langle \mathbf{x}, \mathbf{x} \rangle} = \sqrt{x_1^2 + x_2^2 + \cdots + x_n^2}.$$ For complex vectors in $\mathbf{C}^n$ one cannot use this definition directly. One needs to use complex conjugation in the definition in such a way that $\langle \mathbf{x}, \mathbf{x} \rangle \geq 0$ and $\|\mathbf{x}\| = \sqrt{\langle \mathbf{x}, \mathbf{x} \rangle}$ still make sense. The inner product (or the scalar product) $\langle \mathbf{x}, \mathbf{y} \rangle$ of vectors $\mathbf{x}$ and $\mathbf{y}$ is defined as $\langle \mathbf{x}, \mathbf{y} \rangle = \bar{x}_1 y_1 + \bar{x}_2 y_2 + \cdots + \bar{x}_n y_n$ where $\bar{a}$ is utilized here (and only here) to denote the complex conjugate of a (otherwise we use a*); if a = x + iy, then $\bar{a}$ = a* = x - iy and if a = x - iy, then $\bar{a}$ = a* = x + iy. Notice that $\langle \mathbf{x}, \mathbf{x} \rangle = \bar{x}_1 x_1 + \bar{x}_2 x_2 + \cdots + \bar{x}_n x_n = |x_1|^2 + |x_2|^2 + \cdots + |x_n|^2 \geq 0$. The norm of x is given by $$\|\mathbf{x}\| = \sqrt{\langle \mathbf{x}, \mathbf{x} \rangle} = \sqrt{|x_1|^2 + |x_2|^2 + \cdots + |x_n|^2}.$$ An inner product on a *complex* vector space is a function that assigns to each pair of vectors x and y a complex number denoted by $\langle \mathbf{x}, \mathbf{y} \rangle$, such that the following conditions are satisfied: $\quad$ (C1) $\langle \mathbf{x}, \mathbf{y} \rangle \geq 0$, and $\langle \mathbf{x}, \mathbf{x} \rangle = 0$ if and only if x = 0 $\quad$ (C2) $\langle \mathbf{y}, \mathbf{x} \rangle = \overline{\langle \mathbf{x}, \mathbf{y} \rangle}$ $\quad$ (C3) $\langle a_1 \mathbf{x}_1 + a_2 \mathbf{x}_2, \mathbf{y} \rangle = a_1 \langle \mathbf{x}_1, \mathbf{y} \rangle + a_2 \langle \mathbf{x}_2, \mathbf{y} \rangle$ $\qquad \langle \mathbf{x}, b_1 \mathbf{y}_1 + b_2 \mathbf{y}_2 \rangle = \bar{b}_1 \langle \mathbf{x}, \mathbf{y}_1 \rangle + \bar{b}_2 \langle \mathbf{x}, \mathbf{y}_2 \rangle$
Inner product space	A vector space on which the operation of vector multiplication has been defined. A space where for each two elements, an inner product $(\cdot,) \to [0, \infty)$ is defined.
ℓ_2-norm	(Aka Euclidean norm) is a vector norm defined for a complex vector $$\mathbf{x} = \begin{bmatrix} x_1 \\ x_2 \\ \vdots \\ x_n \end{bmatrix}$$ by $$|\mathbf{x}| = \sqrt{\sum_{k=1}^{n} |x_k|^2},$$ The norm is also denoted by utilizing double vertical lines: $\|\mathbf{x}\|$.

Concept	Description
Linear Operator Op	Linear operators have the following properties: (i) $Op(\lvert A\rangle + \lvert B\rangle) = Op\lvert A\rangle + Op\lvert B\rangle$ and, (ii) $Op(c\lvert A\rangle) = c\,(Op\lvert A\rangle)$ This last expression also implies $$Op\left(\sum_i scalar\ o_i\ \times\ vector\ v_i\right) = \sum_i scalar\ o_i\ \times\ Op\left(vector\ v_i\right)$$ or, $$Op\left(\sum_i o_i \times \lvert v_i\rangle\right) = \sum_i o_i \times Op\left(\lvert v_i\rangle\right)$$ Linear operators can be represented by matrices.
Matrix multiplication	Given a pair of matrices A of size $m \times p$ and B of size $p \times n$, the (traditional) matrix product $C = AB$ is a matrix of size $m \times n$ with entries $$\mathbf{C} = \mathbf{A}\mathbf{B} = \left(c_{ij} = \sum_{k=1}^{p} A_{ik}\,B_{kj} \right)_{\substack{1 \le i \le m \\ 1 \le j \le n}}$$ Namely, (in mathematical notation with round brackets) $$\mathbf{A} = \begin{pmatrix} a_{11} & a_{12} & \cdots & a_{1n} \\ a_{21} & a_{22} & \cdots & a_{2n} \\ \vdots & \vdots & \ddots & \vdots \\ a_{m1} & a_{m2} & \cdots & a_{mn} \end{pmatrix}, \quad \mathbf{B} = \begin{pmatrix} b_{11} & b_{12} & \cdots & b_{1p} \\ b_{21} & b_{22} & \cdots & b_{2p} \\ \vdots & \vdots & \ddots & \vdots \\ b_{n1} & b_{n2} & \cdots & b_{np} \end{pmatrix}$$ $m \times n$ m = rows n = columns $n \times p$ n = rows p = columns $$\mathbf{C} = \mathbf{A}\mathbf{B} = \begin{pmatrix} a_{11}b_{11} + \cdots + a_{1n}b_{n1} & a_{11}b_{12} + \cdots + a_{1n}b_{n2} & \cdots & a_{11}b_{1p} + \cdots + a_{1n}b_{np} \\ a_{21}b_{11} + \cdots + a_{2n}b_{n1} & a_{21}b_{12} + \cdots + a_{2n}b_{n2} & \cdots & a_{21}b_{1p} + \cdots + a_{2n}b_{np} \\ \vdots & \vdots & \ddots & \vdots \\ a_{m1}b_{11} + \cdots + a_{mn}b_{n1} & a_{m1}b_{12} + \cdots + a_{mn}b_{n2} & \cdots & a_{m1}b_{1p} + \cdots + a_{mn}b_{np} \end{pmatrix}$$ $m \times p$ m = rows p = columns $$\mathbf{A} = \begin{pmatrix} a_{11} & a_{12} \\ a_{21} & a_{22} \end{pmatrix} \quad \mathbf{B} = \begin{pmatrix} b_{11} \\ b_{21} \end{pmatrix} \quad \mathbf{C} = \mathbf{A}\mathbf{B} = \begin{pmatrix} a_{11}b_{11} + a_{12}b_{21} \\ a_{21}b_{11} + a_{22}b_{21} \end{pmatrix}$$ $m \times n$ m = rows = 2 n = columns = 2 2x2 $n \times p$ n = rows = 2 p = columns = 1 2x1 $m \times p$ m = rows = 2 p = columns = 1 2x1 For example, if the first matrix (vector) has m = rows = 2 and n = columns = 1, and the second matrix (vector) has n = rows = 1 and p = columns = 2 and, the resulting matrix is $m \times n$ or 2×2 (this *specific example* also includes the utilization of conjugates—otherwise, general elements can be used): $$\lvert \psi \rangle = \begin{bmatrix} \alpha \\ \beta \end{bmatrix}$$ $$\langle \psi \rvert = \begin{bmatrix} \alpha \\ \beta \end{bmatrix}^{\dagger} = (\alpha^{*}\ \beta^{*})$$ $$\lvert \psi \rangle \times \langle \psi \rvert = \lvert \psi \rangle \langle \psi \rvert = \begin{bmatrix} \alpha \\ \beta \end{bmatrix} \times (\alpha^{*}\ \beta^{*}) = \begin{bmatrix} \alpha\,\alpha^{*} & \alpha\,\beta^{*} \\ \beta\,\alpha^{*} & \beta\,\beta^{*} \end{bmatrix}$$ 2x1 1x2 2x2

Table 2.A1 Basic Mathematical Concepts Applicable to Quacom (*cont.*)

Concept	Description
Model	A mathematical structure utilized to represent some physical phenomenon of interest.
Norm	The norm generalizes the notion of length of a vector in $\mathbf{R}^n$. Say V is a vector space. A function $\alpha : V \rightarrow \mathbf{R}$ is called a norm on V if it has the following properties: (i) $\alpha(x) \geq 0$, $\alpha(x) = 0$ only for $x = 0$ (positivity); (ii) $\alpha(rx) = \lvert r \rvert\, \alpha(x)$ for all $r \in \mathbf{R}$ (homogeneity); and (iii) $\alpha(x + y) \leq \alpha(x) + \alpha(y)$ (triangle inequality). A normed vector space is a vector space endowed with a norm. Example: $V = \mathbf{R}^n$, $x = (x_1, x_2, \ldots, x_n) \in \mathbf{R}^n$. $\lvert\lvert x \rvert\rvert = \max(\lvert x_1 \rvert, \lvert x_2 \rvert, \ldots, \lvert x_n \rvert)$. A vector $\lvert v \rangle$ has norm 1 if the inner product $\langle v \vert v \rangle$ (which is by nomenclature also $\langle v \vert v \rangle =$ $\langle v \vert\, \lvert v \rangle$, $\langle\langle v \vert\, \lvert v \rangle\rangle$) is equal to the scalar 1. ℓ_2-norm (aka Euclidean norm) is a vector norm defined for a complex vector $$\mathbf{x} = \begin{bmatrix} x_1 \\ x_2 \\ \vdots \\ x_n \end{bmatrix}$$ by $$\lvert \mathbf{x} \rvert = \sqrt{\sum_{k=1}^{n} \lvert x_k \rvert^2}\,,$$ The norm is **also** denoted by utilizing double vertical lines: $\lvert\lvert x \rvert\rvert$.
Operator O, $\mathrm{Op}\lvert A \rangle$	A mapping of a vector space (say vector space A) onto itself: it takes any vector $\lvert B \rangle$ in a space onto another vector $\lvert B' \rangle$ also in that same space—namely, $\mathrm{Op}\lvert B \rangle = \lvert B' \rangle$. In QM, one deals extensively with *linear* operators. Linear operators can be expressed as a matrix multiplication of the operator-specific matrix onto the vector in question. $$\mathrm{Op} = \begin{bmatrix} a & b \\ c & d \end{bmatrix} \qquad v = \begin{bmatrix} x \\ w \end{bmatrix}$$ $$\text{2x2} \qquad\qquad \text{2x1}$$ $$\mathrm{Op} = \begin{bmatrix} a & b \\ c & d \end{bmatrix} \bullet \begin{bmatrix} x \\ w \end{bmatrix} = \begin{bmatrix} ax + bw \\ cx + dw \end{bmatrix} = v'$$ $$\text{2x2} \quad \text{2x1} \qquad\qquad \text{2x1}$$ $$\mathrm{Op}\colon v \rightarrow v'$$ More generally the (linear) operator can map between two spaces, V and W.
Operator O, $\mathrm{Op}\lvert A \rangle$, matrix representation	A linear operator on a vector space can be represented in a column notation by N^2 numbers $$\mathrm{Op} = \begin{bmatrix} O_{11} & O_{12} & \cdots & O_{1N} \\ O_{21} & O_{22} & & \\ \cdots & & & \\ O_{N1} & & & O_{NN} \end{bmatrix}$$

(continues)

Concept	Description
Operator O, Op$\|A\rangle$, matrix representation, (cont.)	where $O_{ij} = \langle A_i \| Op \| A_j \rangle$ with $i, j \leq N$, with A_N the basis vectors (the first "$\|$" here representing the dot product). The application of operator Op on 2×1 vector $\|B\rangle$ in two-dimensional space (as illustrative example) is $$Op\|B\rangle = \begin{bmatrix} O_{11} & O_{12} \\ O_{21} & O_{22} \end{bmatrix} \times \begin{bmatrix} b_1 \\ b_2 \end{bmatrix}$$ $$= \begin{bmatrix} (O_{11}b_1 + O_{12}b_2) \\ (O_{21}b_1 + O_{22}b_2) \end{bmatrix}$$ $$= (O_{11}b_1 + O_{12}b_2)\|A_1\rangle + (O_{21}b_1 + O_{22}b_2\|A_2\rangle$$ $$= \|B'\rangle$$ Note: in the matrix multiplication, an $m \times p$ matrix multiplied by a $p \times n$ matrix gives rise to a matrix of size $m \times n$; here a 2×2 matrix operator and a 2×1 vector (matrix) gives rise to a 2×1 vector (matrix).
Orthogonal vector basis	A collection of N mutually orthogonal vectors (of length 1 per the inner product) in an N-dimensional vector space. When $\|A_1\rangle, \ldots, \|A_N\rangle$ is such a collection of unit vectors, then every vector in the space can be uniquely expressed as a sum of the form: $$\|B\rangle = b_1\|A_1\rangle + b_2\|A_2\rangle + \ldots + b_N\|A_N\rangle \text{ where } b_i = \langle B\|A_i\rangle.$$ The b_i's are known as B's expansion coefficients in the context of the A-basis. For any vector $\|P\rangle$ and $\|Q\rangle$ described by an A-basis $$\|P\rangle + \|Q\rangle = \sum_{i=1}^{N} (p_i + q_i)\|A_i\rangle$$ and $$\langle P\|Q\rangle = \sum_{i=1}^{N} p^*_i q_i$$
Orthogonality	$\|A\rangle$ and $\|B\rangle$ are mutually perpendicular, or orthogonal, if, and only if $\langle A\|B\rangle = 0$. Orthogonality is defined using the inner product over C^k: given two vectors $\|A\rangle = \|\phi\rangle = \Sigma_i\alpha_i\|i\rangle$ and $\|B\rangle = \|\psi\rangle = \Sigma_i\beta_i\|i\rangle$, then the inner product is $\langle A \| B \rangle = \langle\phi\|\psi\rangle = \Sigma_i\alpha^*_i\beta_i$, or, in a two-dimensional space, $\langle A \| B \rangle = (a^*_1)(b_1) + (a^*_2)(b_2)$. Two vectors are orthogonal if they are perpendicular to each other—namely, the dot product of the two vectors is zero. Elements of an orthogonal basis are orthogonal by definition.
Projection-Valued Measure (PVM)	A projection-valued measure on a Hilbert space H is a finite family of projections $H -^{P_i}\rightarrow H$, which are complete and orthogonal. A PVM is nondegenerate when $\text{Tr}(p_i) = 1$ for all i. A finite family of linear maps $H -^{f_i}\rightarrow H$ is complete when the following holds: $$\sum_i f_i = \text{id}_H$$ where id_H is the identity—namely, $H -^{id}_H\rightarrow H$—namely, any Cauchy sequence of vectors in the space converges to a vector in the space. Also See *Complete vector space*. A family of linear maps $H -^{f_i}\rightarrow H$, $i \geq 2$, is orthogonal when for any $i \neq j$, the following holds: $f_i \circ f_j = \langle f_i, f_j\rangle = 0$ (orthogonality equates to zero inner product).

Table 2.A1 Basic Mathematical Concepts Applicable to Quacom (cont.)

Concept	Description
Representation of vectors, matrix	Relative to a basis A_i, the matrix notation is $$\lvert Q\rangle = \begin{bmatrix} q_1 \\ q_2 \end{bmatrix}$$ with $q_i = \langle Q\lvert A_i\rangle$. Namely, any vector in an N-dimensional space can be represented by a column of N numbers, specific to a choice of basis for the space.
Representation of vectors, wave function	For practical nomenclature in vector spaces (especially those of infinite dimension being that one cannot list the infinite set of expansion coefficients in the matrix column) one utilizes a "wave function" $\psi(i)$ for vector $\langle Q\lvert$—namely, $\psi(i) = q_i = \langle Q\lvert A_i\rangle$. Important operations on vectors often are basic algebraic operations on their wave functions. In a vector space, given any vector and given any basis, one can obtain the wave function of the vector in that basis. Conversely, given a wave function for a vector in a given basis, one can construct the vector whose wave function it represents.
Rotation	Altering the state of a qubit with a single-qubit quantum gate changes the representation of the state; altering the state vector of a qubit is referred to as a "rotation". This change can be viewed (i) as a rotation of the Block-sphere itself, thus resulting in a change of the basis vectors of the state vector; or (ii) as a rotation of the state vector within the Block sphere, resulting in a change of the coefficients of the state vector [BOC23]. A rotation, state change, or single-qubit quantum-gate operation is represented mathematically by a unitary 2×2 matrix with complex elements: $$U = \begin{bmatrix} a & b \\ -b^* & a^* \end{bmatrix}$$ where a and b are complex numbers and "x^*" is the complex conjugate of x. A unitary 2×2 matrix U with complex elements is defined as a 2×2 matrix U, with the following property: $UU^\dagger = U^\dagger U = I$—namely, $$U = \begin{bmatrix} a & b \\ -b^* & a^* \end{bmatrix} \quad U^\dagger = \begin{bmatrix} a^* & -b \\ b^* & a \end{bmatrix} \quad U^\dagger U = UU^\dagger = I = \begin{bmatrix} 1 & 0 \\ 0 & 1 \end{bmatrix}$$ $$U^\dagger U = UU^\dagger = \begin{bmatrix} a^* & -b \\ b^* & a \end{bmatrix}\begin{bmatrix} a & b \\ -b^* & a^* \end{bmatrix} = \begin{bmatrix} a^*a + bb^* & a^*b - ba^* \\ ab^* - ab^* & bb^* + aa^* \end{bmatrix} = \begin{bmatrix} 1 & 0 \\ 0 & 1 \end{bmatrix}.$$ given that $a^*a + bb^* = \lVert a\rVert^2 + \lVert b\rVert^2 = 1$. Thus, the operation of a quantum gate on a qubit with state $\lvert\psi\rangle$, where $\lvert\psi\rangle$ is expressed in vector form as $$\lvert\psi\rangle = \begin{bmatrix} \alpha \\ \beta \end{bmatrix},$$ can be expressed as left-hand multiplication of the state vector by the unitary matrix corresponding to the unitary operation: $$\begin{bmatrix} a & b \\ -b^* & a^* \end{bmatrix}\begin{bmatrix} \alpha \\ \beta \end{bmatrix} = \begin{bmatrix} a\alpha + b\beta \\ -b^*\alpha + a^*\beta \end{bmatrix} = \begin{bmatrix} \alpha' \\ \beta' \end{bmatrix}.$$
Scalar multiplication of a vector	Multiplying a vector $\lvert A\rangle$ by k, where k is a scalar constant, results in a vector that is the same direction as $\lvert A\rangle$ but whose length is k times $\lvert A\rangle$'s length.

Concept	Description
State-space of a composite system	The state-space of a composite system HC is obtained from the "tensor product" $H_C = H_A \otimes H_B$ of the individual state-spaces, H_A and H_B.
State-space of a system	The space formed by the set of the space's possible states—namely, the physically possible ways of combining the values of quantities that characterize the system internally.
Structure, general	A set of elements on which specified operations and relations are defined.
Structure, mathematical	A structure in which the elements are mathematical objects (for example, vectors) and the operations are mathematical operations.
Trace of a matrix M	The trace of M is the sum of its diagonal elements: $$tr(M) \equiv \sum_i M_{ii}$$ With T and R arbitrary matrices and z a complex number, one has (i) $tr(TR) = tr(RT)$ (it is cyclic) (ii) $tr(T+R) = tr(T)+tr(R)$, $tr(z^*T) = z^*tr(T)$ (it is linear)
Transposed vector	The transpose of a vector changes a column vector to a row vector or vice versa: $$x = \begin{bmatrix} x_1 \\ x_2 \\ x_3 \\ . \\ . \\ . \\ x_n \end{bmatrix} \qquad x^T = \begin{bmatrix} x_1 & x_2 & x_3 & ...x_n \end{bmatrix}$$
Unit vector	A vector in the orthogonal basis of a space.
Unitary matrix	A matrix U is said to be unitary if $U^\dagger U = I$.
Unitary operator	(Also known as *unitary*; also known as *gate*.) An operator U is unitary if $U^\dagger U = UU^\dagger = I$, that is, an operator is unitary if and only if each of its matrix representations is unitary. A unitary operator U satisfies the requirement $U^\dagger = U^{-1}$—that is, the adjoint operator is equal to its inverse. The following holds in matrix notation: $[U^\dagger]_{ij} = [U^*]_{ij} = [U^T]^*_{ij}$. Operators or transformations that are unitary in a (complex) vector space are of interest in QM (one example is a rotation of a vector by some angle).
Vector	A vector, say vector A, $|A\rangle$, is a mathematical object described by a length, $|A|$, and a direction. A normalized vector is a vector of length 1 ($|A| = 1$). By nomenclature, vectors are represented by round or square parenthesis; the latter is more common in QM. $$v = \begin{pmatrix} a \\ b \end{pmatrix} = |v\rangle \quad \text{or} \quad v = \begin{bmatrix} a \\ b \end{bmatrix} = |v\rangle$$
Vector addition	The mapping of any pair of vectors onto another vector with the parallelogram law: adding vectors $|A\rangle$ and $|B\rangle$|generates vector $|C\rangle (= |A\rangle + |B\rangle)$ as in the diagram: IA> IB>

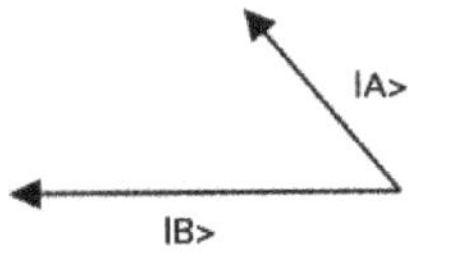

(continues)

Table 2.A1 Basic Mathematical Concepts Applicable to Quacom (cont.)

Concept	Description
Vector addition (cont.)	 (intuitively: moving the second vector in such a manner that its tail coincides with the tip of the first, without altering the length or direction of either vector, then joining the tail of the first vector to the tip of the second vector).
Vector Basis	Representing a vector as the sum of unit vectors. Also see *Orthogonal basis*. In two-dimensional space two unit vectors, $\lvert A_1\rangle$ and $\lvert A_2\rangle$, can be a basis (under the right condition). $\lvert A_1\rangle$ and $\lvert A_2\rangle$ must be vectors of length 1 ("unit vectors") and need the property that $\langle A_1\lvert A_2\rangle = 0$ (the angle between these two unit vectors is 90°, and the $\cos\theta$ is 0). Then, any two-dimensional vector $\lvert B\rangle$ can be expressed in terms of the unit vectors: $\lvert B\rangle = b_1\lvert A_1\rangle + b_2\lvert A_2\rangle$, mainly the *sum of the two unit vectors* $\lvert A_1\rangle$ and $\lvert A_2\rangle$
Vector Length	The length of $\lvert A\rangle$ is the square root of inner product of $\lvert A\rangle$ with itself, $$\lvert A\rvert = \sqrt{\langle A \mid A\rangle}$$
Vector space	A set of vectors closed under the operation of vector addition and vector multiplication by constants.
Zero operator	An operator maps all vectors to the zero vector.

Table 2.A2 More Advanced Mathematical Concepts Applicable to Quacom

Concept	Description
Composite systems/ entanglement	Consider two quantum systems A and B that interact with each other. Let H_A and H_B be the Hilbert spaces as the state spaces of these systems, respectively, and let $\lvert x\rangle_A$ and $\lvert y\rangle_B$ be the respective basis, so that states $\lvert\phi\rangle_A$ and $\lvert\zeta\rangle_B$ are described as $$\lvert\phi\rangle_A = \Sigma\, c_x \times \lvert x\rangle_A$$ $$\lvert\zeta\rangle_B = \Sigma\, d_y \times \lvert y\rangle_B$$ for appropriate c_x's and d_y's. Then the state space of the combined interacting system is described by the tensor product $H_{AB} = H_A \otimes H_B$ with basis comprising the set $\lvert x\rangle_A \otimes \lvert y\rangle_B$: $$\lvert\psi\rangle_{AB} = \sum_{x,y} q_{x,y} \lvert x\rangle_A \otimes \lvert y\rangle_B$$ where $q_{x,y}$ are the appropriate parameters from the tensor product. Also see *Tensor product*. The composite state $\lvert\psi\rangle_{AB}$ is said to be *separable* if $q_{x,y} := c_x \times d_y$, in which case $\lvert\psi\rangle_{AB} = \lvert\phi\rangle_A \otimes \lvert\zeta\rangle_B$. If the composite state is not separable, then it is *entangled*. Also see *Entanglement/separable states*
Density matrix	A quantum state $\lvert\psi\rangle$ can (also) be represented (for convenience) as a density matrix $$\rho = \lvert\psi\rangle\langle\psi\rvert. \qquad \text{(continues)}$$

Concept	Description
Density matrix *(cont.)*	The density matrix is obtained by taking the outer product (see *Outer product*) of the *ket* and the *bra* of the state—namely, $|\psi\rangle$ and $\langle\psi|$.
Density operator ρ	A mechanism for an alternate formulation to describe QM. Say a quantum system is in one of a number of states $|\psi_i\rangle$ with probabilities $p_i - \{p_i|\psi_i\rangle\}$ represents a collection of pure states. Then ρ for the system is the outer product of the $|\psi_i\rangle$ and its corresponding conjugate $\langle\psi_i|$.
Eigenstate	Given an operator, $A(\cdot)$, acting on a general wave function (or state, or vector) ψ, the result is usually a wave function with a completely different shape. However, there are special wave functions ψ_e that are such that when A acts on them, the result is just a multiple of the original wave function (or state, or vector); these special wave functions are called *eigenstates*, and the multiples are called *eigenvalues* [FIT23]. Thus, if $$A(|\psi_e\rangle) = e \times |\psi_e\rangle$$ where e is a complex number, then $|\psi_e\rangle$ *is called an eigenstate of operator A* corresponding to the eigenvalue a. This equation implies that if $|\psi_e\rangle$ is the eigenstate with eigenvalue a, then acting on $|\psi_e\rangle$ with the operator A just returns the same state back, but multiplied by the eigenvalue e. *The quantum state after a measurement becomes the eigenstate corresponding to that measurement.*
Eigenvalue	A value of a parameter that allows the solution of a differential equation considering boundary conditions. As applied to QM: a measurement of an observable yields a value, called an eigenvalue of the observable.
Eigenvector	The vector $|B\rangle$ is an eigenvector of a linear operator Op with eigenvalue a if, and only if, $$\text{Op}|B\rangle = a|B\rangle;$$ a is a complex number known as the eigenvalue of O corresponding to $|B\rangle$. The eigenvector-to-operator relation is invariant under the space basis: the eigenvector-to-operator relation depends only on the operator and vectors in question.
Entanglement/ separable states	A combined system A and B is called separable if and only if it can be written as a convex combination of outer products of single-node states, which is $\rho_{AB} = \sum_j p_j \sigma^j_A \otimes \tau^j_B$ for some distribution $\{p_j\}_j$, and states $\{\sigma^j_A\}_j$ in A and $\{\tau^j_B\}_j$ in B. Any state ρ_{AB} that is not *separable* is called *entangled*. Entangled states form the primary building block of most quantum network applications [VAR22]. Also see *Composite systems/entanglement*.
Kronecker product	(Aka matrix direct product.) (Also see *Tensor product*.) An operation (not the usual matrix multiplication) on two matrices that results in a block matrix. It is a particularization of the generic tensor product (and is shown by the same symbol). If A is an $m \times n$ matrix and B is a $p \times q$ matrix, then the Kronecker product $A \otimes B$ is the $pm \times qn$ block matrix: $$\mathbf{A} \otimes \mathbf{B} = \begin{bmatrix} a_{11}\mathbf{B} & \cdots & a_{1n}\mathbf{B} \\ \vdots & \ddots & \vdots \\ a_{m1}\mathbf{B} & \cdots & a_{mn}\mathbf{B} \end{bmatrix}$$ Properties include: $\mathbf{A} \otimes (\mathbf{B} + \mathbf{C}) = \mathbf{A} \otimes \mathbf{B} + \mathbf{A} \otimes \mathbf{C}$ and $(\mathbf{A} \otimes \mathbf{B})(\mathbf{C} \otimes \mathbf{D}) = (\mathbf{AC}) \otimes (\mathbf{BD})$. Example of product and block matrix:

Table 2.A2 More Advanced Mathematical Concepts Applicable to Quacom *(cont.)*

Concept	Description
Kronecker product *(cont.)*	$$\begin{bmatrix} 1 & 2 \\ 3 & 4 \end{bmatrix} \otimes \begin{bmatrix} a & b \\ c & d \end{bmatrix} = \begin{bmatrix} 1\begin{bmatrix} a & b \\ c & d \end{bmatrix} & 2\begin{bmatrix} a & b \\ c & d \end{bmatrix} \\ 3\begin{bmatrix} a & b \\ c & d \end{bmatrix} & 4\begin{bmatrix} a & b \\ c & d \end{bmatrix} \end{bmatrix} = \begin{bmatrix} 1\times a & 1\times b & 2\times a & 2\times b \\ 1\times c & 1\times d & 2\times c & 2\times d \\ 3\times a & 3\times b & 4\times a & 4\times b \\ 3\times c & 3\times d & 4\times c & 4\times d \end{bmatrix}$$
Outer product	(Aka tensor product.) The outer product of two vectors $u = [u_1, u_2, \ldots, u_m]$ and $v = [v_1, v_2, \ldots, v_n]$, denoted $u \otimes v$, is an m-by-n matrix W of rank 1 such that its coordinates satisfy $w_{ij} = u_i v_j$. The outer product $u \otimes v$ is equivalent to a matrix multiplication uv^* (or uv^T if vectors are real) when u is represented as a column $m \times 1$ vector and v is represented as a column $n \times 1$ vector. Also see *Tensor product*.
Pure state	A state of system is pure if it is representable by (there exists) an orthogonal vector basis v^A for the state-space. In a composite state, if the states of A and B are pure (that is, respectively representable by vectors v^A and u^B), then the state of $(A+B)$ is pure and represented by $v^A \otimes u^B$, that is the pair set $(v^A i, u^B j)$.
	Quantum states for which the density operator ρ is also a projector are called pure states $(\text{tr}(\rho^2) = 1)$, otherwise they are called mixed states $(\text{tr}(\rho^2) < 1)$.
Quantum gate	(Aka quantum logic gate.) A basic quantum circuit that operates on a small number of qubits. Quantum gates form the building blocks of quantum circuits, similar to classical logic gates in classical digital circuits. Quantum gates are unitary operators and are characterized as unitary matrices relative to some basis, say $|0\rangle, |1\rangle, |d{-}1\rangle$. There are many (a large number of) gates, where some are named after various researchers. Quantum logic gates are represented by unitary matrices. A gate that acts on n qubits is represented by a $2^n \times 2^n$ unitary matrix.
Systems interaction, composite system	When two systems interact, say V and U, a new state space for the combined sate W is generated. A composite state defines uniquely the states of its components. If H_V and H_U are Hilbert spaces for V and U, then $H_W = H_V \otimes H_U$ is also a Hilbert space. The state-space of a composite system W is obtained from the *"tensor product"* of the individual state-spaces, H_V and H_U: $H_W = H_V \otimes H_U$. If $\{v_i\}$ is an orthonormal basis for H_V and $\{u_j\}$ is an orthonormal basis for H_U, then the set of pairs (v_i, u_j) can be taken to form an orthonormal basis for the tensor product space $H_V \otimes H_U$. Notation-wise $v_i \otimes u_j$ is used to represent the *pair set* (v_i, u_j). Also see *Composite systems*.
	Note: As described in [ISM21], it turns out that the state of a composite system is not uniquely defined by those of its components: in quantum mechanics, facts about composite systems that do not supervene on facts about their components; this means that there are facts about systems as wholes that do not supervene on facts about their parts and the way those parts are arranged in space.
Tensor	(Simplistically) A multidimensional array of numbers.
Tensor product $V \otimes U$ of two vector spaces V and U	The tensor product is an operation combining two (smaller) vector spaces into one larger vector space. Elements of this larger space are called tensors.
	Say V and U are two finite dimensional spaces over the same field of scalars $\mathbb{F}$. Let $\alpha = \{v_1, v_2, \ldots, v_n\}$ and $\beta = \{u_1, u_2, \ldots, u_m\}$ be their respective bases; then the tensor (outer) product $V \otimes U$ of vector spaces V and U is spanned on the basis $\{v_i \otimes u_j$ with $i = 1, 2, \ldots, n; j = 1, 2, \ldots, m\}$. *(continues)*

Concept	Description
Tensor product $V \otimes U$ of two vector spaces V and U (cont.)	Elements of $V \otimes U$, the tensors, are linear combinations of nm of basis vectors $$\sum_{i,j} \tau^{i,j} v_i \otimes u_j, \qquad \tau^{i,j} \in \mathbb{F},$$ satisfying the following two conditions $$c(v \otimes u) = (cv) \otimes u = v \otimes (cu), \; c \in F \text{ (scalar multiplication)},$$ and $$a \otimes u + b \otimes u = (a+b) \otimes u,$$ $$v \otimes u + v \otimes w = v \otimes (u+w).$$
Tensor product $V \otimes U$ of two vector spaces V and U (cont.)	Another description is as follows: $V \otimes U$, the tensor product space of U and V, is the (Hilbert) (vector) space spanned by the basis elements $|v\rangle \otimes |u\rangle$, where $|v\rangle \in V$ and $|u\rangle \in U$, such that the following relations hold for any $|v\rangle$, $|v'\rangle \in V$ and $|u\rangle$, $|u'\rangle \in U$: $$(|v\rangle + |v'\rangle) \otimes |u\rangle = |v\rangle \otimes |u\rangle + |v'\rangle \otimes |u\rangle$$ $$|v\rangle \otimes (|u\rangle) + |u'\rangle) = |v\rangle \otimes |u\rangle + |v\rangle \otimes |u'\rangle.$$ The (tensor) outer product is somewhat similar, but not identical to the Kronecker Product (see *Kronecker product*)
Value for observable	A system has a value for observable A if, and only if, the vector representing its state is an eigenstate of the A-operator; the value is the eigenvalue associated with that eigenstate.
Value for observable A	A system has a value for observable A if, and only if, the vector representing its state is an eigenstate of the A-operator.

Table 2.A3 Yet More Advanced Mathematical Concepts Applicable to Quacom

Concept	Description
Del	The upside-down capital delta symbol del, ∇, also called "nabla", is used to denote the gradient of a vector (and other related vector derivatives—e.g., ∇^2 being *Laplacian* or *vector Laplacian*).
Hamiltonian ($\hat{H}$) quantum operator	The result of the quantum Hamiltonian $\hat{H}$ operator on a quantum system is the characterization of its total energy of the system, and the eigenvalues of the Hamiltonian operator are the energy of the system. It corresponds to the total energy inside a system, including kinetic and potential energy. The Hamiltonian operator $\hat{H}$ is the sum of two operators. $\hat{H} = -\hbar^2/2m \times \partial^2/\partial x^2 + V(x)$ where $-\hbar^2/2m \times \partial^2/\partial x^2$ (the first operator) is the kinetic energy operator and $V(x)$ (the second operator) is the potential energy operator. The eigenvalues of this operator are the possible outcomes of the total energy of a QM system.
	A Hamiltonian operator is the energy operator for wave function ψ in time-independent Schrödinger's wave theory equation. It combines the operators for kinetic energy and potential energy to constitute a time-independent Hamiltonian operator formula equation. ψ is the wave representing a spatial distribution of particles—for example, electrons in an atom are described by a wave function centered on the nucleus. ψ does not describe the exact location of the electron; it

(continues)

Table 2.A3 Yet More Advanced Mathematical Concepts Applicable to Quacom *(cont.)*

Concept	Description
Hamiltonian ($\hat{H}$) quantum operator *(cont.)*	indicates the probability of finding the electron at a given point. These points are the eigenvalues of the Hamiltonian operator [ARE22]. Example (from [CRE11]): Hamiltonian for the ion O^-_2 ion takes the form: $$\hat{H} \doteq \begin{bmatrix} E_0 & -A \\ -A & E_0 \end{bmatrix}$$ where A is a real number. The eigenvalues of $\hat{H}$ are $E_1 = E_0 + A$ and $E_2 = E_0 - A$. The associated eigenstates are: $$|E_1\rangle = \frac{1}{\sqrt{2}}(|+a\rangle - |-a\rangle) \doteq \frac{1}{\sqrt{2}}\begin{bmatrix} 1 \\ -1 \end{bmatrix} \qquad |+a\rangle = \begin{bmatrix} 1 \\ 0 \end{bmatrix} \quad |-a\rangle = \begin{bmatrix} 0 \\ 1 \end{bmatrix}$$ $$|E_2\rangle = \frac{1}{\sqrt{2}}(|+a\rangle + |-a\rangle) \doteq \frac{1}{\sqrt{2}}\begin{bmatrix} 1 \\ 1 \end{bmatrix}$$
Hamiltonian operator (mathematics)	In a Cartesian coordinate system $x = (x_1 \ldots x_n)$ with unit (basis) vectors $e_1, \ldots, e_n$ the Hamilton operator has the form $$\nabla = \sum_{j=1}^{n} e_j \frac{\partial}{\partial x_j}.$$ The application of the Hamilton operator to a scalar function f, which is understood as multiplication of the "vector" ∇ by the scalar $f(x)$ yields the gradient of f: $$\text{grad } f = \nabla f = \sum_{j=1}^{n} e_j \frac{\partial f}{\partial x_j},$$ namely, the vector with components $(\partial f/\partial x_1 \ldots \partial f/\partial x_n)$ [STA23].
Hermitian operator	An operator Op that has the property that there is an orthonormal basis consisting of its eigenvectors, and those eigenvalues are all real. Every Hermitian operator on a Hilbert space associated with a system represents a distinct observable.
Hermitian conjugate of the operator	Say Op is any linear operator on a Hilbert space, V. Then there exists a unique linear operator $Op^\dagger$ on V such that for all vectors $|A\rangle, |B\rangle \in V$, the inner product meets the condition: $$\langle|A\rangle, Op|B\rangle\rangle = \langle\langle Op^\dagger|A\rangle, |B\rangle\rangle.$$ *This linear operator $Op^\dagger$ is known as the* adjoint *or Hermitian conjugate of the operator Op.* An operator Op whose adjoint is also Op is known as a *Hermitian* or *self-adjoint* operator. By nomenclature, if $|A\rangle$ is a vector, one denotes the new vector obtained by applying operator $Op^\dagger$ to this vector as $Op^\dagger|A\rangle = |A\rangle^\dagger$ and then ultimately denotes $|A\rangle^\dagger = \langle A|$. $Op|A\rangle$ is a (new) vector, also by nomenclature when one applies the operator $Op^\dagger$ to the vector $Op|A\rangle$ one denotes the new vector that results, $Op^\dagger(Op|A\rangle) = (Op|A\rangle)^\dagger$, and then ultimately denotes **[(Op|A⟩)†] = Op†⟨A|.**
Hermitian matrix	(Aka self-adjoint matrix) is a complex square matrix that is equal to its own conjugate transpose—namely, the element in the i-th row and j-th column is equal to the complex conjugate of the element in the j-th row and i-th column, for all values of i and j.

Concept	Description
k-local Hamiltonian	A Hermitian matrix H acting on the space of n-state particles (n qubits) that can be expressed as $H = \Sigma_i H_i$, where each H_i acts non-trivially on at most k particles; that is, $H_i = h \otimes \|$ where h is a Hamiltonian on at most k particles (qubits), and $\|$ denotes the identity operator [ALB18].
Planck constant	$\hbar = h/2\pi$ is the cornerstone of quantum mechanics; it enables (i) establishing when classical concepts no longer apply; (ii) specifying the uncertainties introduced by measurement; and (iii) determining the size of atoms.
Positive operator on a complex Hilbert space	A subclass of Hermitian operators. A symmetric operator and has a self-adjoint extension that is also a positive operator. A positive operator Op is one for which the inner product between $\langle\psi\|$ and $Op\|\psi\rangle$ is greater or equal to 0 for all $\|\psi\rangle$ not equal to zero—namely, $\langle\psi \| Op\|\psi\rangle \geq 0$. Furthermore, a positive definite operator Op is one for which $\langle\psi \| Op\|\psi\rangle > 0$ for all $\|\psi\rangle$ not equal to zero.
Projective Hilbert space	A set of equivalence classes where the equivalence relation $\sim$ is $$a \sim b \Longleftrightarrow a = \lambda b,$$ where λ is a complex number.
Schrödinger equation	An equation that prescribes the non-relativistic behavior of the wave function in space and time in terms of a Hamiltonian, which combines the kinetic and potential energies [JAC23]. The Schrödinger equation describing non-relativistic motion of free particles is $$i\hbar\frac{\partial\psi}{\partial t} = -\frac{\hbar^2}{2m}\nabla^2\psi.$$ where $\hbar = h/2\pi$ (Planck constant), mass m (particles), momentum p, and $\psi(x, t)$ describes a state of motion of a particle. When a particle is moving in a potential $V(\mathbf{x})$, the time-dependent Schrödinger equation that governs the motion is $$i\hbar\frac{\partial\psi}{\partial t} = \left[\frac{p^2}{2m} + V(\mathbf{x})\right]\psi$$ Note that $\|\psi\|^2$ is a probability density.
Time-evolution of quantum sate	In general, systems, including quantum systems, are dynamic—namely, they evolve in time. Time evolution is the change of state brought about by the passage of time. $\|\psi(t)\rangle = e^{-i\omega t}\|\psi(0)\rangle$ where ω is a constant. The Hamiltonian operator $\hat{H}$ generates the time evolution of quantum states: if $\|\psi(t)\rangle$ is the state of the system at time t, then the evolution is described by the Schrödinger equation: $$H\|\psi(t)\rangle = i\hbar\frac{\partial}{\partial t}\|\psi(t)\rangle.$$ If H is independent of t, then given the state at some initial time $t = 0$, the unitary time evolution operator $U(t)$ is the exponential operator: $$\|\psi(t)\rangle = U(t)\|\psi(0)\rangle = e^{-iHt/\hbar}\|\psi(0)\rangle$$
Unitary operator, time-evolution	Operator $U = e^{(-i\hat{H}t)}$ where $\hat{H}$ is the Hamiltonian operator of the system. The quantum Hamiltonian operator is the total energy of the system, and the eigenvalues of the Hamiltonian operator are the energy of the system.

Appendix B

Bibliography

This bibliography is extracted from reference [ISM21]; the authors are greatly indebted to Prof. D. Albert, 1994, *Quantum Mechanics and Experience*, Cambridge, MA: Harvard University Press.

L. Ballentine, 1998, *Quantum Mechanics: A Modern Approach*, Singapore: World Scientific Publishing Company.

J. Barrett, 2019, *The Conceptual Foundations of Quantum Mechanics*, New York: Oxford University Press.

J.L. Basdevant, J. Dalibard, 2005, *Quantum Mechanics*, Berlin: Springer.

A. Becker, 2018, *What is Real? The Unfinished Quest for the Meaning of Quantum Mechanics*, New York: Basic Books.

J.S. Bell, 1987, *Speakable and Unspeakable in Quantum Mechanics*, Cambridge: Cambridge University Press.

W. Benenson, J. Harris, H. Stoecker, and H. Lutz, 2006, *Handbook of Physics* (2nd edition), Berlin: Springer.

I.N. Bronshtein, K.A. Semendyayev, 2007, *Handbook of Mathematics* (5th edition), Berlin: Springer.

P. Busch, P. Lahti, and P. Mittelstaedt, 1991, *The Quantum Theory of Measurement*, Berlin: Springer-Verlag.

S. Carroll, 2019, *Something Deeply Hidden: Quantum Worlds and the Emergence of Spacetime*, New York: Dutton.

R.K. Clifton, (Ed.), 1996, *Perspectives on Quantum Reality*, Dordrecht: Kluwer.

C. Cohen-Tannoudji, 2006, *Quantum Mechanics*, New York: Wiley-Interscience.

P.A.M. Dirac, 1930 [1958], *The Principles of Quantum Mechanics*, Oxford: Clarendon Press, 1930; 4th edition, revised, 1958.

B. d'Espagnat, 1995, *Veiled Reality*, Reading, MA: Addison-Wesley.

S. Gasiorowicz, 1995, *Quantum Physics* (3rd edition), New York: Wiley.

D. Griffiths, 1995 (also 2018), *Introduction to Quantum Mechanics* (2nd edition), Upper Saddle River, NJ: Prentice Hall; 3rd edition, 2018.

D. Halliday, R. Resnick, and J. Walker, 2008, *Fundamentals of Physics* (8th edition), Hoboken, NJ: Wiley.

P. Halmos,1957, *Introduction to Hilbert Space* (2nd edition), Providence: AMS Chelsea Publishing.

R.I.G. Hughes, 1989, *Structure and Interpretation of Quantum Mechanics*, Cambridge, MA: Harvard University Press.

J. Ismael for this extensive compilation of quantum-related textbooks.

P. Lewis, 2016, *Quantum Ontology: A Guide to the Metaphysics of Quantum Mechanics*, New York: Oxford University Press.

R. Liboff, 1998, *Introductory Quantum Mechanics* (4th edition), San Francisco: Addison-Wesley.

T. Maudlin, 2019, *Philosophy of Physics: Quantum Theory*, Princeton: Princeton University Press.

E. Merzbacher, 1997, *Quantum Mechanics* (3rd edition), New York: Wiley.

T. Norsen, 2017, *Foundations of Quantum Mechanics: An Exploration of the Physical Meaning of Quantum Theory*, Cham: Springer.

R. Omnès, 1994, *The Interpretation of Quantum Mechanics*, Princeton: Princeton University Press.

H. Primas, 1983, *Quantum Mechanics, Chemistry and Reductionism* (2nd edition), Berlin: Springer.

A. Rae, 1986, *Quantum Physics: Illusion or Reality?* Cambridge: Cambridge University Press.

M.L.G. Redhead, 1989, *Incompleteness, Nonlocality and Realism*, Oxford: Clarendon Press.

J.J. Sakurai, 1993 (also 2021), *Modern Quantum Mechanics* (revised edition), Reading, MA: Addison Wesley; 3rd Ed., coauthor Jim Napolitano, Cambridge: Cambridge University Press, 2021.

J. Schwinger, 2003, *Quantum Mechanics* (corrected edition), Berlin: Springer.

R. Shankar, 1994, *Principles of Quantum Mechanics* (2nd edition), Berlin: Springer.

E. Squires, 1990, *Conscious Mind in the Physical World*, Bristol, New York: Adam Hilger.

L. Susskind, A. Friedman, 2014, *Quantum Mechanics: The Theoretical Minimum (2nd edition)*, New York: Basic Books.

A. Whitaker, 1996, *Einstein, Bohr and the Quantum Dilemma*, Cambridge: Cambridge University Press.

N. Zettili, 2009, *Quantum Mechanics: Concepts and Applications*, Chichester: John Wiley & Sons, Ltd.

Chapter 3

More Advanced Concepts in Quantum Physics—Mathematical Tools

This chapter extends some of the basic general mathematical Quantum Mechanics (QM) machinery discussed in Chapter 2, still aiming to be at an introductory level. We discuss concepts of eigenstate and eigenvalue, noisy quantum states, quantum quality/fidelity measures, photon number operator, Fock state, Bell states, measurements in different basis, and measurement of photons.

The basic mathematics of QM is operator algebra. In particular, an operator (say A) acts on a state vector, and a new vector emerges: $A(|\psi\rangle) = |\phi\rangle$. An operator can be represented as a matrix. In mathematics, a function is analogous to a vector; hence ψ is the implicit representation of a vector $|\psi\rangle$. Table 3.1 captures some of the basic concepts discussed in Chapter 2 and/or in this chapter; some additional machinery is also introduced. Readers should review the table in order to become familiar or refamiliarize themselves with these key concepts.

3.1 Basic Principles of Quantum Mechanics

As already noted, QM describes the behavior of nanoscale, atomic, and sub-atomic entities (particles—for example, electrons and photons). Mathematical models (mathematical abstractions) are used to study their behavior and the "system state". There are three basic principles of interest, captured in the three basic principles of quantum mechanics [WHA09]:

- **The superposition principle:** This principle informs as to what the state of a quantum system looks like. To describe a physical system, one needs identify its "type" and its time-dependent "state"; the former properties are "state-independent", while the latter are "state-dependent".
- **The measurement principle:** This principle establishes how much information about the state one can access. A *physical quantity* is an "*observable*" value that can describe the state of the system.
- **Unitary evolution principle:** This principle describes how the state of the quantum system evolves in time, especially in the presence of noise.

[Text continues on page 107]

**Table 3.1 Basic Concepts Discussed in Previous Chapters
and Some Elaborated in This Chapter**

Concept	Description
'bra' vector of $|\psi\rangle$, $\langle\psi|$	The Hermitian conjugate: $\langle\psi| = (|\psi\rangle)^\dagger = (|\psi*\rangle)^T$
Basis	A collection of N of mutually orthogonal vectors of length 1 in an N-dimensional vector space such that any vector in the space is a unique linear sum (superposition) of these vectors.
Bell-State Measurement (BSM)	A projective measurement where the four elements are the projectors on the Bell states.
Creation operators and annihilation operators	Mathematical operators introduced by Paul Dirac that have widespread applications QM. The annihilation operator (usually denoted by a) lowers the number of particles in a given state by one. A creation operator (usually denoted $a^\dagger$, being the adjoint of the annihilation operator) increases the number of particles in a given state by one.
Density matrix ρ	$\rho = |\psi\rangle\langle\psi|$. Density matrices are Hermitian ($\rho = \rho^\dagger$), positive semi-definite ($\rho > 0$) and with trace of one, namely, $\mathrm{tr}(\rho) = 1$.
Dimension of a vector space	The maximum number of nonzero, mutually orthogonal vectors it contains.
Dirac notation	There is a one-to-one correspondence of Dirac notation to matrix algebra notation. $\langle x| <-> x^\dagger$; $\langle x|y\rangle = \langle x|y\rangle = x^\dagger y$; $\hat{A}|x\rangle <-> \bar{A}x$. $\bar{a}$ is the complex conjugate: if $z = a + bi$, then $\bar{z} = a - bi$
	$\bar{A}x$ is an *implicit* notation; $\Sigma_{ij} A_{ij}x_j$ is an *explicit* notation.
	ψ is a function which is a vector in a functional space. It is denoted as ψ in math notation and $|\psi\rangle$ in Dirac notation. This is also known as the "ket". The conjugate transpose of a vector in Dirac notation is called a "bra", which is denoted as $\langle\psi|$.
Eigenvector of operator Op on $|v\rangle$	The mapping of vector $|v\rangle$ to a vector $|v'\rangle += \mathrm{Op}(|B\rangle)$.
Eigenvector of operator Op with eigenvalue a	$|E\rangle$ is eigenvector when $\mathrm{Op}(|E\rangle) = a|E\rangle$.
Expansion coefficients in a basis	The values of the coefficients that represent a given vector in a vector space when utilizing the specified basis.
Hadamard basis	A basis $|+\rangle := (|0\rangle + |1\rangle)/\sqrt{2}$ and $|-\rangle := (|0\rangle - |1\rangle)/\sqrt{2}$. See Figure 3.3, page 111.
Hermitian operator	An operator Op that has the property that there is an orthonormal basis consisting of its eigenvectors, and those eigenvalues are all real. Every Hermitian operator on a Hilbert space associated with a system represents a distinct observable.
Inner product space	A vector space on which the operation of vector multiplication has been defined
Kraus operators	Kraus operators satisfy the relation $\Sigma_m A^\dagger_m A_m = \mathbb{I}$ with $\mathbb{I}$ being the identity matrix.
Maximally entangled state between two d-dimensional systems A and B	This is the state $|\Psi\rangle = \frac{1}{\sqrt{d}} \sum_j |j\rangle_A \otimes |j\rangle_B$

Table 3.1 Basic Concepts Discussed in Previous Chapters . . . (cont.)

Concept	Description
Mixed states $\|w\rangle$ in a $(A+B)$ aka $(A \otimes B)$ ensemble)	States that are weighted sums of pure states, e.g., $\|w\rangle = w_1 \times \|a\rangle + w_2 \times \|b\rangle$. These states can be utilized to represent the states of ensembles of systems such as A and B whose components are in different pure states, or can be utilized to represent states of individual systems about which one has only partial knowledge.
	Although any such state $\|w\rangle$ is representable by some basis set $\{W_1, W_2, W_3, ..., W_i,...\}$ in $(A \otimes B)$, it is not representable by the inner product set of basis $v_i^A \otimes u_j^B$ where $\{v_i^A\}$ is an orthonormal basis for H_A and $\{u_j^B\}$ is an orthonormal basis for H_B.
	One needs a way of representing the weighted sum of a set of pure states (equivalently, of the probability functions associated with them) that is distinct from adding the (suitably weighted) vectors that represent them; that means that we need either an alternative way of representing mixed states or a uniform way of representing both pure and mixed states that preserves the distinction between them—this can done by utilizing the density operator: advantageously, both pure and mixed states are representable by density operators [ISM21].
Normalized vectors	Vectors, such that property $\langle v\|\|v\rangle = \langle v\|v\rangle = 1$; such vector has norm 1. For example, given scalars such as v_x and v_y, one can define a normalized vector: $$\|v\rangle = \frac{v_x}{\sqrt{v_x^2 + v_y^2}}\|0\rangle + \frac{v_y}{\sqrt{v_x^2 + v_y^2}}\|1\rangle,$$ $$\langle v\|v\rangle = \sum_i \bar{\alpha}_i \beta_i = \frac{v_x^2}{v_x^2 + v_y^2} + \frac{v_y^2}{v_x^2 + v_y^2} = 1$$
Observable	A 'physical quantity'.
Operator algebra	An operator (say A) acts on a state vector and a new vector emerges: $A(\|\psi\rangle) = \|\phi\rangle$. An operator can be represented as a matrix. The outer (tensor) product of two vectors $\|\psi\rangle$ and $\|\phi\rangle$ results in a matrix or an operator that is written as $\|\psi\rangle \otimes \langle\phi\|$ or $\|\psi\rangle \langle\phi\|$.
	A particular operator is $\underline{P} = \|\psi\rangle \otimes \langle\psi\| = \|\psi\rangle\langle\psi\| = \rho$; it is a projector operator: $\underline{P}(\|\phi\rangle) = \|\psi\rangle \otimes \langle\psi\| (\|\phi\rangle) = \langle\psi\|\phi\rangle\|\psi\rangle$. Note that $\underline{P}^2 = \underline{P}$.
Positive Operator Valued Measure (POVM)	A collection of operators E_m such that $\Sigma_m E_m = \mathbb{I}$, the identity matrix.
	The relation between a POVM and the Kraus operator representation of a measurement is given by the relation $E_m = A^\dagger_m A_m$.
Pure state $\|v\rangle$	A state in Hilbert vector space A, such as $\|v\rangle$, is pure if it representable by/ with a set of basis vectors $\{A_1, A_2, A_3, ..., A_i,...\}$. A pure state is described by a state vector: $\|\psi\rangle$. It is common to see pure states as being states represented by vectors; they can also be seen as being represented by density operators. Quantum states for which the density operator ρ is also a projector are called pure states $(\text{tr}(\rho^2) = 1)$; otherwise they are called mixed states $(\text{tr}(\rho^2) < 1)$.
QM intrinsics	The Hilbert spaces that represent the state-spaces of quantum mechanical systems.
Quantum measurement	A quantum measurement of a quantum state ρ is characterized by Kraus operators A_m, where the index m refers to the measurement outcome. The probability to measure the outcome m is $p_m = \text{tr}(A_m \rho A_m^\dagger)$. The state after the measurement becomes $\rho_m = (A_m \rho A_m^\dagger)/p_m$.

Concept	Description
Separable states; entangled states	Define $expA = d_A \times d_A$ and $expB = d_B \times d_B$. A quantum state $\rho_{AB} \in \mathbf{C}^{expA} \otimes \mathbf{C}^{expB}$ of a combined quantum system of qubits A and B is said to be *separable* when ρ_{AB} can be written as a convex combination of tensor products of single-node states—that is, for some distribution $\{p_j\}_j$, and states $\{\sigma^j_A\}_j$ on A and states $\{\tau^j_B\}_j$ on B, $\rho_{AB} = \sum_j p_j \sigma^j_A \otimes \tau^j_B$. *Any state where ρ_{AB} that is not separable is called entangled.*
State-space of a system	The space formed by the set of its possible states, which are the physically possible ways of combining the values of quantities that characterize it internally
Vector space	A set of vectors closed under addition and multiplication by constants: any addition of vectors (in a Hilbert space) or multiplying them by scalars results in a vector in the space.
Wavefunction	(Especially for infinite-dimensional vector spaces.) (Being that one cannot write out the entire column of expansion coefficients needed to identify the vector) one defines a function ψ instead to represent the vector, here $\vert Q\rangle$ under a basis $\{A_1, A_2, A_3, ..., A_i,...\}$: $$\vert Q\rangle = \begin{bmatrix} q_1 \\ q_2 \\ q_3 \\ \cdot \\ \cdot \\ \cdot \end{bmatrix}$$ $\psi(Q, A_i) = \psi_1(Q, A_1) \times \vert A_1\rangle + \psi_2(Q, A_2) \times \vert A_2\rangle + \psi_3(Q, A_3) \times \vert A_3\rangle + ... = \Sigma_i \psi_i(Q, A_i) \times \vert A_i\rangle$ where $\psi_i(Q, A_i) = q_i = \langle Q \vert A_i\rangle$. In mathematics, a function is analogous to a vector; thus, ψ is the implicit representation of a vector.

To elaborate on these principles, four basic aspects of QM are as follows,[1] per [ISM21]:

- **Physical states:** (i) Every physical system is associated with a Hilbert Space; (ii) every *unit vector* in the space corresponds to a possible pure state of the system, and (iii) every possible pure state corresponds to some vector in the space.
- **Physical quantities:** Hermitian operators in the Hilbert space associated with a system represent physical quantities; their eigenvalues represent the possible results of measurements of those quantities. In particular, the Hamiltonian provides the total energy of the system.
- **Composition:** The Hilbert space associated with a complex system is the tensor (inner) product of those associated with the simple systems (in the standard, non-relativistic theory: the individual particles) of which it is composed.
- **Dynamics:**
  - *Contexts of type 1:* Given the state of a system at time t and the forces and constraints to which it is subject, Schrödinger's equation gives the state at any other time—namely the operator U provides $U\vert vt\rangle \rightarrow \vert vt'\rangle$. U is *deterministic* (it takes the state of a system at one time into a unique state at any other); is *unitary* (it is an automorphism of the Hilbert space on which it acts—namely, it is a mapping of that space onto itself that preserves the

---

[1] The reader may want to consult reference [ISM21] for a more exhaustive treatment of these topics, particularly as related to measure an observable. Among a wide multitude of sources on this topic, the reader may consult [TRA17], [GAS95], [MER98].

linear space structure and inner product); and is *linear* (that if it takes a state $|A\rangle$ onto the state $|A'\rangle$, and it takes the state $|B\rangle$ onto the state $|B'\rangle$, *then* it takes any state of the form $\alpha|A\rangle + \beta|B\rangle$ onto the state $\alpha|A'\rangle + \beta|B'\rangle$).

- ○ *Contexts of type 2 ("Measurement Contexts"):* "The Collapse Postulate": Carrying out a "measurement" of an observable B on a system in a state $|A\rangle$ has the effect of collapsing the system into a B-eigenstate corresponding to the eigenvalue observed. Which particular B-eigenstate it collapses into is probabilistic, and the probabilities are given by Born's Rule: $\Pr(b_i) = |\langle A \mid b_i \rangle|^2$.

3.2 Superposition and States

A point of departure is that "system state" in QM is conceptually different from the concept of state in classical mechanics. In classical mechanics parameters such as positions of a particle, momentum of the particle, spin, electrical charge, and energy characterize the state of the system; these same parameters can be measured (in the context of various defined experiments). Therefore, state and measurements are inextricably intertwined. In QM system states are defined by different mathematical constructs and have a different logical structure. Furthermore, in QM states and measurements are distinct characterizations, and the relationship is mostly devoid of intrinsic intuitiveness.

As noted in Chapter 2, the *state-space* of a system is the space formed by the set of its possible states—namely, the physically possible ways of combining the values of quantities that characterize the system internally. The state-space can be perceived as coordinates in an appropriate (mathematical) vector space. In QM, the state-spaces are representable as vectors in Hilbert vector spaces—Hilbert vector spaces are vector spaces on which an inner product is defined and which are complete. The finite dimensional complex vector space(s) that underpin QM is effectively synonymous to an inner product space. All the physically relevant features of the behaviors of quantum mechanical systems are derivative of the mathematical properties of those relations.

There are a number of theoretical and philosophical interpretations of QM as a descriptive and explanatory model of the actual world; there is extensive literature on the matter, and this topic is not explored further herewith, except to hint to the use of the well-accepted Copenhagen interpretation. According to the Copenhagen interpretation of QM, the quantum state is not to be taken as a description of the physical system, but rather, the role of the quantum state is to summarize what one can expect if one makes measurements on the system. See [LEW23] for a discussion of various interpretations.[1]

We noted in Chapter 2 that the state of a qubit can be perceived as a two-dimensional vector with basis $\{|0\rangle, |1\rangle\}$ and value $|\psi\rangle = \alpha|0\rangle + \beta|1\rangle$. A quantum bit or qubit is a two-state QM system—namely, a qubit (with state) $|\psi\rangle$ can be in a generic *superposition* of the basis states $|0\rangle$ and $|1\rangle$, with $|\psi\rangle = \alpha|0\rangle + \beta|1\rangle$. The complex coefficients α and β are such that $|\alpha|^2 + |\beta|^2 = 1$, which physically corresponds to the requirement that the probability to find the state $|\psi\rangle$ in

---

[1] [LEW23] points to some useful references on this topic (among many others): [ALB92] as a non-technical overview of the various interpretations of quantum mechanics and their problems; [BEL04] with a discussions of various interpretations of quantum mechanics, particularly hidden variable theories; [CUS94] with a comparison of the Copenhagen interpretation and Bohm's theory; [KAS13] describing transactional interpretation and development of a "possibilist" version; and [SAU10] which discusses the many-worlds interpretation.

precisely one of the basis states $|0\rangle$ or $|1\rangle$ is 1. States $|0\rangle$ and $|1\rangle$ are vectors forming a basis of a two-dimensional Hilbert space H. Examples of qubits include:

- Polarization states of a photon
  - Using rectilinear polarization with basis $|0\rangle$ and $|1\rangle$
  - Using diagonal polarization with basis $|+\rangle := (|0\rangle + |1\rangle)/\sqrt{2}$ and $|-\rangle := (|0\rangle - |1\rangle)/\sqrt{2}$
  - Using circular polarization with basis $|+\rangle := (|0\rangle + i|1\rangle)/\sqrt{2}$ and $|-\rangle := (|0\rangle - i|1\rangle)/\sqrt{2}$
- Electron spin with basis $|\uparrow\rangle$ and $|\downarrow\rangle$.

As discussed in Chapter 2, a quantum state $|\psi\rangle$ can (also) be represented (for convenience) as a *density matrix* $\rho = |\psi\rangle\langle\psi|$ (which, as seen just here in the definition is obtained by taking the outer product of the ket and the bra of the state). For example, for $d = 2$.

$$|\psi\rangle = \alpha|0\rangle + \beta|1\rangle = \begin{bmatrix} \alpha \\ \beta \end{bmatrix}$$

$$\langle\psi| = \begin{bmatrix} \alpha \\ \beta \end{bmatrix}^\dagger = [\alpha^* \ \beta^*] = \alpha^*\langle 0| + \beta^*\langle 1|$$

$$|0\rangle = \begin{bmatrix} 1 \\ 0 \end{bmatrix} \quad \langle 0| = [1 \ 0]$$

$$|1\rangle = \begin{bmatrix} 0 \\ 1 \end{bmatrix} \quad \langle 1| = [0 \ 1]$$

$$\rho = |\psi\rangle\langle\psi| = \underset{2\times1}{\begin{bmatrix} \alpha \\ \beta \end{bmatrix}} \underset{1\times2}{[\alpha^* \ \beta^*]} = \underset{2\times2}{\begin{bmatrix} \alpha\alpha^* & \alpha\beta^* \\ \beta\alpha^* & \beta\beta^* \end{bmatrix}}$$

3.3 Eigenstate and Eigenvalue

Consider operator A: if $A(\psi_a) = a \times \psi_a$ where a is a complex number, then ψ_a *is called an eigenstate of operator* A corresponding to the eigenvalue a—this equation implies that if ψ_a is the eigenstate with eigenvalue a, then acting on ψ_a with the operator A just returns the same state back, but multiplied by the eigenvalue a. *For example*, each of the Pauly matrices has two eigenstates with particular eigenvalues; for example, for σ_z, $|0\rangle$, $|1\rangle$ are the eigenstates of the Pauli matrices with eigenvalues 1 and –1, as seen in Figure 3.1. The widely-used Pauli matrices are a set of three 2×2 complex matrices, depicted in the figure, which are Hermitian, involutory, and unitary; they act on single qubit quantum states.

In summary, a state described by an eigenfunction is called an *eigenstate of the operator*. Eigenstates that are possible values of a mechanical quantity are all eigenvalues of its operator. In its own eigenstate, the mechanical quantity takes a certain value—that is, an eigenvalue to which the eigenstate belongs [CHE23]. For a Hamiltonian matrix $\hat{H}$, solutions satisfying the equation $H|\psi\rangle = E|\psi\rangle$ are called *eigenstates* $|\psi\rangle$ of H, having *eigenenergy* E. A ground state corresponds to an eigenstate with the lowest energy of a quantum system.

3.4 Quantum Gates

Quantum information is administered through the application of quantum gates, already mentioned in Chapter 2. Namely, Quacomp and Quanet applications are realized by applying a sequence of quantum gates to one or several qubits and then performing a measurement of the qubits to read out information in the quantum states.

$$\sigma_x = \begin{bmatrix} 0 & 1 \\ 1 & 0 \end{bmatrix} \quad \sigma_y = \begin{bmatrix} 0 & -i \\ i & 0 \end{bmatrix} \quad \sigma_z = \begin{bmatrix} 1 & 0 \\ 0 & -1 \end{bmatrix}$$

$$\sigma_{x+} = \frac{1}{\sqrt{2}}\begin{bmatrix} 1 \\ 1 \end{bmatrix} \quad \sigma_{x-} = \frac{1}{\sqrt{2}}\begin{bmatrix} 1 \\ -1 \end{bmatrix} \quad \sigma_x = |0\rangle\langle 1| + |1\rangle\langle 0|$$

$$\sigma_{y+} = \frac{1}{\sqrt{2}}\begin{bmatrix} 1 \\ i \end{bmatrix} \quad \sigma_{y-} = \frac{1}{\sqrt{2}}\begin{bmatrix} 1 \\ -i \end{bmatrix} \quad \sigma_y = -i\,|0\rangle\langle 1| + i\,|1\rangle\langle 0|$$

$$\sigma_{z+} = 1\begin{bmatrix} 1 \\ 0 \end{bmatrix} \quad \sigma_{z-} = -1\begin{bmatrix} 0 \\ 1 \end{bmatrix} \quad \sigma_z = |0\rangle\langle 0| - |1\rangle\langle 1|$$

Example:

$$\sigma_z = \begin{bmatrix} 1 & 0 \\ 0 & -1 \end{bmatrix} \quad |0\rangle = \begin{bmatrix} 1 \\ 0 \end{bmatrix} \quad |1\rangle = \begin{bmatrix} 0 \\ 1 \end{bmatrix}$$

$$\sigma_z|0\rangle = \begin{bmatrix} 1 & 0 \\ 0 & -1 \end{bmatrix}\begin{bmatrix} 1 \\ 0 \end{bmatrix} = \begin{bmatrix} 1 \\ 0 \end{bmatrix}$$

2×2   2×1   2×1

$$\sigma_z|1\rangle = \begin{bmatrix} 1 & 0 \\ 0 & -1 \end{bmatrix}\begin{bmatrix} 0 \\ 1 \end{bmatrix} = -\begin{bmatrix} 0 \\ 1 \end{bmatrix}$$

Figure 3.1 $|0\rangle$, $|1\rangle$ are the eigenstates of Pauli matrix with eigenvalues 1 and -1.

$$U = \begin{bmatrix} a & b \\ c & d \end{bmatrix} \quad U^{-1} = \frac{1}{ad - bc}\begin{bmatrix} d & -b \\ -c & a \end{bmatrix} \quad UU^{-1} = \begin{bmatrix} a & b \\ c & d \end{bmatrix}\frac{1}{ad-bc}\begin{bmatrix} d & -b \\ -c & a \end{bmatrix} = \begin{bmatrix} 1 & 0 \\ 0 & 1 \end{bmatrix}$$

Unitary: $U^{-1} = U^{\dagger} \Rightarrow U^{-1} = \frac{1}{ad-bc}\begin{bmatrix} d & -b \\ -c & a \end{bmatrix} = U^{\dagger} = \begin{bmatrix} \bar{a} & \bar{c} \\ \bar{b} & \bar{d} \end{bmatrix} = \begin{bmatrix} a^* & c^* \\ b^* & d^* \end{bmatrix}$

$$\Rightarrow \quad a^* = \frac{d}{ad-bc} \qquad c^* = \frac{-b}{ad-bc}$$
$$b^* = \frac{-c}{ad-bc} \qquad d^* = \frac{a}{ad-bc}$$

unitary with
$$\begin{array}{ll} a = 1 & a^* = 1 \\ c = 0 & c^* = 0 \\ b = 0 & b^* = 0 \\ d = 1 & d^* = 1 \end{array}$$

also
unitary with
$$\begin{array}{ll} a = 0 & a^* = 0 \\ c = 1 & c^* = 1 \\ b = 1 & b^* = 1 \\ d = 0 & d^* = 0 \end{array}$$

Unitary: $UU^{\dagger} = I \Rightarrow \begin{bmatrix} a & b \\ c & d \end{bmatrix}\begin{bmatrix} a^* & c^* \\ b^* & d^* \end{bmatrix} = \begin{bmatrix} aa^* + bb^* & ac^* + bd^* \\ ca^* + db^* & cc^* + dd^* \end{bmatrix} = \begin{bmatrix} 1 & 0 \\ 0 & 1 \end{bmatrix}$

$$\Rightarrow \quad \begin{array}{ll} aa^* + bb^* = 1 & ac^* + bd^* = 0 \\ ca^* + db^* = 0 & cc^* + dd^* = 1 \end{array}$$

these hold for the two sets of values from above

- The rows of a unitary U form an orthonormal basis
- The columns of a unitary U form an orthonormal basis
- A unitary U preserves the inner product: $\langle x|y\rangle = \langle U(x)|U(y)\rangle$

Figure 3.2   Unitary example in two dimensions

A quantum logic gate can be described using linear algebra—namely, it can be expressed by a matrix. A single-qubit operation can be a specific rotation on the Bloch sphere; rotations about the axes of the Bloch sphere can be described in terms of the Pauli's Hermitian, unitary operators/matrices. For example, the NOT gate (Pauli's X) maps 0 to 1 and 1 to 0, so that the $|0\rangle$ and $|1\rangle$ states are interchanged—namely, if $|\psi\rangle = \alpha|0\rangle + \beta|1\rangle$ then $X|\psi\rangle = \alpha|1\rangle + \beta|0\rangle$.

A quantum gate G is represented by a matrix $G \in \mathbf{C}^{d \times d}$, where G is unitary (namely, $GG^\dagger = G^\dagger G = \mathbb{I}$, and $\mathbb{I}$ is an identity matrix of dimension d (the siglum U is also used to highlight the unitary nature of the operator). Figure 3.2 provides an example of a unitary gate. Applying a quantum gate G to $|\psi\rangle$ generates the state $|\psi'\rangle$—that is, $|\psi'\rangle = G|\psi\rangle$. Figure 3.3 provides a simple example of a gate; notice that although the quantum logic rotation gate is shown as a single matrix, it can be decomposed into a plurality of matrices of computation. For example, the rotation quantum logic gate can be decomposed into the quantum logic gates depicted at the bottom of Figure 3.3; each of the R_w gates shown is a computation that results in the one-quantum bit rotation gate when executed in succession, as shown in the figure (subdividing an arbitrary computation by creating a circuit that entails a combination of more basic functionality provides practical improvements).

Common quantum gates include the single-qubit Pauli operators and others (see Figure 3.4). Applying gates $G_1, \ldots , G_n$ to n individual qubits results in the application of the overall

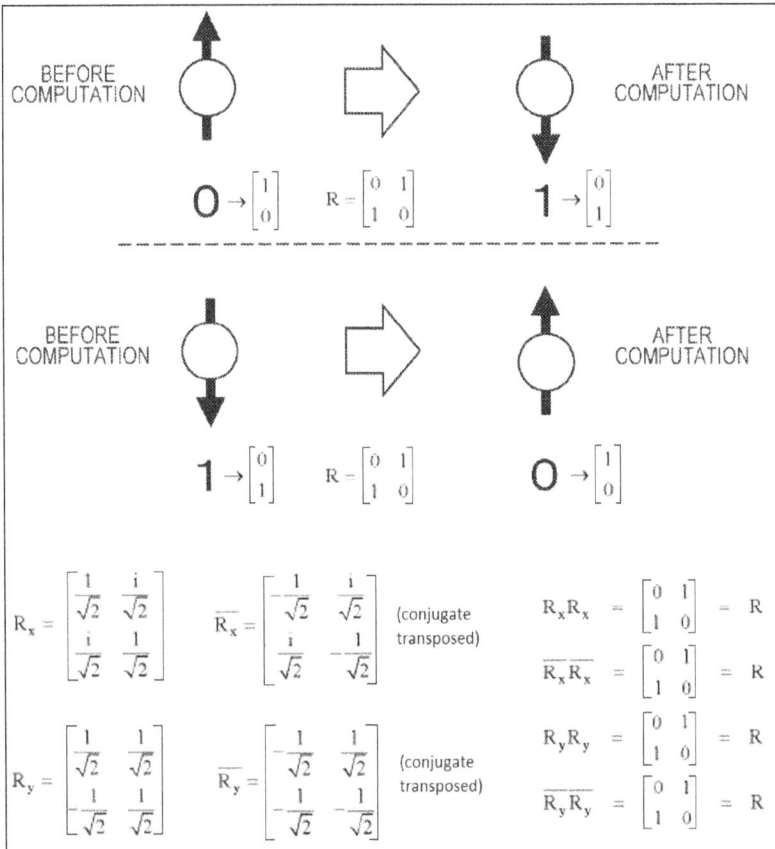

Figure 3.3   Example of the action of a quantum gate

operation $G = G_1 \otimes \cdots \otimes G_n$. Quacomp and Quanet applications are realized by applying a series of quantum gates to one or several qubits and then performing a measurement of the qubits to read out information in the quantum states [VAR22]. Table 3.2 identifies some common quantum gates, while Figure 3.4 depicts some symbology and equivalence relations; see also Figure 3.5 (on page 116).

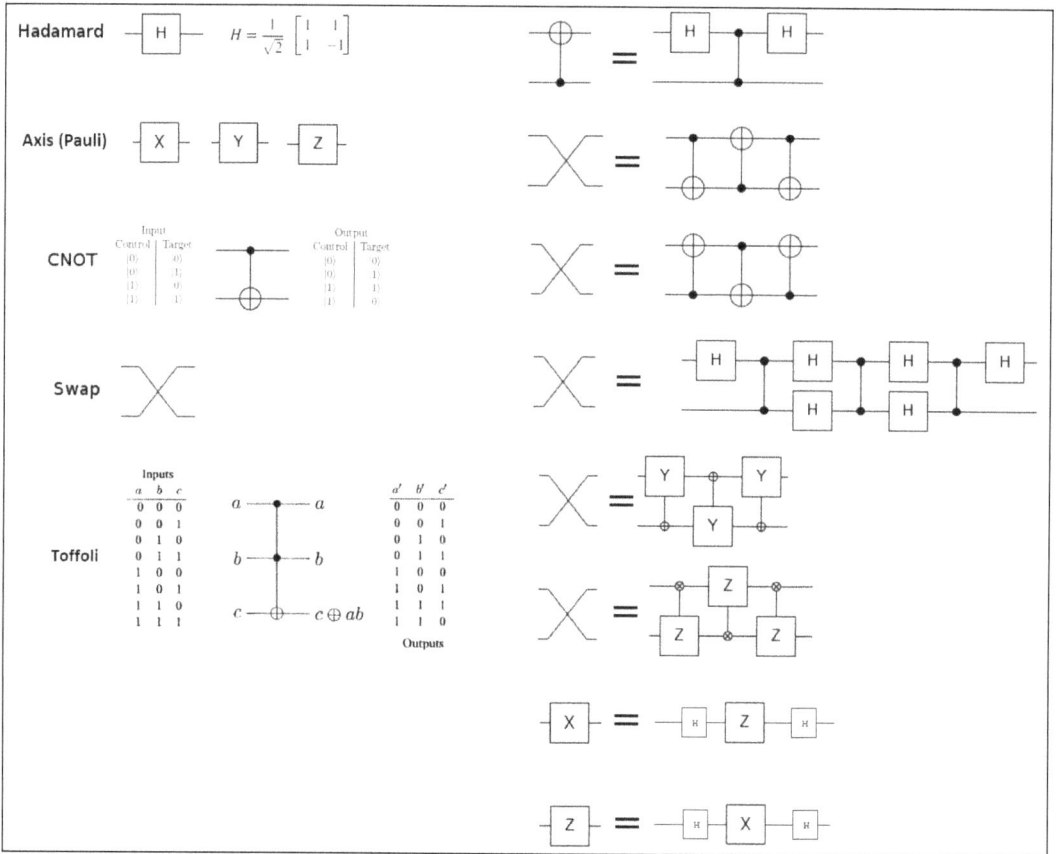

Figure 3.4 Common gates (left) and combined operations (right)

In order to manipulate a qubit $|\psi\rangle = \alpha|0\rangle + \beta|1\rangle$, one must manipulate its state with a unitary/gate; this is achieved by acting on $|\psi\rangle$ with unitary operators (that is, gates) such that $U(|\psi\rangle) = U(\alpha|0\rangle + \beta|1\rangle) = \alpha'|0\rangle + \beta'|1\rangle$; in particular, U is a given 2×2 unitary matrix. The Hadamard gate and the Controlled-Not (CNOT) are two basic gates. For example, Hadamard gate (H)

$$H = \frac{1}{\sqrt{2}} \begin{bmatrix} 1 & 1 \\ 1 & -1 \end{bmatrix}$$

acts on the basis such that $H(|0\rangle) = H|0\rangle = |+\rangle$ and $H(|1\rangle) = H|1\rangle = |-\rangle$, with $|+\rangle := (|0\rangle + |1\rangle)/\sqrt{2}$ and $|-\rangle := (|0\rangle - |1\rangle)/\sqrt{2}$. Hadamard gate corresponds to a reflection of a standard $|0\rangle \times |1\rangle$ plane

(text continues on page 117)

Table 3.2 Some Common Gates

Gate	Qubits	Description
Hadamard Gate (H)	1-qubit gate	$H = \dfrac{1}{\sqrt{2}} \begin{bmatrix} 1 & 1 \\ 1 & -1 \end{bmatrix}$ $$H\lvert 0\rangle = \frac{1}{\sqrt{2}} \begin{bmatrix} 1 & 1 \\ 1 & -1 \end{bmatrix}\begin{bmatrix} 1 \\ 0 \end{bmatrix} = \frac{1}{\sqrt{2}}\begin{bmatrix} 1 \\ 1 \end{bmatrix} = \frac{1}{\sqrt{2}}\left(\begin{bmatrix} 1 \\ 0 \end{bmatrix} + \begin{bmatrix} 0 \\ 1 \end{bmatrix}\right)$$ $$= \frac{1}{\sqrt{2}}\,(\lvert 0\rangle + \lvert 1\rangle)$$ $$= \lvert + \rangle \quad \text{(by definition)}$$ Similarly, $$H\lvert 1\rangle = \frac{1}{\sqrt{2}}\,(\lvert 0\rangle - \lvert 1\rangle) = \lvert - \rangle \ \text{(by definition)}$$ $H^{\dagger} = H$, since H is real and symmetric. $H^2 = \mathbb{I}$. Hadamard gates are a single-qubit operation that rotates 180° around the diagonal $X+Z$ axis, transitioning the $\lvert 0\rangle$ state to the state $(1/\sqrt{2})\lvert 0\rangle + (1/\sqrt{2})\lvert 1\rangle$ and the $\lvert 1\rangle$ state to the state $(1/\sqrt{2})\lvert 0\rangle - (1/\sqrt{2})\lvert 1\rangle$. In the complex plane H can be visualized as a reflection around $\pi/8$. On the Bloch sphere H can also be visualized as a rotation of π about the axis (the $X+Z$ axis of the sphere) (see Figure 1.2 in Chapter 1).
Rotation gate (by θ)	1-qubit gate	$R = \begin{bmatrix} \cos\theta & -\sin\theta \\ \sin\theta & \cos\theta \end{bmatrix}$
General rotation gate: U(3) (also represented as U_3)	1-qubit gate	U(3) is a universal gate: it can perform any single-qubit rotation. It has three parameters: theta (θ), phi (ϕ), and lambda (λ); each parameter determines a specific rotation that the gate performs on a quantum bit (qubit) [DAS23]: Theta (θ): The angle of rotation around the X-axis. It determines how much the quantum state is rotated towards or away from the X-axis. A value of $\theta = 0$ implies no rotation, and $\theta = \pi$ implies a complete flip around the X-axis. Phi (ϕ): The angle of rotation around the Z-axis. It determines how much the quantum state is rotated clockwise or counterclockwise around the Z-axis. A value of $\theta = \pi$ implies no rotation, and $\phi = \pi$ implies a complete flip around the Z-axis. Lambda (λ): The angle of rotation around the Y-axis. It determines how much the quantum state is rotated towards or away from the Y-axis. A value of $\lambda = 0$ implies no rotation, and $\lambda = \pi$ implies a complete flip around the Y-axis. $$\mathrm{U3}(\theta,\ \phi,\ \lambda) = \begin{bmatrix} \cos(\theta/2) & -\exp(i\lambda)\sin(\theta/2) \\ \exp(i\phi)\sin(\theta/2) & \exp(i(\phi+\lambda))\cos(\theta/2) \end{bmatrix}$$
NOT Gate, aka bit flip gate, or X (Pauli X)	1-qubit gate	$NOT = \begin{bmatrix} 0 & 1 \\ 1 & 0 \end{bmatrix}$ Flips a bit from 0 to 1 and vice versa. An X rotation conjugated by H gates is a Z rotation: $X = HZH$. Conjugation of X by H means pre-multiplying X by H^{-1} and post-multiplying it by H. However, note that $H = H^{-1}$.

Gate	Qubits	Description
Pauli Y	1-qubit gate	$$Y = \begin{bmatrix} 0 & -i \\ i & 0 \end{bmatrix}$$
Phase flip, aka Z (Pauli Z)	1-qubit gate	$$Z = \begin{bmatrix} 1 & 0 \\ 0 & -1 \end{bmatrix}$$ $Z\lvert + \rangle = \lvert - \rangle$ and $Z\lvert - \rangle = \lvert + \rangle$ A Z rotation conjugated by H gates is an X rotation: $Z = HXH$. Phase flips and bit flips are related via conjugation.
General phase gate	1-qubit gate	$$R_z(\delta) = \begin{bmatrix} 1 & 0 \\ 0 & e^{i\delta} \end{bmatrix}$$ Note $Z = R_z(\pi)$ $$S = R_z(\pi/2) = \begin{bmatrix} 1 & 0 \\ 0 & i \end{bmatrix} = TT$$ $$T = R_z(\pi/4) = \begin{bmatrix} 1 & 0 \\ 0 & e^{i\pi/4} \end{bmatrix}$$ T is also known as $\pi/8$ gate.
Unitary operation gates	1-qubit gate	Can be constructed via a combination of X, Z, S, T, $R_z(d)$
Spin operator gates	1-qubit gate	The spin operators for the three components of spin (for a spin-1/2 particle) are $\hat{S}_x, \hat{S}_y, \hat{S}_z$, where $$\hat{S}_x = \frac{\hbar}{2}\begin{bmatrix} 0 & 1 \\ 1 & 0 \end{bmatrix} = \frac{\hbar}{2}\sigma_x$$ $$\hat{S}_y = \frac{\hbar}{2}\begin{bmatrix} 0 & -i \\ i & 0 \end{bmatrix} = \frac{\hbar}{2}\sigma_y$$ $$\hat{S}_z = \frac{\hbar}{2}\begin{bmatrix} 1 & 0 \\ 0 & -1 \end{bmatrix} = \frac{\hbar}{2}\sigma_z$$ and where $\sigma_x, \sigma_y, \sigma_z$ are the Pauli operator (matrices) $$\sigma_x = \begin{bmatrix} 0 & 1 \\ 1 & 0 \end{bmatrix} \quad \sigma_y = \begin{bmatrix} 0 & -i \\ i & 0 \end{bmatrix} \quad \sigma_z = \begin{bmatrix} 1 & 0 \\ 0 & -1 \end{bmatrix}$$ Clearly, $\hat{S}_x = (\hbar/2)\sigma_x$, $\hat{S}_y = (\hbar/2)\sigma_y$, $\hat{S}_z = (\hbar/2)\sigma_z$.
Hadamard gate (H)	2-qubit gate	$$H \otimes H = \frac{1}{\sqrt{2}}\begin{bmatrix} 1 & 1 \\ 1 & -1 \end{bmatrix} \otimes \frac{1}{\sqrt{2}}\begin{bmatrix} 1 & 1 \\ 1 & -1 \end{bmatrix} = \frac{1}{2}\begin{bmatrix} 1\begin{bmatrix}1 & 1 \\ 1 & -1\end{bmatrix} & 1\begin{bmatrix}1 & 1 \\ 1 & -1\end{bmatrix} \\ 1\begin{bmatrix}1 & 1 \\ 1 & -1\end{bmatrix} & -1\begin{bmatrix}1 & 1 \\ 1 & -1\end{bmatrix} \end{bmatrix}$$ $$= \frac{1}{2}\begin{bmatrix} 1 & 1 & 1 & 1 \\ 1 & -1 & 1 & -1 \\ 1 & 1 & -1 & -1 \\ 1 & -1 & -1 & 1 \end{bmatrix}$$ $$H \otimes H\lvert 00 \rangle = \frac{1}{2}\begin{bmatrix} 1 & 1 & 1 & 1 \\ 1 & -1 & 1 & -1 \\ 1 & 1 & -1 & -1 \\ 1 & -1 & -1 & 1 \end{bmatrix}\begin{bmatrix} 1 \\ 0 \\ 0 \\ 0 \end{bmatrix} = \frac{1}{2}\begin{bmatrix} 1 \\ 1 \\ 1 \\ 1 \end{bmatrix} = \frac{\lvert 00 \rangle + \lvert 01 \rangle + \lvert 10 \rangle + \lvert 11 \rangle}{2}$$ *(continues)*

Gate	Qubits	Description
Hadamard gate (H) (cont.)	2-qubit gate	and so on. Or, $$H\lvert 0\rangle = \frac{1}{\sqrt{2}}(\lvert 0\rangle + \lvert 1\rangle) \quad H\lvert 1\rangle = \frac{1}{\sqrt{2}}(\lvert 0\rangle - \lvert 1\rangle)$$ $$H \otimes H\lvert 00\rangle = H\lvert 0\rangle H\lvert 0\rangle = \frac{1}{\sqrt{2}}(\lvert 0\rangle + \lvert 1\rangle)\frac{1}{\sqrt{2}}(\lvert 0\rangle + \lvert 1\rangle)$$ $$= \frac{1}{2}(\lvert 0\rangle\lvert 0\rangle + \lvert 0\rangle\lvert 1\rangle + \lvert 1\rangle\lvert 0\rangle + \lvert 1\rangle\lvert 1\rangle)$$ $$= \frac{1}{2}(\lvert 0\ 0\rangle + \lvert 0\ 1\rangle + \lvert 1\ 0\rangle + \lvert 1\ 1\rangle)$$ $$= \frac{1}{2}(\lvert 00\rangle + \lvert 01\rangle + \lvert 10\rangle + \lvert 11\rangle)$$
Tensored gates (local gates)	2-qubit gate	Any 1-qubit gate can be tensored with itself or another gate to make a 2-qubit gate—e.g., $H \otimes H$. (However, tensor products of 1-qubit gates do not have the ability to generate entanglement.)
Controlled NOT (CNOT) gate	2-qubit gate	$$CNOT = \begin{bmatrix} 1 & 0 & 0 & 0 \\ 0 & 1 & 0 & 0 \\ 0 & 0 & 0 & 1 \\ 0 & 0 & 1 & 0 \end{bmatrix} \quad CNOT^{-1} = CNOT \quad (CNOT)^2 = \mathbb{I}$$ The first bit is the "control bit;" the second is the "target bit." When the control bit is state 1, then target bit flips—namely 00 → 00; 01 → 01; 10 → 11; 11 → 10)
SWAP states gate	2-qubit gate	$$SWAP = \begin{bmatrix} 1 & 0 & 0 & 0 \\ 0 & 0 & 1 & 0 \\ 0 & 1 & 0 & 0 \\ 0 & 0 & 0 & 1 \end{bmatrix}$$ $$SWAP(\lvert \phi_1\rangle \otimes \lvert \phi_2\rangle) = \lvert \phi_2\rangle \otimes \lvert \phi_1\rangle$$ SWAP = CNOT x CNOT x CNOT Example[a]: $\lvert 00\rangle$ swaps to $\lvert 00\rangle$, $\lvert 01\rangle$ swaps to $\lvert 10\rangle$, $\lvert 10\rangle$ swaps to $\lvert 01\rangle$, $\lvert 11\rangle$ swaps to $\lvert 11\rangle$
Toffoli gate	3-qubit gates	The third, target, qubit is flipped when both the first and second qubits are in state 1. This is a 3-qubit generalization of the CNOT gate.
Hadamard Gate (H)	n-qubit gate	$$H^{\otimes n}\lvert 00.....0_n\rangle = \frac{1}{\sqrt{2^n}}\sum_{x=0}^{2^n-1}\lvert x\rangle$$
Tensored gates (local gates)	n-qubit gates	n-qubit gates formed as tensor products of one-qubit gates, e.g., $H^{\otimes n}$

[a] As an example, [GID17] provides the following observations about SWAP = CNOT x CNOT x CNOT:

$\lvert 00\rangle$: Desired output is $\lvert 00\rangle$
    The first CNOT's control is **not** satisfied. We stay in the state $\lvert 00\rangle$.
    The middle CNOT's control is **not** satisfied. We stay in the state $\lvert 00\rangle$.
    The final CNOT's control is **not** satisfied. We stay in the state $\lvert 00\rangle$.

$\lvert 01\rangle$: Desired output is $\lvert 10\rangle$.
    The first CNOT's control is **not** satisfied. We stay in the state $\lvert 10\rangle$.
    The middle CNOT's control is satisfied. The bottom qubit gets toggled. We transition to the state $\lvert 11\rangle$.
    The final CNOT's control is satisfied. The top qubit gets toggled. We transition to the state $\lvert 10\rangle$.

(footnote continues on next page)

(footnote continued from previous page)

$|10\rangle$: Desired output is $|01\rangle$.

The first CNOT's control is satisfied. The top qubit gets toggled. We transition to the state $|11\rangle$.
The middle CNOT's control is satisfied. The bottom qubit gets toggled. We transition to the state $|01\rangle$.
The final CNOT's control is **not** satisfied. We stay in the state $|01\rangle$.

$|11\rangle$: Desired output is $|11\rangle$.

The first CNOT's control is satisfied. The top qubit gets toggled. We transition to the state $|10\rangle$.
The middle CNOT's control is **not** satisfied. We stay in the state $|10\rangle$.
The final CNOT's control is satisfied. The first qubit gets toggled. We return to the state $|11\rangle$.

More generally, one can apply the SWAP = CNOT x CNOT x CNOT rule to say $v_1 = a|0\rangle + b|0\rangle$ and $v_2 = c|0\rangle + d|1\rangle$; then the convolution of two qubits $v_1 \otimes v_2$ is

$$|\psi\rangle = \alpha_{00}|00\rangle + \alpha_{01}|01\rangle + \alpha_{10}|10\rangle + \alpha_{11}|11\rangle$$

where $|00\rangle = [1,0,0,0]^T$, $|01\rangle = [0,1,0,0]^T$, $|10\rangle = [0,0,1,0]^T$, $|11\rangle = [0,0,0,1]^T$, with $\alpha_1 = ac$, $\alpha_2 = ad$, $\alpha_3 = bc$, $\alpha_2 = bd$. Note that clearly $|\psi\rangle \in \mathbf{C}^4$

$$|\psi\rangle = ac\begin{bmatrix}1\\0\\0\\0\end{bmatrix} + ad\begin{bmatrix}0\\1\\0\\0\end{bmatrix} + bc\begin{bmatrix}0\\0\\1\\0\end{bmatrix} + bd\begin{bmatrix}0\\0\\0\\1\end{bmatrix} = \begin{bmatrix}ac\\ad\\bc\\bd\end{bmatrix}$$

If one started with the two qubits in state $|y_2y_1\rangle$, after the three CNOTs the qubits are in state $|y_1y_2\rangle$.

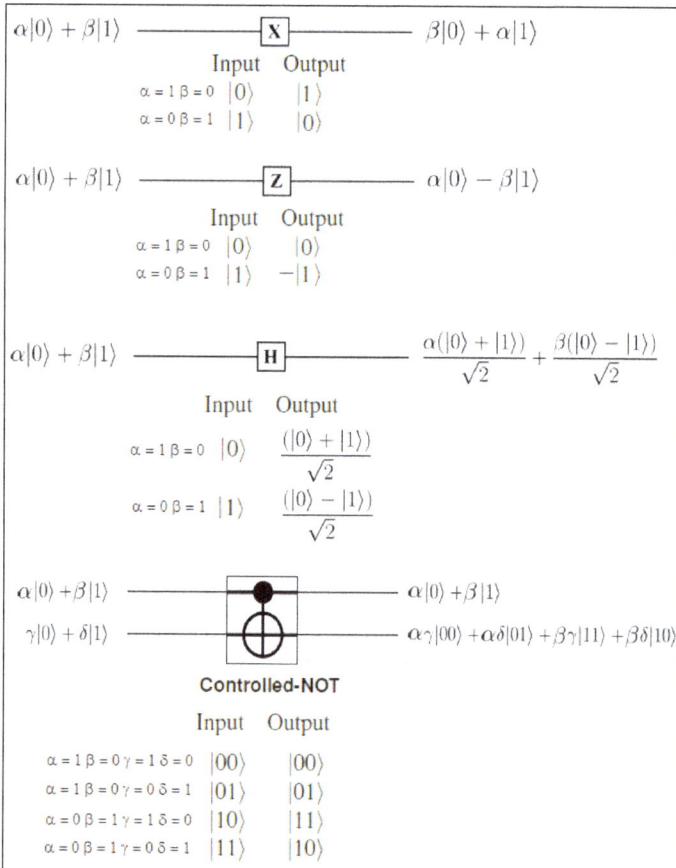

Input	Output
$\alpha = 1\,\beta = 0$ $|0\rangle$	$|1\rangle$
$\alpha = 0\,\beta = 1$ $|1\rangle$	$|0\rangle$

Input	Output
$\alpha = 1\,\beta = 0$ $|0\rangle$	$|0\rangle$
$\alpha = 0\,\beta = 1$ $|1\rangle$	$-|1\rangle$

Input	Output
$\alpha = 1\,\beta = 0$ $|0\rangle$	$\frac{(|0\rangle + |1\rangle)}{\sqrt{2}}$
$\alpha = 0\,\beta = 1$ $|1\rangle$	$\frac{(|0\rangle - |1\rangle)}{\sqrt{2}}$

Controlled-NOT

Input	Output
$\alpha = 1\,\beta = 0\,\gamma = 1\,\delta = 0$ $|00\rangle$	$|00\rangle$
$\alpha = 1\,\beta = 0\,\gamma = 0\,\delta = 1$ $|01\rangle$	$|01\rangle$
$\alpha = 0\,\beta = 1\,\gamma = 1\,\delta = 0$ $|10\rangle$	$|11\rangle$
$\alpha = 0\,\beta = 1\,\gamma = 0\,\delta = 1$ $|11\rangle$	$|10\rangle$

Figure 3.5   Another view of the action of common gates

$$|0\rangle = \begin{bmatrix} 1 \\ 0 \end{bmatrix} \quad |1\rangle = \begin{bmatrix} 0 \\ 1 \end{bmatrix}$$

$$|+\rangle = (|0\rangle + |1\rangle)/\sqrt{2} = \begin{bmatrix} 1 \\ 1 \end{bmatrix}/\sqrt{2}$$

$$|-\rangle = (|0\rangle - |1\rangle)/\sqrt{2} = \begin{bmatrix} 1 \\ -1 \end{bmatrix}/\sqrt{2}$$

$$H(|0\rangle) = H|0\rangle =$$

$$\frac{1}{\sqrt{2}} \begin{bmatrix} 1 & 1 \\ 1 & -1 \end{bmatrix} |0\rangle =$$

$$\frac{1}{\sqrt{2}} \begin{bmatrix} 1 & 1 \\ 1 & -1 \end{bmatrix} \begin{bmatrix} 1 \\ 0 \end{bmatrix} = \frac{1}{\sqrt{2}} \begin{bmatrix} 1 \\ 1 \end{bmatrix} = |+\rangle = (|0\rangle + |1\rangle)/\sqrt{2}$$

2x2   2x1   2x1

$$H(|1\rangle) = H|1\rangle =$$

$$\frac{1}{\sqrt{2}} \begin{bmatrix} 1 & 1 \\ 1 & -1 \end{bmatrix} |1\rangle =$$

$$\frac{1}{\sqrt{2}} \begin{bmatrix} 1 & 1 \\ 1 & -1 \end{bmatrix} \begin{bmatrix} 0 \\ 1 \end{bmatrix} = \frac{1}{\sqrt{2}} \begin{bmatrix} 1 \\ -1 \end{bmatrix} = |-\rangle = (|0\rangle - |1\rangle)/\sqrt{2}$$

2x2   2x1   2x1

Figure 3.6   Applying the Hadamard gate

around the line $\theta = \pi/8$. This reflection maps the x-axis to the 45° line, and the y-axis to the −45° line. See Figure 3.6 (and Figure 3.17 on page 132).

Make note of the graphical symbology shown in Figure 3.4 for this gate in a quantum circuit.

Starting with $|\psi\rangle$ being in the state of either $|0\rangle$ or $|1\rangle$, when measured (that is when H is applied), $H(|\psi\rangle)$ will result in 0 and 1, thus there is no longer any distinguishing information in the bit since the information has transferred to the phase (in the Hadamard basis).

Various types of 2-qubit computations are known. The 2-qubit CNOT gate achieves an EXOR (exclusive OR logic function[1]) of the first qubit into the second qubit; the unitary 4×4 matrix, the CNOT gate is

$$G = CNOT = \begin{bmatrix} 1 & 0 & 0 & 0 \\ 0 & 1 & 0 & 0 \\ 0 & 0 & 0 & 1 \\ 0 & 0 & 1 & 0 \end{bmatrix}$$

CNOT supports the following mapping: $|a,b\rangle \rightarrow |a, a \otimes b\rangle$; specifically, $|00\rangle$ is mapped to $|00\rangle$; $|10\rangle$ is mapped to $|11\rangle$; $|01\rangle$ is mapped to $|01\rangle$; and $|11\rangle$ is mapped to $|10\rangle$. The CNOT graphical symbology for this gate was shown in Figure 3.4 (page 112): the lower wire acts as the Target Bit (TB) and upper wire acts as the Control Bit (CB).

Another example of a two-quantum bit logic gate is the exchange logic gate (SWAP gate): it exchanges the states of two qubits. See Figure 3.7, inspired by [SHI23]. The figure illustrates

---

1   Classically the Exclusive OR Gate, also known as EX OR Gate or XOR Gate, acts as follows: the output is High if and only if one of the inputs is High. If both the inputs are Low or High, then the output is Low.

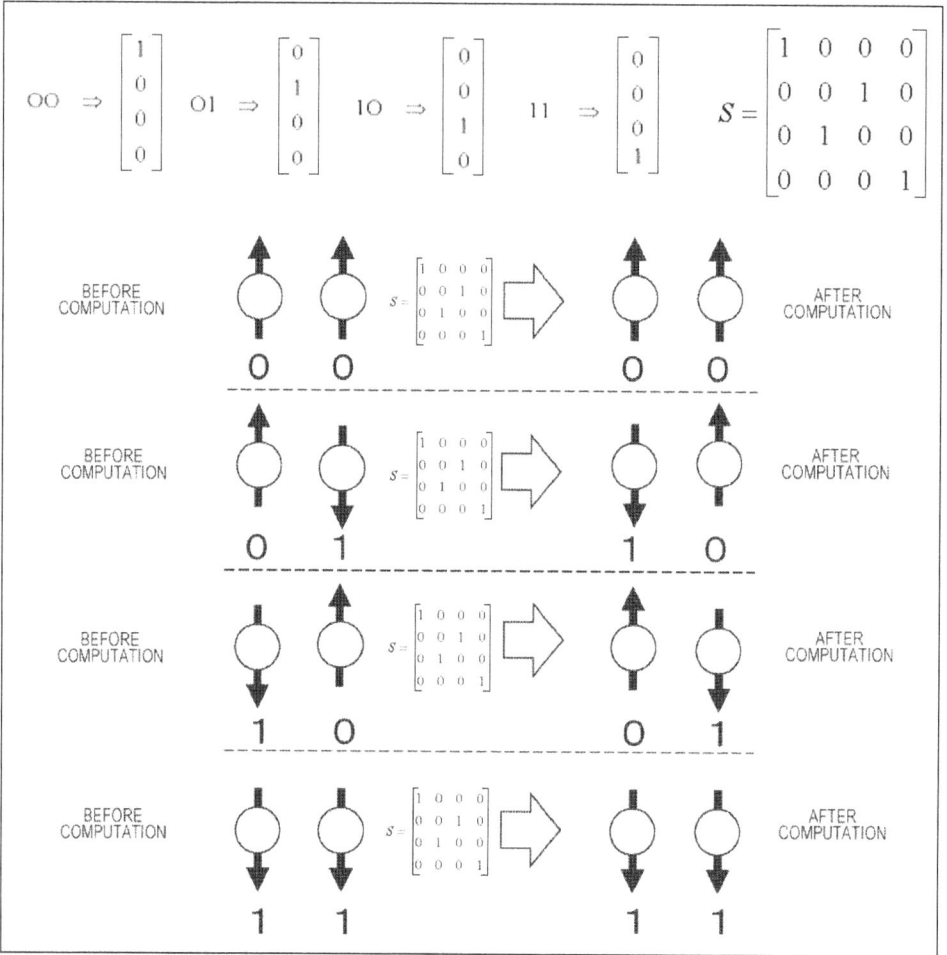

Figure 3.7   Swap operator/operation (inspired by [SHI23])

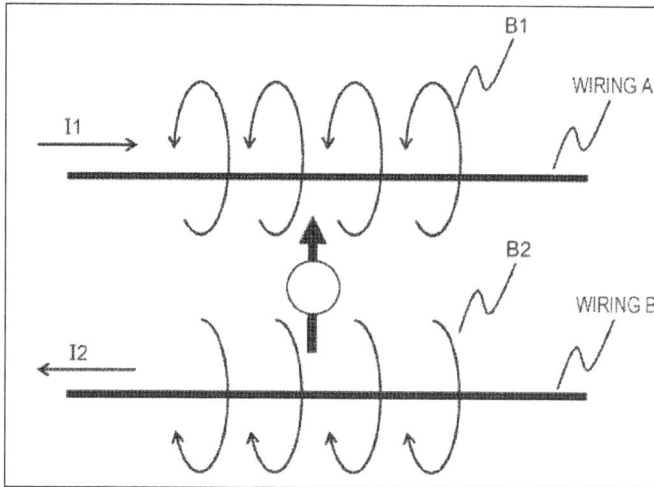

Figure 3.8   Simplified magnetic field approach to alter the spin state of a qubit [SHI23]

the states of electrons and the states of qubits before and after a computation in the SWAP gate. The SWAP gate is a quantum logic gate acting on two qubits act and exchanges the states of the two qubits. The SWAP gate simply exchanges the state and does not exchange two electrons.

Figure 3.8 illustrates the principle of applying a magnetic field to a qubit by controlling a current; when there are two wirings, A and B, on both sides of the qubit, reverse currents I1 and I2 can flow through the respective wirings. In this way, the magnetic fields B1 and B2 generated by the currents are strengthened in the region sandwiched between the two wirings, and the magnetic field is weakened in other regions; this enables application of a selective magnetic field to the qubit in the drawing and alters its spin state as needed.

Yet another example is a control phase rotating gate (CZ gate). See Figure 3.9, which illustrates qubit states before and after the action of the selective quantum logic gate. This quantum logic gate changes the quantum phase of a right qubit TB—referred to as a *first qubit* (target bit)—depending on the state of a left qubit CB—referred to as a *second qubit* (control bit). Only when the second qubit is 1, a phase rotation gate (Z gate) is caused to act on the first quantum bit.

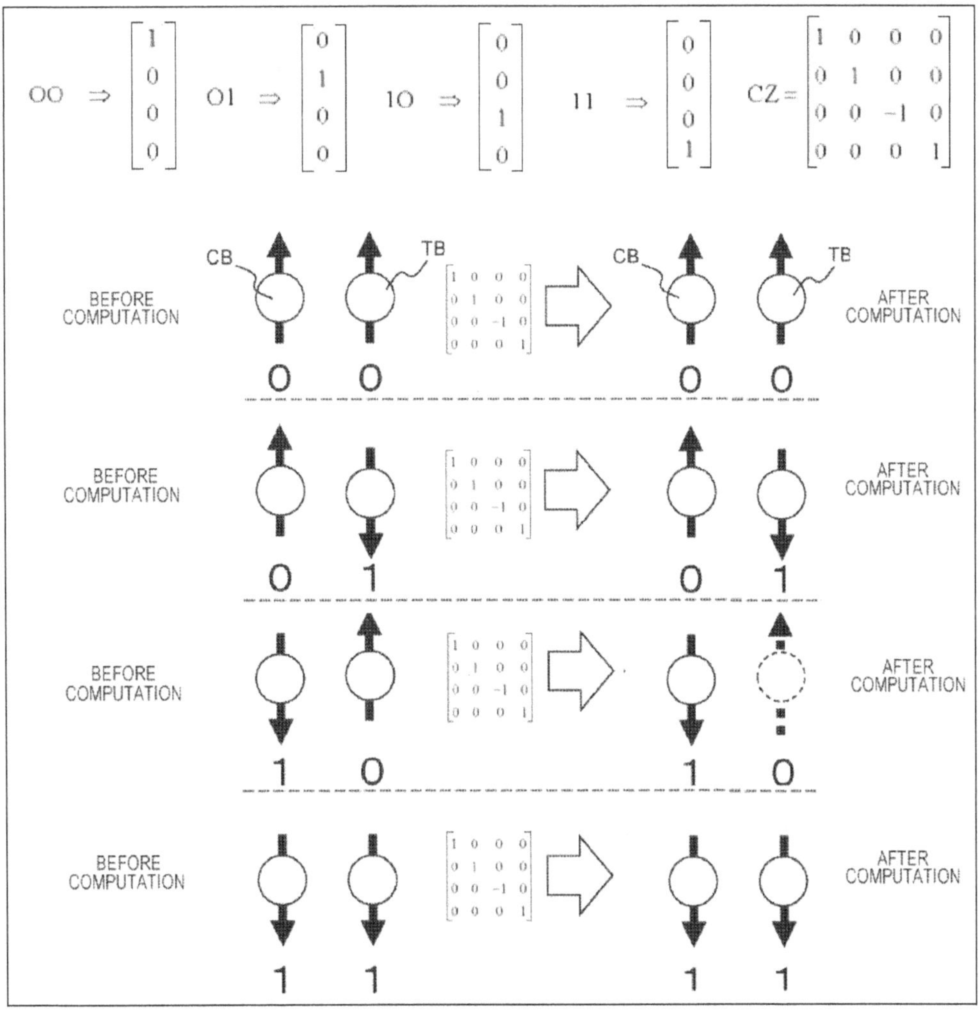

Figure 3.9 Phase-rotating CZ operator/operation (inspired by [SHI23])

Figure 3.10 illustrates a method for realizing the quantum logic gate of Figure 3.9. A-1 as a gate procedure diagram illustrates a CZ gate; A-2 is a procedure for actually realizing the CZ gate; thus, the quantum logic gate of A-1 can be effectively realized by combining and using the quantum logic gates of the 1-quantum bit and the 2-quantum bit described above.

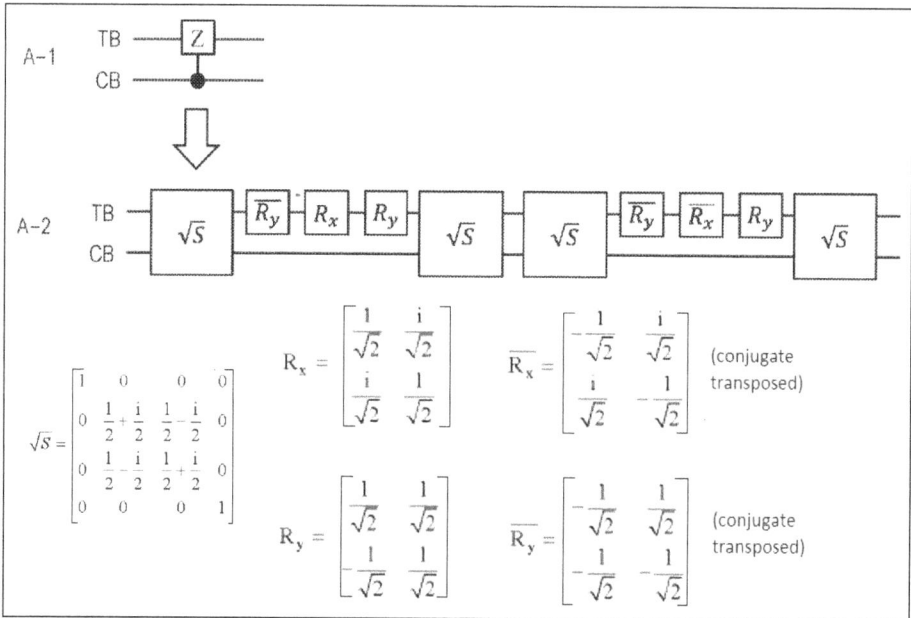

Figure 3.10   Method for realizing the CZ quantum logic gate

3.5 Noisy Quantum States

We focus next on noisy quantum states and operations. Noise has an effect on a quantum state: qubits are susceptible to noise signals in the environment that can alter their state. The impact of environmental interaction on quantum states over time is referred to as *decoherence*.

For example, consider a probabilistic process that aims at preparing a desired state $|0\rangle \langle 0|$ with probability $1-p$, but fails in some manner and prepares instead the state $|1\rangle \langle 1|$ with probability p. This results in a noisy quantum state $\rho = (1-p)^*|0\rangle \langle 0| + p^*|1\rangle \langle 1|$. Among other situations, noise can arise due to incomplete shielding of qubits from external factors; another example could be noise that occurs because implementations of quantum gates that have some design or implementation deficiency; yet in another example some noise process degrades a quantum state ρ_0 that is placed into a quantum memory at time $t = t_0$ into a noisy quantum state ρ_1 at time $t = t_1$.

The density matrix ρ representation of quantum states mentioned above allows for the expression of noisy quantum states. Let S be the set of (matrices of) all quantum states on a d-dimensional quantum system (including noisy states); this corresponds to the set of matrices $S = \{\rho \in \mathbf{C}^{d \times d}, \rho > 0$ with $\mathrm{tr}(\rho) = 1\}$ where $\mathrm{tr}(.)$ is the trace function. Then, formally, the set of all possible noise processes corresponds to the set of Completely Positive Trace Preserving Maps (CPTPM) Λ, with $\Lambda: S \to S$. The effect of environmental interaction is modeled utilizing noise models that describe the mapping Λ.

An example of a noise model is $\Lambda = \mathcal{D}$ being depolarizing noise [VAR22],

$$\mathcal{D}_t(\rho) = (1 - 3p)\rho + pX\rho X + pY\rho Y + pZ\rho Z = (1 - 4p)\rho + 4p\frac{\mathbb{I}}{2},$$

which drives a quantum state towards the maximally noisy state—the quantum equivalent of white noise (the depolarizing noise model is often used as a worst-case estimate, when the physical noise process is not accurately characterized). The impact of time t as it relates to the degradation of the quality of quantum state stored in a quantum memory for a given interval T of expected operation ($t \leq T$) is often taken using $p = \frac{1}{4}(1 - e^{-\frac{t}{T}})$. For some specific implementations of quantum memories, additional information about the noise process of the quantum memory device is available, providing an improved view of the impact of noise.[1]

Noisy quantum gates can impact overall performance. Using the density matrix ρ representation of quantum states, the impact of applying a gate G on a quantum state can be expressed as $G(\rho) = G\rho G^{\dagger}$. The effect of a noisy quantum gate G needs to be considered by modeling a noisy implementation $\mathcal{E}$ as being the ideal gate, followed by time-dependent potential noise $\mathcal{N}_t$, with $\mathcal{E} = \mathcal{N}_t \circ G$, where G is the ideal implementation of the gate.

3.6 Quantum Quality/Fidelity Measure

Distributed quantum applications require quality assurance on the quantum states that they utilize because when such applications are executed on systems or architectures with limitations, such as imperfect gates, the overall performance is greatly impacted. *Fidelity*, briefly introduced in Chapter 2, is an important quantity used to assess how a fixed noise process (e.g., during the preparation of the state) affects its quality. Fidelity is also used to assess how the quality of an already prepared state degrades as a function of a waiting time t that this state spends in a quantum memory.

Thus, fidelity characterizes the quality of a quantum state, describing how close a quantum state is to the state one (say, the sender) tried to create. Several quantum applications (algorithms) do not need a fidelity of one to be able to operate meaningfully, but rather require the fidelity to be above a given threshold—otherwise, the applications (algorithms) just operate at lower rates. Higher fidelity can be achieved in two ways: (i) using quantum hardware that produces states of better fidelity (typically trading off computing rate for higher fidelity); or (ii) utilizing quantum error detection and correction mechanisms.

---

[1] [VAR22] notes that noise can be modeled as *dephasing* and *damping* noise (or a combination of both). *Dephasing* noise is expressed as $\mathcal{P}_t(\rho) = (1 - p)\rho + pZ\rho Z$, again with $p = \frac{1}{4}(1 - e^{-\frac{t}{T}})$ being used to capture time-dependence, for a fixed T characterizing the memory. *Damping* noise is expressed as $\mathcal{A}_t(\rho) = M_0\rho M_0^{\dagger} + M_1\rho M_1^{\dagger}$ where

$$M_0 = \begin{bmatrix} 1 & 0 \\ 0 & \sqrt{1-\gamma} \end{bmatrix}, \quad M_1 = \begin{bmatrix} 0 & \sqrt{\gamma} \\ 0 & 0 \end{bmatrix}, \text{ and } \gamma = (1 - e^{-\frac{t}{T_1}}) \text{ for a fixed } T_1$$

(this models a noisy channel of one-sided error which preserves $|0\rangle\langle 0|$ but damps $|1\rangle\langle 1|$ to $|0\rangle\langle 0|$ with an error probability γ). In most physical implementations of quantum devices, both $\mathcal{P}$ and $\mathcal{A}$ occur, and the noise is described by a composite model $\mathcal{C}_t(\rho) = \mathcal{P}_t(\mathcal{A}_t(\rho)) = (1 - p)\mathcal{A}_t(\rho) + pZ\mathcal{A}_t(\rho)Z$.

More precisely, the *fidelity F of a quantum state* ρ, $F(\rho, |\psi\rangle)$, measures how well this state ρ approximates a specific target state $|\psi\rangle$. F lies in the interval $[0, 1]$, and larger values of F indicate that ρ is closer to the target state $|\psi\rangle$ and is defined as

$$F(\rho, |\psi\rangle) = \langle\psi|\rho|\psi\rangle$$

where $F = 1$ only when ρ is identical to the target state $|\psi\rangle$ (note that $\langle.\rangle$ is the inner product). F is also computed as ([SHE22])

$$|\psi\rangle) = \left(\text{tr}\sqrt{\sqrt{\rho}\,|\psi\rangle\sqrt{\rho}}\right)^2$$

and this simplifies to $F(\rho, |\psi\rangle) = \text{tr}\,(\rho|\psi\rangle)$ when either ρ or $|\psi\rangle$ is a pure state.

Gate

The average gate fidelity $F_{\text{orig}}(\mathcal{E}, G)$ measures how well a pragmatic (a real-world) implementation $\mathcal{E}$ approximates a desired target gate G—namely, how well the implementation approximates the target gate when applied to a specific input state $|\psi\rangle$, averaged over all possible input states.

Entanglement fidelity

We discussed entanglement in Chapter 2. We revisit the topic here for completeness before defining entanglement fidelity. A bipartite quantum system composed of states $|\phi_A\rangle \in H_A$ and $|\zeta_B\rangle \in H_B$, where

$$|\phi_{A,1}\rangle = \alpha_{A,1}|0\rangle_A + \beta_{A,1}|1\rangle_A \text{ and } |\phi_{A,2}\rangle = \alpha_{A,2}|0\rangle_A + \beta_{A,2}|1\rangle_A \ (|0\rangle_A \text{ and } |1\rangle_A \text{ being a basis})$$

$$|\zeta_{B,1}\rangle = \gamma_{B,1}|0\rangle_B + \delta_{B,1}|1\rangle_B \text{ and } |\zeta_{B,2}\rangle = \gamma_{B,2}|0\rangle_B + \delta_{B,2}|1\rangle_B \ (|0\rangle_B \text{ and } |1\rangle_B \text{ being a basis})$$

has states with basis $|0\rangle_A|0\rangle_B = |00\rangle$, $|0\rangle_A|1\rangle_B = |01\rangle$, $|1\rangle_A|0\rangle_B = |10\rangle$, $|1\rangle_A|1\rangle_B = |11\rangle$ and is expressed as

$$|\psi_{AB}\rangle = |\phi_A\rangle\,|\zeta_B\rangle \in H_{AB} = H_A \otimes H_B,$$

$$|\psi\rangle_{AB} = \sum_{x,y} q_{x,y}\left(|x\rangle_A \otimes |y\rangle_B\right)$$

where $q_{x,y}$ are the appropriate parameters from the outer (tensor) product. These quantum states are called *separable*, being that the composite system is (by definition) an outer (tensor) product of quantum states, each belonging to a separate Hilbert space. However, by the superposition principle, which states that any linear combination of valid quantum states is also a valid quantum state, the composite space also contains superpositions of (said) separable quantum states. For example, the two-qubit composite quantum state

$$(1/\sqrt{2})(|0\rangle_A|0\rangle_B + |1\rangle_A|1\rangle_B) = (1/\sqrt{2})(|00\rangle + |11\rangle)$$

cannot be expressed as a product of two 1-qubit quantum states. In other words, each qubit in the composite system cannot be described independently from one another. This property is

Consider two quantum systems A and B that interact with each other. Let H_A and H_B be the Hilbert spaces as the state spaces of these systems respectively, and let $|x\rangle_A$ and $|y\rangle_B$ be the respective basis, so that states $|\phi\rangle_A$ and $|\zeta\rangle_B$ are described as

$$|\phi\rangle_A = \Sigma\, c_x \times |x\rangle_A$$
$$|\zeta\rangle_B = \Sigma\, d_y \times |y\rangle_B$$

for appropriate c_x's and d_y's.

Then the state space of the combined interacting system is described by the tensor product $H_{AB} = H_A \otimes H_B$ with basis comprised of the set $|x\rangle_A \otimes |y\rangle_B$:

$$|\psi\rangle_{AB} = \sum_{x,y} q_{x:y} \left(|x\rangle_A \otimes |y\rangle_B\right)$$

where $q_{x:y}$ are the appropriate parameters from the tensor product.

The composite state $|\psi\rangle_{AB}$ is said to be *separable* if $q_{x:y} := c_x \times d_y$, in which case $|\psi\rangle_{AB} = |\phi\rangle_A \otimes |\zeta\rangle_B$. If the composite state is not separable, then it is *entangled*.

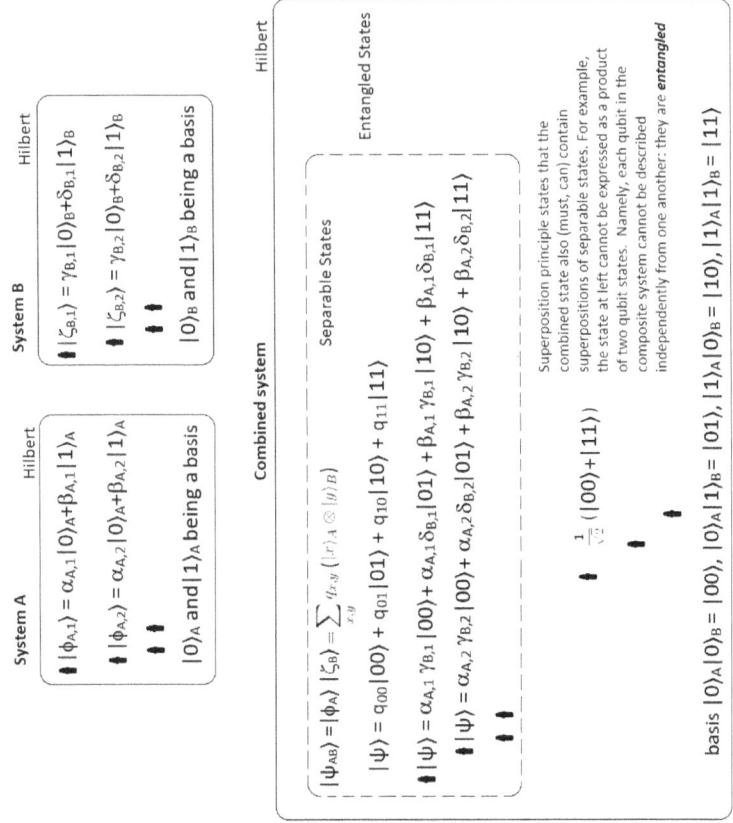

System A Hilbert

➧ $|\phi_{A,1}\rangle = \alpha_{A,1}|0\rangle_A + \beta_{A,1}|1\rangle_A$

➧ $|\phi_{A,2}\rangle = \alpha_{A,2}|0\rangle_A + \beta_{A,2}|1\rangle_A$

$|0\rangle_A$ and $|1\rangle_A$ being a basis

System B Hilbert

➧ $|\zeta_{B,1}\rangle = \gamma_{B,1}|0\rangle_B + \delta_{B,1}|1\rangle_B$

➧ $|\zeta_{B,2}\rangle = \gamma_{B,2}|0\rangle_B + \delta_{B,2}|1\rangle_B$

$|0\rangle_B$ and $|1\rangle_B$ being a basis

Combined system Hilbert

$|\psi_{AB}\rangle = |\phi_A\rangle\, |\zeta_B\rangle = \sum_{x,y} q_{x:y}\left(|x\rangle_A \otimes |y\rangle_B\right)$

Separable States   **Entangled States**

$|\psi\rangle = q_{00}|00\rangle + q_{01}|01\rangle + q_{10}|10\rangle + q_{11}|11\rangle$

➧ $|\psi\rangle = \alpha_{A,1}\gamma_{B,1}|00\rangle + \alpha_{A,1}\delta_{B,1}|01\rangle + \beta_{A,1}\gamma_{B,1}|10\rangle + \beta_{A,1}\delta_{B,1}|11\rangle$

➧ $|\psi\rangle = \alpha_{A,2}\gamma_{B,2}|00\rangle + \alpha_{A,2}\delta_{B,2}|01\rangle + \beta_{A,2}\gamma_{B,2}|10\rangle + \beta_{A,2}\delta_{B,2}|11\rangle$

➧ $\frac{1}{\sqrt{2}}\left(|00\rangle + |11\rangle\right)$

Superposition principle states that the combined state also (must, can) contain superpositions of separable states. For example, the state at left cannot be expressed as a product of two qubit states. Namely, each qubit in the composite system cannot be described independently from one another: they are *entangled*

basis $|0\rangle_A|0\rangle_B = |00\rangle$, $|0\rangle_A|1\rangle_B = |01\rangle$, $|1\rangle_A|0\rangle_B = |10\rangle$, $|1\rangle_A|1\rangle_B = |11\rangle$

Figure 3.11 Separable and entangled states (➧ represents a vector) *[Figure is described on next page]*

known as *entanglement*. The composite state $|\psi\rangle_{AB}$ is said to be *separable* if $q_{x,y} := c_x d_y$; then $|\psi\rangle_{AB} = |\phi\rangle_A \otimes |\zeta\rangle_B$. If the composite state is not separable, then it is *entangled*.[1] See Figure 3.11.

The *entanglement fidelity* $F_e(\mathcal{N}_t)$ measures the quality of an initially maximally entangled state after it was stored in a noisy memory on node B (or A) for time t. For a noise process $\mathbb{I}_A \otimes \mathcal{N}_t$ (no noise on A, and time-dependent noise $\mathcal{N}_t$ on system B), the entanglement fidelity is defined as:

$$F_e(\mathcal{N}_t) = (\langle\psi|\mathbb{I}_A) \otimes \mathcal{N}_t (|\psi\rangle\langle\psi|) |\psi\rangle$$

where $|\psi\rangle$ is the maximally entangled state defined above.

The gate fidelity and the entanglement fidelity are related as

$$F_{\text{orig}}(\mathcal{E}, G) = (d \times F_e(\mathcal{E}) + 1) / (d + 1)$$

where d is the dimension of A and B (for qubits, $d = 2$).

3.7 Photon Number Operator

It is generally accepted that light consists of discrete particles, or photons, each carrying one quantum of energy, and the photon number *operator* described next, gives the number of photons in one mode—say, how many photons with an energy $\hbar\omega$ are in the fundamental mode.[2,3] The *photon number n* (an integer) refers to the number of photons in a photonic quantum state; the mean

---

[1] More generally, a state $|\psi\rangle$ in the composite (Hilbert) space in $\mathbf{C}^n$,

$$\bigotimes_{k=1}^{n} {}^{H_{A_k}}$$

is separable if for all k there exists states

$$|\psi_{A_k}\rangle \in H_{A_k}$$

such that state $|\psi\rangle$ is expressible by the outer product (and appropriate coefficients)—namely,

$$|\psi\rangle = \bigotimes_{k=1}^{n} |\psi_{A_k}\rangle;$$

(otherwise, if not, $|\psi\rangle$ is an entangled state of n qubits). One example of a 'highly' entangled n-qubit state with many applications is the Greenberger-Horne-Zeilinger (GHZ) state ([SHE22])

$$|\psi_{\text{GHZ}}\rangle = \frac{1}{\sqrt{2}}(|0\rangle^{\otimes n} + |1\rangle^{\otimes n})$$

[2] Photons are bosons. All known elementary particles are either bosons (with integer spin, 0, 1, 2, . . .) or fermions (with odd half-integer spin, $\frac{1}{2}$, $\frac{3}{2}$, $\frac{5}{2}$, . . .).

[3] Some of the Physics play a role in these definitions. Schrödinger equation for quantum harmonic oscillator can be written as

$$\frac{1}{2} \left(\hat{a}^\dagger \hat{a} + \hat{a} \hat{a}^\dagger \right) \psi = \left(\hat{a}^\dagger \hat{a} + \frac{1}{2} \right) \psi = \frac{E}{\hbar\omega_0} \psi$$

In Dirac's notation this is

$$\left(\hat{a}^\dagger \hat{a} + \frac{1}{2} \right) |\psi\rangle = \frac{E}{\hbar\omega_0} |\psi\rangle$$

If one denotes a photon number state by $\psi_n(x)$ in explicit notation, ψ_n in math notation or $|\psi_n\rangle$ in Dirac notation, then one the has

photon number is the average number of photons per optical pulse. *Photon number statistics* represent the probability distribution to find a certain number of photons in a field. Fock states such as $|n\rangle$ (more below) are quantum states with well-defined numbers of photons associated with them; these states are eigenstates[1] of the photon number *operator* $\hat{n}$, with $\hat{n} = \hat{a}^\dagger\hat{a}$; the eigenvalue equation reads

$$\hat{n}|n\rangle = |n\rangle n$$

By notational definition $|n\rangle = |\psi_n\rangle$; thus, put slightly differently,

$$\hat{n}|\psi_n\rangle = n|\psi_n\rangle, \text{ or}$$

$$\hat{a}^\dagger\hat{a}\,|\psi_n\rangle = n|\psi_n\rangle,$$

where n is the eigenvalue of $\hat{n}$ and $|\psi_n\rangle$ the corresponding eigenvector. The number operator $\hat{n} = \hat{a}^\dagger\hat{a}$ is an Hermitian operator, therefore its eigenvalues n are real and its eigenvectors $|\psi_n\rangle$ form a complete set of orthogonal states. $|\psi_n\rangle$ is an eigenvector of $\hat{n} = \hat{a}^\dagger\hat{a}$ operator with eigenvalue n. It can be shown that

$$\hat{a}|\psi_n\rangle = \sqrt{n}|\psi_{n-1}\rangle \quad \Leftrightarrow \quad \hat{a}|n\rangle = \sqrt{n}|n-1\rangle$$

$$\hat{a}^\dagger|\psi_n\rangle = \sqrt{n+1}|\psi_{n+1}\rangle \Leftrightarrow \hat{a}^\dagger|n\rangle = \sqrt{n+1}|n+1\rangle$$

$$|n\rangle = \frac{1}{\sqrt{n!}}\left(\hat{a}^\dagger\right)^n|0\rangle$$

The annihilation operator (usually denoted by $\hat{a}$) lowers the number of particles in a given state by one (thus it is also known as a lowering operator). A creation operator (usually denoted $\hat{a}^\dagger$, being the adjoint of the annihilation operator) increases the number of particles in a given state by one (thus it is also known as a raising operator).

3.8 Fock State

Quanets/Quareps involve the notion of Fock states. The quantization of the electromagnetic (EM) field leads to the existence of Fock states $|n\rangle$ alluded to just above. This quantization is expressed by the Fock states $|n\rangle$ ($n = 0, 1, 2, ...$). Individual photons are the fundamental quantum excitation of the EM field. Fock states, or photon-number states, being eigenstates of the number

$$\left(\hat{a}^\dagger\hat{a} + \frac{1}{2}\right)|\psi_n\rangle = \frac{E_n}{\hbar\omega_0}|\psi_n\rangle = \left(n + \frac{1}{2}\right)|\psi_n\rangle$$

with $E_n = (n + \frac{1}{2})\,\hbar\omega0$. By comparing terms, one gets

$$\hat{a}^\dagger\hat{a}\,|\psi_n\rangle = n|\psi_n\rangle$$

$$\hat{n} = \hat{a}^\dagger\hat{a}$$

---

[1] We noted in Chapter 2 that there are (special) wave functions ψ such that when an operator acts on them, the result is just a multiple of the original wave function—these wave functions are called *eigenstates* and the multiples are called *eigenvalues*. Thus, if $\hat{\boldsymbol{n}}\psi = \mathbf{dd}^\dagger\psi = c \times \psi$ where c is a complex number, then ψ is called an eigenstate of $\hat{\boldsymbol{n}} = \mathbf{dd}^\dagger$ corresponding to the eigenvalue c. More pedantically, $\hat{\boldsymbol{n}}|n\rangle = \mathbf{dd}^\dagger|n\rangle = c \times |n\rangle$. The creation operator $\mathbf{d}^\dagger$ has this property: $\mathbf{d}^\dagger|n\rangle = (\sqrt{(n+1)})|n+1\rangle$; the annihilation operator $\mathbf{d}$ has this property: $\mathbf{d}|n\rangle = (\sqrt{n})|n-1\rangle$ [ABR13].

operator, are the generalization of single-photon states to higher excitation numbers. The ability to create Fock states of a well-defined single EM field mode with a prescribed excitation number is of foundational interest and also has significant practical impact in QM applications [COO13].

Fock states $|n\rangle$ are eigenstates of the (photon) number operator $\boldsymbol{n} = \mathbf{dd}^\dagger$, where $\mathbf{d}$ is the annihilation operator of a quantized mode of the EM field (the *annihilation operator* $\mathbf{d}$ and the *creation operator* $\mathbf{d}^\dagger$ are operators such that $\mathbf{dd}^\dagger + \mathbf{d}^\dagger\mathbf{d} = \mathbb{I}$, with $\mathbb{I}$ the identity matrix). Then define the *number operator* as $\boldsymbol{n} = \mathbf{dd}^\dagger$ and denote its eigenstates as $|n\rangle$ (above we equivalently defined the operator $\hat{n}$ as $\hat{n} = \hat{a}^\dagger\hat{a}$). A generic Fock state is a state of the form

$$|\phi\rangle = \sum_{n=0}^{\infty} \alpha_n |n\rangle \quad \text{with} \quad \sum_{n=0}^{\infty} |\alpha_n|^2 = 1$$

and α_n are complex coefficients.

3.9 Bell States

These states related to two qubits in a combined $A \otimes B$ space were mentioned in passing in Chapter 2. Bell pairs have the property that the qubits are entangled (correlated); in fact, they *are maximally entangled states on two qubits*. They provide a set of entangled and normalized basis vectors where the overall probability of the particle being in one of these states is $\langle\psi|\psi\rangle = 1$. The four states are (also see Figure 3.12):

$$|\phi^+\rangle_{AB} := \frac{1}{\sqrt{2}}(|00\rangle_{AB} + |11\rangle_{AB})$$

$$|\phi^-\rangle_{AB} := \frac{1}{\sqrt{2}}(|00\rangle_{AB} - |11\rangle_{AB})$$

$$|\psi^+\rangle_{AB} := \frac{1}{\sqrt{2}}(|01\rangle_{AB} + |10\rangle_{AB})$$

$$|\psi^-\rangle_{AB} := \frac{1}{\sqrt{2}}(|01\rangle_{AB} - |10\rangle_{AB})$$

$|\psi^+\rangle$ and $|\psi^-\rangle$ are also known as an Einstein-Podolsky-Rosen (EPR) pair. Bell's states come into play in Quacom in the context of techniques such as superdense coding and quantum teleportation. Because the qubits are entangled, measurement of one qubit will "collapse" the second qubit to a new state whose measurement yield one of two possible values with a given probability, where the new state depends on which Bell's state the two qubits are in initially. Notice that Bell's states (such as $|\psi^+\rangle$ and $|\psi^-\rangle$) are linear combinations of two states $|01\rangle_{AB}$ and $|10\rangle_{AB}$.

Consider $|\psi^-\rangle$ for argument's sake. Measuring the first qubit of $|\psi^-\rangle$ in the standard basis yields a 0 with probability 0.5, and 1 with probability 0.5; similarly, measuring the second qubit of $|\psi^-\rangle$ results in the same outcomes. Hence, measuring one qubit of this state yields what is a random outcome (50% chance either way). The relevant observation is that determining either qubit, by consequence, *exactly* determines the other. Here when one measures the first qubit, the second qubit is completely determined.[1] For example, if the first qubit is measured and

---

[1] The entanglement means the following: the qubit held by Alice is in a superposition of $|01\rangle_{AB}$ and $|10\rangle_{AB}$. If Alice measured her qubit in the standard basis, the outcome would be either 0 or 1, each with probability ½. Now if Bob also measured his qubit, the outcome would be the same as for Alice. Thus, although

$$|\psi\rangle = \alpha_{00}|00\rangle + \alpha_{01}|01\rangle + \alpha_{10}|10\rangle + \alpha_{11}|11\rangle$$

$$|\psi\rangle = \alpha_{00}\begin{bmatrix}1\\0\\0\\0\end{bmatrix} + \alpha_{01}\begin{bmatrix}0\\1\\0\\0\end{bmatrix} + \alpha_{10}\begin{bmatrix}0\\0\\1\\0\end{bmatrix} + \alpha_{11}\begin{bmatrix}0\\0\\0\\1\end{bmatrix} = \begin{bmatrix}\alpha_{00}\\\alpha_{01}\\\alpha_{10}\\\alpha_{11}\end{bmatrix}$$

$$|\phi^{+}\rangle = \frac{1}{\sqrt{2}}(|00\rangle \qquad\qquad\qquad + |11\rangle) = \begin{bmatrix}1/\sqrt{2}\\0\\0\\1/\sqrt{2}\end{bmatrix}$$

$$|\phi^{-}\rangle = \frac{1}{\sqrt{2}}(|00\rangle \qquad\qquad\qquad - |11\rangle) = \cdots$$

$$|\psi^{+}\rangle = \frac{1}{\sqrt{2}}(\qquad |01\rangle \quad + \quad |10\rangle \qquad) = \begin{bmatrix}0\\1/\sqrt{2}\\1/\sqrt{2}\\0\end{bmatrix}$$

$$|\psi^{-}\rangle = \frac{1}{\sqrt{2}}(\qquad |01\rangle \quad - \quad |10\rangle \qquad) = \cdots$$

Can also be expressed as below, where A and B are two *QM* systems

$$|\phi^{\pm}\rangle = \frac{1}{\sqrt{2}}(|0\rangle_A|0\rangle_B \pm |1\rangle_A|1\rangle_B) \qquad |\psi^{\pm}\rangle = \frac{1}{\sqrt{2}}(|0\rangle_A|1\rangle_B \pm |1\rangle_A|0\rangle_B)$$

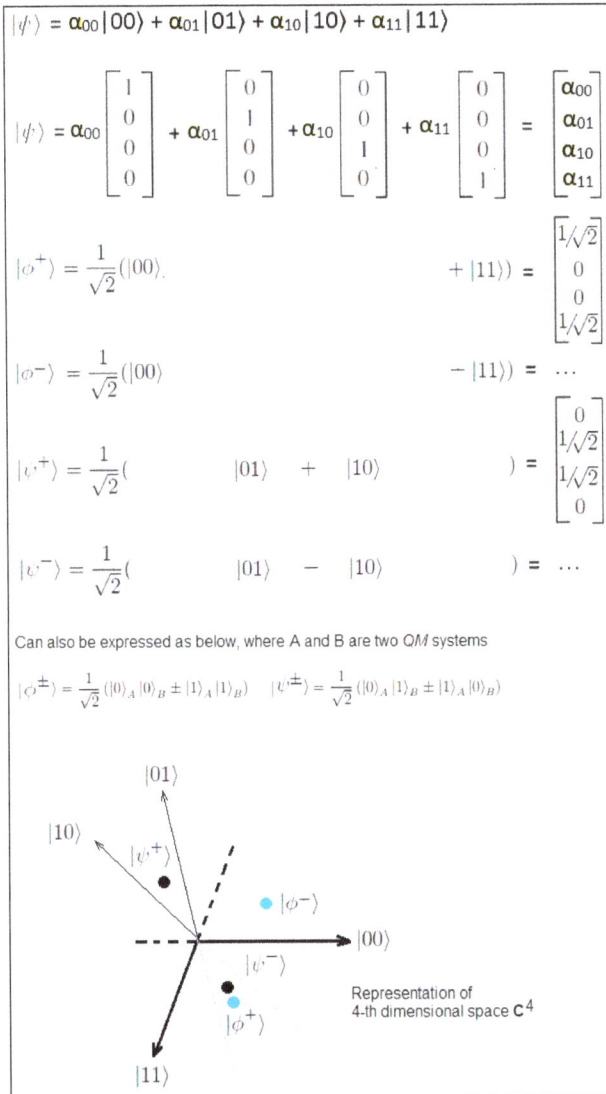

Representation of 4-th dimensional space $\mathbf{C}^4$

Figure 3.12 Bell states

yields a 0, this throws the Bell state onto the state $|01\rangle$; the second qubit is then a 1 for certain. Additionally, measurement of $|\psi^{-}\rangle$ in any basis yields opposite outcomes for the two qubits.

One can generate the Bell states with a Hadamard gate and a CNOT gate, as shown for example in Figure 3.13 (other implementations with unitary operators are possible). Also see Figure 3.14. Bell (EPR) pairs can be constructed from photon polarization states.

Alice and Bob would each seemingly experience random outcome, upon joint consultation they would discover that, in spite of the fact that their individual outcomes appeared random, these were in fact exactly correlated.

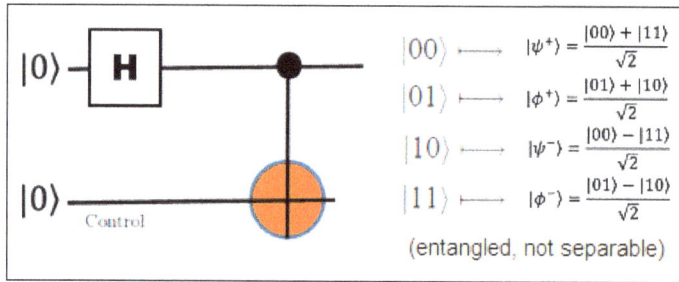

If the top qubit starts with a state of $|0\rangle$, the Hadamard gate puts it into a superposition of $|0\rangle$ and $|1\rangle$. The CNOT gate flips the bottom qubit depending on the state of the top qubit, but since the top qubit is in a superposition of $|0\rangle$ and $|1\rangle$, this makes the final state of bottom qubit dependent on what the final state of the top qubit turns out to be—hence, the two qubits are part of an entangled state.

Figure 3.13   Bell states generated with a Hadamard gate and a CNOT gate

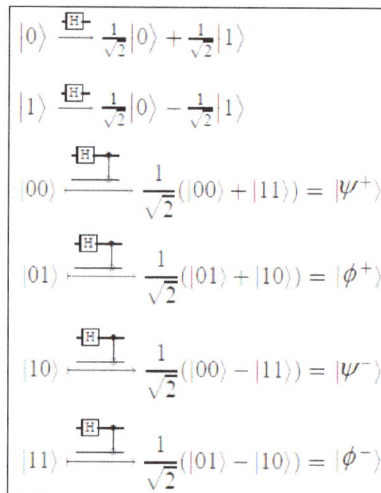

Figure 3.14   Bell states, circuit realization

3.10 Quantum channels

A quantum channel $\mathcal{E}$ is a CPTPM, a completely positive and trace preserving map(ping) introduced earlier. *Completely positive* means that for any density operator ρ, $\mathcal{E}(\rho)$ and $\mathbb{I} \otimes \mathcal{E}(\rho)$ are positive; *trace preserving* means that $\mathrm{tr}(\mathcal{E}(\rho)) = \mathrm{tr}(\rho)$. Some examples include the depolarizing channel and the depolarizing noise for a quantum gate, and the quantum erasure channel. See Table 3.3.

Many physical channels, such as optical fibers, free-space spectrum, or metallic links can operate as conduits that can carry both classical binary (or analog) signals and, under appropriate conditions, quantum states.

Table 3.3 Depolarization

Concept	Definition
Depolarizing channel	For any n-qubit density operator ρ and $0 \leq p \leq 1$ define $$\mathcal{E}(\rho) := p\rho + \frac{1-p}{2^n}\,\mathbb{I}$$
Depolarizing noise for a quantum gate	Given a quantum gate U acting on n-qubits, one can model an imperfect realization of such a gate utilizing the quantum channel $$\mathcal{E}_U(\rho) = pU\rho U^\dagger + \frac{1-p}{2^n}\,\mathbb{I}$$ and $0 \leq p \leq 1$.

3.11 Measurements

As discussed in previous chapters, there are various physical implementations of qubits, *including but not limited to* electrons, photon-based qubits in crystals, and solid-state superconducting qubits. In cases where qubits are implemented in electrons, the electron can transition to a higher (or lower) quantum energy state by applying (or removing) EM energy at an appropriate frequency to the electron in an atom. Thus, by applying a field with the appropriate frequency, one can affect the energy level and spin of the electron to generate a logical "one" state, a logical "zero" state, or a superposition of the two states. As discussed, QM postulates that the electron is in superposition of the two states at the same time, based on a weighted linear combination of the two: here the state of superposition of the qubit is the vector addition of the "upper" and "lower" states. See Figure 3.15. However, the state of superposition only can exist if one does

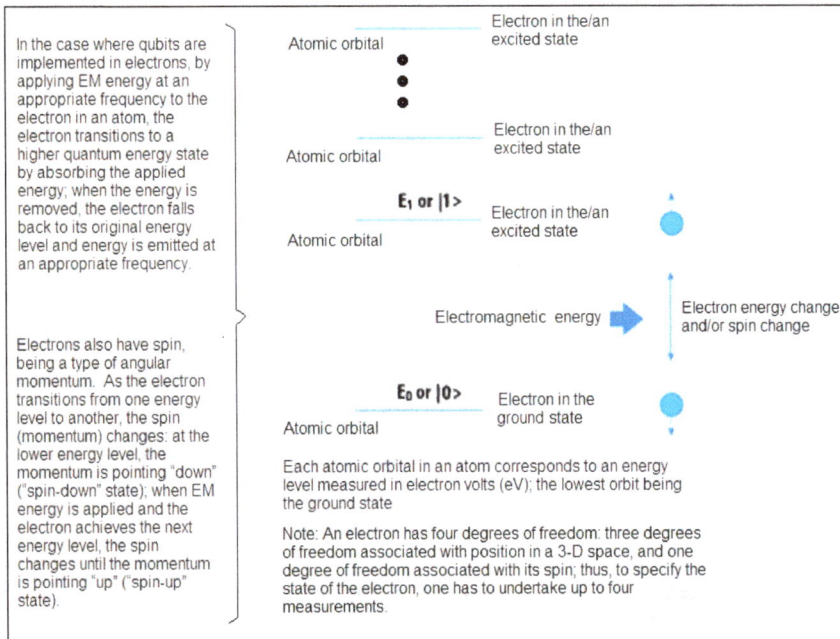

In the case where qubits are implemented in electrons, by applying EM energy at an appropriate frequency to the electron in an atom, the electron transitions to a higher quantum energy state by absorbing the applied energy; when the energy is removed, the electron falls back to its original energy level and energy is emitted at an appropriate frequency.

Electrons also have spin, being a type of angular momentum. As the electron transitions from one energy level to another, the spin (momentum) changes: at the lower energy level, the momentum is pointing "down" ("spin-down" state); when EM energy is applied and the electron achieves the next energy level, the spin changes until the momentum is pointing "up" ("spin-up" state).

Atomic orbital — Electron in the/an excited state

Atomic orbital — Electron in the/an excited state

E_1 or $|1\rangle$ — Atomic orbital — Electron in the/an excited state

Electromagnetic energy — Electron energy change and/or spin change

E_0 or $|0\rangle$ — Atomic orbital — Electron in the ground state

Each atomic orbital in an atom corresponds to an energy level measured in electron volts (eV); the lowest orbit being the ground state

Note: An electron has four degrees of freedom: three degrees of freedom associated with position in a 3-D space, and one degree of freedom associated with its spin; thus, to specify the state of the electron, one has to undertake up to four measurements.

Figure 3.15   Example of qubit operation

not "observe" it; by applying a measurement pulse (an operator) to a qubit to "observe" it, the state of that qubit collapses to one of the two quantized energy levels, and the measurement causes the particle to be projected (in the mathematical sense) to one of its eigenstates [ELO22]. A ground state corresponds to an eigenstate with the lowest energy (the quantum state after a measurement becomes the eigenstate corresponding to that measurement).

Say $|\psi\rangle = \Sigma_j \, \alpha_j |\alpha_j\rangle = \Sigma_j \, \alpha_j |j\rangle$. To gain information on the state of the qubit (e.g., by a photon, an electron spin, and so on), one must undertake measurements. To extract physical information from a quantum system one needs to measure an observable. An observable O is a linear and Hermitian ($O = O^\dagger$) operator. O can be decomposed into a set of *projectors* $\{P_i\}$ satisfying $P_i P_j = P_i \delta_{i,j}$ and $\Sigma_i P_i = \mathbb{I}$, in conjunction with a corresponding set of real-values eigenvalues $\{o_i\}$ such that $O = \Sigma_i o_i P_i$; the index i signifies different measurement outcomes. If the quantum state ρ is measured, then outcome i is observed with probability $\mathrm{tr}(P_i \rho)$. This quantum measurement is known as a *Projection-Valued Measurement* (PVM) and is the simplest description of a quantum measurement.[1] After a measurement is undertaken, the quantum state undergoes a non-unitary transformation, referred to as the "collapse of the wave function" (the collapse of the wave function is a postulate of the Copenhagen interpretation; this interpretation is the most commonly used model for quantum measurements, although some debate on the matter persists). If a PVM is performed on the state ρ and outcome i is observed, then

$$\rho \longrightarrow \frac{P_i \rho P_i}{\mathrm{tr}(P_i \rho)}.$$

To reiterate the above in other terms, measurement in QM can be perceived as a projection onto the (or a) basis of the subtending vector space. A quantum measurement of a quantum state ρ is described via Kraus operators P_i where the index i refers to the measurement outcome. The probability of measuring the outcome i is $p_i = \mathrm{tr}(P_i \rho \, P_i^\dagger)$. The state after the measurement becomes $\rho_i = (P_i \rho \, P_i^\dagger)/p_i$.

It was already noted that the measurement process in a quantum system alters its state: the impact of the measurement is that the new state is precisely the outcome of the measurement—namely, if the outcome of the measurement is j, then following the measurement, the qubit is in state $|j\rangle$; for example, if the outcome of the measurement is 1, then following the measurement, the qubit is in state $|1\rangle$. This implies that one cannot secure any additional information about the amplitudes α_j by repeating the measurement on the resulting state: one needs to make many identical measurements on a set (ensemble) of equivalent states.

One can measure the qubit in any orthonormal basis by projecting $|\psi\rangle$ onto the (two) basis vectors (see Figure 3.16). Consider the state $|\psi\rangle$ of a single qubit written as a superposition over the possibilities $|0\rangle$ and $|1\rangle$, with $|\psi\rangle = \alpha|0\rangle + \beta|1\rangle$, where α and β are complex numbers with $|\alpha|^2 + |\beta|^2 = 1$ ($\alpha = |\langle 0|\psi\rangle|$, $\beta = |\langle 1|\psi\rangle|$), and orthonormal basis

$$|0\rangle = \begin{bmatrix} 1 \\ 0 \end{bmatrix} \text{ and } |1\rangle = \begin{bmatrix} 0 \\ 1 \end{bmatrix}$$

---

[1] A quantum measurement can be further generalized by abandoning the notion that measurement outcomes are orthogonal [SHE22]. This generalization is called a *Positive-Operator-Valued Measure* (POVM). More general operators are used in the measurement. PVM as a special case of a POVM when the set of measurement operators are all projectors.

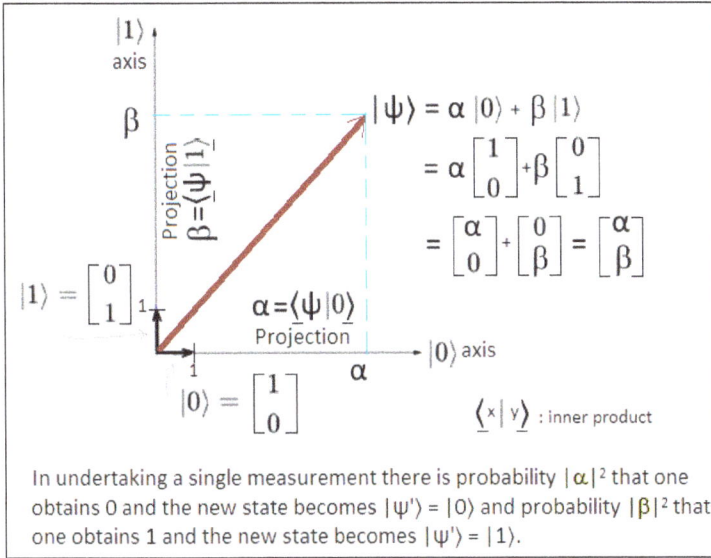

Figure 3.16   Projections as measurements

Undertaking a single measurement in the standard basis, then, there is probability $|\alpha|^2$ that one obtains 0 and the new state becomes $|\psi'\rangle = |0\rangle$ and probability $|\beta|^2$ that one obtains 1 and the new state becomes $|\psi'\rangle = |1\rangle$.

Say $|\psi\rangle = \alpha_1|a_1\rangle + \alpha_2|a_2\rangle + \alpha_3|a_3\rangle + \alpha_4|a_4\rangle = \alpha_{00}|00\rangle + \alpha_{01}|01\rangle + \alpha_{10}|10\rangle + \alpha_{11}|11\rangle$, with $\sum_i|\alpha_i|^2 = 1 = \sum_{ij}|\alpha_{ij}|^2$ (that is, $|\alpha_{00}|^2 + |\alpha_{01}|^2 + |\alpha_{10}|^2 + |\alpha_{11}|^2 = 1$). The α's are complex amplitudes in $\mathbf{C}^4$ that parametrize the basis "vectors" $|00\rangle$, $|01\rangle$, $|10\rangle$ and $|11\rangle$ (here $n = 2$ and the dimension of the Hilbert space is $k = 2^n$, or 4).

Undertaking a single measurement of the observable A on $|\psi\rangle$ results in the outcome a_i with probability $P_{\psi(ai)} = |\alpha_i|^2$, and the resulting state will be $|a_i\rangle$—that is, the measurement of the observable has "collapsed" the state $|\psi\rangle$ to a single eigenstate $|a_i\rangle$—namely, a measurement of the state of the two qubits yields a result whereby the probability that the first qubit is in state i and the second qubit is in state j is $Pr(i, j) = |\alpha_{ij}|^2$. After the measurement, the state of the two qubits is $|\psi'\rangle = |ij\rangle$.

On the other hand, if one were to just measure the first qubit, one would find that $Pr\{1st\ bit = 0\} = |\alpha_{00}|^2 + |\alpha_{01}|^2$; thus the state vector can be seen or interpreted as a projection (i) onto the subspace spanned by $|00\rangle$ and $|01\rangle$ with probability equal to the square of the norm of the projection, or (ii) onto the subspace spanned by $|10\rangle$ and $|11\rangle$ with the remaining probability; the new state is given by the (normalized) projection onto the respective subspace [WHA09]. Thus, although the Hilbert space associated with a system of n qubits is an element in $\mathbf{C}^k$ with $k = 2^n$, when one undertakes a measurement of a state in that combined environment, one can recover only n bits of information.

As stated, measurement can be perceived as a *projector*; earlier (in Table 3.1 on page 105) it was noted that $\underline{P} = |\psi\rangle \otimes \langle\psi|$ is a projector operator: in general, $\underline{P}(|\phi\rangle) = |\psi\rangle \otimes \langle\psi| (|\phi\rangle) = \langle\psi|\phi\rangle|\psi\rangle$. It follows that $P^2 = P$. Define the projector $\underline{P}_i = |i\rangle \otimes \langle i|$ (now abbreviating $|a_i\rangle$ as $|i\rangle$), such that one has the representation $|\psi\rangle = \sum_j \alpha_j|j\rangle$ as given above.

$$|\psi\rangle = \alpha\,|0\rangle + \beta\,|1\rangle = \alpha\begin{bmatrix} 1 \\ 0 \end{bmatrix} + \beta\begin{bmatrix} 0 \\ 1 \end{bmatrix} = \begin{bmatrix} \alpha \\ 0 \end{bmatrix} + \begin{bmatrix} 0 \\ \beta \end{bmatrix} = \begin{bmatrix} \alpha \\ \beta \end{bmatrix}$$

with basis $\quad |0\rangle = \begin{bmatrix} 1 \\ 0 \end{bmatrix} \quad |1\rangle = \begin{bmatrix} 0 \\ 1 \end{bmatrix}$

$$\langle\psi| = [\alpha\ \beta]$$

$$|\psi\rangle\langle\psi| = \begin{bmatrix} \alpha \\ \beta \end{bmatrix}[\alpha^*\ \beta^*] = \begin{bmatrix} \alpha\alpha^* & \alpha\beta^* \\ \beta\alpha^* & \beta\beta^* \end{bmatrix} = \underline{P}$$
$$\qquad\quad {\scriptstyle 2\times1}\quad {\scriptstyle 1\times2}\qquad\qquad {\scriptstyle 2\times2}$$

$$\underline{P_i}(|\psi\rangle) = c_i \times |i\rangle$$

$$\underline{P_1}|\psi\rangle = |0\rangle\langle0||\psi\rangle = \begin{bmatrix} 1 & 0 \\ 0 & 0 \end{bmatrix}\begin{bmatrix} \alpha \\ \beta \end{bmatrix} = \begin{bmatrix} \alpha \\ 0 \end{bmatrix} = \alpha\,|0\rangle = \alpha\begin{bmatrix} 1 \\ 0 \end{bmatrix}$$
$$\qquad\qquad\qquad {\scriptstyle 2\times2}\quad {\scriptstyle 2\times1}\qquad {\scriptstyle 2\times1}$$

$$\underline{P_2}|\psi\rangle = |1\rangle\langle1||\psi\rangle = \begin{bmatrix} 0 & 0 \\ 0 & 1 \end{bmatrix}\begin{bmatrix} \alpha \\ \beta \end{bmatrix} = \begin{bmatrix} 0 \\ \beta \end{bmatrix} = \beta\,|1\rangle = \beta\begin{bmatrix} 0 \\ 1 \end{bmatrix}$$

$$|0\rangle\langle0| = \begin{bmatrix} 1 \\ 0 \end{bmatrix}[1\ 0] = \begin{bmatrix} 1 & 0 \\ 0 & 0 \end{bmatrix}$$

$$|1\rangle\langle1| = \begin{bmatrix} 0 \\ 1 \end{bmatrix}[0\ 1] = \begin{bmatrix} 0 & 0 \\ 0 & 1 \end{bmatrix}$$

$$\mathbb{I} = |0\rangle\langle0| + |1\rangle\langle1| = \begin{bmatrix} 1 & 0 \\ 0 & 1 \end{bmatrix}$$

Figure 3.17   Illustration of measurement as a projector

Since, as just noted, $\underline{P}(|\phi\rangle) = |\psi\rangle \otimes \langle\psi|\,(|\phi\rangle) = \langle\psi|\phi\rangle|\psi\rangle$, one has (by simple substitution and rearrangement) $\underline{P_i}(|\psi\rangle) = |i\rangle \otimes \langle i|\,(|\psi\rangle) = |i\rangle\,\langle i|\psi\rangle = \alpha_i \times |i\rangle$ and because of the orthogonality of the set of $a_i = \langle i|$ one has $\langle i|\psi\rangle = c_i$ as the inner product. Then, $\underline{P_i}(|\psi\rangle) = \alpha_i \times |i\rangle$. Namely, $\underline{P_i}$ takes $|\psi\rangle$ and maps it by its basis component $|i\rangle$ with amplitude $\langle i|\psi\rangle = \alpha_i$. The spectral resolution $\mathbb{I}$ of the identity defines a set of projectors: $\mathbb{I} = \sum_k |k\rangle\langle k| = \sum_k \underline{P_k}$.

For two states the sum has just two components and $\mathbb{I} = |0\rangle\langle0| + |1\rangle\langle1|$; Figure 3.17 illustrates these algebraic concepts for this case.

More generally, on can measure the qubit in *any* orthonormal basis by projecting $|\psi\rangle$ onto the (two) selected basis vectors; after the measurement, the state will migrate to $|\psi'\rangle$. In addition to the use of a standard basis (that is, $|0\rangle$ and $|1\rangle$), measurement in QM can be perceived as a projection onto some (other) appropriate basis of the subtending vector space: as noted above, given the basis $\{a_i\}$, of eigenstates of A, a single measurement of the observable A on a state $|\psi\rangle$ will yield the value a_i, with probability $P_{\psi(ai)} = |\langle a_i|\psi\rangle|^2$; the normalized state after the measurement will be [WHA09]:

$$|a_i\rangle\left\{\langle i|\psi\rangle \Big/ \sqrt{\langle\psi|M_i^\dagger M_i|\psi\rangle}\right\} \quad \text{with } M_i = |a_i\rangle\langle a_i|$$

For example, certain applications take advantage of the Hadamard basis already mentioned in Chapter 2 and above, where the basis is: $|+\rangle := (|0\rangle + |1\rangle)/\sqrt{2}$ and $|-\rangle := (|0\rangle - |1\rangle)/\sqrt{2}$. See

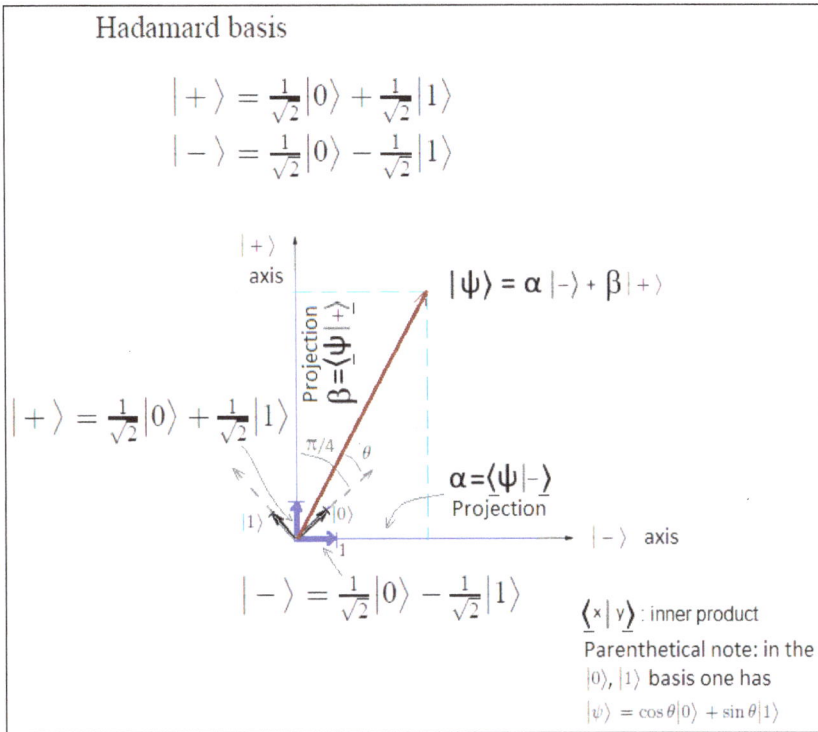

Hadamard basis

$$|+\rangle = \tfrac{1}{\sqrt{2}}|0\rangle + \tfrac{1}{\sqrt{2}}|1\rangle$$
$$|-\rangle = \tfrac{1}{\sqrt{2}}|0\rangle - \tfrac{1}{\sqrt{2}}|1\rangle$$

$$|\psi\rangle = \alpha\,|-\rangle + \beta\,|+\rangle$$

$$|+\rangle = \tfrac{1}{\sqrt{2}}|0\rangle + \tfrac{1}{\sqrt{2}}|1\rangle$$

$$\alpha = \langle\psi|-\rangle$$
Projection

$$|-\rangle = \tfrac{1}{\sqrt{2}}|0\rangle - \tfrac{1}{\sqrt{2}}|1\rangle$$

$\langle x | y \rangle$: inner product
Parenthetical note: in the
$|0\rangle, |1\rangle$ basis one has
$|\psi\rangle = \cos\theta|0\rangle + \sin\theta|1\rangle$

Figure 3.18   Measurements in the Hadamard basis

Figure 3.18. However, if the measurement in that "rotated" basis is impractical for some reason, one can rotate the system (in the opposite direction by θ where θ is the angle between the state vector and the $|0\rangle$, $|1\rangle$ basis) and then measure the state in the original, standard basis.

In two dimensions, the Hadamard rotation equates to a positive $\pi/4$ rotation of the space defined by the basic $|0\rangle/|1\rangle$ coordinates; a Heisenberg (or Schrödinger) rotation/basis corresponds to a more general rotation by ϕ.

3.12 Measurement of a Photon

To elaborate some concepts related to measurements and provide an illustration, consider a photon with a superposition state $|\psi\rangle = \alpha|0\rangle + \beta|1\rangle$, ($|0\rangle$ and $|1\rangle$ being the standard basis), or normalizing the coefficients as $|\psi\rangle = \cos(\theta)|0\rangle + \sin(\theta)|1\rangle$.

$$|\psi\rangle = \begin{bmatrix} \cos(\theta) \\ \sin(\theta) \end{bmatrix} = \cos(\theta)|v\rangle + \sin(\theta)|h\rangle = \cos(\theta)\begin{bmatrix} 1 \\ 0 \end{bmatrix} + \sin(\theta)\begin{bmatrix} 0 \\ 1 \end{bmatrix} \text{ with } |\uparrow\rangle = \begin{bmatrix} 1 \\ 0 \end{bmatrix} = |v\rangle = |0\rangle \quad |\rightarrow\rangle = \begin{bmatrix} 0 \\ 1 \end{bmatrix} = |h\rangle = |1\rangle$$

The polarization of a photon can be measured using a polarizing filter because the material acts as a filter that selects only one component of the state vector; for example, a vertical polarizing filter (suitably oriented) allows/transmits photons $|\uparrow\rangle$ and absorbs y-polarized photons $|\rightarrow\rangle$. Measuring $|\psi\rangle$ in the standard basis $|0\rangle$ and $|1\rangle$ results in the state $|0\rangle$ with probability $|\alpha|^2$, and state $|1\rangle$ with probability $|\beta|^2$. A polarizing filter (film) oriented in the vertical direction in this basis can be represented by the following matrix operator:

$$\hat{V} = \begin{bmatrix} 1 & 0 \\ 0 & 0 \end{bmatrix} \quad \text{Vertical filter}$$

A polarizing filter (film) oriented in the horizontal direction in this basis can be represented by the following matrix operator:

$$\hat{H} = \begin{bmatrix} 0 & 0 \\ 0 & 1 \end{bmatrix} \quad \text{Horizontal filter}$$

In Figure 3.19, note the results of measurement operators $\hat{V}$ and/or $\hat{H}$ (namely, applying these operators) on three polarization states; also see the top portion of Figure 3.20.

When the photon is transmitted through the vertical polarizing filter,

$$|\uparrow\rangle\langle\uparrow| = \begin{bmatrix} 1 & 0 \\ 0 & 0 \end{bmatrix} = \hat{\Theta} = \hat{V}$$

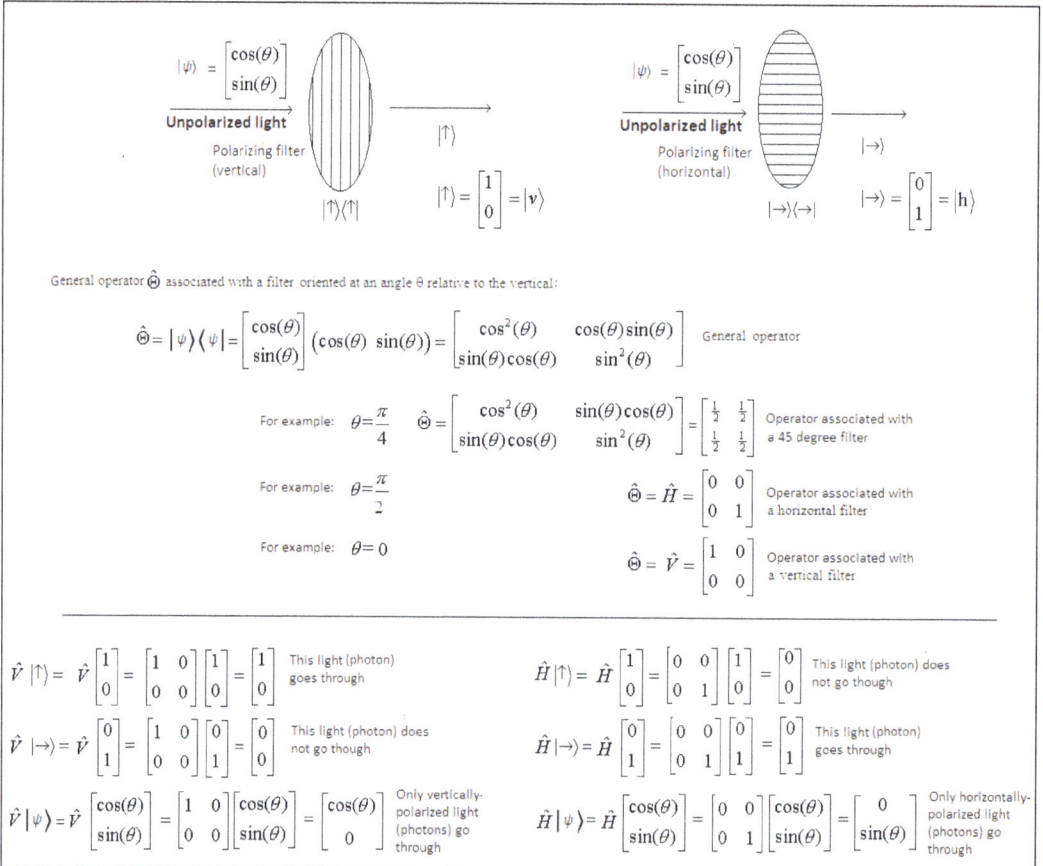

Figure 3.19   Results of the $\hat{V}$ and $\hat{H}$ measurements on three photonic states

$$|\psi\rangle = \begin{bmatrix}\cos(\theta)\\\sin(\theta)\end{bmatrix} = \cos(\theta)|v\rangle + \sin(\theta)|h\rangle = \cos(\theta)\begin{bmatrix}1\\0\end{bmatrix} + \sin(\theta)\begin{bmatrix}0\\1\end{bmatrix}$$

$|\uparrow\rangle = \begin{bmatrix}1\\0\end{bmatrix} = |v\rangle = |0\rangle$ $|\rightarrow\rangle = \begin{bmatrix}0\\1\end{bmatrix} = |h\rangle = |1\rangle$ $|\nearrow\rangle = \frac{1}{\sqrt2}[|\uparrow\rangle+|\rightarrow\rangle] = \frac{1}{\sqrt2}[|\uparrow\rangle+\rightarrow\rangle] = \frac{1}{\sqrt2}\left[\begin{bmatrix}1\\0\end{bmatrix}+\begin{bmatrix}0\\1\end{bmatrix}\right] = \frac{1}{\sqrt2}[1\ 1] = \begin{bmatrix}\frac{1}{\sqrt2}\\\frac{1}{\sqrt2}\end{bmatrix}$

$\langle\uparrow| = [1\ 0] = \langle v|$ $\langle\rightarrow| = [0\ 1] = \langle h|$ $|\nwarrow\rangle = \frac{1}{\sqrt2}[|\uparrow\rangle-|\rightarrow\rangle] = \frac{1}{\sqrt2}[|\uparrow\rangle+|\rightarrow\rangle^T] = \frac{1}{\sqrt2}[1\ 1] = \begin{bmatrix}-\frac{1}{\sqrt2}\\\frac{1}{\sqrt2}\end{bmatrix}$

$|\uparrow\rangle\langle\uparrow| = \begin{bmatrix}1\\0\end{bmatrix}[1\ 0] = \begin{bmatrix}1&0\\0&0\end{bmatrix} = \hat V$ $|\rightarrow\rangle\langle\rightarrow| = \begin{bmatrix}0\\1\end{bmatrix}[0\ 1] = \begin{bmatrix}0&0\\0&1\end{bmatrix} = \hat\Theta = \hat H$ $|\nearrow\rangle\langle\nearrow| = \begin{bmatrix}\frac12&\frac12\\\frac12&\frac12\end{bmatrix} = \hat\Theta$ $|\nearrow\rangle = |+\rangle$ $|\nwarrow\rangle = |-\rangle$

Probability that a polarized photon passes a vertically-oriented filter:

$\hat\Theta = \hat V = \begin{bmatrix}1&0\\0&0\end{bmatrix}$ $\text{Prob} = \frac{1}{\pi}\int_0^\pi\left\|\begin{bmatrix}1&0\\0&0\end{bmatrix}\begin{bmatrix}\cos(\theta)\\\sin(\theta)\end{bmatrix}\right\|^2 d\theta = 0.5$ for $|\uparrow\rangle$ for $\begin{bmatrix}1\\0\end{bmatrix}$

$\text{Prob} = \frac{1}{\pi}\int_0^\pi\left\|\begin{bmatrix}1&0\\0&0\end{bmatrix}\begin{bmatrix}\cos(\theta)\\\sin(\theta)\end{bmatrix}\right\|^2 d\theta = 0$ for $|\rightarrow\rangle$ for $\begin{bmatrix}0\\1\end{bmatrix}$

with $\int_0^\pi|\cos(\theta)|^2 = \pi/2$

Probability that a polarized photon passes a horizontally-oriented filter:

$\hat\Theta = \hat H = \begin{bmatrix}0&0\\0&1\end{bmatrix}$ $\text{Prob} = \frac{1}{\pi}\int_0^\pi\left\|\begin{bmatrix}0&0\\0&1\end{bmatrix}\begin{bmatrix}\cos(\theta)\\\sin(\theta)\end{bmatrix}\right\|^2 d\theta = 0$ for $|\uparrow\rangle$ for $\begin{bmatrix}1\\0\end{bmatrix}$

$\text{Prob} = \frac{1}{\pi}\int_0^\pi\left\|\begin{bmatrix}0&0\\0&1\end{bmatrix}\begin{bmatrix}\cos(\theta)\\\sin(\theta)\end{bmatrix}\right\|^2 d\theta = 0.5$ for $|\rightarrow\rangle$ for $\begin{bmatrix}0\\1\end{bmatrix}$

with $\int_0^\pi|\sin(\theta)|^2 = \pi/2$

Probability that unpolarized light passes two crossed polarizing films (vertical followed by horizontal) is zero:

$\hat\Theta = \hat H = \begin{bmatrix}0&0\\0&1\end{bmatrix}$ $\text{Prob} = \frac{1}{\pi}\int_0^\pi\left\|[0\ 1]\begin{bmatrix}0&0\\0&1\end{bmatrix}\begin{bmatrix}1&0\\0&0\end{bmatrix}\begin{bmatrix}\cos(\theta)\\\sin(\theta)\end{bmatrix}\right\|^2 d\theta = 0$

When a polarizing film oriented at a 45° angle is inserted between the crossed polarizers light gets through the final horizontal filter, with probability 0.125:

$\hat\Theta = \begin{bmatrix}\frac12&\frac12\\\frac12&\frac12\end{bmatrix}$

$\text{Prob} = \frac{1}{\pi}\int_0^\pi\left\|[0\ 1]\begin{bmatrix}0&0\\0&1\end{bmatrix}\begin{bmatrix}\frac12&\frac12\\\frac12&\frac12\end{bmatrix}\begin{bmatrix}1&0\\0&0\end{bmatrix}\begin{bmatrix}\cos(\theta)\\\sin(\theta)\end{bmatrix}\right\|^2 d\theta = 0.125$ with $\int_0^\pi|\cos(\theta)|^2 = \pi/2$

Unpolarized light → $|\uparrow\rangle$ $|\uparrow\rangle\langle\uparrow|$

Unpolarized light → $|\uparrow\rangle$ → $|\nearrow\rangle$ $|\uparrow\rangle\langle\uparrow|$ $|\nearrow\rangle\langle\nearrow|$ $|\rightarrow\rangle\langle\rightarrow|$

Unpolarized light → $|\uparrow\rangle$ → $|\rightarrow\rangle$ $|\uparrow\rangle\langle\uparrow|$ $|\rightarrow\rangle\langle\rightarrow|$

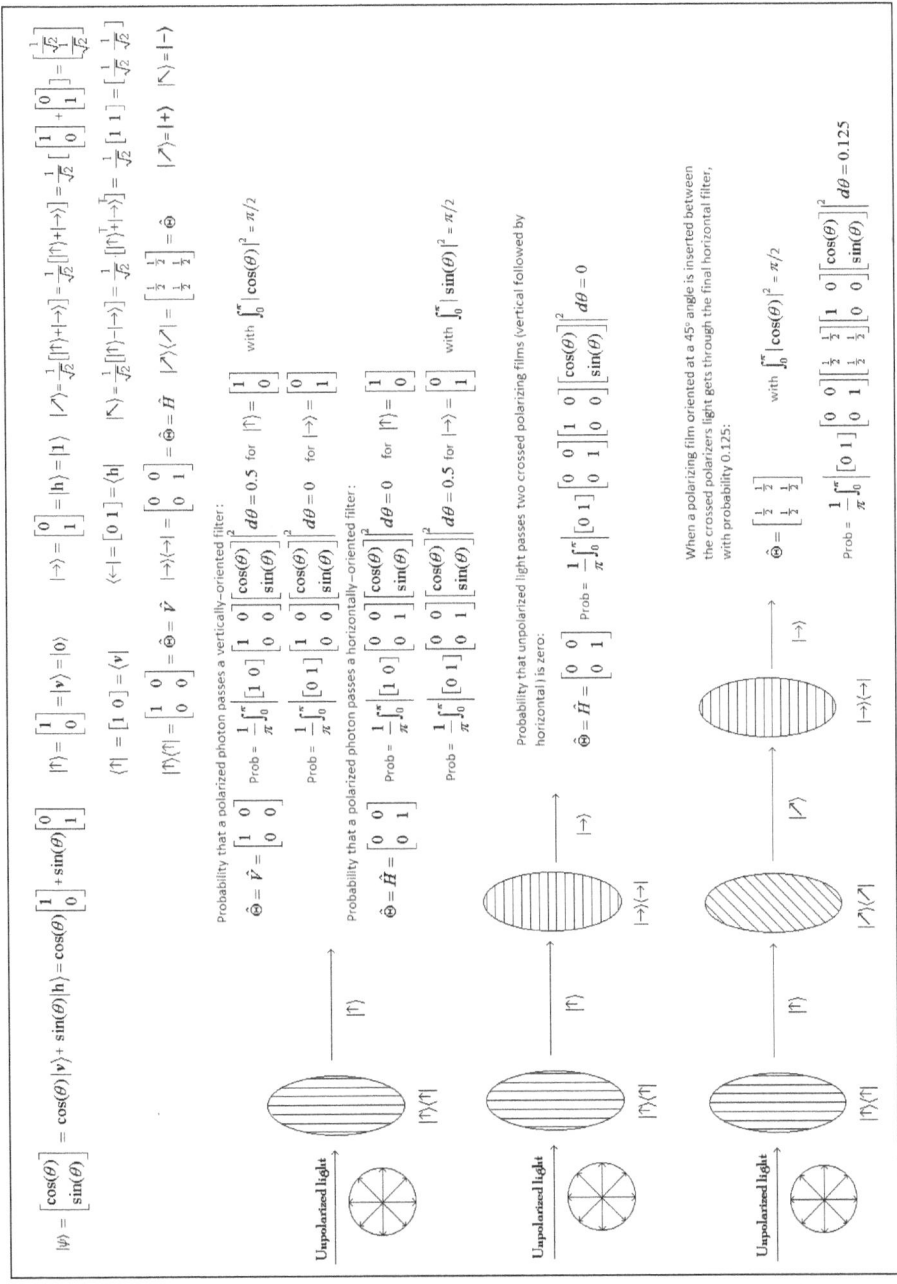

Figure 3.20 Probabilities that photons pass through polarizing filters (inspired in part by [RIO23])

(see the bottom left of Figure 3.19 and the applicable section of Figure 3.20) the photon can be received behind the filter with probability $|\alpha|^2$ (= 0.5 for the normalized case) when the polarizing filter is oriented vertically, allowing it to transmit $|\uparrow\rangle$ (the final state will be $|\uparrow\rangle$); horizontally oriented photons are absorbed (have probability 0).

When the photon is transmitted through the horizontal polarizing filter,

$$|\rightarrow\rangle\langle\rightarrow| = \begin{bmatrix} 0 & 0 \\ 0 & 1 \end{bmatrix} = \hat{\Theta} = \hat{H}$$

(see the bottom right of Figure 3.19 and the applicable section of Figure 3.20) the photon can be received behind the filter with probability with probability $|\beta|^2$ (also 0.5 for the normalized case) when the polarizing filter is oriented horizontally, allowing it to transmit $|\rightarrow\rangle$ (the final state will be $|\rightarrow\rangle$); vertically oriented photons are absorbed (have probability 0).

Consider what happens when adding additional filters. If two polarizing filters are aligned in the same direction, clearly, light from the first polarizer passes through the second filter. If the polarizer filters are opposed at a 90° angle, the polarized light transmitted from the first polarizing filter is absorbed by the second filter.

If a third polarizer is interjected between the two opposed polarizing filters at a 45° angle, some light gets through the third polarizer. Consider the setup depicted in Figure 3.20 (inspired in part by reference [RIO23]); it shows various probabilities of the light passing through filters: (i) the probability that a polarized photon passes a vertically oriented filter; (ii) the probability that unpolarized light (or light of any polarization) will pass two crossed polarizing films (vertical followed by horizontal, for example); and (iii) the probability when a polarizing film oriented at a 45° angle is inserted between the crossed polarizers.

After the first filter one has the state $|\uparrow\rangle$ with probability 0.5 (in the normalized case), but after the second filter the probability drops to 0 (Figure 3.20, middle row). During the transmission through the first filter, the photon was absorbed with probability 0.5; if it actually goes through the filter, it is absorbed (with probability 1) by the second filter.

Consider interjecting a third filter that is polarized at $\pi/4$ (45°); the filter is

$$|\nearrow\rangle\langle\nearrow| = \begin{bmatrix} \frac{1}{2} & \frac{1}{2} \\ \frac{1}{2} & \frac{1}{2} \end{bmatrix} = \hat{\Theta}$$

Thus, it will pass photons with polarization orientation of $\pi/4$, and the polarization of light after that filter is:

$$|\nearrow\rangle = \frac{1}{\sqrt{2}}\left[|\uparrow\rangle + |\rightarrow\rangle\right] = \frac{1}{\sqrt{2}}\left[|\uparrow\rangle + |\rightarrow\rangle\right] = \frac{1}{\sqrt{2}}\left[\begin{bmatrix} 1 \\ 0 \end{bmatrix} + \begin{bmatrix} 0 \\ 1 \end{bmatrix}\right] = \begin{bmatrix} \frac{1}{\sqrt{2}} \\ \frac{1}{\sqrt{2}} \end{bmatrix}$$

(which is also known as $|+\rangle$). For completeness

$$|\nwarrow\rangle = \frac{1}{\sqrt{2}}\left[|\uparrow\rangle - |\rightarrow\rangle\right] = \frac{1}{\sqrt{2}} \cdot \left[|\uparrow\rangle^{\top} + |\rightarrow\rangle^{\top}\right] = \frac{1}{\sqrt{2}}\begin{bmatrix} 1 & 1 \end{bmatrix} = \begin{bmatrix} \frac{1}{\sqrt{2}} & \frac{1}{\sqrt{2}} \end{bmatrix}$$

(which is also known as $|-\rangle$).

The photon that is transmitted by the filter (with probability ½) then makes it through the next two with probability ¼; therefore, the photon that had initially hit the vertical polarizer makes it through the three filters with probability 0.125 (see Figure 3.20, bottom row). The final state of the photon is $|\rightarrow\rangle$.

The operator projections across multiple scenarios discussed above can be perceived as a case of sequential measurements.

3.14 Theoreticals, EPR paradox, Bell's Inequality

This section is a high-level review of some theoretical considerations related to the EPR Paradox and Bell's inequality (summarized from [WHA09]).

It was noted above that for a Bell state, when one measures the first qubit, the second qubit is completely determined. However, since the speed of light is finite, if two qubits in the measurement scenario are spatially far apart, then the second qubit must have had a determined state in some time interval *before* measurement.

A source (say Dan) emits an entity that is in the Bell state $|\psi^-\rangle$ and transmits one qubit to Alice and one qubit to Bob. If Alice measures her qubit in the standard basis ($\{|0\rangle, |1\rangle\}$) and, say, gets a 1, then Bob will get a 0 upon measuring the qubit in the standard basis. On the other hand, if Alice measures her qubit in the Hadamard basis $|+\rangle = (|0\rangle + |1\rangle)/\sqrt{2}$ and $|-\rangle = (|0\rangle - |1\rangle)/\sqrt{2}$), and gets a '+', then Bob will get a '-' in the Hadamard basis. Clearly, the states $|1\rangle$ and $|+\rangle$ are not the same, and states $|0\rangle$ and $|-\rangle$ are not the same, since they differ by a rotation of $\pi/4$ (as seen in Figure 3.18). Which state Bob ends up with depends on the measurement made by Alice.

More generally, (in some cases) measurements of properties (such as spin or polarization, among others) performed on entangled particles are deterministically correlated, implying that with entangled particles a measurement on one particle affects the entire entangled system. For example, if a pair of entangled particles is created, and one particle is found upon measurement to have an anticlockwise spin on a given basis, then the spin of the other particle, measured with the same basis, is found, upon measurement, to be clockwise. Thus, when selecting a measurement and obtaining a result, the quantum state of the other particle appears to collapse into a new state, depending upon that original measurement result.

In QM nothing *can* be known about the quantum state until a measurement is made; and the result of a measurement on one system can instantaneously influence the result of a measurement on the second system. This behavior evokes the theoretical EPR paradox—any measurement of a particle's properties results in the collapse of the state of the "companion" particles and alters the original quantum state of that particle, no matter how far away the other particle is.

Work has been undertaken in the past decades to clarify and/or validate the entanglement phenomenon. In the mid-1960s John Bell developed correlation functions of the measurements that satisfy an inequality under which the assumption of *local realism* holds (local realism postulates that the result of a measurement on one system cannot instantaneously influence the result of a measurement on the second system); that is, he developed mathematical constraint on how the outcomes on the two measurements are correlated [BEL64]. Bell then demonstrated that QM predicts correlations that violate this inequality. *Bell nonlocality* is a counter-intuitive property of quantum correlations. In the early 1980s, experiments were designed and carried out actually demonstrating that polarization or spin of entangled particles measured at separate locations are correlated, consistent with the entanglement concept; this violates Bell's inequality and, so, negating the local realism assumption, and negating the local realism implies that the two particles are able to interact instantaneously no matter how widely they are separated.

In summary, Bell's theorem (Bell's inequalities) shows that the predictions of QM, specifically those involving separated measurements on entangled states, cannot be replicated when one abides by local realism theories. Local realism, which would appear to be a natural physical assumption, embodies two facets: (i) locality requires that spacelike-separated events cannot influence one another (i.e., influences cannot travel faster than the speed of light); and (ii) realism requires that physical systems have pre-existing properties, independent of measurements. That local realism is incompatible with QM (and that QM is, in fact, correct) requires *abandoning local realism* as a fundamental principle of the universe. Some researchers find this abrogation difficult, but recent loophole-free violations of Bell's inequalities leads one in that direction [MEY16].

References

[ABE23] A. J. G. Abelém, G. Vardoyan, and D. Towsley, "Quantum Internet: The Future of Internetworking". April 2023, arXiv:2305.00598v1.

[ABR13] S. Abruzzo, "Long Distance Quantum Key Distribution with Quantum Repeaters". Inaugural-Dissertation zur Erlangung des Doktorgrades der Mathematisch-Naturwissenschaftlichen Fakultat der Heinrich-Heine-Universit at Dusseldorf, Dusseldorf, November 2013.

[ALB92] D. Z. Albert, *Quantum mechanics and experience*. Harvard University Press, 1992.

[BEL64] J. S. Bell, "On the Einstein Podolsky Rosen Paradox". Physics Physique Физика. 1(3): 195–200. November 1964. doi:10.1103/PhysicsPhysiqueFizika.1.195.

[BEL04] J. S. Bell, *Speakable and unspeakable in quantum mechanics: Collected papers on quantum philosophy*. Cambridge University Press, 2004. (Earlier related publication by Cambridge University Press was in 1987 in the book *Speakable and unspeakable in quantum mechanics*, ISBN 9780521368698).

[COO13] M. Cooper, L. J. Wright, *et al*, "Experimental generation of multi-photon Fock states". Optics Express, Vol. 21, Issue 5, pp. 5309-5317, 2013. https://doi.org/10.1364/OE.21.005309

[CUS94] J. T. Cushing, *Quantum mechanics: Historical contingency and the Copenhagen hegemony.* University of Chicago Press, 1994.

[DAS23] Das, "The U3 Gate". Jul 21, 2023. Available online on January 5, 2024 at https://medium.com/ {https://medium.com/@iamsounak01/explain-the-u3-gate-to-you-in-simple-terms-51b6dd6a8622#:~:text=The%20U3%20gate%20is%20a,a%20 quantum%20bit%20(qubit).}

[ELO22] M. Elo, "Understanding Quantum Computing". Microwave Journal, September 14, 2020, available online on January 3, 2024 at https://www.microwavejournal.com/ articles/34557-understanding-quantum-computing?page=1.

[FIT23] R. Fitzpatrick, "Eigenstates and Eigenvalues". Available online on 10/25/2023 at https:// phys.libretexts.org/Bookshelves/Quantum_Mechanics/Introductory_Quantum_Mechanics_(Fitzpatrick)/03%3A_Fundamentals_of_Quantum_Mechanics/3.08%3A_Eigenstates_and_Eigenvalues

[GAS95] S. Gasiorowicz, *Quantum Physics* (3rd edition), 1995, New York: Wiley. ISBN: 978-0-471-05700-0.

[GID17] C. Gidney, "Breaking Down the Quantum Swap". 22 Aug 2017, Available online on December 12, 2023 at https://algassert.com/post/1717.

[GOT04] K. Gottfried, T. M. Yan, *Quantum Mechanics: Fundamental,* 2Ed, Springer 2004, ISBN-10: 0387220232, ISBN-13: 978-0387220239.

[ISM21] J. Ismael, "Quantum Mechanics". Stanford Encyclopedia of Philosophy, Edward N. Zalta (Editor). Fall 2021 Edition. Available online (10/24/2023) at https://plato.stanford.edu/entries/qm/ or https://plato.stanford.edu/archives/fall2021/entries/qm/

[JAC23] J. D. Jackson, *A Course in quantum mechanics,* August 2023, Wiley, ISBN: 978-1-119-88041-7.

[KAS13] R. Kastner, *The transactional interpretation of quantum mechanics: The reality of possibility.* Cambridge University Press, 2013.

[LEW23] P. J. Lewis, "Interpretations of Quantum Mechanics". Internet Encyclopedia of Philosophy (IEP). Available online on 11/3/2023 at https://iep.utm.edu/int-qm/#:~:text=According%20to%20the%20Copenhagen%20interpretation,make%20measurements%20on%20the%20system.

[MER98] E. Merzbacher, *Quantum mechanics* (3rd edition), 1998, New York: Wiley. ISBN: 978-0-471-88702-7.

[MEY16] E. Meyer-Scott, *Heralding Photonic Qubits for Quantum Communication.* Thesis presented to the University of Waterloo in fulfillment of the thesis requirement for the degree of Doctor of Philosophy in Physics (Quantum Information) Waterloo, Ontario, Canada, 2016.

[MIN05] D. Minoli, *Nanotechnology applications to telecommunications and networking,* New York, Wiley, 2005.

[NIE10] M. Nielsen, *Quantum computation and quantum information: 10th Anniversary Edition.* Cambridge University Press, 2010, New York, NY.

[RIO23] F. Rioux, "Matrix Mechanics Approach to Polarized Light". Department of Chemistry, St. John's University/College of St. Benedict, St. Joseph, MN. Available online on December 3, 2023 at https://faculty.csbsju.edu/frioux/polarize/polarize.htm

[SAU10] S. Saunders, J. Barrett, *et al, Many Worlds?: Everett, Quantum Theory, & Reality.* Oxford University Press, 2010.

[SHE22] N. Shettell, "Quantum Information Techniques for Quantum Metrology". Sorbonne Université, France, Submitted for the degree of Doctor of Philosophy, December 2021, arXiv:2201.01523v1 [quant-ph] 5 Jan 2022. Available online on December 19, 2023 at https://www.researchgate.net/publication/357618914_Quantum_Information_Techniques_for_Quantum_Metrology

[SHI23] G. Shinkai, R. Tsuchiya, Y. Kanno, Control Method of Quantum Bit and Quantum Computer. U.S. Patent Application 20230297873, 2023-09-21. Uncopyrighted material.

[SUS14] L. Susskind, A. Friedman, *Quantum mechanics: the theoretical minimum* (2nd edition), 2014, 2014, New York: Basic Books.

[TRA17] T. Norsen, *Foundations of quantum mechanics: an exploration of the physical meaning of quantum theory,* 2017, Springer.

[UTK24] The University of Tennessee, Department of Physics and Astronomy, "Quantum Mechanics - Elements of Physics II". Available online on January 4, 2024 at https://labs.phys.utk.edu/mbreinig/phys222core/modules/m10/quantum_mechanics.html.

[VAR22] G. Vardoyan, M. Skrzypczyk, S. Wehner, "On the Quantum Performance Evaluation of Two Distributed Quantum Architectures". Performance Evaluation 153, 2022. https://doi.org/10.1016/j.peva.2021.102242.

[WHA09] B. Whaley, D. Gorman, "C/CS/Phys191: Qubits, Quantum Mechanics and Computers - Fall 2009". Available online on 11/27/2023 at https://inst.eecs.berkeley.edu/~cs191/fa09/ (e.g., https://inst.eecs.berkeley.edu/~cs191/fa09/lectures/lecture1_fa09.pdf, etc.)

Basic Glossary of Key Concepts in Chapter 3

This Glossary is based in totality on various industry sources, including in particular [GOT04], [MEY16], [UTK24], and [WHA09]. This Glossary should be used not only as a repository of terms and terminology, but as a tool that further elaborates, explains, and expands on various fundamental quantum concepts.

Fundamental concepts	Description
Bell inequality	An inequality that can be utilized to test if the system exists Bell nonlocality; a Bell test is a test to test Bell inequality.
Bell states	(Aka EPR states or EPR pairs.) Quantum states of two qubits that represent an example of quantum entanglement: when one of the two qubits is measured, it acquires a specific value, and the second qubit is forced to also take on a specific value, being that the entangled state collapses. There are four Bell states (first digit in the brackets refers to the first qubit, and the second digit refers to the second qubit): $$\lvert\beta_{00}\rangle = \frac{\lvert 00\rangle + \lvert 11\rangle}{\sqrt{2}}$$ $$\lvert\beta_{01}\rangle = \frac{\lvert 01\rangle + \lvert 10\rangle}{\sqrt{2}}$$ $$\lvert\beta_{10}\rangle = \frac{\lvert 00\rangle - \lvert 11\rangle}{\sqrt{2}}$$ $$\lvert\beta_{11}\rangle = \frac{\lvert 01\rangle - \lvert 10\rangle}{\sqrt{2}}$$ The Bell states can be generated by a quantum circuit consisting of a Hadamard gate and a CNOT gate.
Bell theorem	Bell's theorem encompasses several related results (a family of results) in physics that establish that QM is incompatible with local hidden-variable theories. The terminology applies to various descriptions of the concept, the first of which was introduced by Bell [BEL64]. Accompanying inequalities show that the predictions of QM cannot be replicated when one abides by local realism theories. Bell noted that *if a hidden-variable theory is local it will not agree with QM, and if it agrees with QM it will not be local.* Local realism embodies two facets: (i) locality requires that spacelike-separated events cannot influence one another; and (ii) realism requires that physical systems have pre-existing properties, independent of measurements. That local realism is incompatible with QM (and that QM is, in fact, correct) requires abandoning local realism as a fundamental principle of the universe [MEY16].
Compatible/incompatible quantum observables/measurements	The wave function immediately after a measurement is an eigenfunction or eigenstate of the operator associated with the measurement. If the operator (measurement) associated with a different observable does not change this eigenfunction, then the two measurements are said to be compatible; if the operator associated with a different observable changes the eigenfunction of the first observable, then the two observables are incompatible [UTK24]. When dealing with incompatible observables, the exact values of the state cannot be known to the observer at any specific given time: a follow-on measurement of an observable that is incompatible with the first one changes the state of the system to an eigenstate of the follow-on (second) observable and obliterates the information about the value of the first observable; thus, the observer is unable to know the value of both incompatible observables simultaneously with arbitrary precision; this is related to the uncertainty principle.

Fundamental concepts	Description
Compatible/incompatible quantum observables/ measurements *(cont.)*	In summary, ([UTK24] further elaborates that) although a state can be a simultaneous eigenstate of several compatible observables—and thus the observer can exactly know the values of several properties of the system at the same time and make exact predictions about the outcome of measurements of those properties—a state cannot be a simultaneous eigenstate of incompatible observables; thus, if a system is in an eigenstate of one of the incompatible observables and the value of this observable is known, then QM gives only the probabilities for measuring each of the different eigenvalues of the other incompatible observables. The outcome of a measurement of any of the other incompatible observables is uncertain; a measurement of one of the other incompatible observables changes the state of the system to one of its eigenstates and destroys the information about the value of the first observable.
Local realism	Postulating that the result of a measurement on one system cannot instantaneously influence the result of a measurement on the second system.
Observables	The physically meaningful entities of classical mechanics, such as position, momentum, and so on, are represented by Hermitian operators. These are known as "observables". In QM, one can generalize these observables, say A, to also include those observables that have no classical correspondence (for example, intrinsic spin).
Probabilistic outcomes and measurements	(Information summarized from [WHA09], Lecture 12) **Probabilistic outcomes:** A set of N replicas of a quantum system S described by a state $\lvert\Psi\rangle$, when subjected to measurements for a physical observable A, will yield in each measurement one of the eigenvalues $\{a_1, a_2, ...\}$ of operator $\hat{A}$, $\hat{A}$ here being the measurement operator; as $N \to$ the eigenvalues will appear with probability $P_\Psi(a_1), P_\Psi(a_2), ..., P_\Psi(a_i) = \lvert\langle a_i\lvert\Psi\rangle\rvert^2$, where $\lvert a_i\rangle$ is the eigenvector associated with the eigenvalue a_i. This defines the probability in terms of specific outcomes in a sequence of identical tests on copies of S, provided that $\sum_i P_\Psi(a_i) = \sum_i \lvert\langle a_i\lvert\Psi\rangle\rvert^2 = \langle a_i\lvert\Psi\rangle = 1$ (this is automatically satisfied for states that are normalized to unity). The expectation value of an observable A in an arbitrary state $\lvert\Psi\rangle$ appears as an average over a probability distribution: $\langle A\rangle_\Psi = \langle\Psi\lvert\hat{A}\lvert\Psi\rangle = \sum_i a_i P_\Psi(a_i)$. If the state $\lvert\Psi\rangle$ is an eigenstate of $\hat{A}$, then $\langle\Psi\lvert\hat{A}\lvert\Psi\rangle = a_j$, where a_j is the corresponding eigenvalue (thus, only a single term contributes to the expectation value of an observable A); this equates to projection onto eigenstates. One can generalize to projection onto an arbitrary state $\lvert\phi\rangle$. The probability to find a quantum system S that is in state $\lvert\Psi\rangle$ in another state $\lvert\phi\rangle$ is $P_\Psi(\phi) = \lvert\langle\phi\lvert\Psi\rangle\rvert^2$ This projection of the ket $\lvert\Psi\rangle$ onto another state—be it (i) an eigenfunction of some operator $\lvert a_i\rangle$, (ii) a basis function for the Hilbert space $\lvert v_i\rangle$, or (iii) an arbitrary state $\lvert\phi\rangle$)—is referred to as a "probability amplitude" (its square modulus is a probability). As seen in the equation above, the probability amplitude is specified both by $\lvert\Psi\rangle$ *and* the other state: the latter specifies the "representation" of $\lvert\Psi\rangle$, which realizes the quantum state in a measurable basis. **A single measurement** of the observable A on a state $\lvert\Psi\rangle$ in the basis (representation) of eigenstates of the measurement operator $\hat{A}$ yields the value a_i, with probability $P_\psi(a_i) = \lvert\langle a_i\lvert\Psi\rangle\rvert^2$. *(continues)*

Fundamental concepts	Description
Probabilistic outcomes and measurements (*cont.*)	Example: Consider the linear superposition $|\Psi\rangle = \alpha_1|1\rangle + \alpha_2|2\rangle + \ldots + \alpha_k|k\rangle$. Making a single measurement of the observable A on $|\Psi\rangle$ will result in the outcome a_i with probability $P_\psi(a_i) = |\alpha_i|^2$, and the resulting state after the measurement is equal to $|i\rangle$ $(\alpha_i/|\alpha_i|)$—namely, the measurement of the observable has "collapsed" the state $|\Psi\rangle$ to a single eigenstate $|i\rangle = |a_i\rangle$ of the measurement operator $\hat{A}$.
	More generally, the measurement operator $\hat{M}_i = |a_i\rangle\langle a_i|$ acts on the state $|\Psi\rangle$. The normalized state after measurement is equal to
	$$\frac{\hat{M}_i|\Psi\rangle}{\sqrt{\langle\Psi|M_i^\dagger M_i|\Psi\rangle}}.$$
	For a measurement in the $|a_i\rangle$ basis, this state is
	$$|i\rangle\frac{\langle i||\Psi\rangle}{\sqrt{\langle\Psi|M_i^\dagger M_i|\Psi\rangle}}.$$
	Note: Many observables only admit quantized eigenvalues—that is, a measurement can only result in one of a discrete set of values.
States, superposition	In QM, the "characterization" of the *state* $|\Psi\rangle$ of a physical system S at time t is provided by, or is perceived as, a vector $|v\rangle$ in a Hilbert space H that is appropriate to that system. $|\Psi\rangle$ is also perceived as a function, or more specifically a "wave function". QM provides a statistical prediction of the outcomes of all observables on the system.
	It is important that every linear combination of such state vectors $|\Psi\rangle$ represents a possible physical state of S—this is the *superposition principle*.
	Classical mechanics is deterministic; in a classical system, the state description is achieved with the specification of the positions and momenta of all particles, each of which can be precisely measured at any time; particle positions and momenta can be specified for all times using the classical equations of motion.
	In contrast, the quantum description is specified by the wave function $|\Psi\rangle$ that belongs to an (abstract) Hilbert space—the wave function has no obvious connection to the physical world.

Chapter 4

Basic Concepts in Quantum Physics: Some Physical Technologies

Many practical applications of Quantum Computing (Quacomp) require communication between quantum nodes and Quantum Computers (QCs). This chapter covers some of the basic physical technologies that have applications to Quantum Communications (Quacom) and Quantum Networking/Networks (Quanet[s]); it covers:

- Sources of nonclassical light (e.g., entangled light, squeezed light)
- Single-photon detectors (e.g., trapped ions techniques)
- Quantum memory and repeaters (e.g., Single Photon Avalanche Diodes)
- Transducers (e.g., Electro-Optic EO transducers)
- Simulators

However, it is not the purpose of this chapter to provide a primer on Quantum Mechanics (QM) *per se,* on quantum physics materials, or on specific quantum behavior.

4.1 Introduction

Quacomp and Quanets make use of QM effects such as superposition, entanglement, interference, tunneling, and superconductivity. As covered in previous chapters, *superposition* describes the ability of a quantum particle to be in multiple states at the same time; entanglement establishes the correlation between two quantum particles that forces the particles to behave in the same way even if they are separated by great distances; their behavior is related in predictable ways. Most Quanet applications require long-lived superpositions and entanglement. At the time of this writing, only short-lived superpositions and entanglement were practically achievable. Desktop (lab) devices such as intermediate-scale quantum processors and quantum-enabled sensors for use in measurements have utilized these short-lived phenomena; in the meantime, active research and development (R&D) in the field continues to seek ways of enhancing the durability of these QM phenomena for scalable commercial applications and deployability.

In the context of Quacomp, a QC is any computer that utilizes the principles of QM just listed to undertake computational operations. A number of variations of QC design exist, including photonic quantum computing, superconducting quantum computing, nuclear magnetic resonance quantum computing, and/or ion-trap quantum computing (more on this in the next chapter). Regardless of the particular type of QC implementation and underlying physical technology, all QCs encode data onto *quantum bits (qubits)*. Whereas classical computers encode bits into ones and zeros, as discussed in previous chapters, QCs encode data by placing a qubit into one of two identifiable quantum states in a two-state QM system. Qubits are subject to QM phenomena that cause them to behave very differently than classical bits; the quantum behavior enables the QC to process a large number of calculations simultaneously.

Qubits (or qudits) may be implemented in a variety of physical quantum systems; examples of qubits include the spin of an electron, the polarization states of photons, or the energy states of molecules, atoms, ions, nuclei, or photons. Other examples include quantum systems that utilize flux qubits, phase qubits, or charge qubits (e.g., in a superconducting Josephson junction), topological qubits (e.g., Majorana fermions—that is, fermions that are their own antiparticles), or spin qubits formed from vacancy centers (for example, nitrogen vacancies in diamond—see Figure 4.1) [MUK23], [LIT23], [DAY19]. Most qubits in use at the time of this writing require cryogenic temperatures. Superconducting qubits (e.g., such as Superconducting Quantum Interference

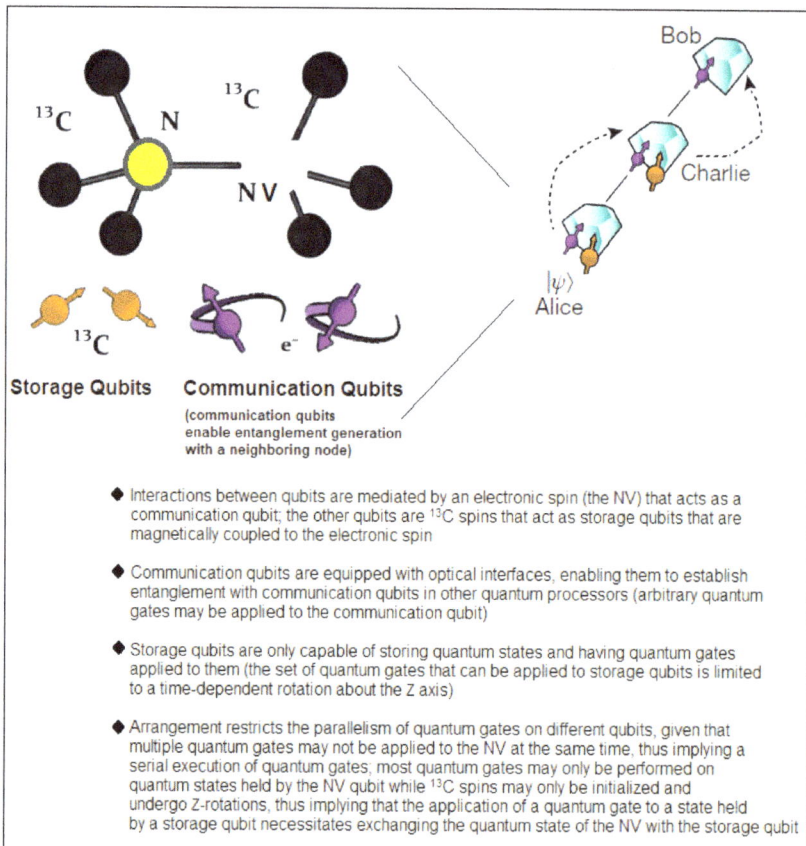

Interactions between qubits are mediated by an electronic spin (the NV) that acts as a communication qubit; the other qubits are ^{13}C spins that act as storage qubits that are magnetically coupled to the electronic spin

Communication qubits are equipped with optical interfaces, enabling them to establish entanglement with communication qubits in other quantum processors (arbitrary quantum gates may be applied to the communication qubit)

Storage qubits are only capable of storing quantum states and having quantum gates applied to them (the set of quantum gates that can be applied to storage qubits is limited to a time-dependent rotation about the Z axis)

Arrangement restricts the parallelism of quantum gates on different qubits, given that multiple quantum gates may not be applied to the NV at the same time, thus implying a serial execution of quantum gates; most quantum gates may only be performed on quantum states held by the NV qubit while ^{13}C spins may only be initialized and undergo Z-rotations, thus implying that the application of a quantum gate to a state held by a storage qubit necessitates exchanging the quantum state of the NV with the storage qubit

Figure 4.1   Nitrogen-Vacancy centers (text observations generated from Reference [VAR22])

Devices—SQUIDs) are electronic circuits that can be cooled to milli-Kelvin temperatures to exhibit quantized energy levels (e.g., due to quantized states of electronic charge or magnetic flux); superconducting qubits can be Josephson junction-based, such as transmon qubits; furthermore, superconducting qubits can be compatible with microwave control electronics and can be utilized with gate-based technology or integrated cryogenic controls [CHE23].

Qubit-utilizing molecular materials have a number of advantages: the ability to create a specific qubit structure, the ability to control qubit properties by altering the chemical structure, the opportunity of scalability in order to support many qubits, and the possibility of enabling Quacomp at room temperature [YAM24]. Trapped atoms[1] are one of the common implementations for Quantum Information Science and Technology (QIST) and Quacomp systems. Qubits based on trapped atomic ions enjoy attractive attributes, such as (i) having good coherence properties, (ii) being able to be prepared and measured with nearly 100% efficiency, and (iii) being easily entangled with each other by modulating them with external control fields such as microwave or optical fields. Atomic-based qubits can be used as quantum memories or as quantum gates in QCs and simulators, and they can act as nodes for Quacom networks.

An example of using ions is an early experiment at the National Institute of Standards and Technology (NIST) that entangled a set of three ions, then destroyed the quantum state in one ion and teleported the state to another ion [WIN04]. This experiment entailed laser-beam manipulations to transfer quantum states of one beryllium atom to another within a set of microscale traps. The properties that were teleported included the "spin state" of the ion (up, down, or a superposition of the two) and the "phase" (which has to do with the relative positions of the peaks and troughs of an ion's wave properties). An innovative approach was required because measurements always alter quantum states (for example, causing superpositions to collapse). Therefore, the experiment teleported the quantum state without measuring it. Lasers are used to manipulate the ion's spin and motion and to entangle the ions by linking their internal spin states to their external motion. A key goal of the experiment was the capability to entangle ions and then separate them in the trap, but maintaining entanglement.

In the context of Quanets, particular technologies are needed for transmitting qubits over a distance, given that arbitrary qubits cannot be precisely duplicated, and quantum information cannot be amplified over long distances; typically, technological efforts are aimed at minimizing and addressing channel loss. In developing Quanets, one needs to keep in mind that some qubits exist in systems operating at *microwave* frequencies, while other qubits exist in systems operating at *optical* frequencies. Networking and reading of different quantum computing devices in Quacomp/Quacom can be accomplished in a variety of ways, depending on the type of qubit involved. *Some* of the technologies or subsystems that are needed for transmitting qubits over a distance are discussed in the sections that follow.

---

[1] Quacomp with trapped atomic ions (qubits) use the ions' combined motion to create the entangling gates. Because the initial motional state affects the gate operation, the ions are typically first cooled to near the motional ground state at the beginning of or during the quantum computation. The number of motional modes that need to be cooled is proportional to the number of ions. Traditionally, the motional modes are cooled sequentially—that is, a next-motional mode is cooled only after the previous motional mode has been cooled. As the number of atomic ions increases, this causes the total cooling time to lengthen. As the cooling process becomes longer, heating of the motional modes from electric field fluctuations in the ion trap electrodes can cause noise that can then overwhelm the cooling process. It thus becomes advantageous to implement faster cooling methods—for example, as described in [AMI23].

4.2 Sources of Nonclassical Light for Quanet Application

Classical optical communications entail (i) the generation of a group of photons by an appropriate source, such as a laser diode (LD) or a light emitting diode (LED), at a specified frequency, (ii) transmission of the group of photons along an appropriate low-loss waveguide (such as an optical fiber), and (iii) the collection of the photons at the receiving end by a photodetector. In nearly all commercial cases there is no modulation, just binary encoding schemes, the simplest (illustrative) case being no photons/light seen as a 0 bit and presence of photons/light seen as a 1 bit[1]. Considerable distances (hundreds of unrepeated miles) and bandwidth (gigabits to terabits per second) can be achieved.

4.2.1 Entangled Light

Quacom relies on a different technological mechanism/model. One such technology is nonclassical (quantum) 'light'. Nonclassical light is 'light'—that is, a stream of photons at a specified frequency (which may, or more commonly may not, be in the visible domain), that can only be described by QM; namely, when classical mechanisms are sufficient. *Entangled light (photons)* and *squeezed light* are two types of nonclassical light. Using nonclassical properties of light can improve a number of tasks in communication, computing, and metrology—these applications all require high detection efficiencies in support of the generation and manipulation of underlying quantum states. There are extensive, ongoing R&D efforts in the nonclassical 'light' field.

Sources of *entangled photons* are needed for photonic quantum information processing since, as noted above and earlier in this text, many quantum photonic-based technologies require the efficient generation of entangled pairs of photons. As noted in earlier chapters, in *entangled light* the quantum state of one photon cannot be described independently of the quantum state of one or more other photons, and when some physical properties of these photons are measured, the photons are always found to be correlated. Entangled light is used for numerous applications in QIST, including superdense coding, quantum error correction, quantum cryptography, quantum teleportation, one-way Quacomp (e.g., via cluster states or graph states), and linear optical Quacomp. Make note that multimode squeezed light (described further below) is also entangled.

At the time of this writing, there are a number of ways to produce entangled pairs of photons reliably. Since the 1980s one has been able to generate pairs of entangled photons through the process of Spontaneous Parametric Down Conversion (SPDC), where a beam of photons is split into entangled pairs by passing it through a crystal. Unfortunately, there is an intrinsic disadvantage in these sources: to avoid excessive noise (that is, multiple photon-pair events), the photon pairs can only be produced at a low per-pulse emission probability (p), typically $p < 0.1$ per excitation pulse [CHE18]. Sources based on parametric down-conversion operate at very low efficiency per pulse due to the probabilistic generation process; see Figure 4.2. This inefficiency has been somewhat of a bottleneck, but ongoing technological progress aims at addressing the issue(s).

In spite of the limitations, SPDC is a commonly used technique for generating single photons (SPs). SPDC is a nonlinear process wherein a photon from a classical pump laser is converted, as it travels through a nonlinear optical crystal, into a photon pair—the emerging

---

[1] Most fiberoptic digital communication systems use Non-Return to Zero Inverted (NRZI) encoding: when one has a transition from light to no light or from no light to light, then a "1" bit is signaled; when there are two successive pulses of light or two successive periods of dark then a "0" bit is signaled.

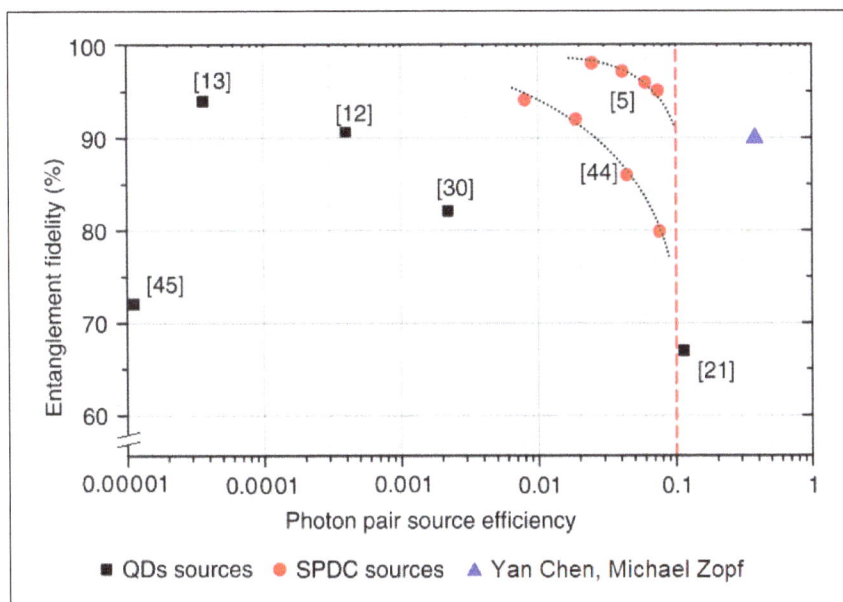

Figure 4.2   Parameters of generated entangled photons (synthetized from [CHE18]). [Refer to source for definition of references cited in graph. Entanglement fidelity was defined in Chapter 3.]

photons are typically entangled. While the down-conversion is a random process, a detection of one photon of the pair heralds the presence of the other; therefore, SPDC can be used as a probabilistic source of SPs.

Because SPDC is easy to implement, the technology lends itself to a straightforward integration, and SPDC-based SPs are available at practically any optical wavelength. SPDC-based entanglement sources are extensively used for Quacom, particularly for sharing entanglement between distant nodes. Combining SPDC-generated photon pairs with optical elements, such as beam splitters and detectors, generates quantum light with so-called *exotic quantum states* that can be useful for multipartite protocols, quantum error-correction, and Quantum Repeaters (Quareps) [NIS23]. In particular, one can generate narrow-linewidth SP pair sources based on cavity-enhanced SPDC. By placing the SPDC non-linear crystal inside a cavity, the generated pairs are restricted to the modes of the cavity, which can be orders of magnitude narrower and brighter than the naturally broadband SPDC process. Each photon in the pair has a different wavelength, and one of the photons is compatible with quantum memory (in bandwidth and wavelength), while the second photon is optimized for fiber transmission [SLA15], [SLA19].

There is ongoing interest in a compact, bright, single-mode, and narrow-linewidth photon source for quantum (as well as both classical) applications. Such sources, as noted, are often based on broadband SPDC where line narrowing then typically relies on the Optical Parametric Oscillator (OPO) solution; optical parametric amplification is triggered by initial SPDC events carried out by a nonlinear optical material—for example, PPLN (Periodically Poled Lithium Niobate)—inside an optical cavity [LAI23]. In classical applications OPOs are operated above a certain level of input light (the pump power threshold) to achieve amplification (cavity enhancement) and narrow-line outputs at those optical frequencies simultaneously compatible with the nonlinear crystal's energy and phase matching conditions, the cavity resonances, and the application at hand. In the context of quantum applications, the output beam must also retain the quantum features that characterize a SP source (i.e., neither thermal nor coherent, but instead anti-bunched); this can be achieved below the pump power threshold, where the quantum nature of the initial SPDC events still prevails. Figure 4.3, synthetized from [LAI23], depicts one recently proposed waveguided solution that capitalizes on the inherent stability to operate at precise pump levels and achieve a compact, single-mode light source with a linewidth in the low GHz range and compatible with both classical and quantum applications[1] (the solution is purported to offer a brightness equal or superior to the sources currently envisioned for quantum applications).

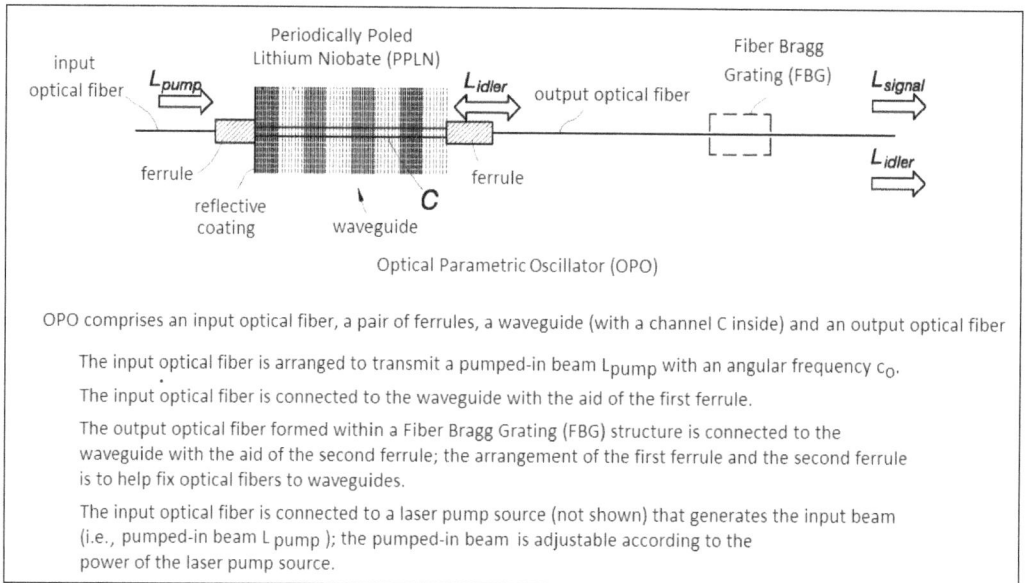

Figure 4.3   Single-mode and narrow-linewidth photon source using an OPO (synthetized from [LAI23])

4.2.2 Squeezed Light

In *squeezed light,* the variance of the amplitude quadrature is reduced relative to the variance of the phase quadrature (or vice versa)—by comparison, coherent light (say, as emitted by a

---

[1] [LAI23] notes that the state of the art of the OPO's resonant cavity is based either on (i) bulky mirrors and etalon (linear or ring configurations), (ii) a combination of mirrors and volume Bragg grating for frequency selection, or (iii) a fiber-loop ring (that includes the wave-guided nonlinear element) knotted by a fiber Bragg grating. It notes that mirrors-based cavities deliver single modes with 1-to-100 MHz linewidths, while Bragg grating solutions are limited to low GHz linewidths; the latter, in their waveguided version, are also a step closer to photonic integrated circuits and could even see their linewidth reduced to 1 kHz if a fibered saturable-absorber is inserted. Among the various cavity-enhanced SPDC implementations, the mirrors-based solution is well established but is not without flaws: alignment and stability issues that go with a free-space and discrete-components solution are expected. For quantum applications, to benefit from some cavity enhancement and conserve at the same time a quantum output, the device is generally operated well below the pump power threshold and results in a relatively weak brightness.

laser) has equally sized variances in the quadratures. Squeezing is characterized by a squeezing factor that quantifies how much the variance of one quadrature has been reduced with respect to the vacuum state or ideal coherent light; a design goal of squeezed-light systems is to achieve high squeezing factors [ZHA23]. Squeezed states have been successfully achieved utilizing SPDC inside an asymmetric cavity. Applications of squeezed light include optical interferometry, radiometry, and coding of quantum information. One challenge that arises in squeezed-light devices is that loss degrades the squeezing factor; the loss can arise from any mechanism (i.e., absorption, scattering, reflection, and so on). To prevent this degradation, it is important that T, the transmissivity, be made as close to 1 as possible, say, using anti-reflection coatings, minimizing changes in the refractive index, and using materials with low absorption.

One type of squeezed light is *squeezed vacuum*. Reference [ZHA23] describes an exemplary Photonic Integrated Circuit (PIC) that generates squeezed and entangled light, and in particular a two-mode squeezed-vacuum state. Squeezed-vacuum injection can be used, for example, to "quantum enhance" a variety of optical interferometers, including Mach-Zehnder interferometers, Fabry-Perot interferometers, and Sagnac interferometers. Applications of quantum-enhanced metrology include optical engineering, timekeeping, inertial navigation, seismology, biology, optical coherence tomography, microscopy, among others. PIC technology is being used to develop compact, robust platforms for squeezed and entangled light generation outside of laboratory environments. PICs can be utilized to generate squeezed light around 1550 nm, which lies within the telecom C-band; for many applications in Quacom, squeezed light may need to travel several kilometers over low-loss optical fibers (e.g., less than 1 dB/km) already used for telecommunications—however, PICs may be used to generate squeezed light at any wavelength with appropriate choice and engineering of materials.

The PIC described in reference [ZHA23] includes a first microresonator[1] that generates a two-mode squeezed vacuum using spontaneous four-wave mixing. The first microresonator has a nonlinear optical medium that converts two pump photons into a pair of entangled signal and idler photons. Due to imperfect conversion efficiency, some of the pump light may co-propagate with the signal light and idler light. This "unconverted" pump light can be spectrally filtered since it has a different wavelength than the signal light and idler light. In existing chip-based squeezed-light generation, pump filtering is typically done off-chip with a wavelength division multiplexer or interference filter; however, such off-chip filters introduce additional losses that degrade squeezing.

By incorporating a pump filter on the PIC, the PIC proposed in the reference reduces loss, and, therefore, preserves squeezing. Specifically, the PIC includes a second microresonator that is tuned to resonate with only the pump light (i.e., not the signal light or idler light). The second microresonator is located after the first microresonator and couples the pump light into a

---

[1] Optical resonators are fundamental devices not only in laser applications, but also as etalons (devices employed for measuring small differences in the wavelength of light), utilizing the interference it produces for optical filtering and as tools for accurate measurements. Bulk optical resonators have a number of limitations due to their size, weight, alignment, and stability problems. An emerging class of miniaturized optical resonators, also exploiting the confinement of light beams, is constituted by dielectric structures having circular symmetry, which sustain the so-called Whispering Gallery Modes (WGMs) (electromagnetic waves that circulate and are strongly confined within the structure). In terms of geometric optics, the confinement is described by the optical rays, which are totally internally reflected and focused by the surface itself [RIG11].

waveguide that brings the light off the PIC. Thus, the second microresonator essentially acts as a notch filter. Integrating this pump filter onto the PIC adds negligibly to the path length of the squeezed light, and therefore may save the propagation losses incurred when using a much larger off-chip filter. In addition, removing the off-chip filter advantageously eliminates the insertion loss incurred when coupling light into and out of an additional optical component. See Figure 4.4 (synthetized from [ZHA23]).

4.2.3 Other Source

NIST has a number of research efforts underway in the nonclassical (quantum) light arena (among many other academic and commercial R&D entities that continuously advance the science). For example, researchers have developed a nanophotonic visible-telecom entangled photon pair source using high-quality factor silicon nitride resonators; the highly non-degenerate and narrow-band photon pairs enjoys purity and brightness supporting entanglement distribution over 20 km in fiber [LU19]. A source of polarization-entangled photon pairs using SPDC in conventional telecom bands (for example, 780 nm or 1500 nm) is useful to have and has recently been documented by NIST; the source can be used to verify point-to-point entanglement distribution in a quantum network (the entanglement visibility is above 93%)—producing narrow band pairs more suitable for most applications [KUO20]. A non-degenerate sequential time-bin entangled photon-pair source has been previously demonstrated using a Periodically Poled Potassium Titanyl Phosphate (PPKTP) waveguide at a clock rate of 1 GHz; the wavelengths of the signal and idler are 895 nm and 1310 nm, which are suitable for local and long-distance optical communications, respectively [MA09]. Rydberg-excited atomic ensembles can also serve as a source for single photons; for example, see [ORN20].

An *isolated single quantum particle* can be used as a deterministic single-photon source. These sources commonly include Quantum dots (QDs), Nitrogen-Vacancy Centers (NVs) mentioned earlier, single atoms, and ions (QDs are single "artificial atoms" comprising a few hundreds of real atoms within a semiconductor crystal).

QD-based single-photon sources can generate up to a billion photons per second and can be optically integrated. However, one of the technical challenges is ensuring that the excitation of the quantum particle produces a photon in a well-defined spatial mode. Furthermore, because emission wavelengths of QDs are different from those used in telecommunications, transducers may be needed, as described in [NIS23]. Semiconductor QDs are a promising alternative to SPDC sources, given that they are more efficient at producing entangled photon pairs: semiconductor QDs can emit single pairs of entangled photons deterministically, but they currently suffer from extremely low-extraction efficiency. Strategies for extracting single photons from QDs, such as embedding them in narrowband optical cavities, are difficult to apply to entangled photons. Researchers have made progress recently in solving several key problems of QDs as sources. One of the well-known problems is that the small energy difference between the intermediate states, or Fine Structure Splitting (FSS), reduces the quality of the entanglement. To obtain a high entanglement fidelity (F) in time-integrated measurements, one needs to cherry-pick a QD with very small FSS [CHE18].

There are many applications (e.g., Quacomp, quantum simulations, atomic and molecular experiments, spectroscopy, magnetic sensors, atomic clocks) *that require light of two or more specific "correct" frequencies*, each with high absolute frequency stability. For example, in the

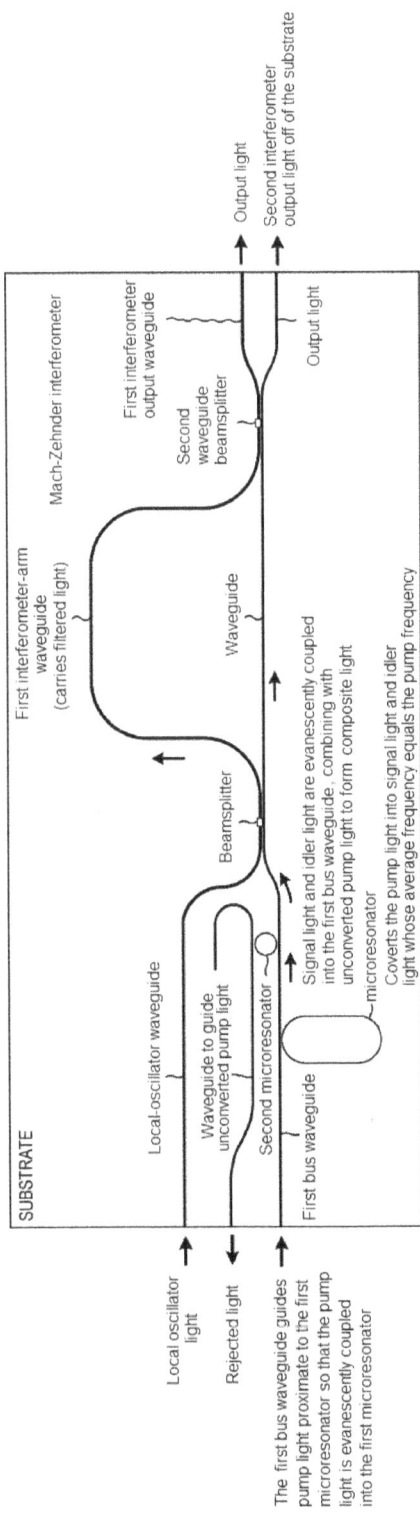

Figure 4.4 Photonic integrated circuit (PIC) that generates entangled light which under certain conditions is also a two-mode squeezed vacuum state (synthetized from [ZHA23])

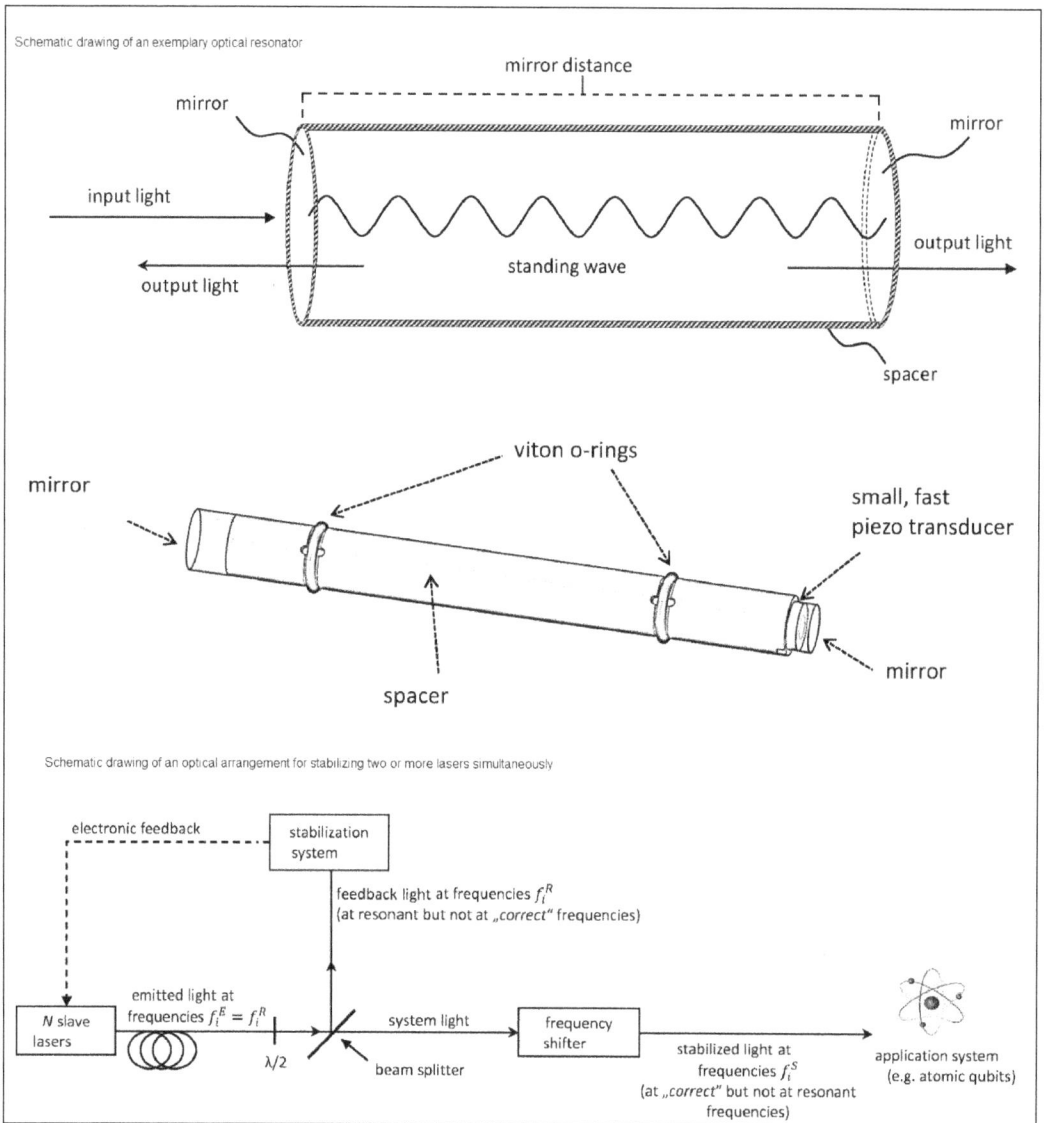

Schematic drawing of an exemplary optical resonator

mirror distance

mirror

mirror

input light

output light

standing wave

output light

spacer

viton o-rings

mirror

small, fast
piezo transducer

mirror

spacer

Schematic drawing of an optical arrangement for stabilizing two or more lasers simultaneously

electronic feedback

stabilization
system

feedback light at frequencies f_i^R
(at resonant but not at „*correct*" frequencies)

emitted light at
frequencies $f_i^E = f_i^R$

N slave
lasers

$\lambda/2$

system light

frequency
shifter

beam splitter

stabilized light at
frequencies f_i^S
(at „*correct*" but not at resonant
frequencies)

application system
(e.g. atomic qubits)

Figure 4.5 Laser stabilization (Synthetized from reference [TAK23])

field of Quacomp, light of lasers with high-frequency stability (e.g., 1 MHz over 1 s) is required
to implement quantum gates when using trapped ions to represent qubits. The light may be
provided by two or more "slave lasers". Each of such slave lasers is required to deliver light at
a "correct" frequency and may have to be stabilized in order to maintain the correct frequency
over a possibly long time period.

Some application systems require many single-frequency lasers, each with high absolute frequency stability. As soon as there is a plurality of lasers, they all have to be separately stabilized
using stabilization schemes. These stabilization schemes may become complex when independent references (e.g., frequency combs) are used for each laser. An optical cavity may be used to
stabilize a plurality of laser sources at the same time. In particular, the length of the cavity is set to

that at which frequencies of all respective laser sources are resonant or nearly resonant. Reference [TAK23] presents an exemplary method for stabilizing simultaneously N slave lasers in order to output stabilized light of N mutually different predetermined frequencies f_i^S, $i = 1, \ldots, N$; see Figure 4.5. The method uses an optical resonator formed by two mirrors and includes a step of adjusting the distance between the two mirrors to a stabilization length. At the stabilization length, there is, for each predetermined frequency f_i^S, a resonant frequency f_i^R of the optical resonator for which a difference between the predetermined frequency f_i^S and the resonant frequency f_i^R is smaller than an established target value. The method also includes a step of feeding light from each of the N lasers to the optical resonator, thereby generating N respective error signals.

4.3 Quantum Memory and Quantum Repeaters

Optical quantum memories are important elements for QIST applications such as Quanets, Quareps, and linear optical quantum computing. Quantum memories store quantum states (qubits) while also preserving entanglement, which is fundamental for data protection. In a classical setting digital information can be copied and stored in a memory; such data can be read and recalled at a desired time. This is not possible with qubits (e.g., with quantum states of light) because the "No-cloning theorem" establishes that an arbitrary unknown quantum state cannot be copied [WOO82]. It prevents quantum information from being transmitted to more than a single destination, such as via re-routing or re-directing. Thus, researchers have endeavored to create a "quantum memory" in which a quantum state of light, such as a single photon, is temporarily localized, or reversibly transferred into a stationary excitation before being later released when needed. Rare earth–doped crystals are candidates for quantum memories: they provide a compact platform comprising a large number of atoms trapped in a crystalline structure; moreover, they feature good coherence properties at cryogenic temperatures.

In existing memory technologies,[1] researchers and developers have assessed the process of mapping a quantum state of light, such as a single photon, into another type of excitation in matter to temporarily store it before reversing the mapping at a later time (effectively, the signal photon is annihilated in a controlled way to create a storage excitation, and then the storage excitation is annihilated in a controlled manner to create a new signal photon). Some techniques utilize a single-emitter approach in which a single entity such as an atom or quantum dot is driven to an excited state and then subsequently emits a single photon when it relaxes; the single emitter approach has the advantage that the source can never emit two photons simultaneously—however, it is a significant technical challenge to isolate and control a single emitter [BUS23], [LAG23]. An alternative approach is to use a photon pair source such as spontaneous parametric down-conversion or spontaneous four-wave mixing [COH09], [COU18].

These sources, however, have significant drawbacks because they only operate in a stochastic, or probabilistic, manner; nonetheless, it is a fundamental property of photon pair sources that they always create pairs of "daughter" photons. Thus, even though they emit photon pairs stochastically, a photon pair source can be used as a source of single photons because detection of one of the daughter photons, conventionally called the "herald photon", is used to herald the fact that a second daughter photon, conventionally called the "signal photon" is available

---

[1] Additional discussion about memories is provided in Chapter 6.

to be used as a resource. The probabilistic emission of photons from photon pair sources is a significant limitation for practical use: technologies such as quantum photonic computers may require hundreds of single photons to be produced simultaneously from independent sources.

To negate this problem and make photon pair sources useful for practical applications, there has been research aimed at multiplexing photon pair sources—that is, combining multiple independent probabilistic photon pair sources together such that their overall success probability as a single photon source is higher. Ideally, independent sources can be multiplexed to create a deterministic single photon source that always emits a single photon on demand. Researchers have investigated the use of various properties of light for multiplexing including frequency, spatial, and temporal degrees of freedom. Quantum memories could be employed for multiplexing photon pair sources because photons created at different times could be stored temporarily and then released together when required [BUS23], [EIS11], [NUN13].

The storage time for a quantum memory depends on the application: memory usage in long-distance communication requires relatively long storage times (ms to s), while other applications can operate with shorter storage times. Memory operation in the telecommunications band (for example, 780 nm or 1500 nm) is desirable for long-distance Quacom because of the practical compatibility with existing optical fiber networks and infrastructure (the majority of existing quantum memory technologies, however, are not tunable in frequency or bandwidth because of the energy-level structure of the storage medium).

SP sources, such as those based on SPDC, have proved reliable emitters of such qubits as noted earlier. Typical implementations of optical quantum memories are based on materials with narrow transition lines and thus require light sources with comparable linewidths. Optically controlled quantum memories of the Raman type or of the Atomic Frequency Comb (AFC) type have extended the range of compatible source linewidths from MHz to GHz [LAI23]. To address and mitigate the loss of photons during their travel inside optical fibers, one needs to establish an efficient physical link between two end nodes in order to achieve higher input rates; in addition, applications that use quantum memories also rely on heralding processes such as detection stages (i) that test anti-bunching, or (ii) that confirm entanglement—the higher rates can mitigate delays incurred as a result of these measurements.

As is the case in classical networks, long-distance Quanets need to use (some kind of) repeaters to mitigate exponential loss of the signal associated with transmission over a distance. As noted in earlier chapters, unknown quantum states cannot be copied with perfect fidelity; in addition, measurement consumes quantum properties such as entanglement. It follows that Quareps need to operate on physical principles different than the principles of classical repeaters. Research has shown that Quareps can be built with quantum memory and/or they can be built utilizing exotic quantum states—although there are various types of not-uniquely-defined "exotic states"; the most common characterization of exotic states refers to certain unique and unusual states (behaviors) of matter that certain materials exhibit under extreme conditions, such as low temperatures or high pressures (one example is an exotic state of matter called a *Bose–Einstein condensate*[1]).

One of the practical issues in developing quantum memories is the efficient coupling of photonic qubits with the material systems used for the quantum storage. Consequently, the implementation of scalable quantum memories, even at the stage of just having lab-grade systems, is still challenging; but R&D is ongoing at a good pace. In order to enable effective long-distance

---

[1] Others (not further discussed here) include, but are not limited to, fermionic condensates, nuclear matter, quantum spin liquid, supercritical fluid, color-glass condensate, quark–gluon plasma, Rydberg matter, and photonic matter.

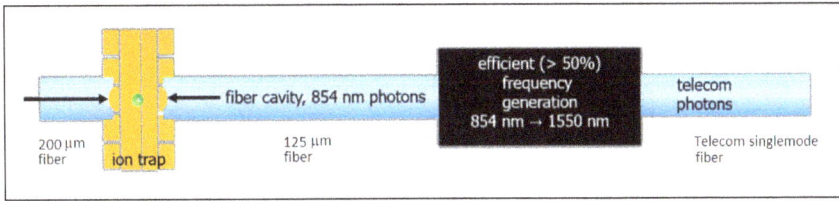

Figure 4.6   Conceptual design of a trapped ion-based quantum memory (*Source:* NIST [NIST23])

Quacom and support advanced quantum applications, there is a need to generate long-lived entanglement between two distant quantum nodes that can store and manipulate qubits. In recent years short-lived entanglement has been demonstrated over short distances of about 60 miles on the ground by transmitting photons over standard telecom fiber, as well as from space over about 800 miles from a satellite [DAY19]. The near-term goal is to enable entanglement over terrestrial distances of hundreds or thousands of miles that can persist for a few hours.

4.3.1 Trapped Ions Techniques and Electromagnetically Induced Transparency

Material systems that can and have been utilized for quantum memories include trapped ions and neutral atoms, as note in [NIS23]:

- *Trapped ions techniques,* alluded to earlier, have been used to develop lab-grade quantum memory with lifetimes exceeding 30 minutes. In addition, low ($< 10^{-3}$) errors per entanglement swapping operation can be reached, which is fundamental for implementing efficient Quareps. The practical challenge is to efficiently couple ions to telecom photons to realize entanglement distribution between repeater stations separated by 20 miles or so.[1] See Figure 4.6.
- Coherent interactions of light with multilevel atoms can modify their optical response via quantum interferences between various excitation pathways or via strong-coupling-field-induced energy level splitting. *Electromagnetically Induced Transparency* (EIT) is a coherent optical nonlinearity that renders a medium transparent within a spectral range around an absorption line; it creates a narrow transparency window by eliminating a resonant absorption. EIT has a variety of applications such as ultraslow light propagation, light storage, dissipation-free light transmission, and nonlinear optics with weak light [PEN14]. EIT in cesium atomic ensembles has been used to experimentally store photons in warm ensembles for several μs before being released on demand [MA17], [BHU20]. Cold ensembles using Magneto-Optical Traps (MOTs) can achieve longer storage times; the MOT traps the atoms, enabling better interaction between the single photon (flying[2] qubit which transports the quantum state) and the atom (stationary qubit that store the quantum state).
- *Atomic ensemble-based quantum memory and repeaters.* This entails (in a specific R&D effort) Furthermore, ensemble-based quantum memories based on arrays of neutral atoms can significantly improve the interfacing of the memory with light (systems) [GOR07], [GOR11], [ORN20].

Being able to individually control each ion in an ion trap is a design requirement, as dictated by a specific quantum gate or quantum operation to be implemented by the ion trap; typically,

---

[1] Classical repeaters had been placed every 2–5 miles for twisted pair/T1 systems, 25–30 miles for microwave systems, and 10–200+ miles for fiberoptic systems. Similar topological designs would be advantageous.

[2] Predicated on the fact that photons carry quantum states from location to location at high speed, they are referred to as "flying qubits".

there is a need to control the phase, frequency, and amplitude, as well as the polarization. At the same time, there is a need to manage or eliminate crosstalk among individual channels, given that the accuracy with which atomic-based qubits are manipulated is critical in order to avoid errors during quantum operations.

4.3.2 Acousto-Optic Modulators

Acousto-Optic Modulators (AOMs) can be used to control optical beams, which in turn are used to control the atomic-based qubits [BEC23]. However, in a multi-channel AOM, in which each channel may correspond to an individual optical beam and its atomic-based qubit, the application of radio frequency (RF) signal in one channel may inadvertently impact one or more other channels; for example, turning on one channel may result in an adjacent or nearby channel being unintentionally turned on. A portion of the RF energy and/or acoustic energy associated with one channel in a multi-channel AOM may "leak" into one or more neighboring channels. The leak contributes to crosstalk noise or errors and typically impacts the operation and/or control of the AOM in a negative way (e.g., causing unintentional changes to the states of the trapped ions). It is important to minimize the impact of crosstalk noise on neighboring channels to avoid the unintended effects of such crosstalk noise on non-targeted atomic-based qubits.

To reduce or eliminate crosstalk among individual channels, additional compensation tones may be applied to each channel of the AOM such that the compensation tones may cancel/reduce crosstalk signals leaked from neighboring channels. Reference [BEC23] discusses an exemplary AOM system with a first waveform generator configured to apply a first RF tone to generate a first acoustic wave in a first channel of the multi-channel AOM. The first acoustic wave interacts with a second channel of the multi-channel AOM to cause a crosstalk effect on the second channel, but a second waveform generator is configured to apply a second RF tone to generate a second acoustic wave in the second channel, such that the second acoustic wave reduces or eliminates the crosstalk effect caused by the portion of the first acoustic wave.

One or more optical beams may be used to individually address the trapped ions, and in some instances, a global optical beam may be applied to all of the trapped ions. These counter-propagating optical beams (known as Raman beams) may be produced by using multi-channel AOMs. A multi-channel AOM is a crystal that has various piezo-electric transducers patterned with radio-frequency antennas, traces, or electrodes on one side and a radio-frequency absorber on the other side that deflect optical beams and shift their frequencies.

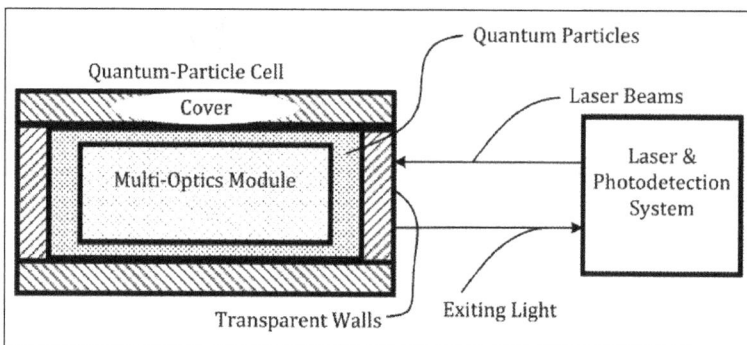

Figure 4.7   Example of quantum particle cell [HUG23]

4.3.3 Quantum Particle Cells

Quantum particle cells (including vacuum cells, cold/ultra-cold matter cells, vapor cells, and channel cells) may be of interest in Quanets. They provide hermetically sealed environments for the study of quantum phenomena such as superposition and entanglement. The quantum particles—for example, neutral or charged atoms or molecules—can be controlled using lasers located outside the cell; cells may have transparent walls or windows to allow light to enter and exit the cell. In some cases, light must be redirected or otherwise modified (e.g., focused, collimated, diffracted, split, polarized) within the cell. For example, optical and magneto-optical traps can require multiple orthogonal laser beams; here mirrors need to be precisely positioned, co-aligned, and attached within the cells to ensure the necessary directivity for the laser beams; however, the challenge of such precision, especially in the confines of a compact quantum-particle cell can contribute to the high cost associated with the finished product.

Reference [HUG23] describes a purportedly more efficient and reliable exemplary process for incorporating precision optical elements within a quantum-particle cell. See Figure 4.7. The quantum-particle cell can be managed (monitored and controlled) using a laser and photo-detection system that generates laser beams that enter the cell via its transparent walls or windows—laser beams can be used to heat dispensers, so they release quantum particles as a vapor.

With the process described in [HUG23], optical attachment modules can modify incoming laser beams; for example, incoming laser beams can be redirected to define an optical trap or to help define a magneto-optical trap to trap or guide quantum particles; incoming laser beams may further be redirected inside the cell to be incident upon a particle source for heating such that the generated plume or spray of particles is minimally or not incident upon the transmissive window of the cell but only upon the redirecting mirror to prevent "mirroring" of the transparent window, especially for corrosive or low vapor pressure species. In addition, incoming laser beams can be used to at least conditionally cause quantum particles to transition between quantum states. The states and transitions can be monitored using photodetection to track light exiting the cell to measure laser-beam absorption and/or quantum-particle fluorescence.

4.4 Single-Photon Detectors (SPDs)

SP detection is critical in the development of Quanets. Well-established technology exists for detection of visible light photons, but detectors at telecom wavelengths (e.g., 780 nm, 890 nm, 1310 nm, or 1500 nm) still require technical improvements. Desiderata for SP detection include very high detection efficiency (> 95%), low latency, low timing jitter, maximum count rate, and minimal afterpulsing. SP detection at telecommunication wavelength is of interest owing to the numerous applications of that technology in optical wireless communication (including free space optical communication and quantum key distribution).

4.4.1 Single Photon Avalanche Diodes

Applications of SP detectors that focus on the timing of a very weak optical signal mostly use *Single Photon Avalanche Diodes* (SPADs). SPADs are widely manufactured by many companies and have numerous applications.

Afterpulsing is a critical non-ideal factor of SPADs operating at telecommunication wavelength, that adversely affects any application that needs to measure the number or timing of detection events. An ideal single-photon detector generates one and only one electric pulse for every incident photon; however, in a real detector, it is possible that a single incident photon results in more than one electrical pulse per incident photon. This is known as *afterpulsing*—the term can also be defined as any pulse in addition to and following on after an isolated detection event and its subsequent dead time, regardless of its origin [ZIA18]. Afterpulsing can be linked to charges trapped in the deep levels of the semiconductor's band structure and released at a later time than the intended signaling; afterpulsing is also a problem in the context of the measurement of photon arrival times in Quacom protocols (as well as in other settings). To efficiently and accurately make use of these diodes, one must account for all non-ideal behavior of the detectors used.

4.4.2 Superconducting Detectors and Transition Edge Sensor Detectors

Types of SP detectors and some development efforts include, but are not limited to, the following, as noted in [NIS23]:

- *Superconducting detectors.* Superconducting Nanowire Single-Photon Detectors (SNSPDs) have become important to quantum optics in recent years because of their high performance: they enjoy low timing jitter (< 3ps), high detection efficiency ($\geq$ 98%), very low dark count rates (<10^{-3}/second), and competitive dead times. A SNSPD is made of a thin film of superconducting material that covers an active detection area as a meandering nanowire. One of the limits of scaling SNSPD systems is the heat load and electrical noise of the input and output connections from ambient temperatures to the SNSPDs in a cryostat [THI22] (this issue is not limited to SNSPDs: many quantum technologies require electronic control in a cryogenic environment). See Figure 4.8 for an example.

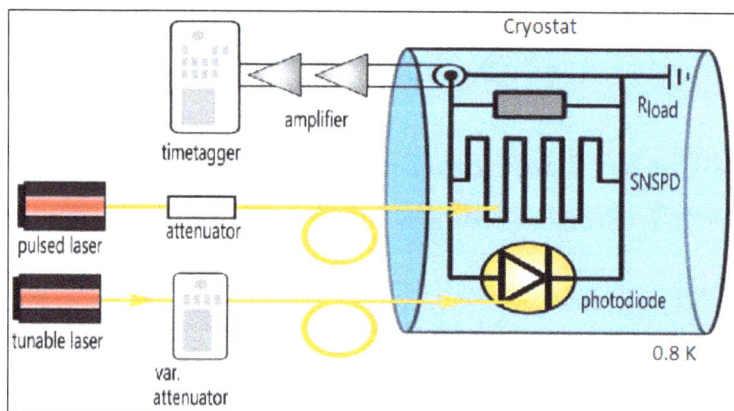

Figure 4.8   SNSPD (modeled after [THI22])

- *Transition Edge Sensor (TES) detectors* take advantage of a sharp dependence of resistance on temperature near the superconducting phase transition. The temperature change resulting from the energy of just one absorbed photon can be resolved; the temperature change due to absorption of several photons can be distinguished from that caused by one photon.

Therefore, TES detectors are photon-number resolving. TES detectors have high detection efficiency (> 98%); however, these detectors have a large jitter and a long recovery time and require very low temperature (~100 mK) to operate.

- *Semiconductor detectors.* Semiconductor-based single-photon detectors typically rely on the avalanche effect in reverse-biased semiconductors. SPADs, mentioned earlier, are convenient, room-temperature, low-cost, and high-performing detectors. Currently, they offer a reasonable detection efficiency (~50%), although dark counts and afterpulsing are still challenges to be addressed; these issues are particularly important for Quanets.

4.5 Transducers

Quantum transduction deals with the conversion of quantum signals from *one form of energy to another form of energy*. To support distributed Quacomp with Quanets, there is typically a need to transfer and convert quantum information between different types of physical qubits implemented via multiple quantum technologies. Transduction is the mechanism to achieve efficient conversion of quantum information from one material system to another without the loss of quantum coherence; thus, it deals with the conversion of quantum signals from one form of energy to another form of energy. In a typical example, a quantum system may generate entangled photons and may need a transducer to convert the given photon to a *photon that fits the well-established telecom band* and vice versa (as seen earlier in Figure 4.6). This can be accomplished using nonlinear frequency conversion. A key design goal is to minimize insertion loss and background noise (experimental transducers with noise measured as being less than 10^7 photons/hour have been demonstrated, but the goal is to achieve less than 10^2 photons/hour[1]).

For example, quantum transduction may involve conversion of energy associated with *microwave photons to a different form of energy associated with optical photons*. Networking of superconducting QCs requires transduction of quantum information between the microwave frequencies at which the QC operates and the optical frequencies at which low loss, quantum-preserving transmission can occur (e.g., in standard optical fibers). For example, quantum frequency transducers are used for interconnecting different quantum devices utilizing optical fibers by applying coherent conversion between superconducting microwave photons and optical photons at telecommunication wavelengths. An approach being studied uses *Vibrating Membrane Convertors* (VMC) by which microwave signals in the circuit cause the membrane to vibrate, altering the phase or amplitude of the laser light; conversely, changes in the phase or amplitude of the laser light cause the membrane to vibrate, producing an electrical signal that encodes the information in a microwave signal [NIS23]. Several other approaches are also being investigated (for instance, using quantum dots, e.g., [IMA22], [AUT20]).

Transduction between electromagnetic frequencies that are far in the electromagnetic spectrum—for instance, from telecom to microwave domains—is more challenging than optical-to-optical conversion (while the term "light" is used in the context of "optical systems", only the signal in the 400–700 nm range is in the visible portion).

---

[1] For example, [NIS23] reports demonstrations of low-noise transduction, with <104 photons/second background noise between the telecom band (1554 nm) and the Near Infrared (NIR) band (837 nm); but the noise figure can in principle be lowered to under ~100 photons per hour.

Near-infrared optical transducers may utilize PPLN crystals; advantageously, lowering the temperature in the crystal reduces the noise count rates. Integrated approaches to transducers are also being developed; one example is *silicon photonic platforms* where wavelength-scale light confinement enables nonlinear processes to be controlled by geometry. For example, see [BUR17], [KUO18], [KUO13]. Transducers have limitations, including losses in conversion from flying qubits to stationary qubits, typically dealing with light–matter interactions; performance is highly dependent on the implementation.

4.5.1 Electro-Optic (EO) Transducers

Among several approaches for quantum transduction (including opto-mechanics and piezo-mechanics), the *Electro-Optic (EO)* quantum transducers based on Pockels effect[1] may be promising for large-scale integration because these devices may be mechanically and thermally stable, electrically tunable, and may be integrated with superconducting circuits [KUD23]. Figure 4.9 is a generic example of an EO transducer.

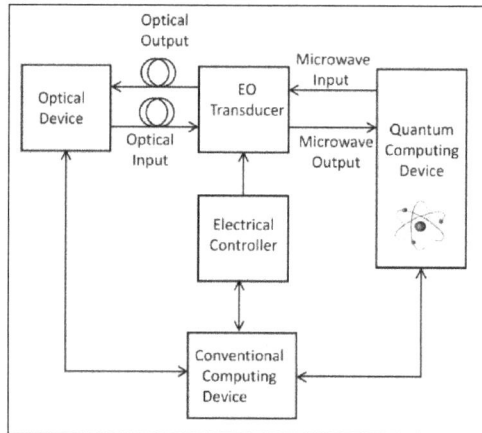

Figure 4.9   Example of use of transducer

4.5.2 Whispering Gallery Mode (WGM) Resonators

EO frequency mixing generally employs a superconducting microwave resonator (resonance at Ω_M) coupled to an optical resonator with large EO coefficient (such as lithium niobate, aluminum nitride, and barium titanate). The optical resonators may be designed to support both pump (ω_p) and generated sideband (sum and/or difference) frequencies ($\omega_s = \omega_p \pm \Omega_M$) where Ω_M is the resonance frequency [KUD23]. Other approaches are based on *coupled ring resonator devices* exhibiting double optical resonance owing to EIT, cited earlier. Such devices may be capable of suppressing undesired sidebands—for example, capable of suppressing the down

---

[1]   The Pockels effect (first described in 1906 by Friedrich Pockels) is the linear electro-optic effect in which the refractive index of a medium is modified in proportion to the applied electric field strength. This effect can occur only in non-centrosymmetric materials [PAS23a]. Materials exhibiting Pockels effects include crystal materials such as lithium niobate (LiNbO3) and lithium tantalate (LiTaO3), among others.

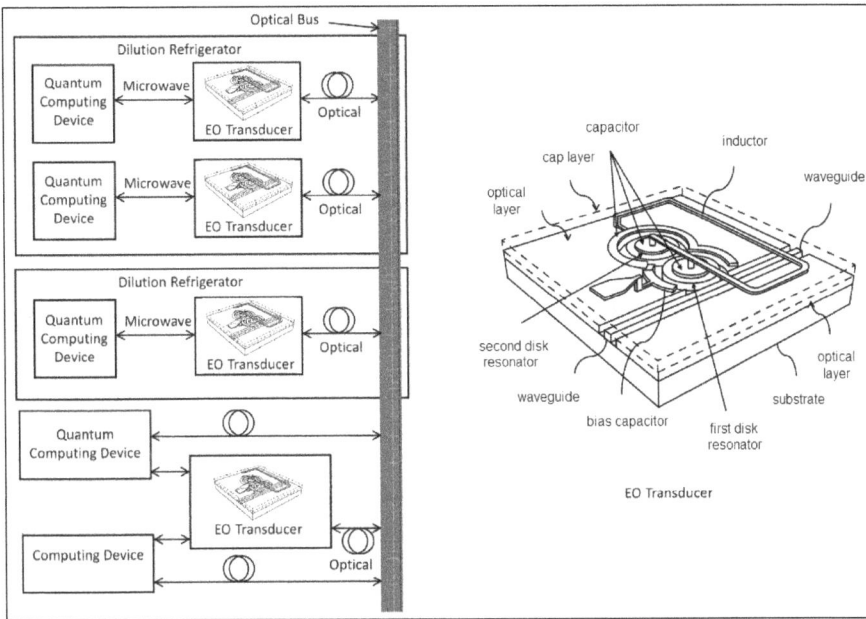

Figure 4.10   EO Transducer based on Whispering Gallery Mode (WGM) resonators (synthetized from [KUD23])

converted signal in the up-conversion (microwave-to-optical) process. However, these devices typically occupy a large footprint, say, up to 1 mm^2, and may have conversion efficiency below 2%; they may also exhibit large optical insertion loss stemming from the sidewall roughness of the ring waveguides. In contrast to larger ring resonators, *Whispering Gallery Mode* (WGM) resonators are compact and may provide a larger Q-factor because of their smaller mode volume. Reference [KUD23] provides an excellent example of an EO transducer based on two optically coupled WGM disks and a resonator. See Figure 4.10.

As noted, a class of transducers entails OE quantum transduction. Optomechanical coupling involves an interaction between a mechanical vibrational mode of a mechanical subsystem and an optical radiation field of an optical subsystem. For example, when there is an optomechanical coupling between a mechanical subsystem (such as a mechanical resonator) and an optical subsystem, vibration in the mechanical subsystem can cause a corresponding change in an optical radiation field of the optical subsystem.

Proposals have been made for utilizing a silicon-germanium (SiGe)/silicon (Si) electro-optomechanical quantum transducer, comprising an SiGe/Si optical ring resonator and a capacitor, that can be associated with (e.g., electronically and communicatively connected to) a qubit of a QC; the transducer has a photoelastic coupling with the silicon-germanium optical waveguide [TOK23]. Figure 4.11 is a diagram of a top view and a side view of an example of such a device that can comprise a quantum transducer associated with a qubit (the figure encompasses only one qubit, but in general it can be one of a group of qubits of a QC—e.g., a superconducting quantum computer). For example, the qubit can be a transmon qubit[1] comprising one or more Josephson junctions and a shunt capacitor that can be associated with one or more Josephson

---

[1] A transmon (transmission line shunted plasma oscillation qubit) is a type of solid-state qubit.

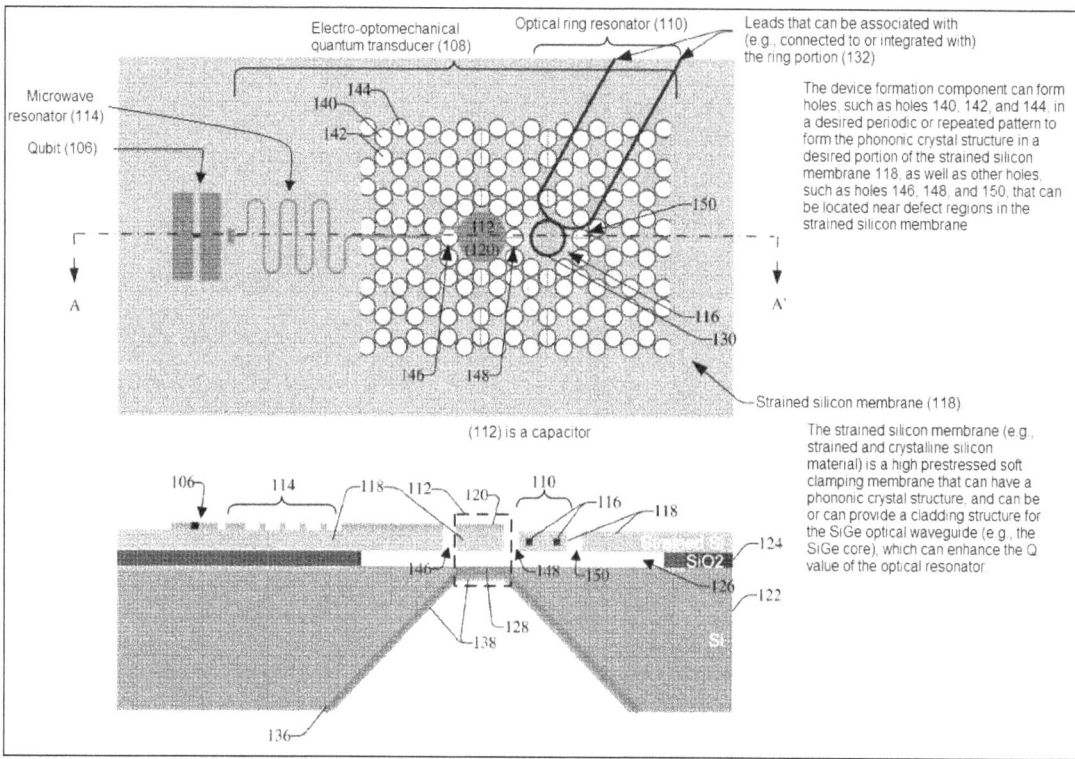

Figure 4.11   Silicon-germanium (SiGe)/silicon (Si) electro-optomechanical quantum transducer (synthetized form [TOK23])

junctions. The SiGe/Si electro-optomechanical quantum transducer can be used for a number of applications, including, but not limited to, state-sensing or charge-sensing applications or applications that involve sensing microwave photons and transducing or converting microwave photons to generate corresponding optical photons.

The SiGe/Si electro-optomechanical quantum transducer described in [TOK23] has an optical ring resonator, a mechanical resonator (e.g., silicon membrane), and a microwave resonator; the optical ring resonator can be coupled to the mechanical resonator, and the mechanical resonator can be coupled to the microwave resonator, to facilitate performing desirable quantum transduction. In some cases, the QC can comprise a group of qubits, comprising a desired number of qubits, that can be respectively associated with SiGe/Si electro-optomechanical quantum transducers. The SiGe/Si electro-optomechanical quantum transducer, comprising the SiGe/Si optical ring resonator, the capacitor, and the microwave resonator, as well as the qubit and associated components and circuitry, can be formed on a single die (e.g., a single integrated circuit [IC] chip). The SiGe/Si electro-optomechanical quantum transducer has a number of advantages over existing technology, including the following:

  i. The SiGe/Si electro-optomechanical quantum transducer can have a (e.g., significantly) higher Q than existing transducers (Q is a measure of the strength of the damping of its oscillations or for the relative linewidth).

  ii. The SiGe/Si electro-optomechanical quantum transducer also has low mechanical dissipation at low temperature.

iii.   The SiGe/Si electro-optomechanical quantum transducer is compact, as such quantum transducer, associated qubit, and associated circuitry can be formed on a single IC chip, and can be compatible with superconducting QCs.

In connection with the quantum operations being performed, it can be desirable to sense or read charges of qubits. When the SiGe/Si electro-optomechanical quantum transducer is being used to sense or read the state of the qubit (e.g., to sense the charge of the qubit, where the qubit charge can indicate or correspond to the state of the qubit), the state of the qubit can correspond to the frequency (e.g., the relatively higher frequency) of the up-converted optical photons produced as an output from the transducer [TOK23]. For instance, the change or difference in frequency between the relatively higher (or lower) frequency of the up-converted (or down-converted) optical photons and the frequency of the pump photons can correspond to or indicate the state or charge of the qubit. The up-converted or down-converted optical photons can exist in a coherent or quantum state in a room temperature environment, in contrast to the microwave photons (e.g., sensed from the qubit), which, while able to exist in a coherent or quantum state in the very low temperature environment of a cryostat in which the QC can be located, can rapidly and undesirably decohere when exposed to a room temperature environment.

Operation of qubits in QCs can involve a microwave field comprising photons that can be generated at microwave frequency levels. These photons can be encoding the state of the quantum computer. To further scale up superconducting QCs, microwave to optical quantum transduction using optomechanical coupling has been explored as a potential technology for use in quantum transduction. However, existing approaches for optomechanical resonators and quantum transduction have *a number of deficiencies*; reference [TOK23] describes some of the limitations of existing approaches, as follows:

i.   One existing approach relates to forming nanomechanical resonators, where a thin silicon nitride film is deposited on a silicon wafer; the film is patterned with a honeycomb lattice of air holes over a square region, the silicon nitride membrane being patterned with a phononic crystal structure. However, this technology can be inefficient, deficient with regard to bandwidth, deficient with regard to optical and microwave properties (e.g., can have relatively poor optical and microwave properties), unreliable, difficult to implement or scale up, bulky with regard to the overall structure that would have to be used to implement the resonator, and incompatible for use with superconducting quantum computers.

ii.   Another existing approach involves an optomechanical resonator that is formed by an optical resonator, a mechanical element for being set in rapid oscillations, by an optical signal confined in the optomechanical resonator, and in slow oscillations, by an acoustic wave generated in the photoacoustic cavity. The optomechanical resonator can be part of a photoacoustic sensor. Such existing approach, however, can be deficient with regard to optical quality and mechanical quality and be undesirable for use for quantum transduction purposes, even if it ostensibly may be used for sensing acoustic modes.

iii.   Still another existing approach involves a nano-optomechanical transducer, comprising a suspended membrane; a transducing optical waveguide, mechanically coupled to the suspended membrane; and an input optical signal that is coupled into the transducing optical waveguide, wherein the input optical signal generates an acoustic wave that propagates in the suspended membrane. The optical waveguide can be embedded in a low-Z phononic crystal membrane. The transducing optical waveguide can also be part of a signal delay system with an array of receiving waveguides, wherein an encoded input optical signal is injected in the transducing optical waveguide that generates an acoustic wave comprising the encoded information. The acoustic signal propagates in a phononic membrane and reaches a receiving optical waveguide after a time delay; this existing system, however, simply uses a membrane to couple a transducing optical waveguide to another optical waveguide for the purpose of creating a signal delay.

4.6 Quantum Simulators

Understanding the behavior of quanta in materials (or newly-designed materials) is important. QM effects are generated through interactions between a large number of particles; but while a mathematical description of a phenomenon is possible in principle, QM behaviors with a plurality of particles (quantum many-body problem) is rather complicated, and predicting the behavior theoretically and numerically can be considered impractical or impossible in actuality. Quantum simulators have received attention in recent years as a method for studying the QM many-body problem. A quantum simulator prepares a model system including physical characteristics of an object under study and actually drives the model system to observe what phenomenon occurs [SAK23]. For example, as noted in the reference, when studying a QM phenomenon in a crystal, a model system is prepared in which appropriate atoms are arranged according to spatial arrangement in accordance with a crystal structure. In an actual crystal, an interatomic distance is small, and observing the behavior of the atoms is difficult; however, by arranging atoms at intervals of about micrometers, it is possible to prepare a model system of a size in which a quantum phenomenon can be easily controlled and observed.

A quantum simulator can control the positions of arranged atoms and applies some stimulus to each of the arranged atoms, so as to be able to detect an influence which appears in an entire

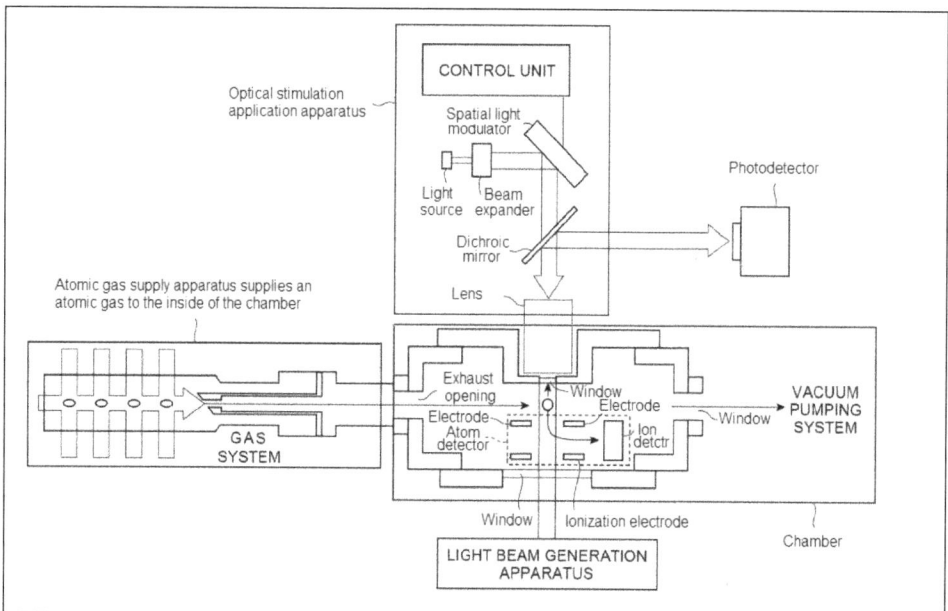

Figure 4.12 Quantum Simulator example (synthetized from [SAK23])

system. A quantum simulator can use an optical trap technique in which light is focused to trap atoms at a focusing spot as a means for arranging atoms; the quantum simulator uses a technique of generating a light pattern having a predetermined shape and irradiating arranged atoms as a means for applying a stimulus to the atoms. Controllability and reproducibility are required for both the means of arranging atoms and the means of applying a stimulus to the atoms. By repeating a detection process multiple times under identical conditions, the existence probability of an electron can be established empirically.

A quantum simulator (say, the one discussed in reference [SAK23], Figure 4.12) may include (i) a chamber having a window; (ii) a light beam generation apparatus for causing light to enter the chamber through the window, forming and regularly arranging a plurality of focusing spots for trapping atoms one-dimensionally or two-dimensionally on an image plane in the chamber; and (iii) a detector for detecting a state of the atoms trapped in the focusing spots in the chamber. From a method perspective the simulation includes (i) an optical trapping step of causing light to enter a chamber through a window of the chamber, and forming and regularly arranging a plurality of focusing spots for trapping atoms one-dimensionally or two-dimensionally on an image plane in the chamber; and (ii) a detection step of detecting a state of the atoms trapped in the focusing spots in the chamber.

References

[AMI23] J. M. Amini, Efficient Cooling of Ion Chains for Quantum Computation. U.S. Patent Application 20230351233, 2023-11-02. Uncopyrighted material.

[AUT20] T.M. Autry, S. Berweger, et al., "Surface Acoustic Wave Cavities and InAs Quantum Dots for Quantum Transduction". CLEO: Fundamental Science, FTh4D.6, 2020. https://doi.org/10.1364/CLEO_QELS.2020.FTh4D.6.

[BEC23] K. M. Beck, M. Cetina, M. L. Goldman, Crosstalk Reduction in Multi-Channel Acousto-Optic Modulators. U.S. Patent Application 20230254047, 2023-08-10. Uncopyrighted material.

[BHU20] S. Bhushan, O. Slattery, et al., "Terahertz Electromagnetically Induced Transparency in Cesium Atoms". Frontier in Optics, virtual conference, 2020. ISBN: 978-1-943580-80-4.

[BUR17] I. A. Burenkov, S. V. Polyakov, "Quantum Frequency Bridge: High-Accuracy Characterization of a Nearly-Noiseless Parametric Frequency Converter". Optics Express, 25, 907–917, 2017. https://doi.org/10.1364/OE.25.000907.

[BUS23] P. J. Bustard, D.G., England, et al., A Fibre Optic Integrated Quantum Memory for Light. U.S. Patent Application 20230420046, 2023-12-28. Uncopyrighted material.

[CHE18] Y. Chen, M. Zopf, et al., "Highly-Efficient Extraction of Entangled Photons from Quantum Dots Using a Broadband Optical Antenna". Nature Communications, 9, Article number: 2994, July 2018, https://doi.org/10.1038/s41467-018-05456-2

[CHE23] E. H. Chen, A. W. Cross, et al., Calibrated Decoders for Implementations of Quantum Codes. U.S. Patent 11803441, 2023-10-31. Uncopyrighted material.

[COH09] O. Cohen, J. S. Lundeen, et al., "Tailored Photon-Pair Generation in Optical Fibers". Physical Review Letters 102, 123603; 2009. https://doi.org/10.1103/PhysRev Lett. 102.123603.

[COU18] C. Couteau "Spontaneous Parametric Down-Conversion". arXiv:1809.00127. https://doi.org/10.48550/arXiv.1809.00127. Also Contemporary Physics 59, 291–304; 2018. https://doi.org/10.1080/00107514.2018.1488463.

[DAY19] A. Dahlberg, M. Skrzypczyk, et al., "A Link Layer Protocol for Quantum Networks". arXiv:1903.09778, https://doi.org/10.48550/arXiv.1903.09778. SIGCOMM '19 Proceedings of the ACM Special Interest Group on Data Communication (2019) 159–173. https://doi.org/10.1145/3341302.3342070.

[EIS11] M. D. Eisaman, J. Fan, et al., "Invited Review Article: Single-Photon Sources and Detectors". Rev. Sci. Instrum. 82, 071101; 2011. https://doi.org/10.1063/1.3610677.

[FUR23] W. Furtner, Multi Dimensional Electrode Controller for Quantum Computing. U.S. Patent Application 20230325698, 2023-10-12. Uncopyrighted material.

[GOR 11] A.V. Gorshkov, J. Otterbach, et al., "Photon-Photon Interactions via Rydberg Blockade". Physical Review Letters, 107, 133602, 2011, https://doi.org/10.1103/Phys RevLett. 107.133602

[GOR07] A.V. Gorshkov, A. André, et al., "Universal Approach to Optimal Photon Storage in Atomic Media". Physical Review Letters, 98, 123601, 2007. https://doi.org/10.1103/PhysRevLett.98.123601

[HUG23] S. M. Hughes, C. R. Sheridan, Drop-In Multi-Optics Module for Quantum-Particle Cell. U.S. Patent Application 20230260670, 2023-08-17. Uncopyrighted material.

[IMA22] P. Imany, Z. Wang, et al., "Quantum Phase Modulation with Acoustic Cavities and Quantum Dots". Optica, 9, 501, 2022. https://doi.org/10.1364/OPTICA.451418.

[KUD23] R. Kudalippalliyalil, S. Chandran, et al., Electro-Optic Frequency Transducer Using Coupled Microdisk Resonators. U.S. Patent Application 20230314717, 2023-10-05. Uncopyrighted Material.

[KUO13] P. Kuo, J. Pelc, et al., "Dual-channel, Single-Photon Upconversion Detector at 1300 nm". Nonlinear Photonics 2012, Colorado Springs, CO, 2012. https://doi.org/10.1364/NP.2012.NM3C.6.

[KUO18] P. Kuo, J. Pelc, et al., "Using Temperature to Reduce Noise in Quantum Frequency Conversion". CLEO: Science and Innovations, San Jose, CA, 2018. https://doi.org/10.1364/CLEO_SI.2018.SM2D.5.

[KUO20] P. Kuo, V. Verma, and S. Nam, "Demonstration of A Polarization-Entangled Photon-Pair Source Based on Phase-Modulated PPLN". OSA Continuum, 3, 295-304, 2020. https://doi.org/10.1364/OSAC.387449.

[LAG23] D. Lago-Rivera, J. V. Rakonjac, et al., "Long Distance Multiplexed Quantum Teleportation from A Telecom Photon to A Solid-State Qubit". Nature Communications, April 2023, 14:1889. https://doi.org/10.1038/s41467-023-37518-5.

[LAI23] J-Y. Lai, C. Le Touze, M-H. Chou, "Cavity-enhanced Frequency Mixer for Classical and Quantum Applications". U.S Patent 11,762,262; 2023-09-19. Uncopyrighted material.

[LIT23] D. Litinski, First-Quantization Block Encoding for Quantum Emulation. U.S. Patent Application 20230297866, 2023-09-21. Uncopyrighted material.

[LU19] X. Lu, Q. Li, et al., "Chip-Integrated Visible–Telecom Entangled Photon Pair Source for Quantum Communication". Nature Physics, 15, 373–381, 2019. https://doi.org/10.1038/s41567-018-0394-3.

[MA09] L. Ma, O. Slattery, et al., "Non-Degenerated Sequential Time-Bin Entanglement Generation Using Periodically Poled KTP Waveguide". Optics Express, 17, 15799-15807, 2009, https://doi.org/10.1364/OE.17.015799.

[MA17] L. Ma, O. Slattery, X. Tang, "Optical Quantum Memory Based on Electromagnetically Induced Transparency". Journal of Optics, 19, 043001, 2017. doi 10.1088/ 2040-8986/19/4/043001.

[MUK23] M. Mukherjee, J. Z. Kim, Quantum Approximation Optimizer for Distributed Register Validation, U.S Patent Application 20230316119, 2023-10-05. Uncopyrighted material.

[NIS23] NIST, "Quantum Networks at NIST". Available on Nov. 11, 2023 at https://www.nist.gov/pml/productsservices/quantum-networks-nist.

[NUN13] J. Nunn, N. K. Langford, et al., "Enhancing Multiphoton Rates with Quantum Memories". Physical Review Letters 110, 133601; 2013. https://doi.org/10.1103/Phys RevLett. 110.133601.

[ORN20] D. P. Ornelas-Huerta, A. N. Craddock, et al., "On-Demand Indistinguishable Single Photons from an Efficient and Pure Source Based on a Rydberg Ensemble". Optica, 7, 813-819, 2020, https://doi.org/10.1364/OPTICA.391485.

[PAS23] R. Paschotta, "Q Factor". Available online on December 23 at https://www.rp-photonics.com/q_factor.html. doi: https://doi.org/10.61835/dyf.

[PAS23a] R. Paschotta, "Pockels Effect". Available online on December 27, 2023 at https://www.rp-photonics.com/pockels_effect.html. https://doi.org/10.61835/4j0.

[PEN14] B. Peng, S. Özdemir, et al., "What Is and What Is Not Electromagnetically Induced Transparency in Whispering-Gallery Microcavities". Nature Communications 5, 5082 (2014). https://doi.org/10.1038/ncomms6082.

[RAU01] R. Raussendorf, H. J. Briegel, "A One-Way Quantum Computer". PhysRev Lett.86.5188. Volume 86, Issue 22, May 2001. doi: 10.1103/PhysRevLett.86.5188.

[RIG11] G. C. Righini, P. Fero, et al., "Whispering Gallery Mode Microresonators: Fundamentals and Applications". La Rivista del Nuovo Cimento 34(7):435. July 2011. doi:10.1393/ncr/i2011-10067-2.

[SAK23] H. Sakai, K. Ohmori, et al., Quantum Simulator and Quantum Simulation Method. U.S. Patent Application US-20230317309, 2023-10-05. Uncopyrighted material.

[SLA15] O. Slattery, L. Ma, et al., "Comparing the Linewidths from Single-Pass SPDC and Singly-Resonant Cavity SPDC". Quantum Communications and Quantum Imaging XII, San Diego, CA, 9615, 961507, 2015 https://doi.org/10.1117/12.2187059

[SLA19] O. Slattery, L. Ma, et al., "Background and Review of Cavity-Enhanced Spontaneous Parametric Down-Conversion". J. Res. Natl. Inst. Stand. Technol., 124, 124019, 2019, https://doi.org/10.6028/jres.124.019.

[TAK23] T. Takekoshi, Methods and Apparatuses for Laser Stabilization. U.S. Patent Application 20230275394, 2023-08-31. Uncopyrighted material.

[THI22] F. Thiele, T. Hummel, et al., "Opto-electronic Bias of a Superconducting Nanowire Single Photon Detector Using a Cryogenic Photodiode". APL Photonics 7, 081303, 2022. https://doi.org/10.1063/5.0097506.

[TOK23] M. Tokunari, R.D. Schilling, Electro-Optomechanical Quantum Transduction. U.S. Patent Application 20230204861, 2023-06-29. Uncopyrighted material.

[VAR22] G. Vardoyan, M. Skrzypczyk, S. Wehner, "On the Quantum Performance Evaluation of Two Distributed Quantum Architectures". Elsevier, Performance Evaluation 153, 2022. https://doi.org/10.1016/j.peva.2021.102242.

[WIN04] D. Wineland, "NIST Demonstrates 'Teleportation' of Atomic States for Quantum Computing". June 16, 2004. Available online December 15, 2023, at https://www.nist.gov/news-events/news/2004/06/nist-demonstrates-teleportation-atomic-states-quantum-computing.

[WOO82] W. Wootters, W. Zurek, "A Single Quantum Cannot Be Cloned". Nature 299, 802–803 (1982). https://doi.org/10.1038/299802a0.

[XU23] C. Xu, "Exotic State". Available online on December 18, 2023 at https://web.physics.ucsb.edu/~xucenke/exotic.html#:~:text=Exotic%20states%20usually%20refer%20to,by%20local%20Landau%20order%20parameters.

[YAM24] A. Yamauchi, K. Tanaka, et al., "Room-Temperature Quantum Coherence of Entangled Multiexcitons in A Metal-Organic Framework". Science Advances, 3 Jan 2024, Vol 10, Issue 1. doi: 10.1126/sciadv.adi3147.

[ZHA23] Z. Zhang, S. Liu, B-H Wu, Photonic Integrated Circuit with Squeezed and Entangled Light Generation, And Associated Methods. U.S Patent Application, 20230221616, 2023-07-13. Uncopyrighted material.

[ZIA18] A. W. Ziarkash, S.K. Joshi, et al., "Comparative Study of Afterpulsing Behavior and Models in Single Photon Counting Avalanche Photo Diode Detectors". Sci Rep 8, 5076, 2018. https://doi.org/10.1038/s41598-018-23398-z.

Basic Glossary of Key Concepts in Chapter 4

This Glossary is based on various industry sources, including [BUS23], [DAY19], [FUR23], [KUD23], [NIS23], [PAS23], [PAS23a], [RAU01], [RIG11], [TOK23], and [ZHA23]. This Glossary should be used not only as a repository of terms and terminology, but as a tool that further explains and elaborates on various fundamental quantum concepts.

Term/Concept	Description
Acousto-Optic Modulators (AOMs)	Devices used to control optical beams which in turn are used to control the atomic-based qubits
Cluster state	An entangled state of multiple qubits. May be utilized in Quareps. A Quacomp model based on a special type of entangled states: they refer to a family of quantum states of n-qubit two- or three-dimensional lattice wherein each vertex corresponds to a qubit; it represents a highly entangled state that has applications in measurement-based quantum computing applications [RAU01].
Detection efficiency	The probability that a photon incident at the optical input of the detection system induces an output signal.
Detector signal jitter	Variation in the detection latency, where latency measured as the time between when a photon arrives at the detector input port and when a signal is output from the detector.
Electromagnetically Induced Transparency (EIT)	A coherent optical nonlinearity that renders a medium transparent within a spectral range around an absorption line; it creates a narrow transparency window by eliminating a resonant absorption.
Etalons	Devices consisting of two reflecting glass plates, employed for measuring small differences in the wavelength of light.
Heralding	In endeavoring to generate long-lived entanglement between two distant quantum nodes, there is a need to be able to confirm entanglement. This can be achieved by undertaking heralded entanglement generation, in which a heralding signal is utilized to indicate the successful generation of entanglement. Currently heralded entanglement has been achieved at a distance of about one mile using a solid-state platform of Nitrogen-Vacancy (NV) centers (which are photoluminescent point defects) in diamond. This platform is a QC with several qubits (up to eight at the time of this writing, with lifetimes of around 1.5 s) that supports a variety of quantum gates; an optical interface is used for initialization, measurement, and entanglement generation. Other hardware platforms on which heralded long-lived entanglement generation has been demonstrated include ion traps, neutral atoms, quantum dots, rare earth ion-doped crystals, atomic gases, and superconducting qubits; however, development work is still needed to achieve the goal of generating entanglement faster than it is lost [DAY19].
Heralded sources	(Particle) source wherein measuring one (particle, photon) of the pair heralds the presence of the other.
Mean photon number	The average number of photons per optical pulse.
Multi-photon signal	An optical signal containing more than one photon.

Basic Glossary of Key Concepts in Chapter 4 *(cont.)*

Term/Concept	Description
Nitrogen-vacancy center	(Also known as NV-centers). A crystal defect in diamond where a nitrogen atom substitutes for a carbon atom in a lattice next to a lattice vacancy. Diamonds with NV-centers can be used to store and process qubits. Transitions between states can generate single photons.
Nonclassical (quantum) light	'Light' that can only be described by QM mechanisms.
Pockels effect	The electro-optic effect in which the refractive index of a medium is modified in proportion to the applied electric field strength.
PPLN (Periodically Poled Lithium Niobate)	A crystal that is able to switch polarity in a periodic fashion; used for achieving quasi-phase-matching in nonlinear optics. It is a highly efficient physical medium for nonlinear wavelength conversion processes.
Q factor (quality factor) of a resonator	A measure of the strength of the damping of its oscillations or for the relative linewidth. The term was originally developed for electronic circuits, microwave cavities, and mechanical resonators, but later also applied to optical resonators. $Q = \lambda/\Delta\lambda$ where λ is the wavelength at which a resonance occurs and $\Delta\lambda$ the linewidth of the resonant wavelength. There are two common definitions of the factor of a resonator: (i) Definition via energy storage: The factor is 2π times the ratio of the stored energy to the energy dissipated per oscillation cycle, or equivalently the ratio of the stored energy to the energy dissipated per radian of the oscillation (for a microwave or optical resonator, one oscillation cycle corresponds to the field oscillation period); (ii) Definition via resonance bandwidth: The factor is the ratio of the resonance frequency and the full width at half-maximum (FWHM) bandwidth $\delta\nu$ of the resonance [PAS23].
Quantum dot (QD)	A nanoscale semiconductor particle with optical and electronic properties. Quantum dots commonly have discrete electronic states that can be used to process quantum information, and transitions between states may result in the generation of a single photon.
Quantum frequency transduction	Methods and devices used for interconnecting of different quantum devices using optical fibers—for example, for the coherent conversion between superconducting microwave photons and optical photons in telecommunication wavelengths
Rydberg atom	An atom whose electrons are excited to a high quantum state. Rydberg interactions among atoms in the ensemble can be used to generate single photons.
Single-photon source	A source generating single photons. Single-photon sources are typically characterized by measuring photon number statistics and showing that the probability of two or more photons in the field is small or negligible.
Single-Photon Detector (SPD)	A detector that can distinguish between a vacuum state (no input photons) and a faint state of light (one photon or more) at its input. A device that maps a single-photon into a detectable signal with non-zero probability. Some single-photon detectors can also measure photon number statistics by being able to distinguish between one or more photons. SPDs are hardware used for photon counting. The SPD emits a pulse of signal for each detected photon. Counting efficiency (high level of sensitivity upon absorption of just a single photon) is an important parameter and is determined by the quantum efficiency and the system's electronic losses. *(continues)*

Term/Concept	Description
Single-photon detector (SPD) *(cont.)*	An example is the Superconducting Nanowire Single-Photon Detector (SNSPD) operated at 2.5 Kelvin, which consists of a thin film of superconducting material shaped into a long nanowire. This design enables it to cover a wide surface area, collecting the whole output of an optical fiber, while providing a single path for the current.
Single-photon source	A light source which emits precisely one photon at a given time.
Source timing jitter	Uncertainty in the emission time of an optical pulse.
Spin-photon entanglement	The entanglement between a single photon and an electric spin.
Spontaneous Parametric Down Conversion (SPDC)	Approach to generating pairs of entangled photons, where a beam of photons is split into entangled pairs by passing it through a crystal.
Squeezed light	Light where the variance of the amplitude quadrature is reduced relative to the variance of the phase quadrature (or vice versa). Squeezed states exhibit a squeezed uncertainty of electric field strength—for example, the beam of squeezed light has a lower quantum uncertainty than a beam without photons. Squeezed light enables better measurement accuracy and facilitates secure communication. Squeezing is characterized by a squeezing factor that quantifies how much the variance of one quadrature has been reduced with respect to the ideal coherent light; a design goal of squeezed-light systems is to achieve high squeezing factors.
Time-bandwidth product	The product of the memory lifetime multiplied with the bandwidth of the stored signal photon; it measures how many distinct photon wave packets could be stored in the memory. This metric is important for use in repeat-until-success protocols, such as multiplexing the operation of one or more photon pair sources as a single photon source.
Transmon	(Transmission line shunted plasma oscillation qubit): A type of solid-state qubit.
Trapped Ion Quantum Computing (TIQC)	A system that utilizes ions (atoms or molecules with a net electrical charge) as qubits for computation, with the excitation state of an electron indicating a logical value or logic state—qubits are stored in the electronic states of each ion, and information can be transferred through the aggregate quantized motion of the ions in a trap.
Whispering Gallery Modes (WGMs) microresonators (WGMRs)	Optical dielectric resonators, being a class of cavity devices with exceptional properties, such as extremely small mode volume, very high power density, and very narrow spectral linewidth. WGMs can be perceived as electromagnetic waves that circulate and are strongly confined within the structure. These resonators have a Q factor up to 10^{11}, compared to values around 10^5 for the best Fabry-Perot resonators. WGMs have been known for over 100 years (initially studied by Lord Rayleigh), but their importance to study quantum electrodynamics and for application to very low-threshold microlasers and high-sensitivity microsensors has been accentuated only in recent years.

Chapter 5

Quantum Computers

This chapter provides a basic overview of Quantum Computers (QCs) used in Quantum Computing (Quacomp) applications. Classical computers utilize silicon-based semiconductor devices. In recent years integrated circuit lateral feature sizes have been reduced to 10 nm, increasing chip density; however, as these sizes progress to be reduced to 5 nm and smaller, silicon-based circuits become impractical. QCs make use of quantum phenomena such as entanglement and superposition to increase and improve the computation speed as compared to classical computers, enabling them to solve some computationally complex problems in practical time windows. At this juncture QCs are not able to act as replacements for general-purpose computers, but they can outperform classical computers in a set of specific computational problems, such as optimization problems or for identifying the best solution from a set of feasible solutions. This chapter covers three QC approaches: the *circuit* model (or gate model), the *adiabatic quantum computation* model, and the *categorical quantum* model.

Quacomp has experienced relatively rapid development in recent years, starting with the basic ideation in the early 1980s, then continuing with early proof of concepts for hardware in the next two decades, and now, in the 2020s, with QC hardware that can implement over one-thousand qubits (e.g., [TIL22], [FEY82], [LEU00], [NEG06], [PLA07], [MON11], [MON16], [CAS23]). Figure 5.1 (courtesy of IBM) depicts recent illustrative achievements in practical QC development at IBM, a key player in this arena (in the fall of 2023, IBM® announced the Condor prototype, the world's first universal QC with 1,121 qubits[1]).

The generation of QCs at the time of this writing is known as Noisy-Intermediate Scale Quantum (NISQ) devices (some proponents cite this as being the third generation). These NISQ QCs are typically cooled to cryogenic temperatures (around –273° C) in order to operate efficiently; these environmental conditions are challenging to maintain in other than laboratory settings. In practical terms, QCs remain of somewhat limited and specialized functionality and are prone to computational errors. Considering these limitations, Hybrid QC (HQC)

---

[1] A June 2024 press report noted that IBM and Japan's National Institute of Advanced Industrial Science and Technology (AIST) were reportedly seeking to produce a QC containing 10,000 qubits by 2029; the 10,000-qubit machine would go beyond IBM's official QC roadmap that targeted a 1,000–2000 qubit QC as a commercial product by 2033 [GRI24].

(text continues on page 175)

2016–2019

Run quantum circuits on the IBM Quantum Platform

2020

Release multi-dimensional roadmap publicly with initial aim focused on scaling

2021

Enhancing quantum execution speed by 100x with Qiskit Runtime

2022

Bring dynamic circuits to unlock more computations

2023

Enhancing quantum execution speed by 5x with quantum serverless and Execution modes

2024

Improving quantum circuit quality and speed to allow 5K gates with parametric circuits

Data Scientist
- Platform
- Code assistant

Researchers
- Middleware
- Quantum Serverless
- Transpiler Service

Quantum Physicist
- IBM Quantum Experience
- Qiskit Runtime
- QASM3
- Dynamic circuits
- Execution Modes

Early
- Canary 5 qubits
- Albatross 16 qubits
- Penguin 20 qubits
- Prototype 53 qubits

Falcon
- Benchmarking 27 qubits

Eagle
- Benchmarking 127 qubits

Heron (5K)
- Error Mitigation
- 5k gates
- 133 qubits
- Classical modular
- 133x3 = 399 qubits

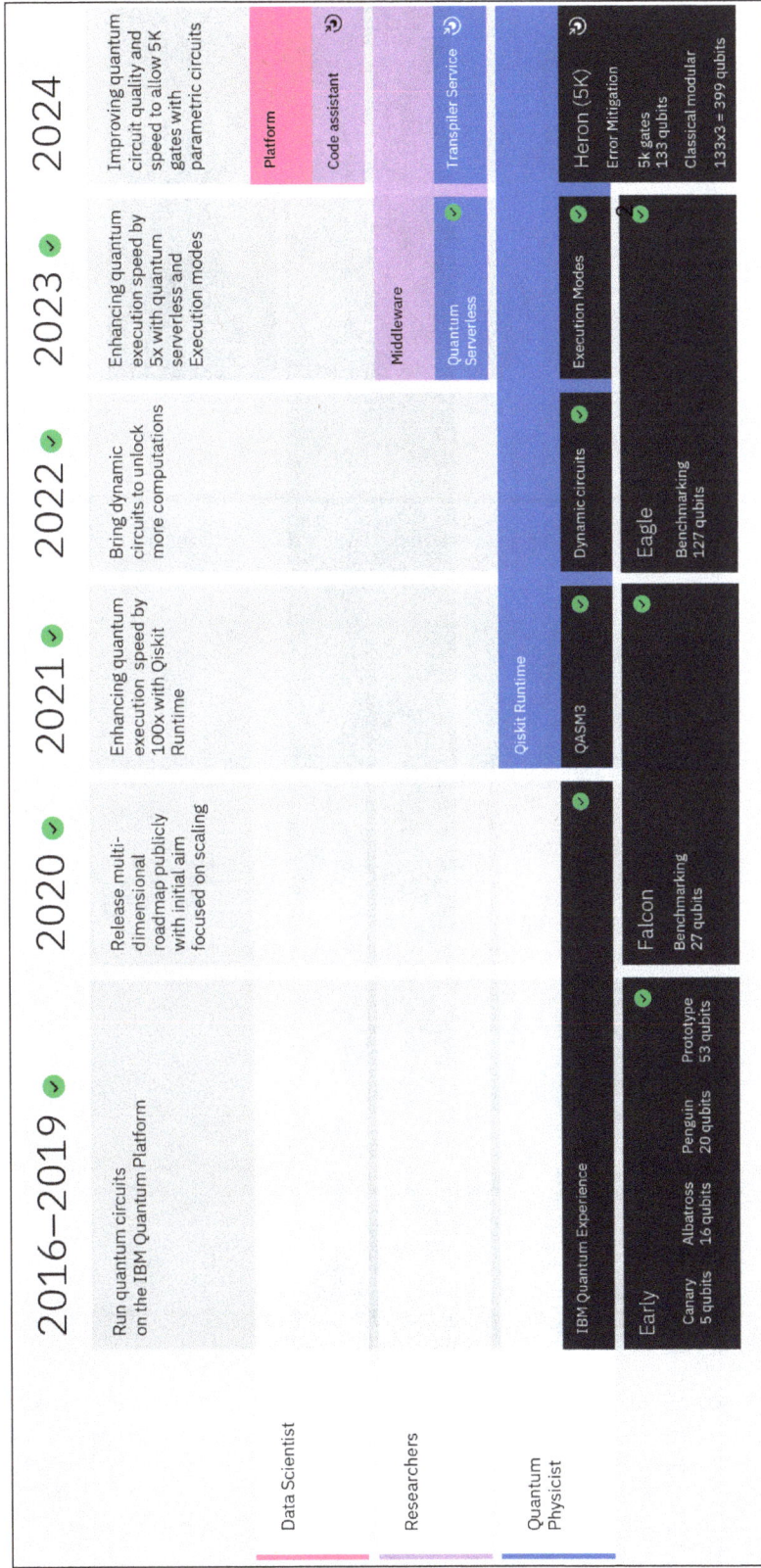

Figure 5.1 Recent developments in practical QC development at IBM (courtesy of IBM) (*this page and next*)

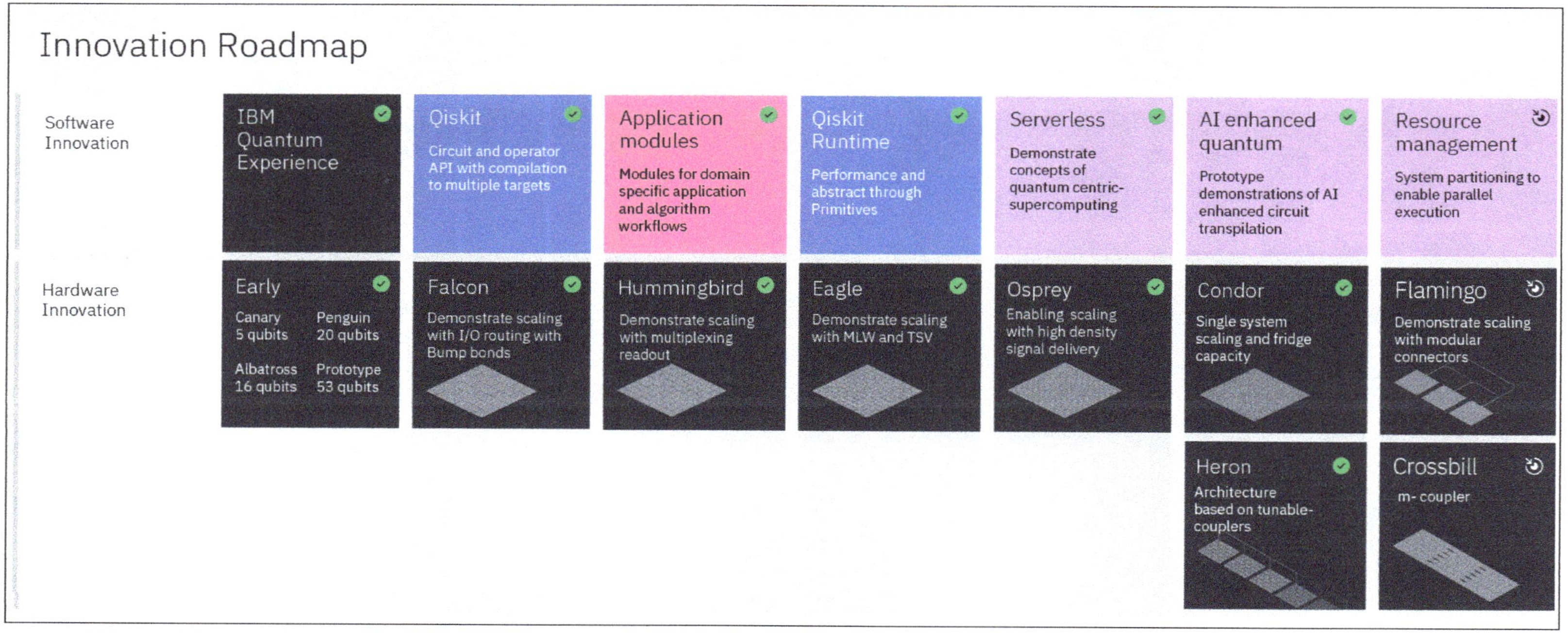

Innovation Roadmap

Software Innovation

IBM Quantum Experience

Qiskit
Circuit and operator API with compilation to multiple targets

Application modules
Modules for domain specific application and algorithm workflows

Qiskit Runtime
Performance and abstract through Primitives

Serverless
Demonstrate concepts of quantum centric-supercomputing

AI enhanced quantum
Prototype demonstrations of AI enhanced circuit transpilation

Resource management
System partitioning to enable parallel execution

Hardware Innovation

Early
Canary 5 qubits
Albatross 16 qubits
Penguin 20 qubits
Prototype 53 qubits

Falcon
Demonstrate scaling with I/O routing with Bump bonds

Hummingbird
Demonstrate scaling with multiplexing readout

Eagle
Demonstrate scaling with MLW and TSV

Osprey
Enabling scaling with high density signal delivery

Condor
Single system scaling and fridge capacity

Flamingo
Demonstrate scaling with modular connectors

Heron
Architecture based on tunable-couplers

Crossbill
m- coupler

arrangements and Hybrid Quantum Applications (HQApps) endeavor to combine classical and QC systems in order to undertake calculations directed at various (but specific) optimization or related use cases; tasks such as preparing inputs for quantum algorithms, storing data, or interacting with the user are typically still performed using classical computers.

The number of operations in a QC typically equates with the number of quantum gates and the number of qubits supported; however, the larger the number of gates, the more likely it is that errors, including decoherence, are introduced into the quantum state. NISQ entails a somewhat limited number of logical qubits and associated physical qubits to support error correction schemes. Fortunately, a category of algorithms running on NISQ devices can proceed with only a small number of qubits and exhibit some degree of noise resilience; typically, they are hybrid algorithms, with a portion of the computational steps performed on a quantum device and with other computational steps running on a conventional computer. The bottom line is that with NISQ devices, the number of operations is typically moderate at this juncture, thus the scope and type of algorithms that can be implemented is relatively limited. However, there is research that shows that even these QCs can outperform conventional computers on a set of mathematical problems (e.g., [PRE18], [BRO19], [MAD22]). The next step in the evolution is the interconnection of QCs over a distance network, as discussed in Chapter 6.

A classification of fundamental QC technologies has been proposed the European standards body CEN-CENELEC, specifically the Focus Group on Quantum Technologies (FGQT), which was established in June 2020 to coordinate and support the development of standards relevant for European industry and research [OVA22]. See Table 5.1.

5.1 Introduction

There are many optimization problems whose complexity grows exponentially; these are known as *NP-complete problems* ("NP-complete" means "nondeterministic polynomial-time complete"). The Traveling Salesman problem is a well-known example (prime factorization is another). These

Table 5.1 QC Technologies, as per CEN-CENELEC FGQT [OVA22]

QC technology	Examples
Cryogenic solid-state based	Superconducting solutions (transmons, flux qubits), semiconductor spin qubits, topological qubits, and artificial atoms in solids
Room-temperature solid-state based	Artificial atoms in solids (such as NV centers in diamond) and optical quantum dots A NV platform is a few-qubit processor capable of executing arbitrary quantum gates and measurements (encompassing up to about a dozen qubits at present); NV center hardware has demonstrated entanglement establishment over a distance of about 1 mile and has also demonstrated implementation of quantum protocols required for Quacom, such as purification (distillation) and entanglement swapping.
Trapped ions	Room-temperature and cryogenic (4K) solutions such as optical qubits, Raman qubits, and spin (microwave) qubits
Neutral atoms	Collision-based and Rydberg-based solutions
Photonic quantum computing	Solutions based on continuous variables, cluster states/measurement based, Knill-LaFlamme-Milburn and Boson sampling

problems cannot (easily) be solved in closed form because they tend to have computational complexity that is $O(n^n)$ or $O(n!)$, n being the number of nodes/entities under consideration; brute-force solutions quickly become impractical on classical computers; heuristic approximations have to be used. For example, in a 10-node problem, there are 10! possible solutions and 10! = 3,628,800; and, say, 20! = 2,432,902,008,176,640,000. See Figure 5.2. These problems are called *classically intractable problems.* Numerous practical problems are known to be in the NP class.

A classically intractable (NP) problem cannot be solved simply by utilizing a(ny) classical algorithm whose running time would desirably grow only polynomially as a function of the number of nodes (e.g., problem objects/elements); unfortunately, with classical algorithm the solution complexity grows exponentially with the number of objects/elements. To date no one has succeeded in discovering a polynomial-time algorithm for any NP problem (although no one has been able to prove that a polynomial-time algorithm *cannot* be constructed for an NP-complete problem). For example, all known classical prime factoring algorithms require more than polynomial time when parameterized on the number of digits in the integer to be factored. On the other hand, Shor's quantum algorithm for the factoring problem is able to factor an integer in a time that increases (approximately) as the square of the number of digits. It turns out that QCs are able to solve (some) classically difficult NP problems faster than classical algorithms and/or classical computers can; thus, Quacomp promises to be able to solve classically intractable tasks in polynomial time.

As traditional silicon-based circuit feature sizes approach the sub-decimal nanometer-scale, Heisenberg's uncertainty principle and quantum effects become dominant; in the limit, using single particles (e.g., ions or photons) to represent data in the form of qubits achieves optimal miniaturization and supports enhanced computing performance. A QC manipulates qubits while in operation; superposition and entanglement allow a QC to process a large number of calculations simultaneously. However, current QCs are noisy and have limited quantum resources (limited number of qubit and/or coherence time); thus, they are often referred to as NISQ devices, as defined earlier.

The just-cited Shor's quantum algorithm for factoring integers is a notable case of QC speed-up; several other algorithms that can be sped up have been discovered in recent years, such as, but not limited to, Hallgren's algorithm for solving Pell's equation, Watrous's algorithms for the group black box model, and the Legendre symbol algorithm by van Dam and Hallgren. All of these algorithms use similar (if not even identical) quantum circuitry [AHA07a]. However, it is also still an open question whether a generic NP-complete problem can be solved in polynomial time on a QC. The improvement achieved by Quacomp over classical computers can be characterized according to various grades (as seen in Table 5.2) as "strong", "provable", "limited", or "generic".

Two QC benchmarks are being sought by researchers and implementers [ARU19]: (i) demonstration of quantum advantage—namely, a demonstration that a quantum device can solve a problem faster than classical computers can; and (ii) quantum supremacy—namely, to establish the ability of a (programmable) QC to solve a problem that classical computers are unable to do (in polynomial time).

Currently, Quacomp has three models, as described in [GHO13]:

- The **circuit** model (or gate model, Standard Quantum Circuit model, or Standard Quantum Computation [SQC] model) with QCs using techniques such as superconducting elements and special silicon-based elements. It is considered to be the "standard" model [DEU89].

Start point

A —— B

2x1

AB
BA

Start point

A ——— B
c_1
c_3 c_2
C

$3X2 = 6$

ABC $C_T = c_1 + c_2$
ACB $C_T = c_3 + c_2$

BAC $C_T = c_1 + c_3$
BCA $C_T = c_2 + c_3$

CAB $C_T = c_3 + c_1$
CBA $C_T = c_2 + c_1$

Problem: minimize C_T

Say $c_1 = 10$, $c_2 = 20$, $c_3 = 30$

ABC $C_T = c_1 + c_2 = 30$
ACB $C_T = c_3 + c_2 = 50$

BAC $C_T = c_1 + c_3 = 40$
BCA $C_T = c_2 + c_3 = 50$

CAB $C_T = c_3 + c_1 = 40$
CBA $C_T = c_2 + c_1 = 30$

Optimal: ABC or CBA

Start point

A ——— B

C ——— D

$4X3X2 = 24$

ABDC
ACBD
ACDB
ADBC
ADCD

BACD
BADC
BCAD
BCDA
BDAC
BDCA

CABD
CADB
CBAD
CBDA
CDBA
CDBD

DABC
DACB
DBCD
DBDC
DCAB
DCBA

Start point

A   B   E

C   D

$5X4X3X2 = 120$

ABCDE
ABCED
ACBDE
ACBED
ADCBE
ADCEB
AEDCB
AEDBC

...

Start point

A   B   E

C   D   F

$$n! \sim \sqrt{2\pi n}\left(\frac{n}{e}\right)^n$$

$$n! \sim \sqrt{2\pi n}\left(\frac{n}{e}\right)^n \left(1 + \frac{1}{12n} + \frac{1}{288n^2} - \frac{139}{51840n^3} - \frac{571}{2488320n^4} + \cdots\right)$$

Number of possible paths growths exponentially on n

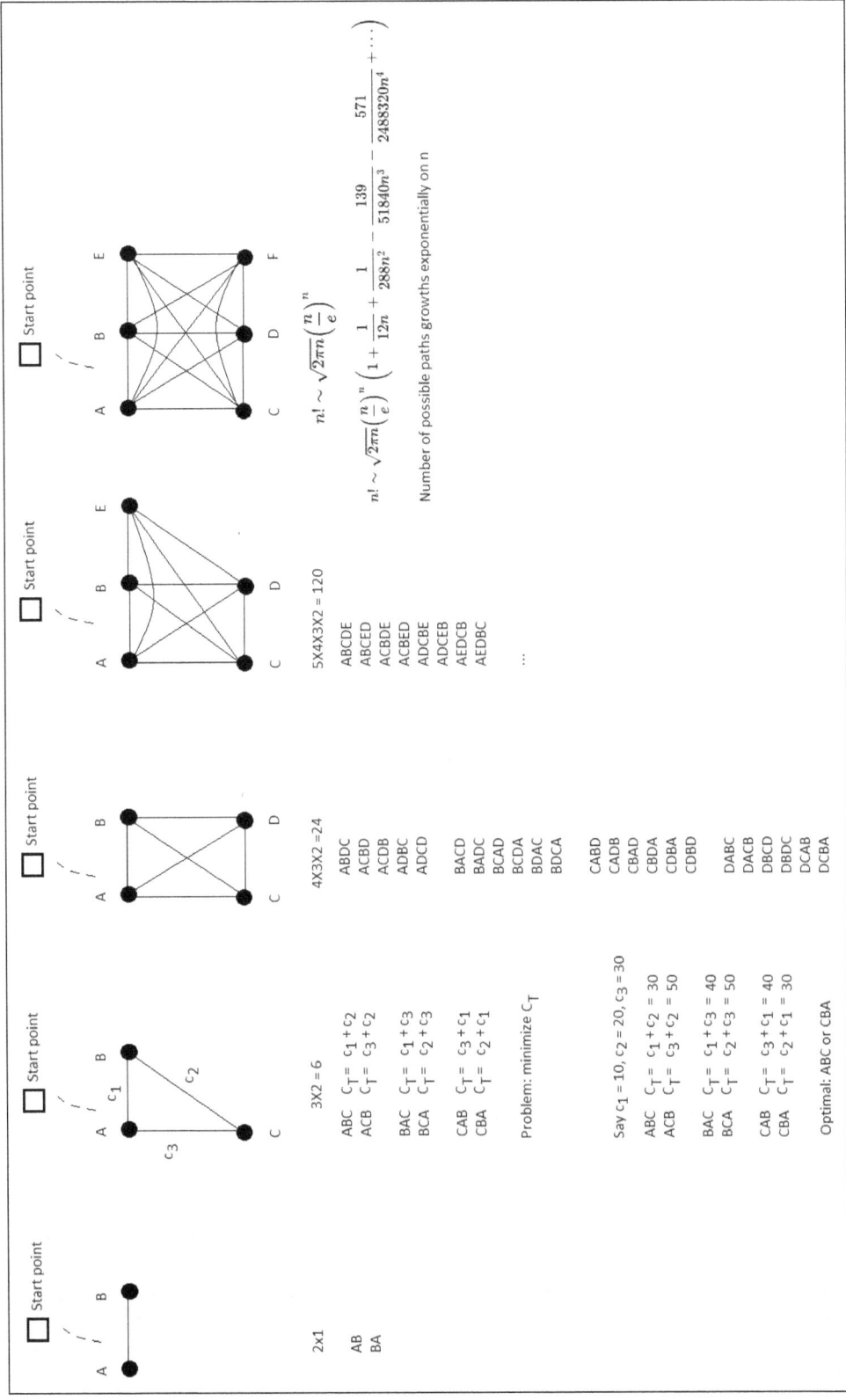

Figure 5.2 Example of NP-complete problem

Table 5.2 Quacomp Improvement over Classical Computers

Quantum speedup characterization	Description
Quantum speedup characterization: "generic"	A speedup against the best available classical algorithm. Such a speedup quantification may be tentative however, given that a better classical algorithm may later be found.
Quantum speedup characterization: "limited"	A speedup obtained when compared specifically with classical algorithms that "are comparable" to the quantum algorithm—that is, the calculations implement the same algorithmic approach but on classical Turing hardware.
Quantum speedup characterization: "provable"	The case where there is a proof that no classical algorithm can outperform a given quantum algorithm. An example is Grover's search algorithm, which enables a provable quadratic speedup over the best possible classical algorithm.
Quantum speedup characterization: "strong"	Contrasts a quantum algorithm against the performance of the "best" existing or yet-to-be discovered classical algorithm. Note, however, that the performance of the "best" possible classical algorithm is unknown in many instances.

Here a computation evolves in Hilbert space and is encoded via a series of unitary quantum logic gates—for example as shown in Figure 5.3. Make note that in a circuit QC the set of bits are manipulated using a classical computer in such a manner that a bit is mapped to a qubit.

- **Adiabatic quantum computation** (AQC) model—for example, using superconducting chips. AQC originated as an approach for solving optimization problems but is now becoming an alternative to the standard circuit model of Quacomp. AQC is a framework for quantum algorithms that, instead of employing unitary gates utilized in the standard quantum circuit model, makes use of the more physical mechanisms of Hamiltonians, spectral gaps, and ground states [AHA07a]. The concept entails encoding the solution to a computational problem in the ground state of a quantum Hamiltonian. In AQC, the computation moves along from an initial Hamiltonian with a given ground state to a final Hamiltonian, whose ground state encodes the solution to the computational problem. Assuming that the Hamiltonian varies relatively slowly, the adiabatic theorem guarantees that the system will track the instantaneous ground state[1] [ALB18].

This approach was called *Quantum Annealing*[2] (QA) by early researchers (specifically [APO88] and also [DEF11]), where it was initially perceived as an algorithm that exploits simulated quantum fluctuations and tunneling. The approach then evolved into the concept

---

[1] Reference [ALB18] provides an extensive, but highly technical, review of the AQC field; readers may want to consult the reference.

[2] QA was initially an optimization process (particularly when the search space is discrete and when there are several local minima) for finding the global minimum of a specified objective function over a specified set of candidate solutions (candidate states) by a process that utilizes quantum fluctuations. Non-stoquastic AQC and other models for Quacomp can simulate one another with at most polynomial resource overhead; thus, the current use of the term *AQC* typically refers to the general, nonstoquastic setting, hence extending AQC beyond optimization computational to any Quacomp [ALB18]. That is, AQC is equivalent to SQC; the adiabatic computation model and the conventional quantum computation model are polynomially equivalent [AHA04], [AHA07], [AHA08].

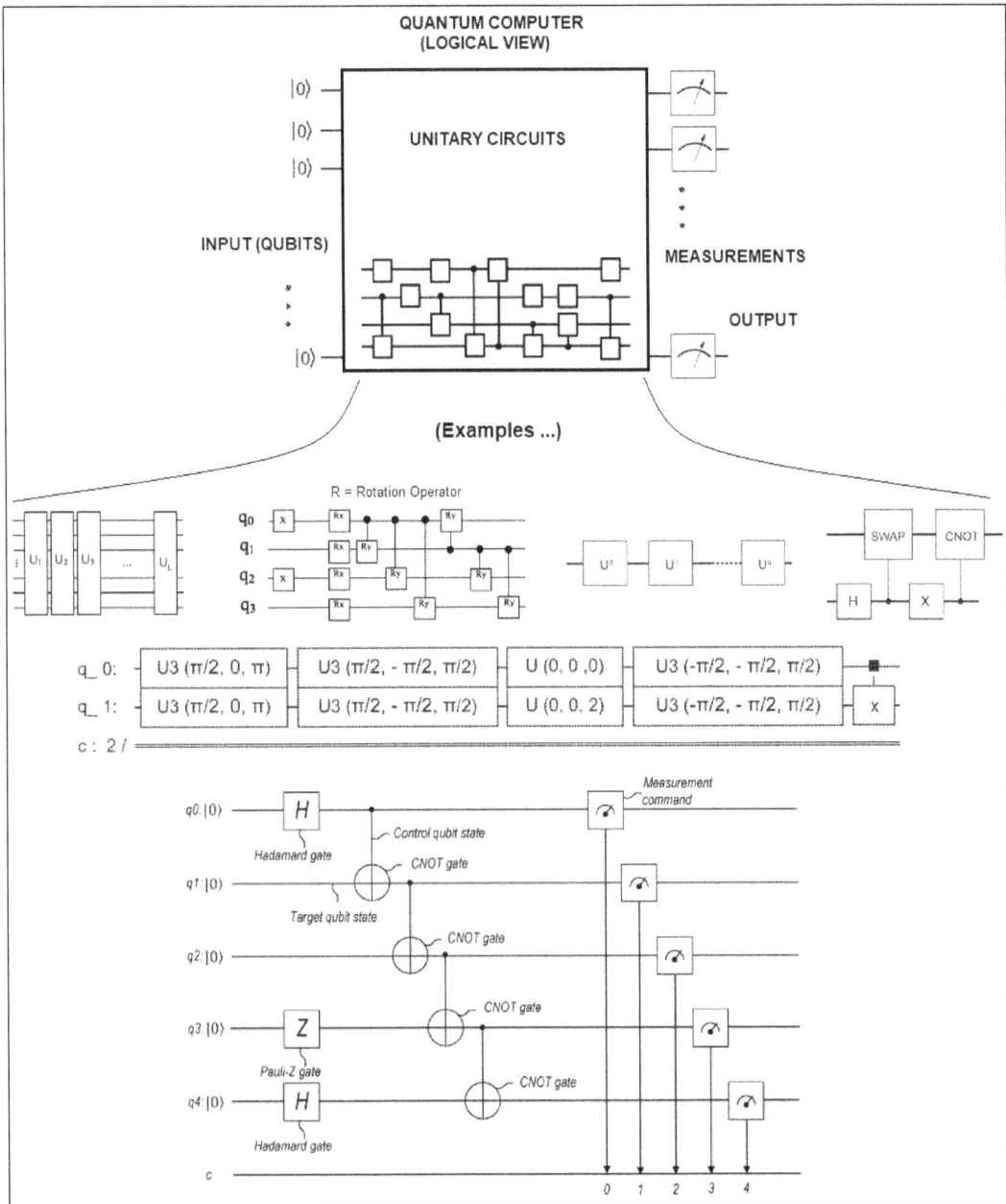

Figure 5.3 Basic elements of a circuit-based QC (The next-to-last row of this diagram is based on [RAS23]; the last row is based on [SHI23].)

of the *Quantum Adiabatic Algorithm* (QAA) (also referred to as *adiabatic quantum optimization*); here, a physical QC solves a combinatorial optimization problem by evolving adiabatically in its ground state [ALB18]. QAA is thus based on quantum adiabatic evolution.

It has been shown that adiabatic Quacomp (which makes use of non-stoquastic Hamiltonians—Hamiltonian operators that have only non-positive off-diagonal elements in the computational basis) is as effective as the circuit model of Quacomp, and in fact they can be interrelated (it has been proved that any standard quantum circuit, specified by a cascade of unitary operators, can be implemented as an adiabatic evolution).

- The **categorical quantum model** is based on category theory; one of its main applications is the study and verification of network and cryptographic protocols. Only a brief discussion at the end of this chapter is included in this text.

The glossary at the end of this chapter summarizes some of the key QC concepts that are discussed in the sections that follow below. A QC operates at a probabilistic level: measurements of algorithmic outputs provide a solution within a confidence interval; the computation is repeated until a satisfactory probabilistic certainty is achieved.

Figure 5.3 depicts a logical diagram of the basic elements (or concept) of a circuit/gate-based QC. Input qubits are fed to unitary circuits that manipulate the qubits; measurements are then made, and output states are generated, upon which some other (more traditional) computations can be made. Naturally, all these elements have to be rendered and supported in physical systems—for example, in superconducting cavities; more generally, QCs can be realized on a variety of physical platforms, including but not limited to photonic systems, continuous spin systems, ion trap systems, nuclear magnetic resonance systems, and molecular magnets systems. Figure 5.4 depicts an example of an implementation of a superconducting quantum circuit.

5.2 Circuit Model QCs

A quantum circuit is a representation of a quantum-based "general-purpose" computer, embodying a hardware implementation of a corresponding quantum algorithm/program under

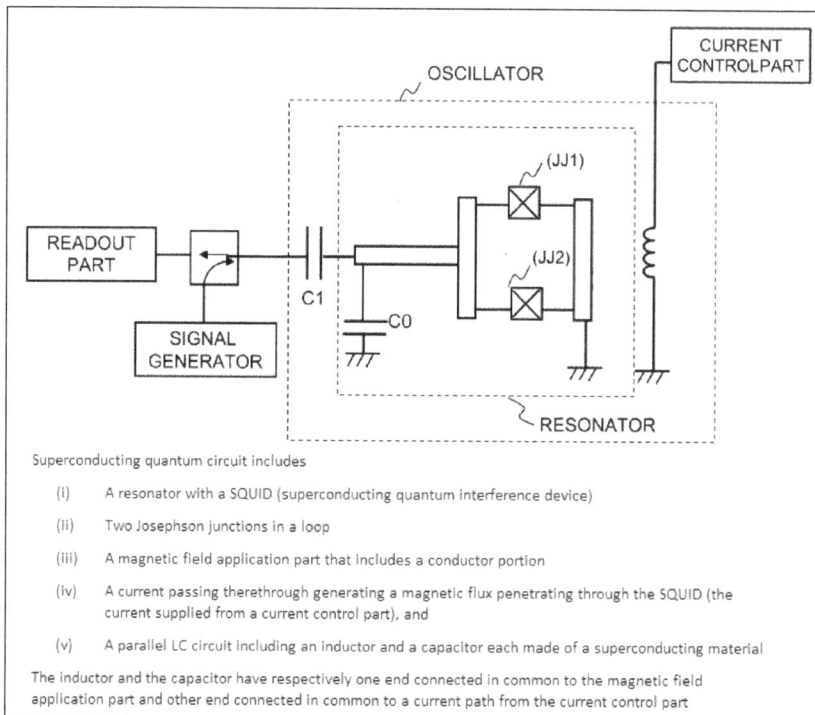

Figure 5.4   Superconducting quantum circuit using a Josephson parametric oscillator (*Source:* [MIY23])

a quantum gate model. If the quantum circuit contains adjustable parameters that control a quantum gate, the quantum circuit is called a *Parameterized Quantum Circuit* (PQC) or a *Variational Quantum Circuit* (VQC), both of which are the same concept [CHE24].

In quantum computation, particularly in a computing model of a quantum circuit, a quantum gate (or quantum logic gate) is a basic quantum circuit that operates a small quantity of qubits. A quantum circuit typically includes:

- Data qubits storing logical quantum data.
- Quantum gates that perform logical operations using the logical quantum data as inputs and generating logical quantum data outputs that are stored in data qubits; gates may include Pauli gates, Hadamard gates, controlled gates (e.g., CNOT, CZ, CX), swap gates, Toffoli gates (e.g., CCNOT, CCX, TOFF), phase gates, T gates, and other types of gates.

Figure 5.3, cited above, depicts an example. Figures 5.5, 5.6, and 5.7 depict a handful of additional examples of quantum circuits; in the figures the symbol at right represents a measurement process. Figure 5.8 depicts an example of a hybrid QC that utilizes both traditional computers and QC elements.

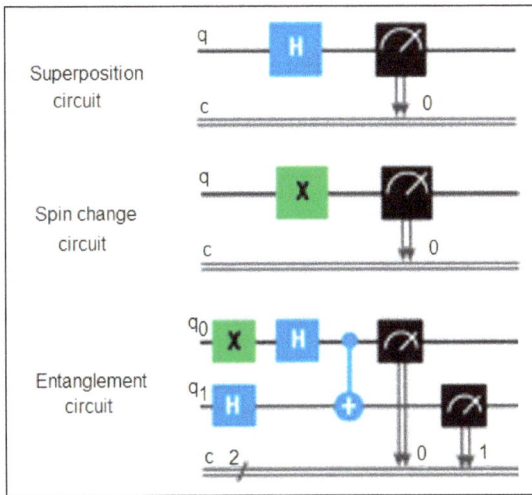

Figure 5.5 Examples of Quantum gate circuits/"programming"

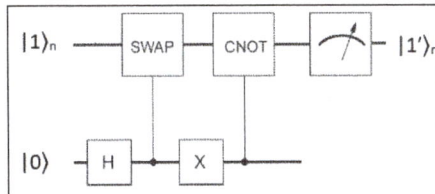

Figure 5.6 Example of quantum circuit ([SOG23])

Quantum gates and quantum circuits are generally single-qubit quantum gates and quantum circuits, respectively, and thus, they have 2×2 complex matrix representations. A quantum circuit can be specified as a cascade, a sequence, of quantum gates in which the quantum gates are represented symbolically, or, equivalently, they are represented numerically by matrices. As

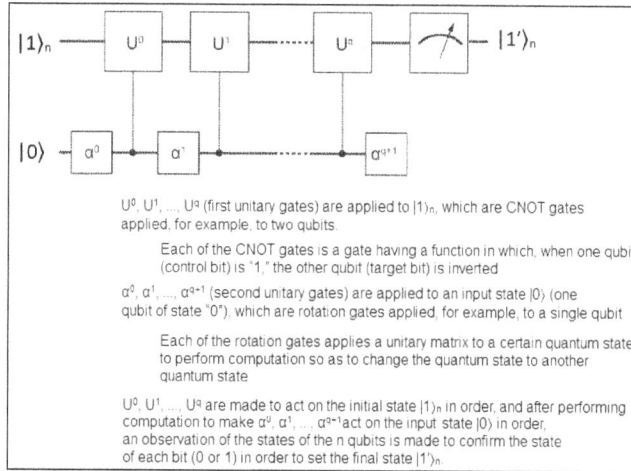

U^0, U^1, ..., U^q (first unitary gates) are applied to $|1\rangle_n$, which are CNOT gates applied, for example, to two qubits.

Each of the CNOT gates is a gate having a function in which, when one qubit (control bit) is "1," the other qubit (target bit) is inverted

α^0, α^1, ..., α^{q-1} (second unitary gates) are applied to an input state $|0\rangle$ (one qubit of state "0"), which are rotation gates applied, for example, to a single qubit

Each of the rotation gates applies a unitary matrix to a certain quantum state to perform computation so as to change the quantum state to another quantum state

U^0, U^1, ..., U^q are made to act on the initial state $|1\rangle_n$ in order, and after performing computation to make α^0, α^1, ..., α^{q-1} act on the input state $|0\rangle$ in order, an observation of the states of the n qubits is made to confirm the state of each bit (0 or 1) in order to set the final state $|1'\rangle_n$.

Figure 5.7   Example of quantum circuit (synthetized from [SOG23])

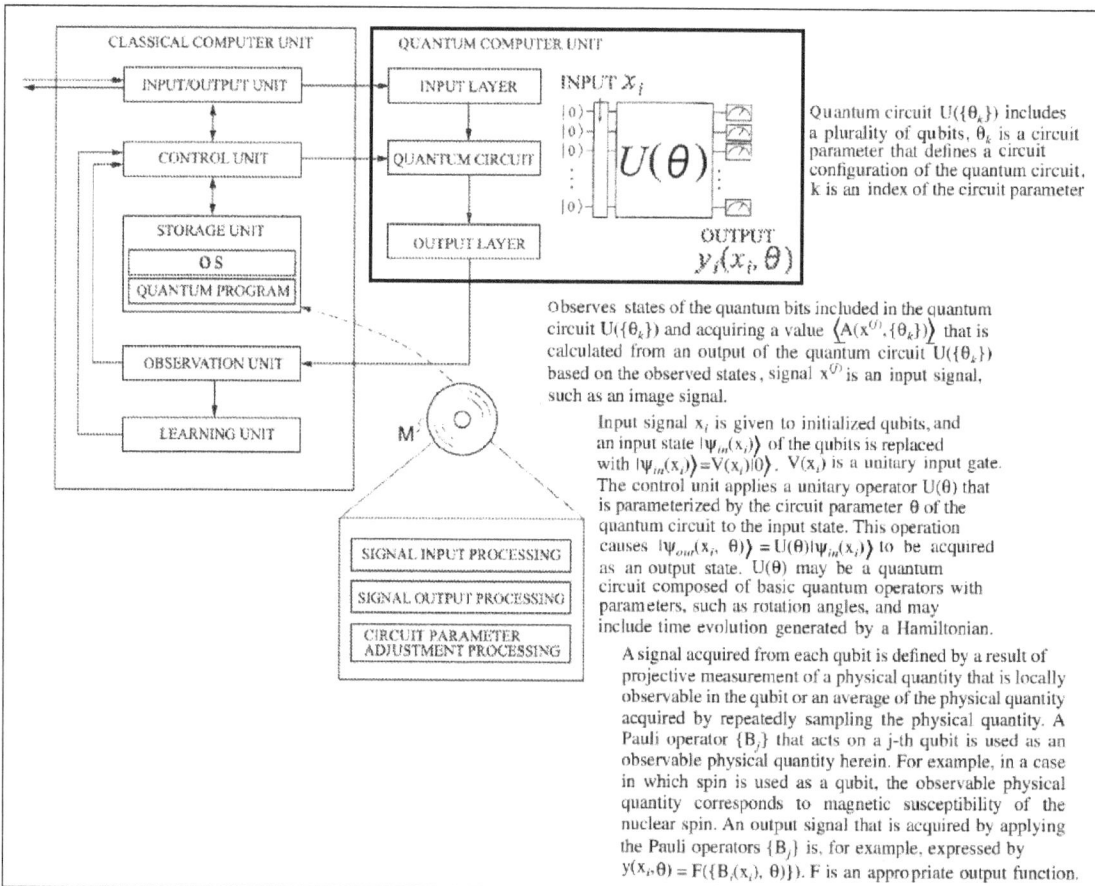

Quantum circuit $U(\{\theta_k\})$ includes a plurality of qubits, θ_k is a circuit parameter that defines a circuit configuration of the quantum circuit, k is an index of the circuit parameter

Observes states of the quantum bits included in the quantum circuit $U(\{\theta_k\})$ and acquiring a value $\langle A(x^{(j)},\{\theta_k\})\rangle$ that is calculated from an output of the quantum circuit $U(\{\theta_k\})$ based on the observed states, signal $x^{(j)}$ is an input signal, such as an image signal.

Input signal x_i is given to initialized qubits, and an input state $|\psi_{in}(x_i)\rangle$ of the qubits is replaced with $|\psi_{in}(x_i)\rangle = V(x_i)|0\rangle$. $V(x_i)$ is a unitary input gate. The control unit applies a unitary operator $U(\theta)$ that is parameterized by the circuit parameter θ of the quantum circuit to the input state. This operation causes $|\psi_{out}(x_i, \theta)\rangle = U(\theta)|\psi_{in}(x_i)\rangle$ to be acquired as an output state. $U(\theta)$ may be a quantum circuit composed of basic quantum operators with parameters, such as rotation angles, and may include time evolution generated by a Hamiltonian.

A signal acquired from each qubit is defined by a result of projective measurement of a physical quantity that is locally observable in the qubit or an average of the physical quantity acquired by repeatedly sampling the physical quantity. A Pauli operator $\{B_j\}$ that acts on a j-th qubit is used as an observable physical quantity herein. For example, in a case in which spin is used as a qubit, the observable physical quantity corresponds to magnetic susceptibility of the nuclear spin. An output signal that is acquired by applying the Pauli operators $\{B_j\}$ is, for example, expressed by $y(x_i,\theta) = F(\{B_j(x_i), \theta)\})$. F is an appropriate output function.

Figure 5.8   Example of hybrid circuit QC (synthetized from [FUJ24])

A quantum circuit comprising two quantum gates: U and V

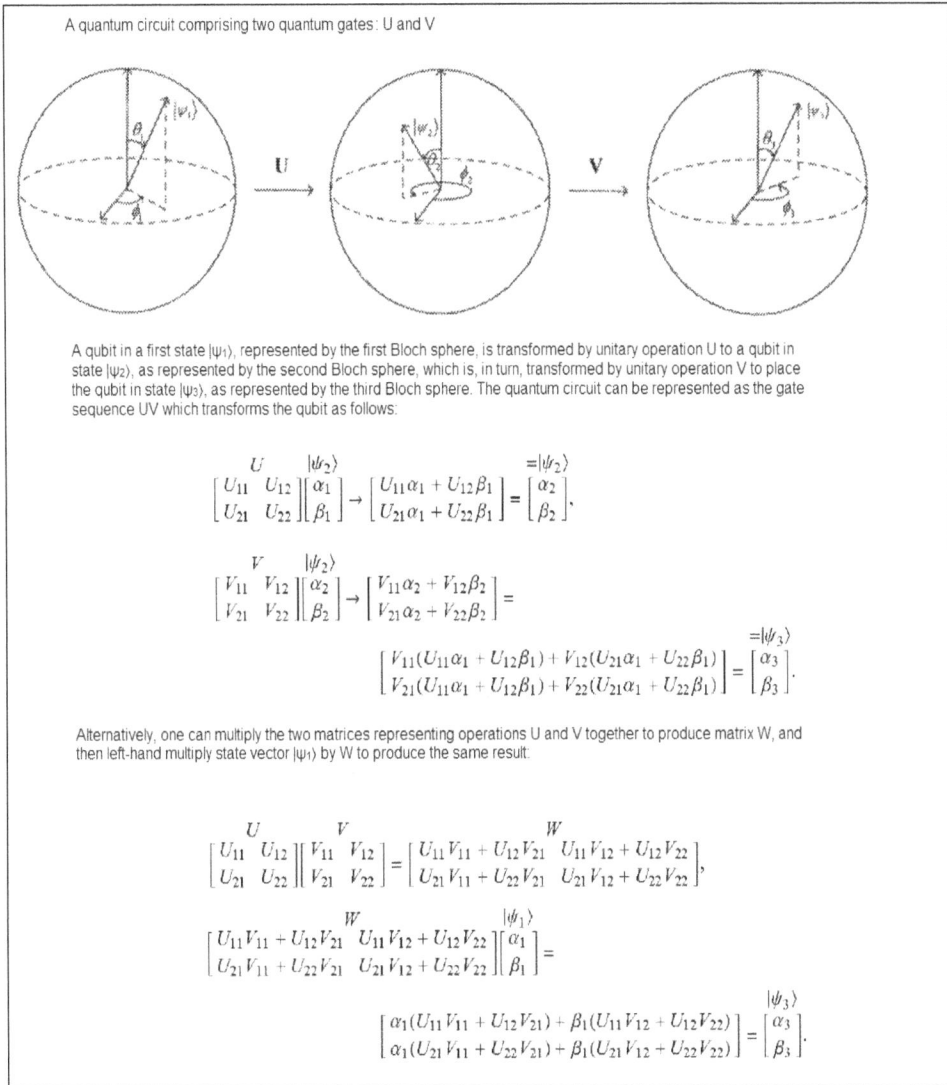

A qubit in a first state |ψ₁⟩, represented by the first Bloch sphere, is transformed by unitary operation U to a qubit in state |ψ₂⟩, as represented by the second Bloch sphere, which is, in turn, transformed by unitary operation V to place the qubit in state |ψ₃⟩, as represented by the third Bloch sphere. The quantum circuit can be represented as the gate sequence UV which transforms the qubit as follows:

$$\overset{U}{\begin{bmatrix} U_{11} & U_{12} \\ U_{21} & U_{22} \end{bmatrix}} \overset{|\psi_2\rangle}{\begin{bmatrix} \alpha_1 \\ \beta_1 \end{bmatrix}} \rightarrow \begin{bmatrix} U_{11}\alpha_1 + U_{12}\beta_1 \\ U_{21}\alpha_1 + U_{22}\beta_1 \end{bmatrix} \overset{=|\psi_2\rangle}{=\begin{bmatrix} \alpha_2 \\ \beta_2 \end{bmatrix}},$$

$$\overset{V}{\begin{bmatrix} V_{11} & V_{12} \\ V_{21} & V_{22} \end{bmatrix}} \overset{|\psi_2\rangle}{\begin{bmatrix} \alpha_2 \\ \beta_2 \end{bmatrix}} \rightarrow \begin{vmatrix} V_{11}\alpha_2 + V_{12}\beta_2 \\ V_{21}\alpha_2 + V_{22}\beta_2 \end{vmatrix} =$$

$$\begin{bmatrix} V_{11}(U_{11}\alpha_1 + U_{12}\beta_1) + V_{12}(U_{21}\alpha_1 + U_{22}\beta_1) \\ V_{21}(U_{11}\alpha_1 + U_{12}\beta_1) + V_{22}(U_{21}\alpha_1 + U_{22}\beta_1) \end{bmatrix} \overset{=|\psi_3\rangle}{=\begin{bmatrix} \alpha_3 \\ \beta_3 \end{bmatrix}}.$$

Alternatively, one can multiply the two matrices representing operations U and V together to produce matrix W, and then left-hand multiply state vector |ψ₁⟩ by W to produce the same result:

$$\overset{U}{\begin{bmatrix} U_{11} & U_{12} \\ U_{21} & U_{22} \end{bmatrix}} \overset{V}{\begin{bmatrix} V_{11} & V_{12} \\ V_{21} & V_{22} \end{bmatrix}} = \overset{W}{\begin{bmatrix} U_{11}V_{11} + U_{12}V_{21} & U_{11}V_{12} + U_{12}V_{22} \\ U_{21}V_{11} + U_{22}V_{21} & U_{21}V_{12} + U_{22}V_{22} \end{bmatrix}},$$

$$\overset{W}{\begin{bmatrix} U_{11}V_{11} + U_{12}V_{21} & U_{11}V_{12} + U_{12}V_{22} \\ U_{21}V_{11} + U_{22}V_{21} & U_{21}V_{12} + U_{22}V_{22} \end{bmatrix}} \overset{|\psi_1\rangle}{\begin{bmatrix} \alpha_1 \\ \beta_1 \end{bmatrix}} =$$

$$\begin{bmatrix} \alpha_1(U_{11}V_{11} + U_{12}V_{21}) + \beta_1(U_{11}V_{12} + U_{12}V_{22}) \\ \alpha_1(U_{21}V_{11} + U_{22}V_{21}) + \beta_1(U_{21}V_{12} + U_{22}V_{22}) \end{bmatrix} \overset{|\psi_3\rangle}{=\begin{bmatrix} \alpha_3 \\ \beta_3 \end{bmatrix}}.$$

Figure 5.9   Example of quantum circuit as a sequence of two gates [BOC23]

shown in Figure 5.9 [BOC23], one can apply the operators sequentially (as the middle of the figure shows), or one can multiply the matrices corresponding to the component quantum gates together in the order specified by the symbol sequence to produce a final 2×2 complex matrix that represents the same state change, in a single operation or quantum gate, corresponding to the state change produced by sequential application of the quantum gates specified in the original symbol sequence (bottom of the figure). As implied by this figure, the quantum circuit can be viewed as a quantum-computing program in which operations corresponding to the sequence of quantum gates represented by the symbol sequence are applied to a qubit in order to produce a final qubit state; or it can viewed as a design for an actual hardware circuit in a QC, where the hardware circuit needs to perform the operation specified by the single-gate representation of the quantum circuit. (Consequently, the term "quantum circuit" is somewhat interchangeable with the term "quantum gate" because a quantum circuit can be expressed as a resultant unitary operator). Symbolic representations of quantum circuits facilitate quantum circuit design for at least two reasons [BOC23]:

i.   In many cases, it is challenging to design and implement arbitrary physical quantum gates, but it is easier to design and implement certain standard quantum gates.

ii. Error-detection-and-correction methods have been worked out and are available for only certain well-known standard quantum gates; thus, designing and implementing quantum circuits as sequences of the well-known standard quantum gates provides a design and implementation to which error detection and error correction can be applied.

It is well known that the circuit model is *universal* for Quacomp—namely, there exist sets of gates acting on a defined number of qubits each that can efficiently simulate a quantum Turing machine (e.g., [YAO93]). In the standard model of Quacomp, the state of n qubits evolves in discrete time steps by unitary operations; this is in contrast with AQC, wherein the underlying physical description of the evolution is continuous (with the Schrödinger equation).

5.3 Adiabatic Quantum Computation (AQC)

We discussed time evolution in Chapter 2. In general, systems, including quantum systems, are dynamic; that is, they evolve in time. Time evolution is the change of state brought about by the passage of time from a start point in state $|\psi(0)\rangle$. Let $|\psi(t)\rangle$ be the time-dependent state vector and H (t) be the time-dependent Hamiltonian operator. The state $|\psi(t)\rangle$ can be derived and shown to be

$$|\psi(t)\rangle = U(t)|\psi(0)\rangle = e^{-iHt/\hbar}|\psi(0)\rangle$$

where the evolution operator $U(t)$ is the exponential operator. This relationship is derived from the fact that all quantum systems evolve in time according to the Schrödinger equation—specifically, the Hamiltonian operator H *generates the time evolution of quantum states*: if $|\psi(t)\rangle$ is the state of the system at time t, then the evolution is described by the Schrödinger equation:

$$H|\psi(t)\rangle = i\hbar\frac{\partial}{\partial t}|\psi(t)\rangle$$

This is the basic expression for the evolution in time of a stationary state of an isolated system (units also can be chosen so that Planck's constant is 1).

The application of time evolution to the matter at hand follows. As noted in [FAR01], a QC algorithm can be viewed as a specification of a Hamiltonian $H(t)$ and an initial state $|\psi(0)\rangle$. More specifically, the elements of the quantum *adiabatic time evolution* algorithm are: (i) an instance-dependent, time-dependent Hamiltonian $H(t)$; (ii) an initial state $|\psi(0)\rangle = |\psi_g(0)\rangle$; (iii) system evolution from 0 to T according to the Schrödinger equation; and (iv) a measurement of $|\psi_g(T)\rangle$. These elements are chosen so that the state at time T, $|\psi(T)\rangle$, encodes the answer to the problem at hand.

In designing a quantum algorithm, one takes advantage of the quantum adiabatic theorem, as follows. At time t, the Hamiltonian $H(t)$ has an instantaneous ground state $|\psi_g(t)\rangle$, which is the eigenstate of $H(t)$ with the lowest energy. Adiabatic evolution refers to the situation where $H(t)$ is slowly varying. Say the quantum system starts at $t = 0$ in the ground state of $H(0)$—that is, $|\psi_g(0)\rangle$); the adiabatic theorem affirms that if $H(t)$ varies slowly enough, then the evolving state vector $|\psi(t)\rangle$ will remain close to the instantaneous ground state $|\psi_g(t)\rangle$. To specify the algorithm, one must provide $H(t)$ for $0 \le t \le T$, where T is the running time of the algorithm. One chooses $H(t)$ so that the ground state of $H(0)$ is known in advance and is easy to construct. For any case of the problem under study, there is a Hamiltonian, H_P whose ground state encodes the solution. Although it is straightforward to construct H_P finding its ground state is computationally difficult. One takes $H(T) = H_P$ which means that $|\psi_g(T)\rangle$ encodes the solution. For intermediate times, $H(t)$ smoothly interpolates between $H(0)$ and $H(T) = H_P$ One starts with the quantum state in the known ground state of $H(0)$. If the running time T is large enough, $H(t)$ will indeed be slowly varying, and by the adiabatic theorem the final state reached, $|\psi(T)\rangle$, will be close to $|\psi_g(T)\rangle$, which encodes the solution to the problem.

As just noted, to specify the quantum algorithm, one must identify the Hamiltonian $H(t)$ that controls the evolution of the quantum system via the time evolution equations (above). Using a n-qubit Hilbert space that can be realized as a system of n spin-½ particles where $|z_i=0\rangle$ corresponds to the ith spin being up in the z-direction, and $|z_i=1\rangle$ corresponds to spin being down in the z-direction, and for $H(0)$ coupling a magnetic field in the x-direction to each quantum spin, [FAR01] derives the following for $H(t)$:

$$H(t) = \left(1 - \frac{t}{T}\right) H(0) + \frac{t}{T} H_P$$

Thus, the final Hamiltonian $H(t)$ is H_P Note that as T gets bigger, $H(t)$ becomes more slowly varying, and by the adiabatic theorem $|\psi(t)\rangle$, which obeys the Schrödinger time evolution equation and begins in $|\psi_g(0)\rangle$, will stay close to $|\psi_g(t)\rangle$. Therefore, $|\psi(t)\rangle$ will be close to $|\psi_g(t)\rangle$, the ground state of H_P Since the ground state of H_P encodes the desired solution, a measurement of the quantum state at time T will with high probability determine if the instance has a satisfying assignment. Refer to [FAR01] for additional details on this QC.

A formal definition of adiabatic Quacomp follows (from [AHA07]):

A k-local adiabatic Quacomp is specified by two k-local Hamiltonians, H_0 and H_1, acting on n p-state particles, $p \ge 2$. The ground state of H_0 is unique and is a product state. The output is a state that is ε-close in ℓ_2-norm to the ground state of H_1. Let $s(t): [0, t_f] \to [0, 1]$ (the "schedule") and let t_f be the smallest time such that the final state of an adiabatic evolution generated by $H(s) = (1-s)H_0 + sH_1$ for time t_f is ε-close in ℓ_2-norm to the ground state of H_1. (H_1 needs not be unique and multiple final ground states are acceptable, for example in a classical optimization problem).

In particular, Adiabatic Theorems (ATs) provide (only) an upper bound on the evolution time t_f required to achieve a certain fidelity between the actual state and the target eigenstate of $H(t)$. There is not a single AT: there are a number of variants that provide different run-time requirements, under different smoothness and differentiability conditions of the Hamiltonian operator. Descriptions of ATs start with the system being initialized in its ground state and the gap is the ground state gap. One then assumes that for all $s \in [0, 1]$ the Hamiltonian $H(s)$ has an eigenprojector $P(s)$ with eigenenergy $\varepsilon_0(s)$, and that the gap never vanishes—that is, $\Delta > 0$. $P(s)$ represents the "ideal" adiabatic evolution. Let $P_{t,f}(s) = |\psi_{t,f}(s)\rangle \langle h\psi_{t,f}(s)|$; this is the projector onto the time-evolved solution of the Schrödinger equation—namely, the "actual" state. ATs are typically assertions about the "instantaneous adiabatic distance" $P_{t,f}(s) - P(s)$ between the projectors associated with the actual and ideal evolutions, or the "final-time adiabatic distance" $P_{t,f}(1) - P(1)$. Typically, ATs provide a bound on $P_{t,f}(s) - P(s)$ of the form $O(1/t_f)$ for the instantaneous case, and a bound of the form $O(1/t^n_f)$ for any $n \in N$ for the final-time case. After squaring, these projector-distance bounds become bounds on the transition probability. One result is that there is a lower bound on the time required to prepare state $|\varepsilon(1)\rangle$ from $|\varepsilon(0)\rangle$ with bounded precision, and $t_f \sim O(\max_{\text{over } s} H(s)/\Delta^2)$. These AT bounds can be complex to derive; the reader may refer to Reference [ALB18] for an extensive discussion of bounds in particular and of ATs in general.

In summary, in AQC, the computation advances by moving from a low-energy eigenstate of the initial Hamiltonian to the ground state of the final Hamiltonian. The process evolves the quantum state towards a user-defined final Hamiltonian of the problem at hand, while at the same time reducing the influence of initial Hamiltonian. Tunneling[1] between various classical states or the eigenstates of the Hamiltonian of the problem at hand is guided by the amplitude of the initial Hamiltonian, and decreasing this amplitude from a large value to zero progresses the system into the ground state of the problem Hamiltonian that corresponds to the optimal solution of the objective function [YOU23].

5.3.1 Example

The Grover algorithm (*aka* quantum search algorithm) informally has the objective to find the marked item (or possibly multiple marked items) in an unsorted database of N items by accessing the database as few times as possible (using just order-of-the-square-root-of-N evaluations of the function—classical methods require order-of-N evaluations) [GRO97]. The algorithm has been implemented with circuit-based operators/Quacomp. The adiabatic Grover algorithm is a canonical example of a provable quantum speedup using AQC. Just to provide a "flavor" of the adiabatic Grover algorithm, make note of the following machinery from reference [ALB18]:

Let $H_0 = \mathbb{I} - |\phi\rangle\langle\phi|$ where $|\phi\rangle$ is the uniform superposition state,

$$|\phi\rangle = \frac{1}{\sqrt{N}} \sum_{i=0}^{N-1} |i\rangle = |+\rangle^{\otimes n}$$

where

$$|+\rangle = \tfrac{1}{\sqrt{2}}(|0\rangle + |1\rangle)$$

Define the time-dependent Hamiltonian to be an interpolation:

$$H(s) = [1 - A(s)] H_0 + A(s)H_1 = [1 - A(s)] (1 - |\phi\rangle\langle\phi|) + A(s)(1 - |m\rangle\langle m|),$$

where $s = t/t_f \in [0, 1]$ is the dimensionless time, t_f is the total computation time, and $A(s)$ is a "schedule" that can be optimized. If the initial state is initialized in the ground state of $H(0)$—that is, $|\psi(0)\rangle = |\phi\rangle$—then the evolution of the system is restricted to a two-dimensional subspace, defined by the span of $|m\rangle$ and $|m^\perp\rangle$ with

$$|m^\perp\rangle = \frac{1}{\sqrt{N-1}} \sum_{i \neq m}^{N-1} |i\rangle.$$

---

[1] See Glossary of Chapter 1.

In this two-dimensional subspace $H(s)$ can be written as

$$[H(s)]_{|m\rangle,|m^\perp\rangle} = \frac{1}{2}\mathbb{I}_{2\times 2} - \frac{\Delta(s)}{2}\begin{pmatrix} \cos\theta(s) & \sin\theta(s) \\ \sin\theta(s) & -\cos\theta(s) \end{pmatrix},$$

$$\Delta(s) = \sqrt{(1-2s)^2 + \frac{4}{N}s(1-s)},$$

$$\cos\theta(s) = \frac{1}{\Delta(s)}\left[1 - 2(1-s)\left(1 - \frac{1}{N}\right)\right],$$

$$\sin\theta(s) = \frac{2}{\Delta(s)}(1-s)\frac{1}{\sqrt{N}}\sqrt{1 - \frac{1}{N}}.$$

The eigenvalues and eigenvectors in this subspace are then given by

$$\varepsilon_0(s) = \frac{1}{2}(1 - \Delta(s)), \quad \varepsilon_1(s) = \frac{1}{2}(1 + \Delta(s))$$

$$|\varepsilon_0(s)\rangle = \cos\frac{\theta(s)}{2}|m\rangle + \sin\frac{\theta(s)}{2}|m^\perp\rangle,$$

$$|\varepsilon_1(s)\rangle = -\sin\frac{\theta(s)}{2}|m\rangle + \cos\frac{\theta(s)}{2}|m^\perp\rangle.$$

The remaining $N-2$ eigenstates of $H(s)$ have eigenvalue 1 throughout the evolution. The minimum gap occurs at $s = \frac{1}{2}$ and scales exponentially with n:

$$\Delta_{\min} = \Delta(s = 1/2) = \frac{1}{\sqrt{N}} = 2^{-n/2}$$

After some extensive math manipulation (and some assumptions), it can be shown that

$$t_f = \frac{N}{c'\sqrt{N-1}}\tan^{-1}\sqrt{N-1} \to \frac{\pi}{2c'}\sqrt{N}$$

which is the expected quadratic quantum speedup. Also, one can derive the optimal scheduling process as

$$A(s) = \frac{1}{2} + \frac{1}{2\sqrt{N-1}}\tan\left[(2s-1)\tan^{-1}\sqrt{N-1}\right]$$

5.3.2 Concluding Observations on AQC

It has been established that the computational power of the circuit model and the adiabatic model of Quacomp are equivalent, up to polynomial overhead. The analog of a set of gates in the circuit model is a Hamiltonian in AQC; furthermore, the circuit model can efficiently simulate AQC. A practical definition of universal AQC is, thus, to efficiently map any circuit to an adiabatic computation using a sufficiently powerful Hamiltonian; the stipulation that the ground state of $H(t_f)$ is equal to the final state at the end of the circuit ensures that the circuit and the adiabatic computation have the same output. Refer to Reference [ALB18] for an extensive discussion on this topic.

5.4 Categorical Quantum Model

Category theory is a general framework, a formalism, an area of applied mathematics. Category theory originated around 1945, with Eilenberg and Lane defining "categories", "functors" between categories, and "natural transformations" between functors [EIL45]. A category has objects (systems, things) and morphisms (processes, ways to go between things). A category C^1 consists of: (i) a collection of **objects**, where if X is an object of C, one writes $X \in C$; and (ii) for every pair of objects (X, Y), a set $\{f_j\}$ of **morphisms** from X to Y, namely, $f_j: X \rightarrow Y$ or $X - {}^{f_j}\!\!\rightarrow Y$, with the additional requirement that for every object X there is an identity morphism $1_X: X \rightarrow X$ (or $X - 1_x \rightarrow X$). In physics, the objects are often physical systems, and the morphisms are processes turning a state of one physical system into a state of another system—possibly the same one. A categorical quantum model is a model that uses category theory. Category theory has only recently been used to model QM; in QM category theory takes Hilbert spaces as objects, and linear operators as morphisms [BAE09], [ABR08]. *Concepts appearing in the gate (circuit) model can be expressed in the language of the category model.* Diagrammatic methods in Quantum Information Science and Technology (QIST) have been used for quite some time; their pragmatic value originates from the fact that they enable one to support mathematical reasoning and even actual calculations using intuitive graphical objects instead of abstract mathematical entities. In particular, Quantum Circuit Diagrams (QCDs) are representable as graphs that are planar, directed, and acyclic. The categorical model subsumes the traditional quantum circuit model; the circuit diagrams themselves now become morphisms in a category, making quantum circuits a special case of a much more general mathematical framework [BER11].

Category theory can be used as a unifying language for mathematics, and in more recent years is used to formulate physical theories. One of the advantages of categorical modeling is that it has a graphical language that is equivalent to the corresponding algebraic notation. One can use this graphical language to define the algebraic properties of the gate networks. Category theory provides a methodology for "diagrammatic reasoning" present in the existing methods to manipulate quantum systems: the string diagrams capture the mathematical properties of how maps (states and operators in the circuit model) can be composed. Utilizing the categorical description of the mathematics used in QM, one essentially obtains and/or recovers the quantum circuits model [BER11]; category theory provides an intuitive graphical interpretation that one can use to manipulate quantum circuits by enabling the creation of diagrams that can then be translated back into a standard, physically implementable quantum circuits.

The discussion herewith is restricted to a basic definition of category theory and a description of well-known concepts in the category model language. The interested reader may refer to [REU19] for an extensive discussion; the observations that follow in this entire subsection are in fact based directly on [REU19].

---

[1] Note that this siglum C does not equate to the complex field **C** defined in earlier chapters.

As stated, categories are formed from two basic structures: objects A, B, C, ..., and morphisms $A - ^f \to B$ going between objects. One can think of an object as a system, and a morphism $A - ^f \to B$ as a process under which the system A becomes the system B. Categories can be constructed from various system and process, for example:

- Physical systems, and the physical processes governing them
- Data types in computer science, and algorithms manipulating them
- Algebraic or geometric structures in mathematics, and structure-preserving functions
- Logical propositions and implications between them

A category is itself just an algebraic structure that focuses entirely on relationships between the objects. Category theory is the study of the patterns formed by these relationships.

Definition. A category C consists of the following data:

- A collection $Ob(C)$ of objects;
- For every pair of objects A and B, a collection $C(A, B)$ of morphisms, with $f \in C(A, B)$ written $A - ^f \to B$;
- For every pair of morphisms $A - ^f \to B$ and $B - ^g \to C$ with common intermediate object, a composite $A - ^{f \circ g} \to C$;
- For every object A an identity morphism $A - ^{idA} \to A$.

These must satisfy the following properties, for all objects A, B, C, D, and all morphisms $A - ^f \to B$, $B - ^g \to C$, $C - ^h \to D$:

- Associativity: $h \circ (g \circ f) = (h \circ g) \circ f$
- Identity: $id_B \circ f = f = f \circ id_A$

The definition of a category refers to collections of objects and morphisms, rather than to sets. From this definition one can see that the morphisms are "more important" than the objects; after all, every object A is canonically represented by its identity morphism id_A. This seems like a basic point, but it is a major departure from much of classical mathematics, in which particular structures (such as groups) play a more important role than the structure-preserving maps between them (such as group homomorphisms). The most basic relationships between sets are given by functions: For sets A and B, a function $A - ^f \to B$ comprises, for each a ∈ A, a choice of element $f(a) \in B$. One writes $f: a \mapsto f(a)$ to denote this choice.

Vector spaces and linear maps form a category: objects are complex vector spaces; morphisms are linear functions. Hilbert spaces are structures that are built on vector spaces. The extra structure enables one to define angles and distances between vectors and is used in QM, as discussed elsewhere, to calculate probabilities of measurement outcomes.

In category theory, Hilbert spaces and bounded linear maps are perceived as follows: (i) objects are Hilbert spaces; (ii) morphisms are bounded linear maps; (iii) composition is composition of linear maps as ordinary functions; (iv) identity morphisms are given by the identity linear maps. Given an element $a \in H$ of a Hilbert space, its ket $\mathbf{C} - ^{|a\rangle} \to H$ is the linear map $s \mapsto sa$; its bra $H - ^{\langle a|} \to \mathbf{C}$ is the linear map $b \mapsto \langle a|b\rangle$.

One can check that $|a\rangle^\dagger = \langle a|$:

$$\left(\mathbf{C} \xrightarrow{|a\rangle} H \xrightarrow{\langle b|} \mathbf{C} \right) = \left(\mathbf{C} \xrightarrow{\langle b| \circ |a\rangle} \mathbf{C} \right) = \left(\mathbf{C} \xrightarrow{\langle b|a\rangle} \mathbf{C} \right)$$

The final expression identifies the number $\langle b|a\rangle$ with the linear map $1 \mapsto \langle b|a\rangle$. Thus, the inner product (or 'bra-ket') $\langle b|a\rangle$ decomposes into a $\langle b|$ and a ket $|a\rangle$; this is the traditional Dirac notation.

The tensor product is a way to make a new vector space out of two given ones. The tensor product of vector spaces U and V is a vector space $U \otimes V$ together with a bilinear function f: $U \times V - f \rightarrow U \otimes V$ (note that $U \times V$ is not itself a vector space). The tensor product of Hilbert spaces H and K is the Hilbert space $H \otimes K$ built by taking tensor product of the underlying vector spaces, giving it the inner product $\langle a_1 \otimes b_1 | a_2 \otimes b_2 \rangle = \langle a_1 | a_2 \rangle_H \times \langle b_1 | b_2 \rangle_K$, then completing it. A qubit is a quantum system with state space $\mathbf{C}^2$. A pure state of a quantum system is given by a vector $v \in H$ in its associated Hilbert space. Such a state is normalized when the vector in the Hilbert space has norm equal to 1: $\langle a | a \rangle = 1$. The joint states of U and V are elements of $U \otimes V$; product states are factorizable states; entangled states are elements of $U \otimes V$ which cannot be factorized. Entangled states represent preparations of $U \otimes V$ which cannot be decomposed as a preparation of U alongside a preparation of V; here there is some essential connection between U and V, which means that they cannot have been prepared independently.

Processing quantum information takes place by applying unitary maps $H - f \rightarrow H$ to the Hilbert space of states. Such a map will take a normalized state $a \in H$ to a normalized state $f(a) \in H$. An example of a unitary map is the X gate represented by the matrix $\begin{bmatrix} 0 & 1 \\ 1 & 0 \end{bmatrix}$, which acts as $|0\rangle \mapsto |1\rangle$ and $|1\rangle \mapsto |0\rangle$ on the computational basis states of a qubit. A postulate of QM states that given two quantum systems with state spaces given independently by Hilbert spaces H and K, as a joint system their overall state space is $H \otimes K$, the tensor product of the two Hilbert spaces; thus a collection of n qubits will have a state space isomorphic to C^N to the $N = 2^n$, requiring 2^n complex numbers to specify its state vector exactly.

For a compound system with state space $H \otimes K$, a product state is a state of the form $a \otimes b$ with $a \in H$ and $b \in K$; an entangled state is a state not of this form. For a quantum system in a pure state, the most basic notion of measurement is a *Projection-Valued Measure* (PVM). A PVM on a Hilbert space H is a finite family of projections $H - p_i \rightarrow H$, which are complete and orthogonal. For a finite-dimensional Hilbert space, nondegenerate PVM corresponds to orthonormal bases, up to phase. A PVM is nondegenerate when $\text{Tr}(p_i) = 1$ for all i. A PVM, when applied to a Hilbert space, has a unique outcome, given by one of the projections; this outcome is probabilistic, with distribution described by the Born rule: for a PVM $\{p_i\}$ on a system in a normalized state $a \in H$, the probability of outcome i is $\langle a | p_i | a \rangle$–note that

$$\sum_i \langle a | p_i | a \rangle \quad = \quad \langle a | (\sum_i p_i) | a \rangle \quad = \quad \langle a | a \rangle \quad = \quad 1$$

A density matrix on a Hilbert space H is a positive map m: $H - m \rightarrow H$. A density matrix is normalized when $\text{Tr}(m) = 1$. A density matrix m: $H - m \rightarrow H$ is pure when $m = |a\rangle \langle a|$ for some $a \in H$; generally, it is mixed. For a finite-dimensional Hilbert space H, the maximally mixed state is the density matrix $(1/\dim(H))(\text{id}_H) - \text{id}_H$ identity morphism.

Decoherence considerations follow. Every nondegenerate PVM $\{p_1, ..., p_n\}$ on a Hilbert space H, $p_i = |i\rangle \langle i|$ corresponds (up to a phase) to an orthonormal basis $\{|1\rangle, ..., |n\rangle\}$, and hence induces n pure states of H. This PVM can be perceived as a controlled preparation: depending on some classical data $i = 1, ..., n$, one prepares state $|i\rangle$. This controlled preparation composes with a measurement in the same basis:

- If one starts with some classical information, one can use it to prepare a quantum system and then immediately measure, then one should end up with the same classical information one started with. According to the Born rule, the probability of getting outcome j after preparing state i is:

$$\langle j|p_i|j\rangle = \langle j|i\rangle\langle i|j\rangle = |\langle i|j\rangle|^2,$$

this being 1 for $i = j$ but 0 for $i \neq j$.

- The other way around is as follows. Assume one measures a quantum system, yielding some classical data, and then immediately uses them to prepare a state of a quantum system, the following situation ensues, supposing that the quantum system starts in a mixed state given by a density matrix

$$H \xrightarrow{m} H \quad \text{with } m = \sum_{ij} c_{ij}|i\rangle\langle j|,$$

the measurement results in outcome $|i\rangle$ with probability $\text{Tr}(p_i m) = \langle i|m|i\rangle$, thus the state eventually prepared is

$$\sum_i c_{ii}|i\rangle\langle i|.$$

The nondiagonal elements of the density matrix m have vanished, and the mixed state has become a convex combination of pure states that no longer cohere. This process is called *decoherence*. Any quantum state undergoes decoherence constantly as it interacts with its environment. To keep a quantum state from decohering rapidly requires good experimental control.

5.5 Examples of QC Elements

A handful of illustrative examples of QCs is shown in this section to illustrate some (graphical) concepts. To begin with, Figure 5.10 is an example of a relatively complex quantum circuit ([SOG23]).

5.5.1 Example of a Hybrid QC

Figure 5.11 depicts an illustrative example of a hybrid QC from reference [WAN23]. In particular, this QC deals with finding solutions for shadow tomography; a brief discussion on this topic follows, from the same reference. Shadow tomography is a process that can be used to estimate properties of an unknown quantum state without resorting to full state tomography. Let ρ denote a quantum state and assume access to multiple copies of the quantum state ρ is possible. Let $\{O_i\}$ represent a collection of M observables; the task is to estimate the expectation values $\text{Tr}(\rho\, O_i)$ (up to some additive error ε) for each operator O_i in the collection of observables. This can be accomplished efficiently by randomly choosing measurement operators from a tomographically complete set. To specify a protocol, an ensemble of unitaries G is chosen; different choices for the ensemble G are possible. Then, unitaries $U_k \in G$ are randomly sampled from the ensemble and the state $U_k \rho U_k^\dagger$ is measured in the computational basis to obtain the basis state $|b_k\rangle\langle b_k|$. Now, consider the state $U_k^\dagger(|b_k\rangle\langle b_k|)U_k$. In expectation, the mapping from ρ to this state defines a quantum channel $M(\rho) := E_k[U_k^\dagger(|b_k\rangle\langle b_k|)U_k]$. It is required that the quantum channel M be invertible, which is true if and only if the collection of measurement operators defined by drawing the unitaries $U \in G$ and measuring in the computational basis is tomographically complete. Assuming that this is true, M^{-1} can be applied to the definition of the quantum channel, yielding $\rho = M^{-1}(E_k[U_k^\dagger(|b_k\rangle\langle b_k|)U_k]) = E_k(M^{-1}[U_k^\dagger(|b_k\rangle\langle b_k|)U_k])$. The collection $\{(M^{-1}[U_k^\dagger(|b_k\rangle\langle b_k|)U_k])\}$ is the classical shadow of the quantum state ρ.

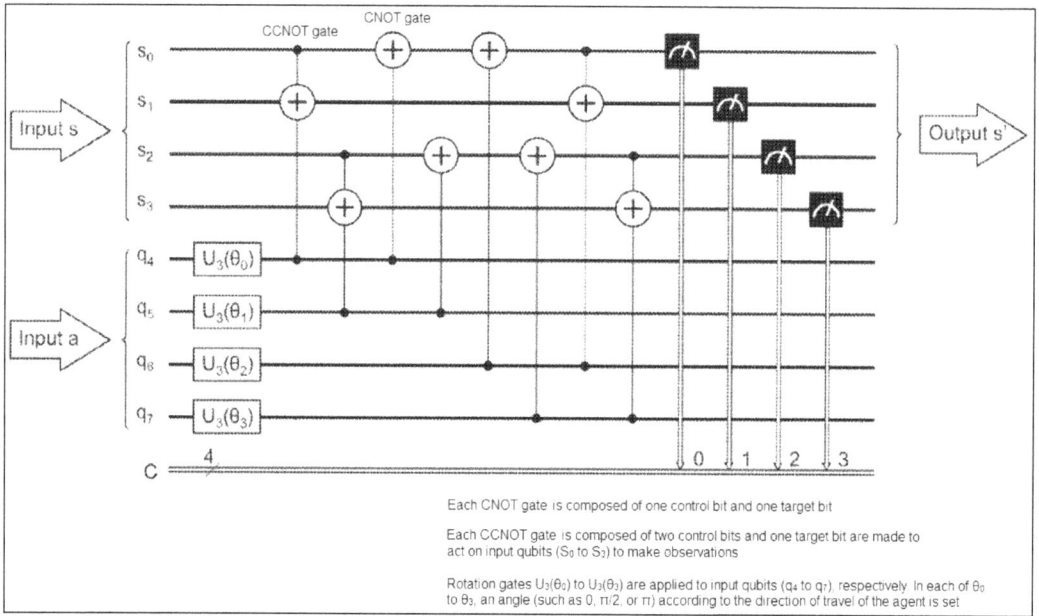

Figure 5.10   Example of quantum circuit ([SOG23])

5.5.2 Examples of Partitioned QC

In some cases, a service executing on a first quantum computing device could utilize a quantum function provided by a second quantum computing device over an application programming interface (API) in order to separate lower-level functions in the downstream system from the more user-related higher-level functions in a more general upstream system. Figures 5.12 and 5.13 depict examples of such an arrangement.

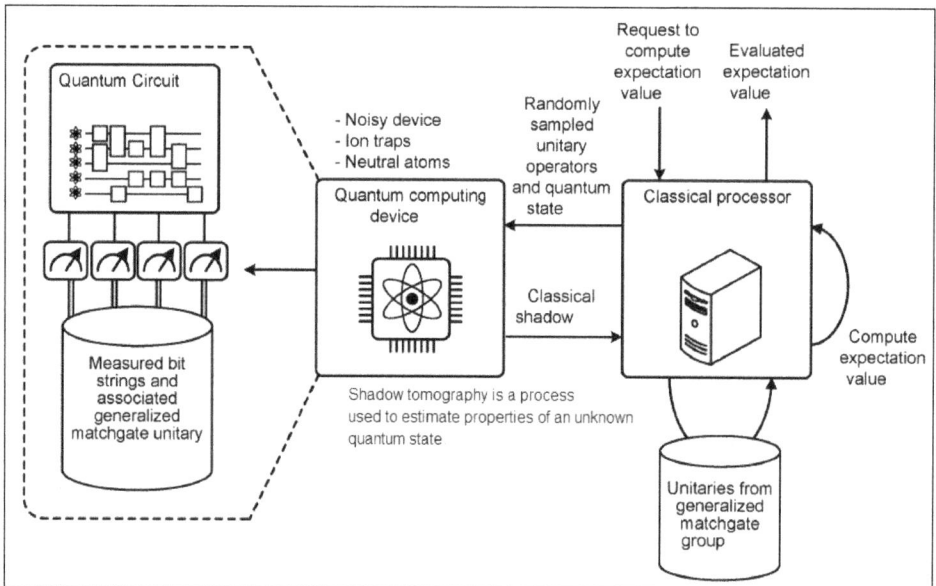

Figure 5.11   Illustrative example hybrid QC used for shadow tomography ([WAN23])

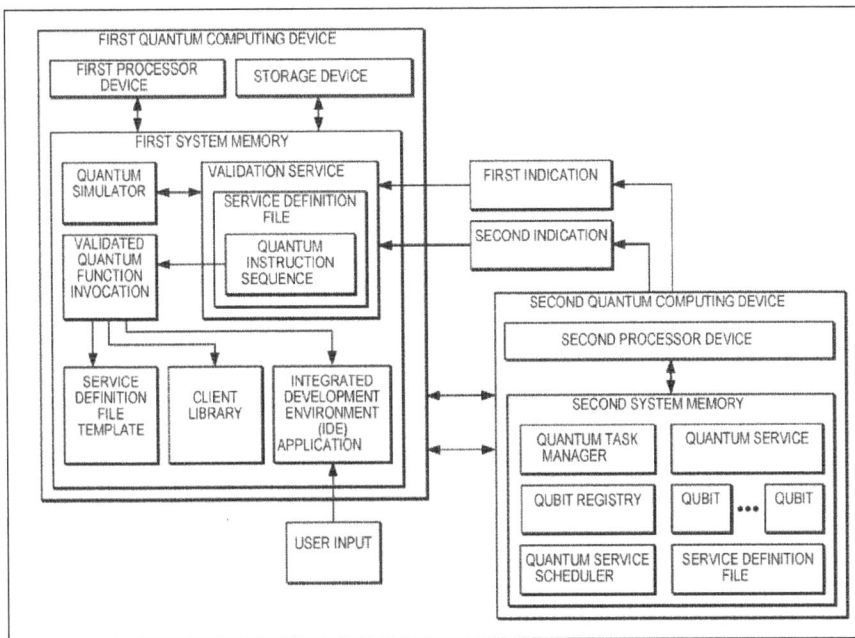

Figure 5.12 Example of partitioning of functionality over two QC systems ([GRI23])

5.5.3 Quantum Circuit Examples

A QC may typically include a classical computer coupled in combination with a quantum computing system; the quantum computing system is configurable to execute program instructions to process input data to generate corresponding output data. For example, [ROS23] describes a quantum computing system configured to execute a *Variational Quantum Amplitude Estimation* (VQAE) algorithm by iteratively executing a sequence of quantum operations including a maximum likelihood amplitude estimation and performing a variational approximation after an iteration to determine a variational quantum state associated with the outcome of the maximum likelihood amplitude estimation (see Figure 5.14). The QC system includes: (i) an array of qubits configured to be provided with initial values depending on input data supplied in use to the quantum computing system; (ii) a quantum computing arrangement and/or a quantum circuit configured to execute a sequence of quantum operations on the array of qubits to generate corresponding processed qubits; and (iii) a measuring device for determining states of the processed qubits to generate measurement data or final measurement.

Naturally, the goal is to reduce the impact of quantum noise that arises when executing the sequence of quantum operations on the qubits[1] and determining the states of the processed

---

[1] Quantum computing devices utilize quantum error correction mechanisms to detect and address quantum errors that may occur while executing quantum computations. These quantum error correction mechanisms may make use of a number of additional qubits to detect quantum errors and to identify and undertake corrective actions to fix or at least mitigate the quantum errors. The number of additional qubits could be high relative to the number of qubits required to perform a quantum operation, and long-running quantum services that generate frequent quantum errors may incur increased performance penalties due to the consumption of quantum resources by quantum error correction services [GRI23a]. More efficient error correction mechanisms are being sought.

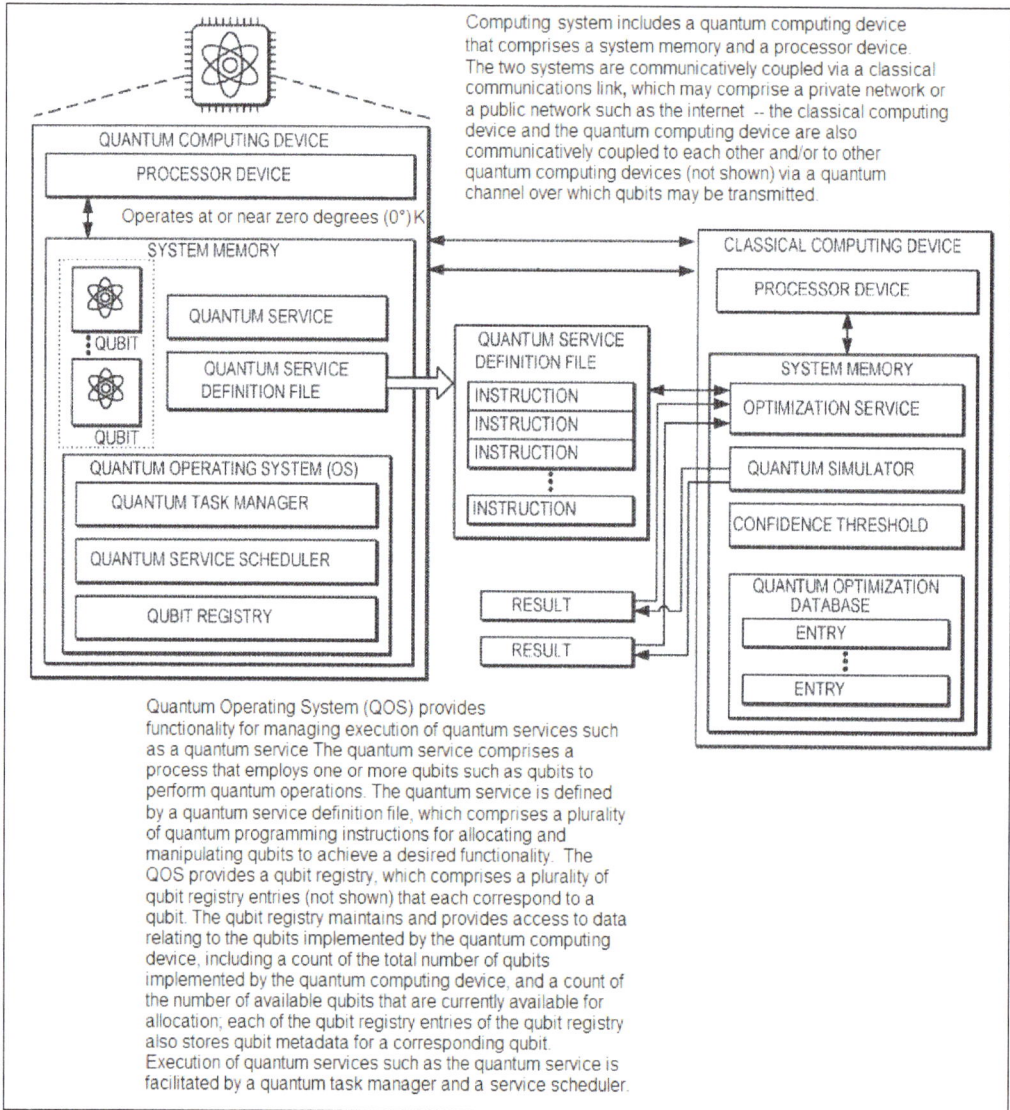

The figure contains the following labels and text:

QUANTUM COMPUTING DEVICE

PROCESSOR DEVICE

Operates at or near zero degrees (0°) K

SYSTEM MEMORY

QUBIT

QUANTUM SERVICE

QUANTUM SERVICE DEFINITION FILE

QUBIT

QUANTUM OPERATING SYSTEM (OS)

QUANTUM TASK MANAGER

QUANTUM SERVICE SCHEDULER

QUBIT REGISTRY

Computing system includes a quantum computing device that comprises a system memory and a processor device. The two systems are communicatively coupled via a classical communications link, which may comprise a private network or a public network such as the internet -- the classical computing device and the quantum computing device are also communicatively coupled to each other and/or to other quantum computing devices (not shown) via a quantum channel over which qubits may be transmitted.

QUANTUM SERVICE DEFINITION FILE

INSTRUCTION
INSTRUCTION
INSTRUCTION

INSTRUCTION

RESULT

RESULT

CLASSICAL COMPUTING DEVICE

PROCESSOR DEVICE

SYSTEM MEMORY

OPTIMIZATION SERVICE

QUANTUM SIMULATOR

CONFIDENCE THRESHOLD

QUANTUM OPTIMIZATION DATABASE

ENTRY

ENTRY

Quantum Operating System (QOS) provides functionality for managing execution of quantum services such as a quantum service The quantum service comprises a process that employs one or more qubits such as qubits to perform quantum operations. The quantum service is defined by a quantum service definition file, which comprises a plurality of quantum programming instructions for allocating and manipulating qubits to achieve a desired functionality. The QOS provides a qubit registry, which comprises a plurality of qubit registry entries (not shown) that each correspond to a qubit. The qubit registry maintains and provides access to data relating to the qubits implemented by the quantum computing device, including a count of the total number of qubits implemented by the quantum computing device, and a count of the number of available qubits that are currently available for allocation; each of the qubit registry entries of the qubit registry also stores qubit metadata for a corresponding qubit. Execution of quantum services such as the quantum service is facilitated by a quantum task manager and a service scheduler.

Figure 5.13   Other illustrative example of hybrid system (synthetized from [COA23])

qubits. In executing a certain calculation, the computing task can be partitioned between execution on the classical computer and execution on the QC: for certain tasks, the classical computer may be more efficient for handling simple computational tasks than the QC, while the QC can in principle solve certain types of computation tasks that would be inefficient to perform on the classical computer. The QC described in the reference includes in a range of 30 to 1,000 qubits and various gates that enable quantum parameters, such as qubit phase, to be modified (namely, rotation operations R), as well as entanglement and superposition operations between qubits to be performed. The QC may be implemented as an ion-trap quantum computing system that benefits from a high connectivity between qubits of the array of qubits.

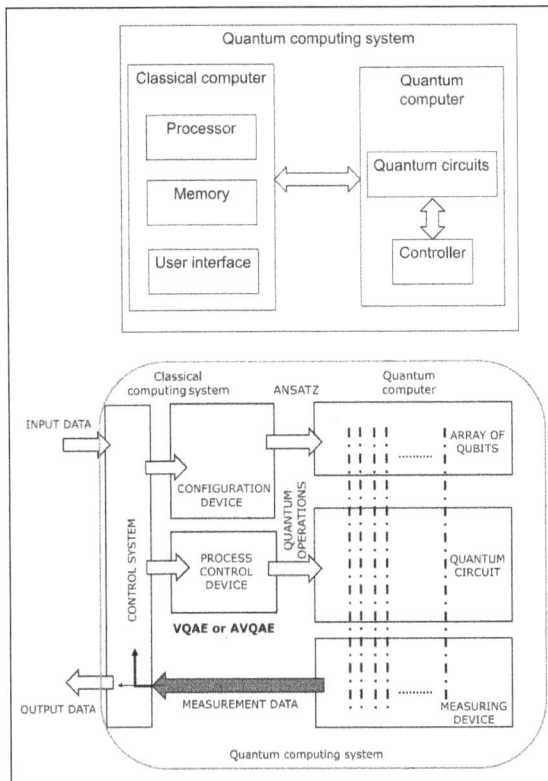

The quantum computing system is a hybrid combination of a classical computer and a quantum computing system (Top: generic view; bottom: more detailed view)

Figure 5.14   Typical quantum computing system at the time of this writing (based on [ROS23])

5.5.4 Parameterized Quantum Circuits

Because current QCs have limited quantum resources, hybrid quantum-classical algorithms have become a prominent user of current quantum hardware, given the fact that these algorithms require fewer quantum resources than do traditional quantum algorithms; in particular, the class of *Variational Quantum Algorithms* (VQAs)—which includes the *Variational Quantum Eigensolver* (VQE) algorithm[1] and the *Quantum Approximate Optimization Algorithm* (QAOA)—is achieving popularity because these algorithms allow quantum speedup to difficult problems [NGU23]. In this context, one of the key factors for success for these types of algorithms is the construction of the parameterized quantum circuit—in particular, the more "expressibility" the parameterized quantum circuit has (i.e., the circuit's ability to generate quantum states), the greater likelihood of identifying the correct solution. Figure 5.15[2] depicts a parameterized quantum circuit; here layers of a circuit template are repeated with the goal of increasing the expressibility of the circuit. However, this type of parameterized quantum

---

[1]   VQE utilizes variational principles to compute the ground state energy of a Hamiltonian; VQE can model complex wave functions in polynomial time, making it a viable near-term application for Quacomp (conventional methods are constrained in their accuracy due to the computational limits) [TIL22].

[2]   The U(3) gate was discussed in Chapter 3.

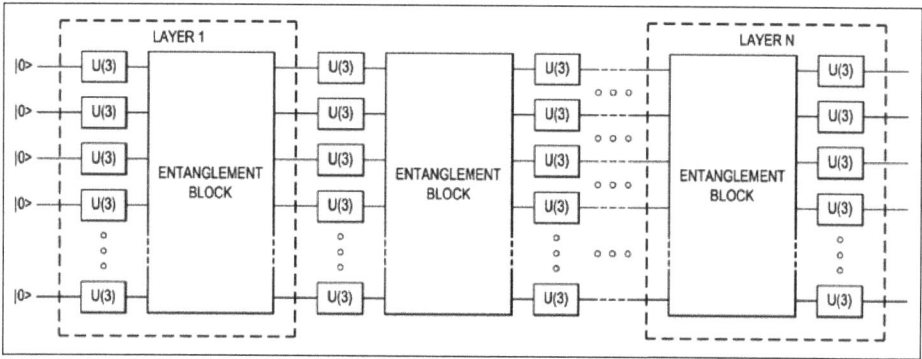

Figure 5.15   Multi-layered parameterized quantum circuit (*Source:* [NGU23])

circuit design can lead to a very long quantum circuit before it is expressible enough to represent the ground state wave function; furthermore, another drawback is that parameterized quantum circuits with arbitrary high expressibility do not scale well in the number of parameters with respect to the dimension of the problem. Hence, a design goal for a parameterized quantum circuit model is to develop a quantum circuit with just high enough expressibility with respect to problems of interest, keeping the quantum resources as low as possible while still achieving high precision results.

Reference [NGU23] proposed a quantum *circuit modeling system* that can be used for variational quantum algorithms such as the VQE algorithm, enabling a reduction of quantum resources in the circuit. Many existing parameterized quantum circuits require a long circuit depth before they are able to generate the correct solution (wave function) sought. The proposed system allows for the design of a parameterized quantum circuit that minimizes the circuit depth while still generating the correct solution. The quantum circuit modeling system generates and fine-tunes parameterized quantum circuit in order generate the ground state wave function to a system; the quantum circuit modeling system progressively fine tunes parameterized quantum circuit until it generates a wave function with the lowest energy (ground state wave function) (see Figure 5.16). The quantum circuit modeling system generates and fine-tunes a parameterized quantum circuit in order generate the ground state wave function to a system. It progressively fine-tunes parameterized quantum circuit until it generates a wave function with the lowest energy (ground state wave function).

The quantum circuit modeling system begins by defining a mean-field initial state (i.e., Hartree-Fock state) of the system in question, which is characterized by the Hamiltonian. The Hamiltonian can be decomposed into a number of Pauli strings; each Pauli string comprises a sequence of operators that are made from Pauli matrices and the 2×2 identity matrix, which defines a basis for the Hamiltonian. Operators may be divided into identity operators and non-identity operators. The sequence of operators can be characterized by a non-identity operator path length and an identity operator path length. The non-identity operator path length is the length of the Pauli string (total number of operators) minus any exterior identity operators (the shorter the non-identity operator path length, the fewer quantum resources associated with implementing the Pauli string in parameterized quantum circuit). The identity operator path length is defined by the number of exterior identity operators within the Pauli string. The model can run on general-purpose CPU or other type of processor.

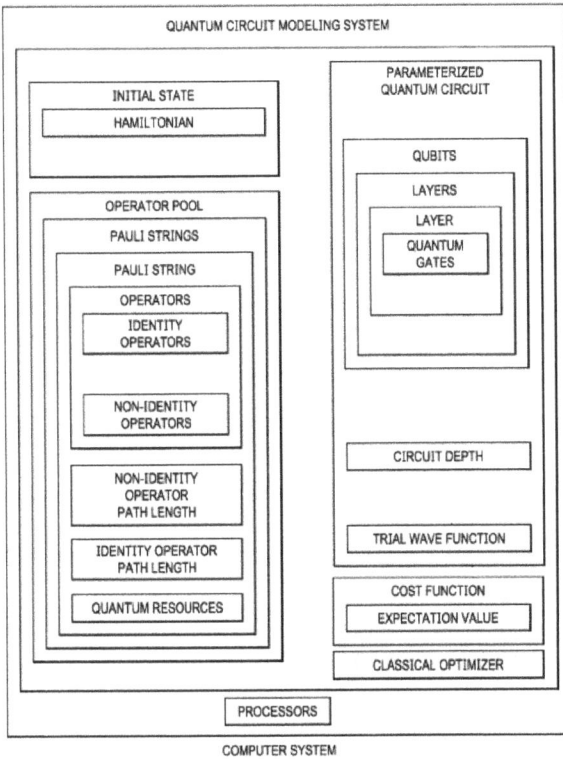

The parameterized quantum circuit is comprised of a number of layers building from the exponentiation of Pauli strings.

Each layer has a corresponding Pauli string which is selected based on its contribution to energy lowering (reducing the cost function value) as well as its structure, i.e., the identity and non-identity path length.

Each layer can be thought of as a sub-circuit in the form of quantum gates applied to qubits within the parameterized quantum circuit according to the non-identity operators within a Pauli string.

The circuit depth is determined by the number of quantum gates used in the longest path of the parameterized quantum circuit, which determines the required coherence time, a quantum resource, for the parameterized quantum circuit.

The parameterized quantum circuit generates a trial wavefunction, which is used to calculate an expectation value (energy) for a cost function associated with Hamiltonian of the system.

Figure 5.16   Illustration of a block diagram for modeling a quantum circuit (synthetized from [NGU23])

Furthermore, Figure 5.17 from [ZHA23] is an illustrative schematic diagram of a different numerical simulation approach for a quantum circuit; the goal here is to identify certain parameters embedded in the (logical) circuit to optimize the circuit performance. The quantum circuit simulation includes receiving a primitive function for the quantum circuit simulation and determining some input parameter of the primitive function. The target quantum circuit can implement numerical simulation of a variational quantum algorithm. The process being simulated and computed includes inputting a specified quantum state (which may be in the form of a matrix product state or a vector); then an output state is measured on different bases in the form of a measured Pauli string by a parameter-containing and possibly noise-containing quantum circuit, in such a manner as to obtain an optimized function value and a gradient about a weight for optimization iteration. The input quantum state of the target quantum circuit is represented as $|\psi_0\rangle$, a circuit parameter of the target quantum circuit is represented as U_θ, a measurement result is represented as $\hat{M}$, and an optimization function is represented as

$$L=\sum_{1}\left\langle \psi_0 \left| U_\theta^\dagger \hat{M}_i U_\theta \right| \psi_0 \right\rangle$$

where $\hat{M}_i$ represents an ith measurement result, i is an integer, and $U_\theta^\dagger$ represents conjugate transpose of U_θ.

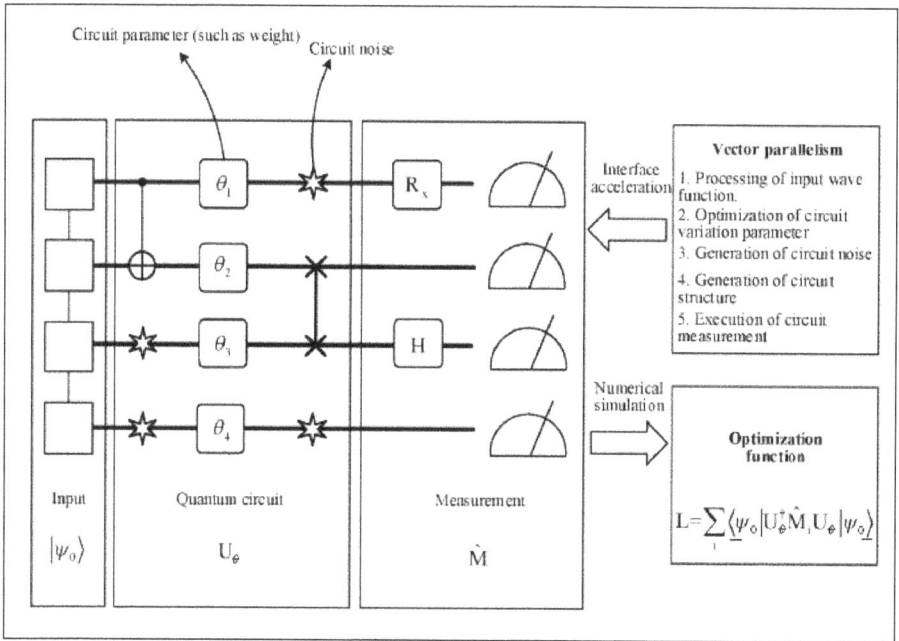

Figure 5.17   Example of a process for Quantum Circuit Simulation ([ZHA23])

5.5.5 Variational Quantum Eigensolver (VQE)

VQE uses the *variational principle* to compute the ground state energy of a Hamiltonian [PER14], [TIL22]. VQE is a promising example of NISQ-supportable algorithms. VQE aims to compute an upper bound for the ground-state energy of a Hamiltonian, which is an intrinsic first step in establishing the energetic properties of electronic structures[1], molecules and materials—applications include chemical engineering, condensed matter physics, material science, and drug discovery, among others. One example is shown in Figure 5.18 from [NGU23]. This particular search algorithm comprises three steps: (i) preparing an initial state, (ii) calculating the expectation value for the Hamiltonian $\hat{H}$ (with a parameterized quantum circuit that generates a trial wave function, and (iii) utilizing a classical optimizer to minimize the cost function $C(\theta) = \langle \psi(\theta)|\hat{H}|\psi(\theta)\rangle$ by updating the parameterized quantum circuit parameters θ in the quantum circuit. One of the advantages of executing variational type calculations on a quantum computer is the efficiency of generating hard trial wave functions on the quantum computer. There are states that can be prepared efficiently on quantum computers but not on classical computers.

VQE is based on the variational principle that optimizes an upper bound for the lowest possible expected value of an observable with respect to a trial wave function: given a Hamiltonian it seeks to find a parametrization of $|\psi\rangle$ such that the expected value of $\hat{H}$ is minimized. Observe that given a Hamiltonian $\hat{H}$ and a trial wave function $|\psi\rangle$, the ground state energy associated with this Hamiltonian, E_0, is bounded $E_0 \leq \langle \psi|\hat{H}|\psi\rangle / \langle \psi|\psi\rangle$; thus, one seeks an expected value that is a tighter (optimized) upper bound for E_0; the goal is to establish an approximation to the eigenvector $|\psi\rangle$ of the Hamiltonian operator $\hat{H}$ corresponding to the lowest eigenvalue E_0.

---

[1]   The study of interactions between electrons requires computation time that scales exponentially in the size of the system studied.

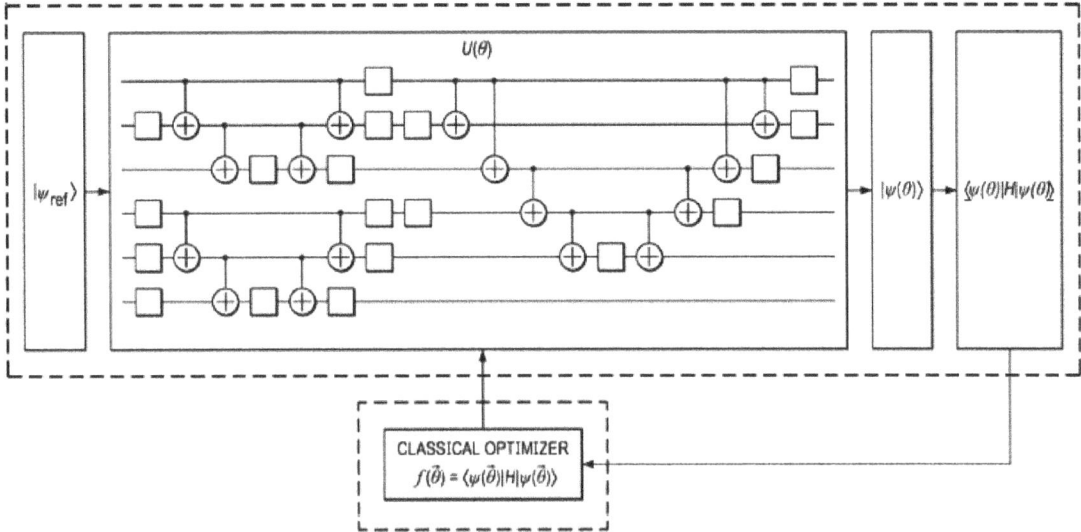

Figure 5.18 Example of VQE search algorithm ([NGU23])

Reference [TIL22] (on which the discussion that follows is based) notes that in order to translate this minimization task into a problem that can be executed on a QC, one must start by defining a so-called *ansatz wave function* that can be implemented on a quantum device as a series of quantum gates (an ansatz is an estimate or guess made to help address or solve a problem; an ansatz is the postulation of some starting equations and/or values describing a physical or mathematical problem). Given that we can only perform unitary operations or measurements on a QC, we do this by using parameterized unitary operations. One can thus express $|\psi\rangle$ as the application of a generic parameterized unitary $U(\theta)$ to an initial state for N qubits, with θ denoting a set of parameters taking values in $(-\pi, \pi]$. The qubit register is generally initialized as $|0\rangle^{\otimes N}$ (written as $|0\rangle$ for simplicity). Noting that $|\psi\rangle$ and $U(\theta)|\psi\rangle$ are a normalized wave function, one can now write the VQE optimization problem as

$$E_{VQE} = \min{}_{\text{over } \theta} \langle 0|[U^\dagger(\theta)\, \hat{H}\, U(\theta)]|0\rangle,$$

(this being the cost function of the VQE optimization problem). One can write the Hamiltonian in a form that is directly measurable on a QC, as a weighted sum of spin operators. Observables suitable for direct measurement on a quantum device are tensor products of spin operators (Pauli operators)—the spin operators for the three components of spin (for a spin-½ particle) are, as noted in Chapter 3, $\hat{S}_x$, $\hat{S}_y$, $\hat{S}_z$, where

$$\hat{S}_x = \frac{\hbar}{2}\begin{bmatrix} 0 & 1 \\ 1 & 0 \end{bmatrix} = \frac{\hbar}{2}\sigma_x$$

$$\hat{S}_y = \frac{\hbar}{2}\begin{bmatrix} 0 & -i \\ i & 0 \end{bmatrix} = \frac{\hbar}{2}\sigma_y$$

$$\hat{S}_z = \frac{\hbar}{2}\begin{bmatrix} 1 & 0 \\ 0 & -1 \end{bmatrix} = \frac{\hbar}{2}\sigma_z$$

and where σ_x, σ_y, σ_z are the Pauli operator[1] (matrices)

$$\sigma_x = \begin{bmatrix} 0 & 1 \\ 1 & 0 \end{bmatrix} = X \quad \sigma_y = \begin{bmatrix} 0 & -i \\ i & 0 \end{bmatrix} = Y \quad \sigma_z = \begin{bmatrix} 1 & 0 \\ 0 & -1 \end{bmatrix} = Z$$

One can define the spin operators as $\hat{P}_x \in \{I, X, Y, Z\}^{\otimes N}$, with N being the number of qubits used to model the wave function. The Hamiltonian can be rewritten as

$$\hat{H} = \sum_a^p w_a \hat{P}_a,$$

with w_a a set of weights, p the *number* of Pauli strings in the Hamiltonian; the cost function of the VQE optimization problem can then be written as

$$E_{VQE} = \min_\theta \sum_a^p w_a \langle 0| U^\dagger(\theta) \hat{P}_a U(\theta) |0\rangle$$

Now the hybrid nature of the VQE becomes discernable: each term

$$\langle 0| U^\dagger(\theta) \hat{P}_a U(\theta) |0\rangle$$

corresponds to the expectation value of a Pauli string $\hat{P}_a$, which can be computed on a quantum device, while the summation and minimization of the just-listed cost function is computed on a conventional computer. Figure 5.19, a simplification from Reference [TIL22], depicts the VQE computation as being decomposed into a number of components. The reader is referred to the cited reference for an extensive discussion of this topic.

5.6 Quantum Error Correction

Algorithms executing on QCs require very low gate error rates (say below about 10^{-10}); managing QC errors is a basic challenge in Quacomp. Quantum information must be protected against errors by encoding it into logical states that are suitable for Quantum Error Correction (QEC). Error management is addressed with QEC methodologies: QCs can utilize one or more QEC codes to identify error-sensitive events that can occur in the underlying quantum circuits, so that the proper corrections can be applied. *QEC codes* are utilized to protect quantum information from errors due to quantum noise, interference, or attenuation, in order to prepare and/or preserve logical quantum states.

QEC algorithms include but are not limited to *Minimum-Weight Perfect-Matching* (MWPM), union-find, and/or maximum likelihood. For example, the stabilizer QEC theory allows one to extend some classical binary error correcting codes for use in Quacomp/Quacom, where to protect qubits, the stabilizer approach appends ancilla qubits to qubits to be protected: a unitary encoding circuit rotates the global state into a subspace of a larger Hilbert space, and the

---

[1] Note: the set $\Pi = \{I, X, Y, Z\}$, where X, Y, and Z are the Pauli operators, forms a mathematical group; operators in Π have eigenvalues ± 1. Π can be utilized in stabilizer formalism to define QEC codes. In Quacomp and Quacom, a stabilizer code is a class of quantum codes for performing QEC to restore a decohered noisy quantum state to a pure quantum state; to protect qubits, a stabilizer QEC code appends ancilla qubits to qubits to be protected [GOT97].

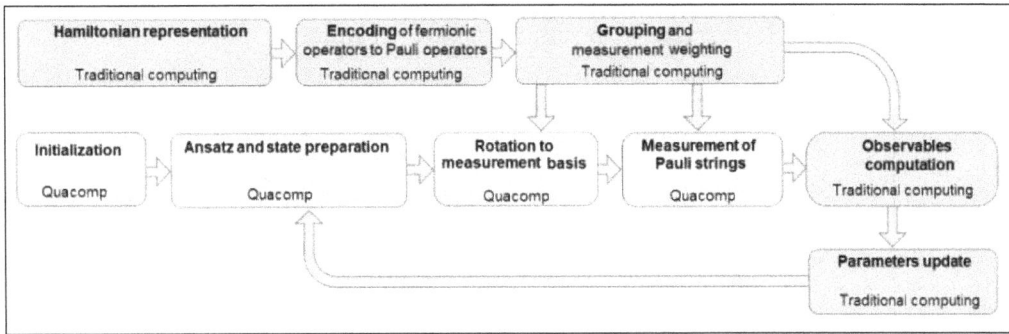

Figure 5.19   VQE calculation as a pipeline (simplified from reference [TIL22])

resulting highly entangled encoded state can be used to correct for *in situ* noisy errors. QEC is an extensive subfield of Quacomp[1]; only a few basic concepts are discussed here.

QEC mechanisms have been developed even taking into account the no-cloning theorem and the incapability to directly measure a quantum state. For example, quantum decoding for QEC codes can operate utilizing decoding graphs and/or decoding hypergraphs. As described in [CHE23], a QEC code can analyze syndrome measurement datasets from the QC in which error-sensitive events can be linear combinations of syndrome measurement bits that would equate to zero in an ideal quantum circuit operation of the QC. A non-zero error-sensitive event can be indicative of an error in the quantum circuit of the QC—for instance, the error can be a Pauli error occurring after a gate of the quantum circuit, after an idle of the quantum circuit, after initialization of the quantum circuit, or before a measurement (see illustrative Figure 20). The Pauli error can be an n-qubit Pauli error if the faulty quantum circuit component acts on "n" qubits (e.g., a 2-qubit Pauli error can occur after 2-qubit gates). Thus, the error-sensitive events can depend on the topology of the quantum circuit of the quantum computers. The effectiveness of quantum decoder algorithms can depend on the edge weights employed in the decoding graph and/or the hyperedge weights employed in the decoding hypergraph.

Another QEC method uses entanglement to delocalize a logical qubit degree of freedom across several redundant physical qubits, such that if any given physical qubit fails, it does not corrupt the underlying logical information. However, the overhead in the realization of error-corrected "logical" qubits, in which information is encoded across many physical qubits, remains challenging.

A physical qubit is represented by actual, physical (quantum) hardware. Current NISQ devices typically operate directly on physical qubits; however, as noted, physical qubits are susceptible to errors. A "logical qubit" is a qubit that is encoded utilizing a collection of physical qubits to protect against errors and noise—a cluster of physical qubits; it is an abstraction used in fault-tolerant Quacomp. Logical qubits are used in error-correction schemes, wherein multiple physical qubits are entangled to encode a single logical qubit. This encoding allows errors in individual physical qubits to be detected and corrected without disturbing the information stored in the logical qubit. By providing a level of protection against errors, logical qubits enable more robust and reliable Quacomp. For example, a common encoding approach may

---

[1] For example, see [DEN02], [DEV13], [FOW10], [GOO23], [GOT97], [HON23], [JIA09], [MUR16], [RYA22], [SEL24], [SIN23], [STE96], [WAN23a] as a short list of references.

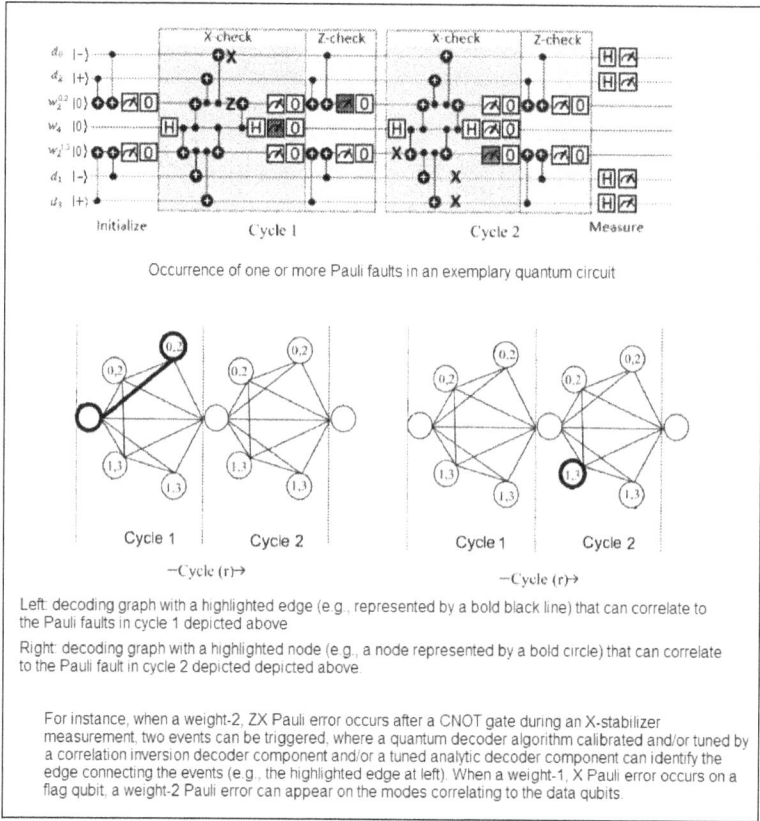

Occurrence of one or more Pauli faults in an exemplary quantum circuit

Left: decoding graph with a highlighted edge (e.g., represented by a bold black line) that can correlate to the Pauli faults in cycle 1 depicted above

Right: decoding graph with a highlighted node (e.g., a node represented by a bold circle) that can correlate to the Pauli fault in cycle 2 depicted depicted above.

For instance, when a weight-2, ZX Pauli error occurs after a CNOT gate during an X-stabilizer measurement, two events can be triggered, where a quantum decoder algorithm calibrated and/or tuned by a correlation inversion decoder component and/or a tuned analytic decoder component can identify the edge connecting the events (e.g., the highlighted edge at left). When a weight-1, X Pauli error occurs on a flag qubit, a weight-2 Pauli error can appear on the modes correlating to the data qubits.

Figure 5.20 QEC example of Pauli fault tracing procedure that can be represented by decoding graphs (synthetized from [CHE23])

use seven physical qubits to represent one logical qubit, allowing for the correction of certain types of errors; in some cases, each logical qubit will require 1,000 physical qubits.

Neutral atoms retain their coherence for relatively long times, and therefore may require relatively few atoms per logical qubit. Implementing logical qubits, however, imposes a non-trivial burden in terms of additional physical qubits and quantum gates, and the complexity of encoding, error detection, and error correction introduces challenges in both hardware and algorithm design; this complexity affects the type of algorithms that can be executed, the algorithms' runtimes, and what computational speedup can be achieved [STA24]. The largest number of logical qubits tested on a QC as of the time of this writing was on a QC that contained *48 logical qubits*. At this time logical qubits utilize too few physical qubits and thus experience relatively high error rates—in the future, direct **logical qubit Quacomp** may become practical. As an illustrative example, QuEra® Computing claimed to have reduced the error rate in qubits with its first commercially available QC with 256 physical qubits and *10 logical qubits*[1] ([AFI2024], [BLU24], [BLU24] provides an extensive discussion on this topic; Figure 5.21 depicts an example of a QC utilizing logical qubits and logical gates).

---

[1]   In 2023, the Google® Quantum AI Lab demonstrated a 2.9% error rate using three logical qubits; QuEra's error rate is 0.5% with 48 logical qubits. IBM also demonstrated error-correction technology in the 127-qubit Heron chip which reduced error rates fivefold compared with its other chips [AFI2024].

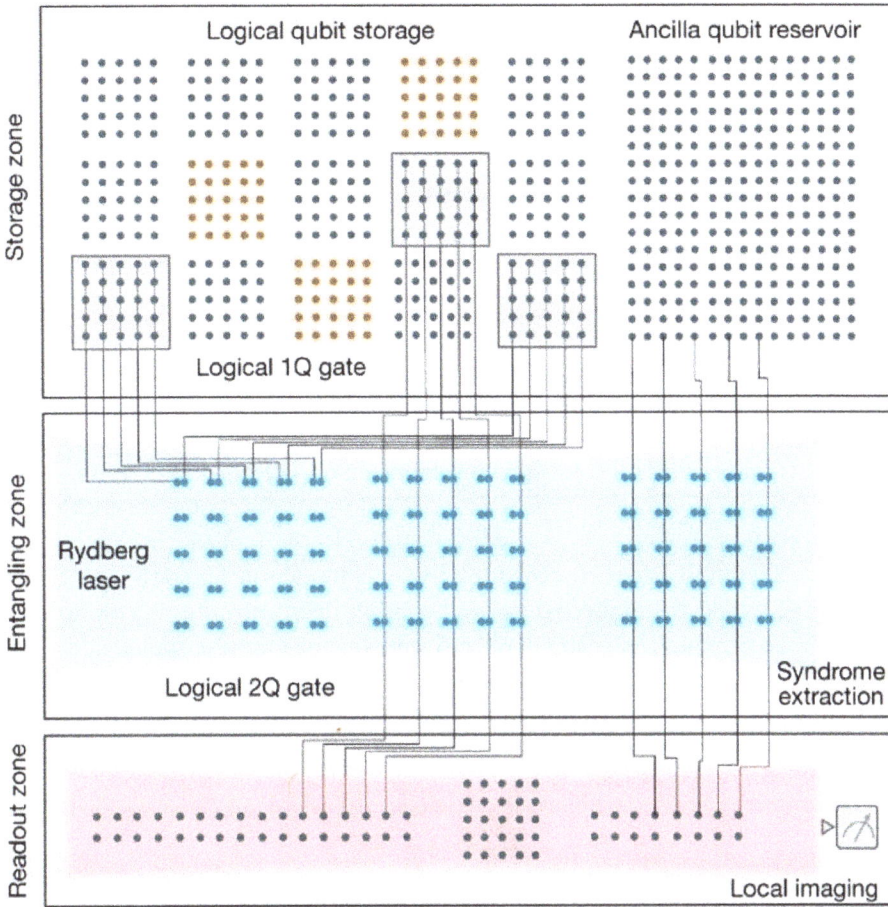

Figure 5.21   Schematic example of a QC that has logical qubits, logical gates, and three processing segments: storage, entanglement, and readout (*Source:* [BLU24])

5.7 QC Technologies

Trapped ions and semiconducting qubit technologies are examples of NISQ QC implementations. Typically, *Trapped Ion Quantum Computing* (TIQC) utilizes ions (atoms or molecules with a net electrical charge) as qubits for computation, with the excitation state of an electron indicating a logical value or logic state—qubits are stored in the electronic states of each ion, and information can be transferred through the aggregate quantized motion of the ions in a trap. Ions such as, but not limited to, magnesium (Mg), calcium (Ca), barium (Ba), and beryllium (Be) may be positively charged, and a single electron in the outer shell of the ion can be used as the logic element. Two or more ions may be entangled, being that changing the state of one qubit causes the entangled qubits to change their state immediately, providing substantial speed and power savings over conventional computing.

However, TIQC requires a well-controlled environment and precise handling of the ions. Ions can be confined and suspended in free space using electromagnetic (EM) fields; for single qubit operations radio signals or lasers can be utilized to achieve coupling between the qubit states, and these signals can also be used to achieve coupling between the internal qubit states and the external states, resulting in entanglement between qubits. For example, ions in a TIQC system are trapped or controlled using a radio frequency (RF) field operating at around, say, 200 volts and 20 megahertz (MHz).

However, ions, once entangled, have a limited duration for which they will stay entangled, requiring rapid handling; furthermore, the ions typically require cryogenic conditions to achieve the maximum benefit of quantum computing. Thus, while photonic Quacomp operates at room temperature, cryogenic TIQC needs to be performed in a sealed cryogenic chamber kept between (approximately) 4 and 10 Kelvin. Figure 5.22 is a logical diagram illustrating a TIQC system. In a TIQC system, electrostatic potentials can be used to move ions between storage and processing locations in a process called *ion shuttling*. In order to control these potentials, hundreds, or even thousands, of electrodes must be simultaneously controlled in order to provide the desired electrical field [FUR23].

Figure 5.23 (simplified and synthetized from Reference [ELO22]) depicts an example of a QC with a transmon qubit in a solid-state implementation (a transmon is a tuned LC circuit that contains an inductor L and a capacitor C attached to a transmission line, which resonates when an appropriate frequency—typically below 10 GHz—is applied); in this case the inductive element in the qubit has been substituted with a Josephson junction. There are a number of circuits in the device connected via transmission lines, providing access for both qubit control and measurement. The energy level and spin change with the application of signals at certain specific frequencies, but only two resonant (or energy) states can result; the resonant frequency corresponds to the equivalent of an electron energy state. The circuit is cooled to very

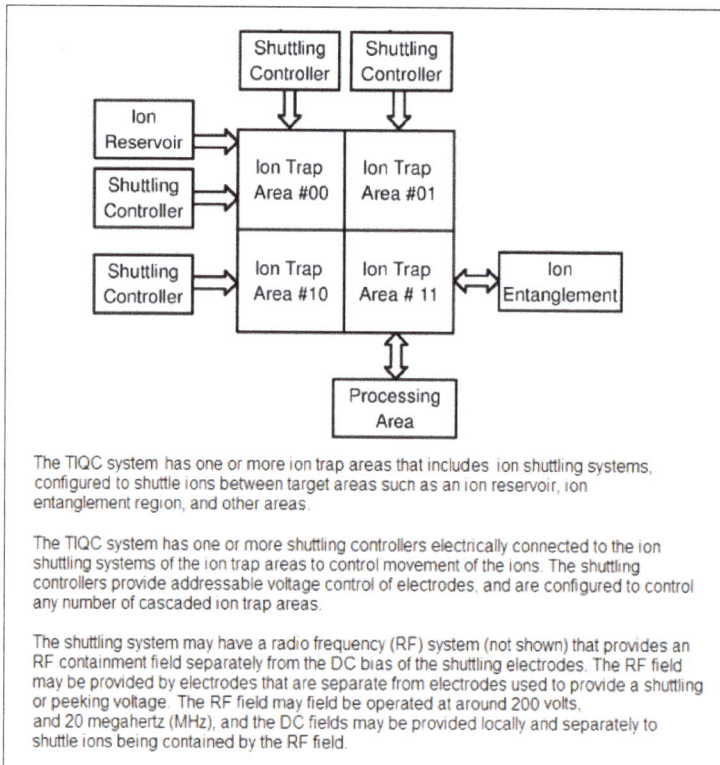

The TIQC system has one or more ion trap areas that includes ion shuttling systems, configured to shuttle ions between target areas such as an ion reservoir, ion entanglement region, and other areas.

The TIQC system has one or more shuttling controllers electrically connected to the ion shuttling systems of the ion trap areas to control movement of the ions. The shuttling controllers provide addressable voltage control of electrodes, and are configured to control any number of cascaded ion trap areas.

The shuttling system may have a radio frequency (RF) system (not shown) that provides an RF containment field separately from the DC bias of the shuttling electrodes. The RF field may be provided by electrodes that are separate from electrodes used to provide a shuttling or peeking voltage. The RF field may field be operated at around 200 volts, and 20 megahertz (MHz), and the DC fields may be provided locally and separately to shuttle ions being contained by the RF field.

Figure 5.22 Logical diagram of a TIQC system (synthesized from [FUR23])

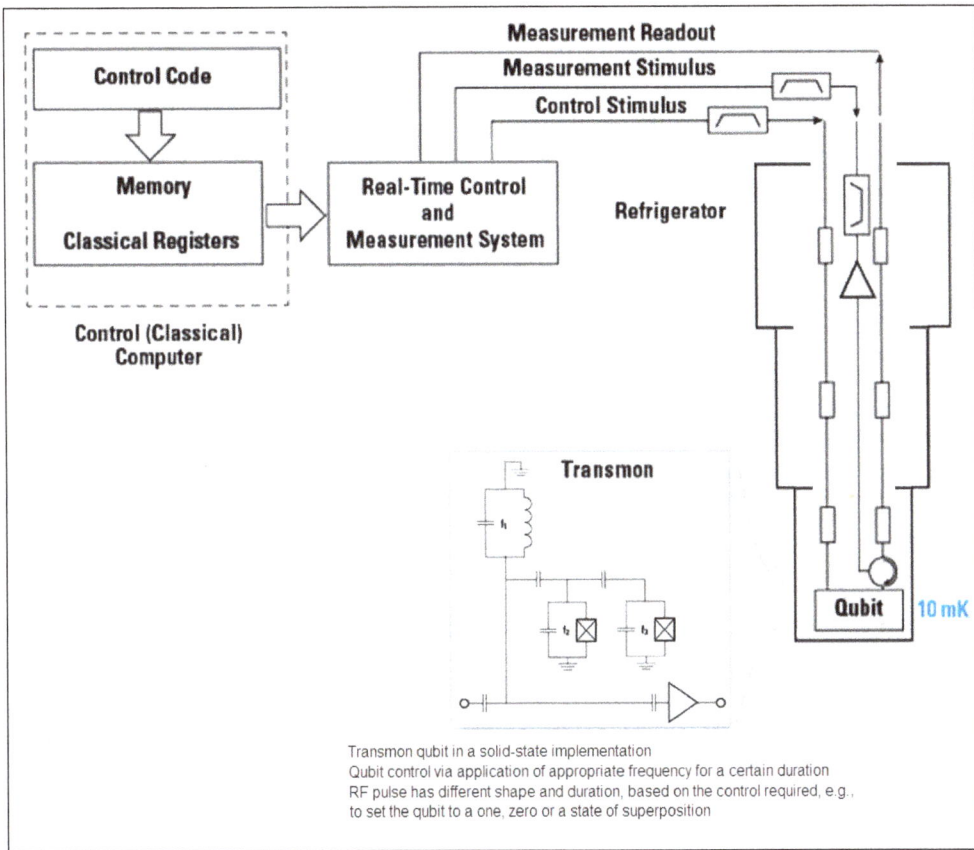

Figure 5.23 Quantum computing system hardware (simplified and synthetized from [ELO22])

low temperatures to reach superconducting properties so that electrons exhibit QM properties. To control the qubit, one applies the appropriate frequency for a certain duration to set the qubit to a one, a zero, or a state of superposition; the applied pulse of RF energy has different durations and shapes, depending on the control required.

The basic operation is as follows [ELO22]: when the classical computer register is set to a value of 1, an appropriate RF frequency pulse is applied to the qubit circuit, which causes the electron to move to the next energy state, changing the magnetic spin and thus setting the qubit to a value of 1. To read the classical register, another RF pulse is applied, and a measurement is undertaken to determine the state of the bit. The value of the measured phase establishes whether the system has transitioned to a logical 1 or a logical 0 (multiple measurements may be required to ascertain the answer is correct—that is, that the probability that a value of 1 will be noted as very high; however, the answer can never be completely certain since there are various interference factors impacting the measurement.

A QC utilizes sets of RF/microwave pulses of different durations and shapes, acting, say, on supercooled semiconducting resonators to instantiate distinct logical operations implemented in operator gates. The NOT, the Hadamard, and the Controlled Not (CNOT) operator gates were discussed in Chapter 3. In a transmon environment, each of these gates is implemented by applying an appropriate RF pulse to the qubit circuit, as follows [ELO22]:

- A *Not gate* entails changing the energy state of the electron and changing the spin. For example, if starting with a bit in a 0 or ground energy state, application of the appropriate frequency causes the electron to transition to the next quantized energy state.
- Applying a *Hadamard gate* results in the qubit's being in a state of superposition—two energy levels at the same time—until the electron (the state) is observed. As discussed elsewhere, when one undertakes a measurement, the quantum state collapses and the qubit returns to a classical quantized energy state representing a 1 or a 0. When undertaking a large number of measurements (say several dozens), the probability (chances) of returning a logical 1 is 50% over the observation interval, and a logical 0 is also 50% over the observation interval.
- The *CNOT gate* connects two bits together, entangling them (physically the superconducting qubits in the cavity oscillator share EM fields and experience the phenomena of quantum entanglement). At the logical level the CNOT gate operates as follows: if the control port is set to a 1, the output of the CNOT gate is the inversion of the input; if the control port is set to 0, the output of the CNOT gate equals the input.

5.8 Quantum Programming

As discussed elsewhere, a QC may be implemented based on different quantum technologies or types of quantum hardware—for example, using qubits built from superconductors, trapped ions, semiconductors, photonics, and so on. Progressing from current QC implementations to larger (or networked) QC systems is often challenging for QCs based on solid-state platforms. Optics-based systems, on the other hand, provide a possibly viable approach to scaling; however, optical systems require engineered photonic states to mitigate for state loss and to operate required QEC overlays. As noted earlier in this text, single photons can serve as physical qubits, being that photons operate more rapidly than solid-state qubits—however, their states are more easily lost; to avoid qubit losses and other errors, one would typically need to couple several single-photon light pulses together to assemble a logical qubit, as in the case of the superconductor-based approach [KON24]. In all cases quantum programs are needed.

A quantum program consists of sequences of quantum gates acting on qubits; it may implement a single quantum circuit or multiple quantum circuits. A quantum program is a computing program that identifies instructions that may be compiled to be executed on a QC in order to execute a given quantum algorithm. Figure 5.24 (inspired by [MAN20]) depicts two approaches for the execution of quantum algorithms: quantum annealers or a gate-model. In the former approach, the problem is encoded into an Ising-type Hamiltonian embedded into a quantum hardware graph (see Glossary at the end of this chapter); then either a quantum annealer or a classical solver is used. In the latter approach, the problem is formulated at a logical level and an adequate quantum algorithm is selected. Thereafter, the quantum algorithm is translated in a quantum circuit (with gates), which is either executed on a quantum processor or simulated with a quantum computer simulator.

A quantum program may include logical loops that instruct a particular quantum circuit to be implemented based on one or more conditions, such as a "for loop", "if loop", "while loop", and so on; it may instruct implementation of a parametric quantum circuit where the angles of one or more quantum gates are not determined [SHI23]. Programmers may utilize Pauli gates, Hadamard gates, controlled gates, and so on, but may also generate their own unique gates, and/or different types of gates may be customized to the capabilities of particular types of

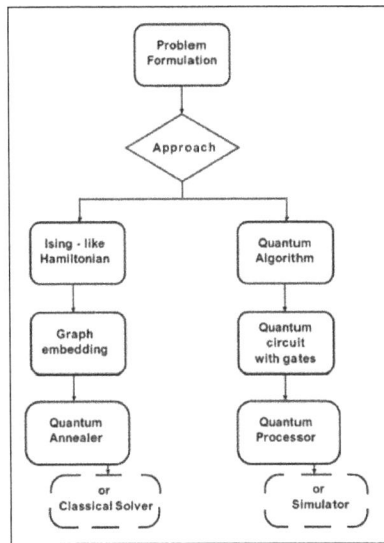

Figure 5.24 Algorithm's implementation choices

QCs or quantum devices. Different QCs may use different types of quantum gates to execute a quantum program to perform a given algorithm.

Some quantum programs may be written using a high-level text-based quantum programming language such as Quil®, OpenQASM, cQASM (see Chapter 8). However, considering the different types of quantum hardware that may be utilized to implement a quantum program, a high-level quantum program needs to be first compiled (translated) into appropriate quantum gates that are native to and specific to a given type of QC, and then further rendered into an executable file—the compiled file of the native quantum gates is called a *low-level quantum circuit* or a *low-level quantum program*. (A programmer with specific knowledge of the native quantum gates of a QC may compose a quantum program or quantum circuit directly using the low-level native quantum gates of the QC, and/or use a mid-level quantum programming language that makes use of non-hardware specific gates—such a mid-level quantum program also eventually needs to be compiled into a low-level quantum circuit before it may be compiled and executed on a QC).

To expedite the development of applications, it would be desirable to have access to tools to assist users to compose quantum programs. Often a complex quantum program and/or quantum circuit may be unintuitive and hard to compose, particularly if the quantum program is not composed using a generic high-level quantum programming language. [SHI23] proposes an assisted composition system to facilitate the composition of quantum objects. The composition system receives a partial portion of a quantum object that is being composed but is not yet fully composed by a user; it determines a first abstract representation of the partial portion of the quantum object being composed. The assisted composition system may be able to determine that the first abstract representation resembles at least a first portion of a second abstract representation of a stored quantum object stored in a library for the quantum programming environment. Then the assisted composition system obtains a second portion of the stored quantum object from the library and provides it to the user as a next portion to the partial portion of the quantum object being composed (see Figure 5.25).

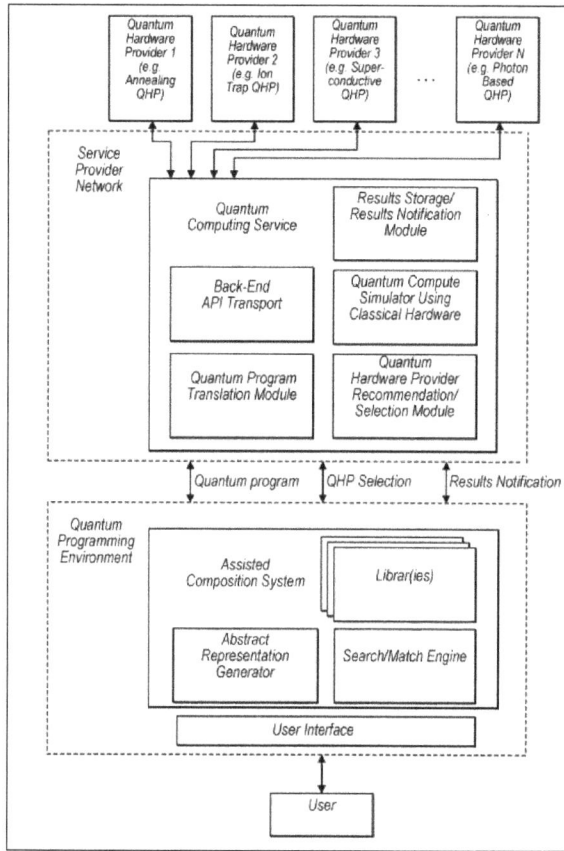

Figure 5.25 Quantum programming environment that includes an assisted composition system ([SHI23])

5.9 Performance Benchmarking

A performance benchmark that is used by some QC developers is the *Algorithmic Qubits* (#AQ) approach. #AQ is an application-based benchmark that aggregates performance across six widely known quantum algorithms that are relevant to the most promising near-term quantum use cases: *Optimization, Quantum Simulation,* and *Quantum Machine Learning.* The *Quantum Fourier Transform* can be used in factoring and hidden subgroup problems; *Hamiltonian Simulations* can be used in quantum chemistry and materials science; and *Monte Carlo Sampling* in quantum finance applications. #AQ is a summary and analysis of multiple quantum algorithms: the higher the #AQ a system offers, the more commercial value the QC can deliver to stakeholders and users [ION24]:

- *Optimization:* problems involving complex routing, sequencing, amplitude estimation, Monte Carlo simulation, and more
- *Quantum Simulation:* understanding the nature of the very small: Hamiltonian simulation, variational quantum eigensolver
- *Quantum Machine Learning:* drawing inferences from patterns in data, at scale: quantum Fourier transform, phase estimation

The AQ metric is derived from algorithmic benchmarking protocols discussed in a vendor-independent study conducted by the Quantum Economic Development Consortium (QED-C) [LUB21], [SWA24b]. In this benchmark the most complex circuits, in terms of the number of qubits and number of gates, determine the resulting #AQ score. For example, at #AQ 35, the IonQ Forte® QC is capable of simultaneously analyzing more than 34 billion different possibilities.

5.10 Implementation and Developments[1]

As alluded to in Chapter 1, IBM, Google, Amazon® and others have already developed lab-oriented QCs. For example, among other initiatives,

- In the Fall of 2023, IBM introduced Condor, the world's first universal quantum computer with 1,121 qubits.
- D-Wave® is developing both annealing QCs and gate-model QCs. In early 2024 D-Wave announced the availability of the Advantage2 prototype that features 1,200+ qubits and 10,000+ couplers, doubling the number of qubits and couplers over the previously released Advantage2 prototype. The prototype QC reportedly is 20 times faster at solving spin glasses, a family of classically hard optimization problems. Benchmarks demonstrate substantial advancements across a number of performance metrics: (i) *qubit connectivity:* increased from 15 to 20-way connectivity to enable solutions to larger problems; (ii) *energy scale:* increased by more than 40% to deliver higher-quality solutions; and (iii) *qubit coherence time:* doubled, which will drive faster time-to-solution [SWA24].
- QuEra Computing reportedly planned to launch several QCs in the second half of the decade, starting with a 30-logical-qubit, 3,000-physical-qubit machine targeted for 2025 and a machine with more than 10,000 physical qubits and 100 logical qubits targeted for 2026 [AFI2024]. QuEra's QCs are based on neutral atoms. The company developed and operates the world's largest publicly-accessible QC available for general use over the Amazon Braket® cloud.
- Current examples of QC performance (for example, with QuantWare's Soprano-D and Contralto-D processors) include measurement results of average coherence times of 60 μs, average single-qubit gate fidelities of 99.9%, and a two-qubit gate fidelity of 99.7% [SWA24a].
- As discussed earlier, atomic ions—individual atoms with one missing electron—are a good resource for storing qubits and utilizing highly-tuned electromagnetic fields. Dozens of ions can be trapped in a long linear chain and be ready for further manipulation; with the aid of precision lasers, the trapped ions can then be entangled using two-qubit quantum gates between any pair of qubits in the chain. This approach is the core technology behind IonQ's QCs. IonQ offered four types of QCs, as follows [ION24]:
    - *IonQ Aria®:* A universally accessible, high-performing flagship quantum system. With an #AQ of 20, IonQ Aria was the best publicly-disclosed quantum computer at the time of this writing. Its trapped-ion architecture offers all-to-all connectivity among the 21 physical qubits in its basic production configuration, producing an average single-qubit gate error rate of 0.05% with a single qubit gate speed of 135 μs; it has an average two-qubit

---
[1] At the time of this writing.

error rate of 0.4% at a gate speed of 600 μs. The average error rate for state preparation and measurement is a mere 0.5%. The lifetime of its qubits (T2 Time) is around 1000ms. In practice, an #AQ of 20 implies that IonQ Aria can successfully execute a quantum circuit with over 20 qubits that contain about 400 entangling gate operations; this makes the system several thousand times more powerful than IonQ Harmony, their previous hardware generation. See Figure 5.26 (courtesy of IonQ).

○ *IonQ Forte:* A software-configurable QC which was the highest performing, commercially available quantum system at the time of this writing; it became broadly available in early 2023. IonQ Forte uses ytterbium ions and integrates highly specialized Acousto-Optic Deflectors (AODs) to direct laser beams at individual qubits in the ion chain to apply logic gates among the qubits. IonQ Forte is designed with a capacity of up to 32 qubits; in a fully connected register of 31 qubits, there are 465 unique qubit pairs between which entangling gates can be performed. In early 2024, IonQ announced that it hit its target technical milestone of 35 #AQ a year ahead of schedule; the milestone was achieved on IonQ Forte through hardware and software improvements [SWA24b].

○ *IonQ Forte Enterprise:* A rack-mounted, on-prem system for production readiness. Built on standard racks, IonQ Forte Enterprise's installation specs are designed to be met by the typical, modern data center (see Figure 5.26).

○ *IonQ Tempo* (Expected, by the vendor, in 2025): A commercial-advantage-capable computer for production applications. At #AQ 64, IonQ Tempo is expected to be capable of commercial advantage for certain applications. Tempo is designed to have faster gate speeds, mid-circuit measurement, and 99.9% fidelity.

Table 5.3 (partially based on [ATT23]) provides a partial list of QC vendors at the time of this writing. As of this time, however, QCs remained limited in available resources and are subject to errors—for example, due to the instability of the generated quantum states; in addition, operations on QCs were often imprecise. Because of these limitations and given the fact that QCs are only superior in certain cases, they cannot fully replace their classical counterparts in the short term (say, to the end of the decade) [VIE21]. Figure 5.27 depicts a roadmap of QC developments as contemplated at Google, a key player in this arena—notice the progression in the number of physical qubits planned to be supported over time; Figure 5.28 (courtesy of IBM) depicts a roadmap of QC developments as contemplated at IBM, also a key player in this arena.

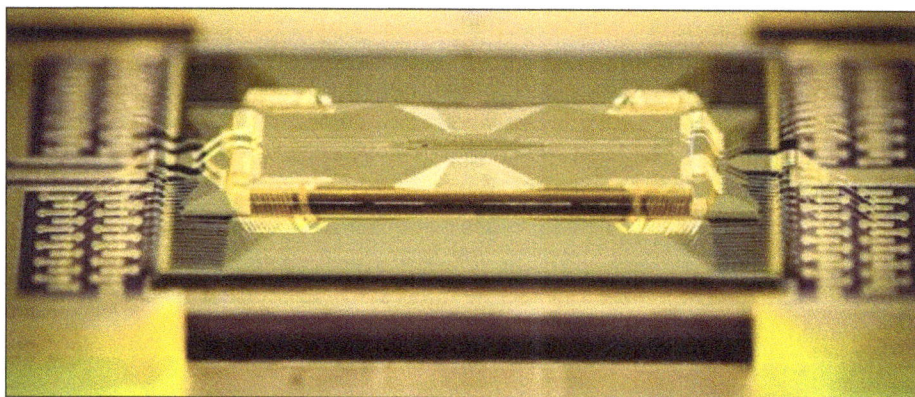

Figure 5.26   Side view of IonQ Forte EGT (evaporated glass ion trap chip) (Image Courtesy of IonQ)

Table 5.3 Subset of QC Vendors at Press Time[a]

	Trapped Ion Qubits	Superconducting Qubits	Silicon Qubits	Photonic Qubits
Vendor Examples	Quantinuum IonQ Alpine Quantum Technologies Oxford Ionics EleQtron	IBM Google D-Wave Anyon Systems Atlantic Quantum Bleximo Oxford Quantum Circuits Rigetti Computing	Quantum Motion Silicon Quantum Computing Equal1 Laboratories Photonic Inc	Xanadu Quantum Technologies ORCA Computing PsiQuantum QuiX Quantum
Physical Qubits	20–100; increasing over time	20–100; increasing over time	Small number	Large number
Coherence Times	~50 sec	~50–200 μsec	~1–10 sec	~150 μsec
Gate Operation Time	~3–50 μsec	~10–50 nsec	~1 nsec	~1–10 nsec
Gate Fidelity	~99%	~99%	~90%	~98%
Scalability	Some	Medium to high	High	High

[a] (partially based on [ATT23])

A typical trade press article at [MAN23] gives a layman's window into some of the possibilities of the QC and Quacom technology, as follows for illustrative purposes:

"The early 1940s saw the first vacuum tube computers put to work solving problems beyond the scope of their human counterparts. These massive machines were complex, specific, and generally unreliable. . . . In many respects, today's quantum systems bear remarkable similarities to early vacuum tube computers in that they're also incredibly expensive, specialized, and not intuitive. Later computers like UNIVAC I in 1951 or the IBM 701 presented the possibility of a competitive advantage for the few companies with the budgets and expertise necessary to deploy, program, and maintain such [systems]. According to Gartner [...] a similar phenomenon is taking place with quantum systems today as companies seek to eke out efficiencies by any means necessary. [...] there is not a single thing that quantum can do today that you can't do classically. However, [...] by combining classical and quantum computing, some early adopters—particularly in the financial and banking industry—have been able to achieve some kind of advantage over classical computing alone. Whether these advantages rise to the level of a competitive edge isn't always clear, but it does contribute to a fear that those who don't invest early may risk missing out. [...] Governments, for example, have poured a significant amount into the possibility that quantum will materialize as a true competitive threat without having the "killer app" for quantum defined yet. [...] DARPA [recently] launched the Underexplored Systems for Utility-Scale Quantum Computing (US2QC) initiative to speed the development and application of quantum systems. [...] companies like Toyota, Hyundai, BBVA, BSAF, ExxonMobil and others have teamed up with quantum computing vendors on

(text continues on page 215)

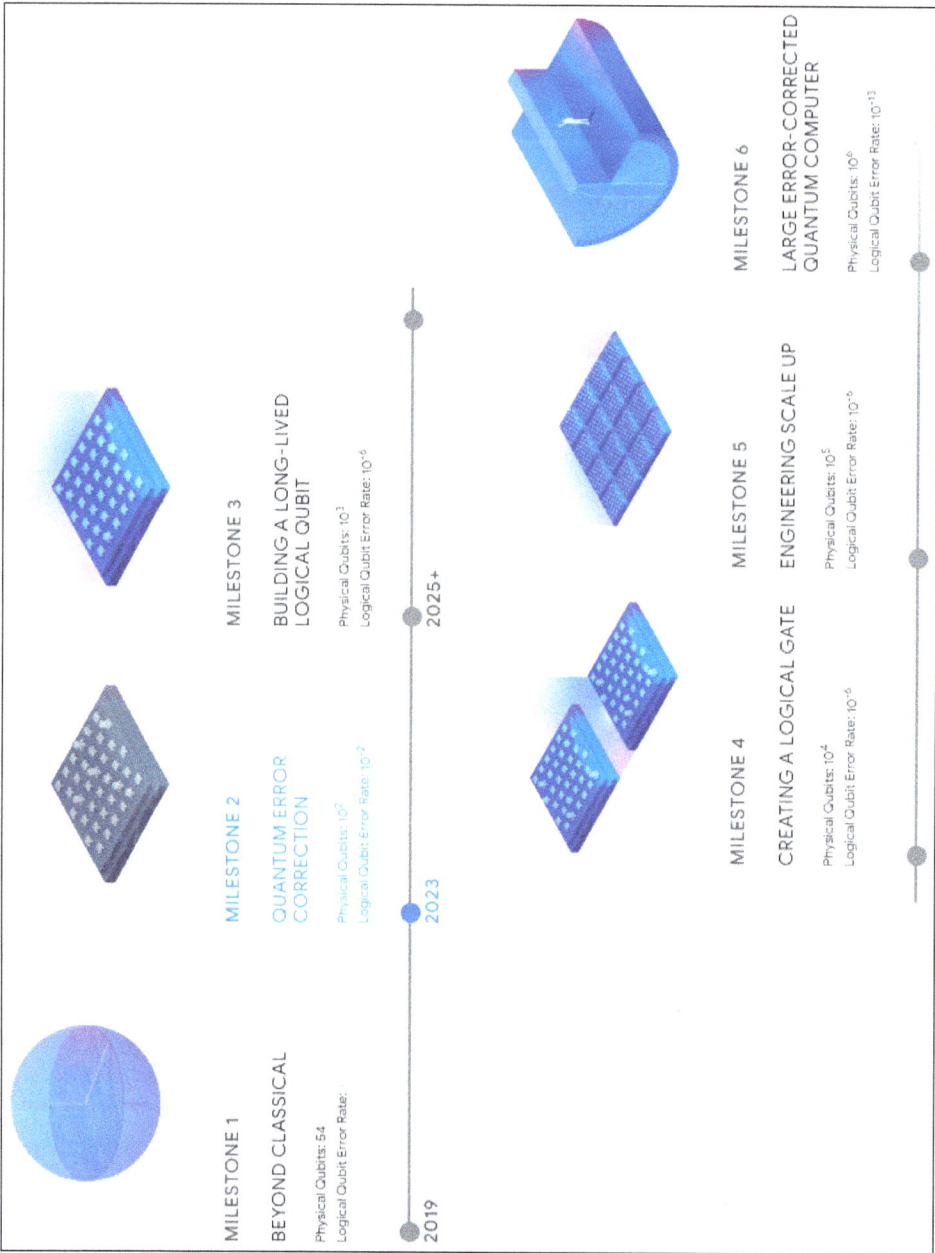

Figure 5.27   Example of planned QC developments at Google (courtesy of Google)

As illustrative of QC trends, at the end of 2024 Google unveiled a new chip called Willow that it says marks a major breakthrough in the field. According to Google, Willow can reduce errors "exponentially" as the number of qubits is scaled up. The vendor reportedly measured Willow's performance using the so-called random circuit sampling (RCS) benchmark, which presents a computational task that is difficult for classical computers to solve. Google asserts that Willow performed a computation in under five minutes that would take one of today's fastest supercomputers 10 septillion years—or 10,000,000,000,000,000,000,000,000 years. The chip has about 100 qubits, but Google is planning to eventually build a system with 1 million qubits.

Figure 5.28 Example of planned QC developments at IBM (courtesy of IBM) (this page and next page)

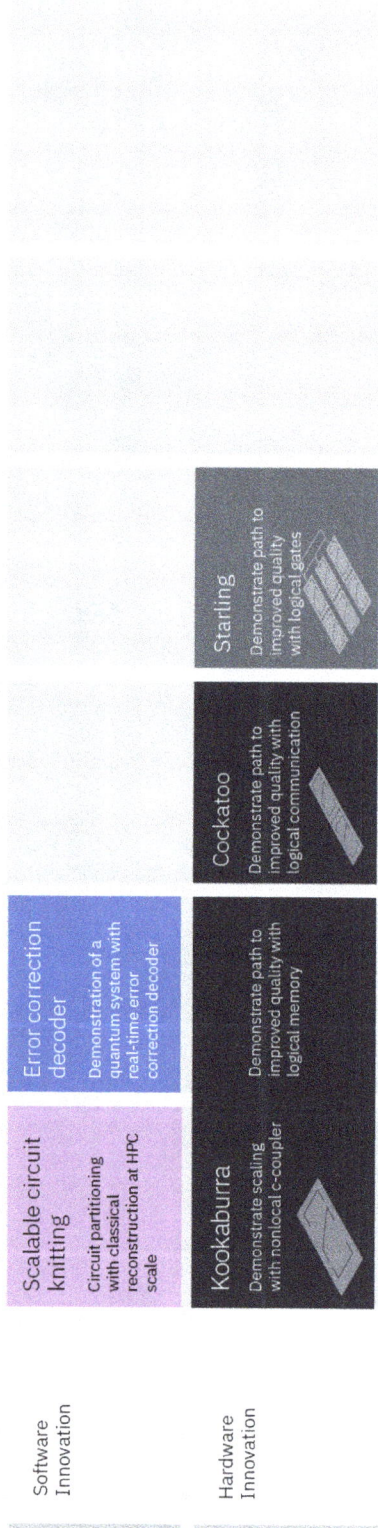

Figure 5.28   *(Continued from previous page)*

the off chance the tech can help develop better batteries, optimize routes and logistics, and/or reduce investment risk. [...] quantum computing remains a long-term investment. [...] There are dozens of vendors claiming to offer quantum services on systems ranging anywhere from a few dozen qubits to thousands of them. While this might seem like an obvious metric to judge the maturity and performance of a quantum system, it really depends on a number of factors— including things like decoherence and the quality of the qubits themselves. We liken this to the "core wars" on modern processors. Those on an Intel CPU are going to have vastly different performance characteristics compared to CUDA (Compute Unified Device Architecture) cores on an Nvidia Graphics Processing Units (GPUs) (CUDA is an Nvidia—proprietary parallel computing platform and application programming interface that enables software to utilize certain types of GPUs for accelerated general-purpose processing, such an approach known as general-purpose computing on GPUs). Depending on what you're doing, a job that might run just fine on a handful of Intel cores might require thousands of CUDA cores—if it runs at all. The same is true of quantum systems, which are often optimized to specific workloads. For example, [...] an IBM system might perform better at computational chemistry, while D-Wave systems may be better tuned for optimization tasks like route planning. [...] The high cost and often exotic conditions—like near-absolute-zero operating temperatures—mean that many quantum systems up to this point have been rented in a cloud-like "as-a-service" fashion. However, some providers, like IonQ, have recently teased rackmount quantum systems that can be deployed in enterprise datacenters. [...] most on-prem deployments will focus on scientific research—likely in conjunction with high performance computing deployments. [...] Today, we're only solving classical problems quantumly, but real innovation is going to come when we solve quantum problems quantumly with quantum algorithms."

References

[ABR08] S. Abramsky and B. Coecke, *Handbook of quantum logic and quantum structures.* (Elsevier, 2008). Also see arXiv:0808.1023.

[AFI24] K. Afifi-Sabet, "World's 1st Fault-Tolerant Quantum Computer Launching This Year Ahead of A 10,000-Qubit Machine In 2026". Live Science, February 1, 2024. Available online on 2/1/24 at https://www.livescience.com/technology/computing/worlds-1st-fault-tolerant-quantum-computer-coming-2024-10000-qubit-in-2026.

[AHA04] D. Aharonov, W. van Dam, et al., "Adiabatic Quantum Computation Is Equivalent to Standard Quantum Computation". Proceedings of the 45th Foundations of Comput. Sci., Rome, Italy. 17–19 October 2004, pp. 42–51.

[AHA07] D. Aharonov, W. van Dam, et al., "Adiabatic Quantum Computation is Equivalent to Standard Quantum Computation". SIAM J. Comput. 37 (1), 166. 2007. https://doi.org/10.1137/S0097539705447323

[AHA07a] D. Aharonov, A. Ta-Shma, "Adiabatic Quantum State Generation". SIAM J. Comput, Vol. 37, No. 1, pp. 47–8, 2007.

[AHA08] D. Aharonov, W. van Dam, et al., "Adiabatic Quantum Computation Is Equivalent to Standard Quantum Computation". SIAM Review, 50(4), 755–787. 2008. http://www.jstor.org/stable/20454175.

[ALB18] T. Albash, D. A. Lidar, "Adiabatic Quantum Computing". Rev. Mod. Phys. 90, 015002, 2018. 2 Feb 2018. https://doi.org/10.48550/arXiv.1611.04471

[APO88] B. Apolloni, N. Cesa-Bianchi, D. de Falco, "A Numerical Implementation of Quantum Annealing". In *Stochastic Processes, Physics and Geometry, Proceedings of the Ascona-Locarno Conference*, 1988. Albeverio et al., Eds, World Scientific (1990), 97–111.

[ARO09] S. Arora, B. Barak, *Computational Complexity: A Modern Approach*. Cambridge University Press. Cambridge, New York, 2009.

[ARU19] F. Arute, K. Arya, et al., "Quantum Supremacy Using a Programmable Superconducting Processor". *Nature*, Vol. 574, No. 7779, October 23, 2019, pp. 505-510. www.nature.com/articles/s41586-019-1666-5.

[ATT23] J. J. Attia, "Quantum Information Technology Standardization—PWI 20153". ISO/IEC JTC 1 Workshop on Quantum Information Technology Standardization, April 2023. Available online on May 16, 2024 at https://jtc1info.org/quantum-information-technology-workshop/

[BAE09] J. C. Baez, M. Stay, "Physics, Topology, Logic and Computation: A Rosetta Stone". Mar. 2009, arXiv:0903.0340.

[BER11] V. Bergholm, J. D. Biamonte, "Categorical Quantum Circuits". J. Phys. A: Math. Theor. 44, 245304, 2011. http://arxiv.org/abs/1010.4840v2, May 29, 2011.

[BLU24] D. Bluvstein, S.J. Evered, et al., "Logical Quantum Processor Based on Reconfigurable Atom Arrays". Nature 626, 58–65, 2024. https://doi.org/10.1038/s41586-023-06927-3.

[BOC23] A. V. Bocharov, E. Kemp, et al., Quantum Bit Prediction. U.S. Patent US-11797872, 2023-10-24. Uncopyrighted material.

[BRA18] S. Bravyi, D. Gosset, R. König, "Quantum Advantage with Shallow Circuits". Science Vol. 362, No. 6412, 2018, pp. 308-311.

[BRO19] M. Brooks, "Beyond Quantum Supremacy: The Hunt for Useful Quantum Computers". Nature; 2019; 574(7776):19–21. doi:10.1038/ d41586-019-02936-3.

[CAS23] D. Castelvecchi, "IBM Releases First-Ever 1,000-Qubit Quantum Chip". 04 December 2023, Nature. Available online on December 5, 2023 at https://www.nature.com/articles/d41586-023-03854-1.

[CHE23] E. H. Chen, A. W. Cross, et al., Calibrated Decoders for Implementations of Quantum Codes. U.S. Patent 11803441, 2023-10-31. Uncopyrighted material.

[CHE24] Y. Chen, S. Zhang, Method and Apparatus for Fabricating Quantum Circuit, Device, Medium, And Product. U.S. Patent Application 20240005192, 2024-01-04. Uncopyrighted material.

[COA23] S. Coady, L. Griffin, Optimizing Execution of Quantum Service Definition Files Using a Quantum Optimization Database. U.S. Patent Application 20230325703, 2023-10-12. Uncopyrighted material.

[DEF11] D. de Falco, D. Tamascelli, "An Introduction to Quantum Annealing". Published online by Cambridge University Press. 15 March 2011. RAIRO—Theoretical Informatics and Applications. Volume 45, Issue 1: ICTCS 09, January 2011, pp. 99 – 116. doi: https://doi.org/10.1051/ita/2011013.

[DEN02] E. Dennis, A. Kitaev, et al., "Topological Quantum Memory". J. Math. Phys. 43, 4452–4505, 2002.

[DES22] A.Deshpande, A. V. Gorshkov, B. Fefferman, "The Importance of The Spectral Gap in Estimating Ground-State Energies". arXiv:2007.11582v2, 9 Dec 2022. Also, PRX Quantum 3, 040327, 2022. https://doi.org/10.48550/arXiv.2007.11582.

[DEU89] D. Deutsch, "Quantum Computational Networks". Proceedings of the Royal Society of London. A. Mathematical and Physical Sciences 425 (1868), 73-90, 1989. http://dx.doi.org/10.1098/rspa.1989.0099.

[DEV13] S. J. Devitt, W. J. Munro, K. Nemoto, "Quantum Error Correction for Beginners". Reports on Progress in Physics Vol. 76, Issue 7, 076001, June 2013. https://iopscience.iop.org/article/10.1088/0034-4885/76/7/076001. doi: 10.1088/0034-4885/76/7/076001.

[EIL45] S. Eilenberg, S. Mac Lane, "General theory of natural equivalences". Trans. Amer. Math. Soc. 58 (1945), 231–294.

[FAR01] E. J. Farhi, J. Goldstone, et al., "A Quantum Adiabatic Evolution Algorithm Applied to Random Instances of an NP-Complete Problem". Science 292 (5516), 2001, 472. https://doi.org/10.48550/arXiv.quant-ph/0104129.

[FEY82] R.P. Feynman, "Simulating Physics with Computers". International Journal of Theoretical Physics 1982;21(6-7):467–488. doi:10.1007/ bf02650179.

[FOW10] A.G. Fowler, D. S. Wang, et al., "Surface Code Quantum Communication". Physical Review Letters Vol. 104, Issue 18, 180503, May 2010, https://journals.aps.org/prl/abstract/10.1103/PhysRevLett.104.180503. doi: 10.1103/PhysRevLett.104.180503.

[FUJ24] K. Fujii, M. Negoro, et al., Quantum Circuit Learning Device, Quantum Circuit Learning Method, And Recording Medium. U.S. Patent 11863164, 2024-01-02. Uncopyrighted material.

[FUR23] W. Furtner, Multi Dimensional Electrode Controller for Quantum Computing. U.S. Patent Application 20230325698, 2023-10-12. Uncopyrighted material.

[GHO13] M. A. Ghonaimy, "Plenary Talk I: An overview of Quantum Information Systems". *2013 8th International Conference on Computer Engineering & Systems (ICCES)*, Cairo, Egypt, 2013, pp. xvii-xix, doi: 10.1109/ICCES.2013.6707154.

[GOO23] Google Quantum AI. "Suppressing Quantum Errors by Scaling a Surface Code Logical Qubit". Nature 614, 676–681. February 2023. https://doi.org/10.1038/s41586-022-05434-1.

[GOT97] D. Gottesman, "Stabilizer Codes and Quantum Error Correction". Thesis in partial fulfillment of the requirements for the Degree of Doctor of Philosophy. California MInstitute of Technology Pasadena, California 2004 (Submitted May 21, 1997). Available online on May 3, 2024 at https://thesis.library.caltech.edu/2900/2/THESIS.pdf.

[GRI23] L. Griffin, S. Coady, Generating Validated Quantum Function Invocations. U.S. Patent Application 20230289648, 2023-09-14. Uncopyrighted material.

[GRI23a] L. Griffin, S. Coady, Performing Just-In-Time (Jit) Error Correction Optimization for Quantum Services Using Quantum Simulators. U.S. Patent Application 20230351240, 2023-11-02. Uncopyrighted material.

[GRI24] D. Grimn, "IBM and AIST Collaborate on 10,000-qubit Quantum Computer — 75x More Qubits Than Rivals". June 17, 2024. Available on June 30, 2024 at https://www.yahoo.com/tech/ibm-aist-collaborate-10-000-151808710.html?guccounter=1

[GRO97] L. K. Grover, "Quantum Mechanics Helps in Searching for a Needle in a Haystack". Phys. Rev. Lett. 79 (2), 325. 1997. https://doi.org/10.1103/PhysRevLett.79.325.

[HON23] D. Honciuc Menendez, A. Ray, M. Vasmer, "Implementing Fault-Tolerant Non-Clifford Gates Using The [[8,3,2]] Color Code". September 2023. https://arxiv.org/abs/2309.08663.

[ION24] IONQ Staff, "The Future is Quantum". Available online on June 7, 2024 at www.ionq.com.

[ISO19] ISO/IEC JTC 1/AG 4, "Meeting report N50—Quantum Computing—Terminology and Vocabulary". 2019. https://www.iso.org/standard/80432.html. International Organization for Standardization, ISO Central Secretariat, Chemin de Blandonnet 8, CP 401, 1214 Vernier, Geneva, Switzerland.

[JAK98] D. Jaksch, C. Bruder, at al., "Cold Bosonic Atoms in Optical Lattices". Physical Review Letters Vol. 81, No. 15,1998, pp. 3108-3111.

[JAN18] D. Janežič, A. Milićević, A., et al., "Topological Complexity of Molecules". In: Meyers, R. (eds) *Encyclopedia of Complexity and Systems Science*. 2018. Springer, Berlin, Heidelberg. https://doi.org/10.1007/978-3-642-27737-5_554-3.

[JIA09] L. Jiang, J. M. Taylor, et al., "Quantum repeater with encoding". Physical Review A Vol. 79, Issue 3, 032325, March 2009, https://journals.aps.org/pra/abstract/10.1103/PhysRevA.79.032325. doi: 10.1103/PhysRevA.79.032325.

[KEM06] J. Kempe, A. Kitaev, O. Regev, "The Complexity of the Local Hamiltonian Problem". SIAM Journal of Computing, Vol. 35(5), p. 1070-1097, 2006. Conference version in Proc. 24th FSTTCS, p. 372-383 (2004). Or arXiv:quant-ph/0406180v2. https://doi.org/10.48550/arXiv.quant-ph/0406180.

[KOK07] P. Kok, W. J. Munro, et al., "Linear Optical Quantum Computing with Photonic Qubits". Reviews of Modern Physics Vol. 79, No. 1, 2007, pp. 135-174.

[KON24] S. Konno, W. Asavanant, et al., "Logical States for Fault-Tolerant Quantum Computation with Propagating Light". Science, 18 Jan 2024, Vol 383, Issue 6680. pp. 289-293. doi: 10.1126/science.adk7560.

[LEU00] D.W. Leung, I. L. Chuang, et al., "Efficient Implementation of Coupled Logic Gates for Quantum Computation". Physical Review A, 2000; 61(4). doi:10.1103/physreva.61.042310.

[LUB21] T. Lubinski, S. Johri, et al., "Application-Oriented Performance Benchmarks for Quantum Computing". arXiv:2110.03137, October 7, 2021. https://doi.org/10.48550/arXiv.2110.03137.

[MAD22] L. S. Madsen, F. Laudenbach, et al., "Quantum Computational Advantage with A Programmable Photonic Processor". Nature; 2022; 606(7912):75–81. doi:10.1038/s41586-022-04725-x.

[MAN20] A. Manzalini, "Quantum Communications in Future Networks and Services". Quantum Rep. 2020, 2(1), 221-232; 11 March 2020. https://doi.org/10.3390/quantum2010014.

[MAN23] T. Mann, "From Vacuum Tubes to Qubits—Is Quantum Computing Destined to Repeat History?" 10/3/23. The Register. Available online on June 10,2024 at https://theregister.com/

[MAZ23] G. Mazzola, P. Ollitrault, I. Tavernelli, Procedure to Speed-up Variational Quantum Eigensolver Calculations in Quantum Computers. U.S. Patent 11803611, 2023-10-31. Uncopyrighted material.

[MIN75] D. Minoli, "Combinatorial Graph Complexity". Att. Accad. Naz. Linc, Rendiconti, Volume LIX, 2nd Semester, Fasc. 6, December 1975, pp. 651-661.

[MIY23] A. Miyata, T. Yamamoto, Superconducting Quantum Circuit Apparatus. U.S. Patent Application 20230363293, 2023-11-09. Uncopyrighted material.

[MON11] T. Monz, P. Schindler, et al., "14-Qubit Entanglement: Creation and Coherence". Physical Review Letters 2011;106(13). doi:10.1103/physrevlett.106.130506.

[MON16] T. Monz, D. Nigg, et al., "Realization of a Scalable Shor Algorithm". Science 2016; 351(6277):1068–1070. doi:10.1126/science.aad9480.

[MUR16] S. Muralidharan, L. Li, et al., "Optimal Architectures for Long Distance Quantum Communication". Scientific Reports Vol. 6, pp. 1-10, February 2016, https://www.nature.com/articles/srep20463. doi: 10.1038/srep20463.

[NEG06] C. Negrevergne, T. S. Mahesh, et al., "Benchmarking Quantum Control Methods on A 12-Qubit System". Physical Review Letters, 2006; 96(17). doi:10.1103/physrevlett.96.170501.

[NGU23] N. H. Nguyen, R. J. Thompson, Reducing Resources in Quantum Circuits. U.S. Patent Application, US 20230289501, 2023-09-14. Uncopyrighted material.

[NIE10] M. Nielsen, L. I. Chuang, *Quantum computation and quantum information: 10th Anniversary Edition*. 2010. Cambridge University Press, New York, NY, USA. ISBN: 9781107002173. http://dx.doi.org/10.1017/cbo9780511976667.016.

[OVA22] O. van Deventer, N. Spethmann, et al., "Towards European Standards for Quantum Technologies". 3 Oct 2022, arXiv:2203.01622. https://doi.org/10.48550/arXiv.2203.01622. EPJ Quantum Technol. 9 (2022) 33. https://doi.org/10.1140/epjqt/s40507-022-00150-1.

[PAP94] C. Papadimitriou. *Computational Complexity*. Addison Wesley, Reading, Massachusetts, 1994. ISBN-13: 978-0201530827.

[PER14] A. Peruzzo, J. McClean, et al., "A Variational Eigenvalue Solver on A Photonic Quantum Processor". Nature Communications 2014; 5(1). doi:10.1038/ncomms5213.

[PLA07] J. H. Plantenberg, P. C. de Groot, et al., "Demonstration of Controlled-NOT Quantum Gates on A Pair of Superconducting Quantum Bits". Nature 2007;447(7146):836–839. doi:10.1038/nature05896.

[PRE18] J. Preskill, "Quantum Computing in the NISQ era and Beyond". arXiv:1801.00862v3. Quantum; 2018; 2:79. doi:10.22331/q-2018-08-06-79.

[REU19] D. Reutter, J. Vicary, *Categorical Quantum Mechanics an Introduction. Lectures and Notes*, Department of Computer Science, University of Oxford, Hilary Term 2019. Available online on January 8, 2024 at https://www.cs.ox.ac.uk/files/10510/notes.pdf

[ROS23] M. Rosenkrantz, M. Lubasch, et al., Method for Reducing Quantum Circuit Depth for Amplitude Estimation, Us. Patent Application 20230385677, 2023-11-30. Uncopyrighted material.

[RUI17] E.C. Ruiz, F. Giacomini, C. Brukner, "Entanglement of Quantum Clocks Through Gravity". In proceedings of the National Academy of Sciences of the United of America (PNAS), March 21. Vol. 114, 2017, No. 12, pp.2303-2309.

[RUS23] V. Rastunkov, F. F. Flother, et al., Method and System of Generating a Classical Model to Simulate a Quantum Computational Model Via Input Perturbation to Enhance Explainability. U.S. Patent Application 20230385682, 2023-11-30. Uncopyrighted material.

[RYA22] C. Ryan-Anderson, N. C. Brownet et al., "Implementing Fault-Tolerant Entangling Gates on The Five-Qubit Code and The Color Code. August 2022. https://arxiv.org/abs/2208.01863

[SEL24] C. N. Self, M. Benedetti, D. Amaro, "Protecting Expressive Circuits with A Quantum Error Detection Code". Nat. Phys. January 2024. https://doi.org/10.1038/s41567-023-02282-2.

[SHI23] Y. Shi, S. Shanmugam Sakthivadivel, "Assisted Composition of Quantum Algorithms". U.S. Patent 11797276, 2023-10-24. Uncopyrighted material.

[SIN23] K. Singh, C. E. Bradley, et al., "Mid-Circuit Correction of Correlated Phase Errors Using an Array of Spectator Qubits". Science 380, 1265–1269. May 2023. doi:10.1126/science.ade5337

[SIP12] M. Sipser, *Introduction to the Theory of Computation*, 3rd Ed. Cengage Learning, Boston, MA, 2012.

[SIU05] M.S. Siu, "From Quantum Circuits to Adiabatic Algorithms". arXiv:quant-ph/0409024v7, 19 Apr 2005, https://doi.org/10.48550/arXiv.quant-ph/0409024.

[SOG23] M. Sogate, C. Chen, K. Shiba, Quantum Circuit and Quantum Computation Method. U.S. Patent Application 20230334356, 2023-10-19. Uncopyrighted material.

[STE96] A. Steane, "Multiple-particle Interference and Quantum Error Correction". Proc. R. Soc. Lond. A Math. Phys. Eng. Sci. 452, 2551–2577 (1996).

[SWA24] M. Swayne, "D-Wave Announces 1,200+ Qubit Advantage2 Prototype in New, Lower-Noise Fabrication Stack, Demonstrating 20x Faster Time-To-Solution on Important

Class of Hard Optimization Problems". The Quantum Insider, January 23, 2024. Available online on January 24, 2024 at https://thequantuminsider.com

[SWA24a] M. Swayne, "QuantWare Say Its Next Generation of Quantum Processors Reach 99.9% Gate Fidelities". The Quantum Insider, February 29, 2024. Available online at https://thequantuminsider.com/2024/02/29/quantware-say-its-next-generation-of-quantum-processors-reach-99-9-gate-fidelities/

[SWA24b] M. Swayne, "IonQ Achieves Technical Milestone One Year Ahead of Schedule". Quantum Computing Business. January 26, 2024. Available online on June 7, 2024 at https://thequantuminsider.com/2024/01/26/ionq-achieves-technical-milestone-one-year-ahead-of-schedule/

[TIL22] J. Tilly, H. Chen, et al., "The Variational Quantum Eigensolver: A Review of Methods and Best Practices". 25 Aug 2022. arXiv:2111.05176.

[VIE21] D. Vietz, J. Barzen, et al., "An Exploratory Study on the Challenges of Engineering Quantum Applications in the Cloud" 2nd Quantum Software Engineering and Technology Workshop, co-located with IEEE International Conference on Quantum Computing and Engineering (QCE21) (IEEE Quantum Week 2021), October 18–22, 2021.

[WAN20] Y. Wang, Z. Hu, et al., "Qudits and High-Dimensional Quantum Computing". Phys., 10 November 2020, Sec. Quantum Engineering and Technology, Volume 8 – 2020. https://doi.org/10.3389/fphy.2020.589504.

[WAN23] K. Wan, W. Huggins, Enhanced Classical Shadows Using Matchgate Quantum Circuits, U.S Patent Application 20230385674, 2023-11-30. Uncopyrighted material.

[WAN23a] Y. Wang, S. Selwyn, et al., "Fault-Tolerant One-Bit Addition with The Smallest Interesting Colour Code". September 2023. https://arxiv.org/abs/2309.09893.

[WIT14] P. Wittek, "Boosting and Adiabatic Quantum Computing" in *Quantum machine learning*, Elsevier, 2014, ISBN 978-0-12-800953-6. doi: https://doi.org/10.1016/C2013-0-19170-2.

[XIA13] Z. L. Xiang, S. Ashhab, et al., "Hybrid Quantum Circuits: Superconducting Circuits Interacting with Other Quantum Systems". Reviews of Modern Physics Vol. 85, No. 2, 2013, pp. 623-653.

[YAO93] A. Yao, "Quantum Circuit Complexity". Proc. 34th IEEE Symp. on Foundations of Computer Science, p. 352. 1993. doi: 10.1109/SFCS.1993.366852. Corpus ID: 2870099.

[YOU23] F. You, A. S. Ajagekar, Quantum Computing Based Deep Learning for Detection, Diagnosis and Other Applications. U.S. Patent Application 20230094389, 2023-03-30. Uncopyrighted material.

[ZHA23] S. Zhang, Z. Wan, S. Zhang, Quantum Circuit Simulation. U.S. Patent Application 20230289640, 2023-09-14. Uncopyrighted material.

Basic Glossary of Key Concepts in Chapter 5

This Glossary is based on various industry sources, including [AFI24], [ALB18], [ARO09], [ARU19], [BAE09], [BER11], [BRA18], [CHE23], [CHE24], [DES22], [ISO19], [JAK98], [KEM06], [KOK07], [MAZ23], [NIE10], [PAP94], [RUI17], [SIP12], [WAN20], [WIT14], [XIA13], and [YOU23]. This Glossary should be used not only as a repository of terms and terminology, but as a tool that further elaborates, explains, and amplifies various fundamental quantum concepts.

Concept	Definition
'Cost' of Quacomp, circuit model	In the circuit model, the cost is equated with the number of gates.
'Cost' of running an algorithm in AQC	The cost of the adiabatic algorithm is defined as the quantity as $\text{Cost} = t_f \max_{\text{over } s} \|H(s)\|$, $\| \cdot \|$ being the ℓ_2-norm. This is seen as being a comparable metric to the circuit gate count. Some researchers use just the run time t_f, which is comparable to the circuit depth of analogous circuit model algorithms.
Adiabatic algorithm time scaling factor	The runtime t_f of an adiabatic algorithm scales, at worst, as $1/\Delta^3$, where Δ is the minimum eigenvalue gap between the ground state H_0 and the first excited state of the Hamiltonian of the adiabatic algorithm. When the Hamiltonian is varied sufficiently smoothly, one can improve this scaling factor [ALB18].
Adiabatic approximation	In broad terms, the adiabatic approximation states that for a system initially prepared in an eigenstate $\lvert\varepsilon_0(0)\rangle$ (namely, in the ground state) of a time-dependent Hamiltonian $H(t)$, the time evolution will, approximately, keep the actual state $\lvert\psi(t)\rangle$ of the system in the corresponding instantaneous ground state $\lvert\varepsilon_0(t)\rangle$ of $H(t)$, provided that $H(t)$ varies "sufficiently slowly"—quantifying the exact nature of this slow variation is the subject of the Adiabatic Theorem (AT), which exists in many variants [ALB18]. Quantum time evolution governed by the Schrödinger equation: $$H\lvert\psi(t)\rangle = i\hbar\frac{\partial}{\partial t}\lvert\psi(t)\rangle$$
Adiabatic process	(From the Greek: *adiábatos*, "impassable".) Relating to or denoting a process or condition wherein heat (or mass) does not enter or leave the system under consideration. An adiabatic process transfers energy to the surrounding environment only as thermodynamic work.
Adiabatic Quantum Computation (AQC)	AQC endeavors to transform or evolve the initial ground state of a quantum system into a final ground state encoding the answer to a computational problem. The basic idea goes as follows as described in [SIU05]: Start with a Hamiltonian whose ground state is easily reachable and prepare our state in the ground state. Change it slowly to a new Hamiltonian that encodes the solution of the problem and maintains a large energy gap between the ground state and the excited state that the evolving state couples to. The Adiabatic Theorem then guarantees that the resultant state will be very close to the ground state of the new Hamiltonian. The original form of the Hamiltonian considered in is a straight-line interpolation: $H(s) = (1-s)H_{\text{initial}} + (s)H_{\text{final}}$. It has been proven that any standard quantum circuit, specified by a sequence of unitary operators, can be implemented as an adiabatic evolution.

Basic Glossary of Key Concepts in Chapter 5 *(cont.)*

Concept	Definition
Adiabatic quantum computer	Computation decomposed into a slow continuous transformation of an initial Hamiltonian into a final Hamiltonian, whose ground states contain the solution [ISO19].
Adiabatic Quantum Optimization (AQO)	The class of procedures for solving optimization problems using a QC. AQO helps escape local minima and overcomes barriers by "tunneling through them rather than stochastically overcoming them" [YOU23].
Adiabatic Theorem (AT)	The AT provides (only) an upper bound on the evolution time t_f required to achieve a certain fidelity between the actual state and the target eigenstate of $H(t)$. But there is not a single AT: there exist different variants that provide different run-time requirements, under different smoothness and differentiability conditions of the Hamiltonian. A criterion commonly used is that the total adiabatic evolution time should be large on the timescale established by the minimum of the square of the inverse spectral gap $\Delta_{ij}(s) = \varepsilon_i(s) - \varepsilon_j(s)$; but since in most cases one is interested in the ground state $\Delta_{ij}(s)$ is replaced by $\Delta \equiv \min \Delta(s)_{\text{over } s\in[0,1]} = \min_{s\in[0,1]} (\varepsilon_1(s) - \varepsilon_0(s))$. Refer to [ALB18] for an extensive discussion on the AT topic.
Category	A category $\mathcal{C}$ consists of (this is the definition given in [BAE09]): • a collection of **objects**, where if X is an object of C, one writes $X \in \mathcal{C}$, and • for every pair of objects (X, Y), a set $\hom(X, Y)$ of **morphisms** from X to Y (some call this set $\hom(X, Y)$ a homset. If $f \in \hom(X, Y)$, then one can write $f: X \to Y$ or $X \xrightarrow{f} Y$). such that: • for every object X there is an identity morphism $1_X: X \to X$ (or $X \xrightarrow{-1}_x X$); Notes: (i) Morphisms are composable: given $f: X \to Y$ and $g: Y \to Z$, there is a composite morphism $g \circ f: X \to Z$; (ii) An identity morphism is both a left and a right unit for composition: if $f: X \to Y$, then $f \circ 1_X = f = 1_Y \circ f$; and (iii) Composition is associative: $(h \circ g) \circ f = h \circ (g \circ f)$ whenever either side is well-defined. A morphism $f: X \to Y$ is an **isomorphism** if it has an inverse—that is, there exists another morphism $g: Y \to X$ such that $g \circ f = 1_X$ and $f \circ g = 1_Y$. A category $\mathcal{C}$ is an algebraic structure that consists of (this is the definition given in [BER11]): (1) $\text{ob}(\mathcal{C}) = \{A, B, C, \ldots\}$, a class of **objects**. (2) $\hom(\mathcal{C})$, being a class of **morphisms** (sometimes called arrows)—that is, maps between the objects. For every pair of objects $A, B \in \text{ob}(\mathcal{C})$ (one can use $\mathcal{C}(A, B) \subset \hom(\mathcal{C})$ to denote the set of morphisms from A to B in the category). (3) Compositions of morphisms—i.e., for every triple of objects A, B, C, the binary operation $\circ: \mathcal{C}(B, C) \times \mathcal{C}(A, B) \to \mathcal{C}(A, C)$. Furthermore, the components of C must fulfill the following axioms: (i) Associativity of composition: $(h \circ g) \circ f = h \circ (g \circ f)$ holds for all morphisms $f \in \mathcal{C}(A, B), g \in \mathcal{C}(B, C), h \in \mathcal{C}(C, D)$. (ii) Existence of identity morphisms: For every object $A \in \text{ob}(\mathcal{C})$ there is an identity morphism $1_A \in \mathcal{C}(A, A)$ such that for every morphism $f \in \mathcal{C}(A, B)$ one has $1_B \circ f = f = f \circ 1_A = f$. (It can readily be shown that the identity morphisms are unique.) *continues)*

Concept	Definition
Category (cont.)	An **isomorphism** is an invertible morphism. The map $f \in \mathcal{C}(A, B)$ is an isomorphism if and only if $\exists g \in \mathcal{C}(B, A)$ for which $g \circ f = 1_A$ and $f \circ g = 1_B$. This makes f and g each other's inverses: $g = f^{-1}$.
Cold atomic quantum computer	A quantum computing model based on the cold atoms, for example optical lattice, Rydberg atom [JAK98].
Complexity theory	Quantum complexity theory aims at formalizing the concept of efficient algorithms. In complexity theory, one is typically interested in the resources required to solve various classes of (decision) problems. In addition, one is interested in how the resource cost scales with the size of the problem to be solved, which is quantified in terms of the length of the input, often denoted n. For example, NP is one example of problem complexity. Several new quantum complexity classes have emerged in the recent past, including QMA (Quantum Merlin Arthur), which is a quantum generalization of NP. There are several complexity classes in Quacomp. Reference [DES22] has an extensive treatment of the topic; also [ARO09], [SIP12], [PAP94] (and an early generic complexity paper by the senior author [MIN75]; also [JAN18]).
Complexity theory, Boolean expression	(Aka propositional logic formula), is an expression assembled from (i) variables (literals); (ii) operators AND (conjunction), OR (disjunction), NOT (negation); and (iii) logical parentheses for establishing a hierarchy of operations. A propositional logic formula is satisfiable if it can be made TRUE by assigning logical values to the variables that it comprises.
Complexity theory, Boolean satisfiability (SAT) problem	(Aka SATISFIABILITY, SAT, or B-SAT) is the problem of determining if there exists an interpretation (a set of values) that satisfies the given Boolean formula. That is, given a Boolean formula, the problem deals with the undertaking of a process to establish whether the formula is satisfiable; hence, the Boolean satisfiability problem seeks to establish whether there exists a truth assignment that renders all clauses true. SAT has been shown to be NP-complete. The 3-SAT problem deals with determining the satisfiability of a formula in conjunctive normal form wherein each clause is limited to at most three literals. It has been shown that 3-SAT is NP-complete (the problem can be re-expressed as a conjunctive statement of literal clauses with three literals, and the satisfiability of the new statement is identical to that of the original formula).
Complexity theory, Maximum Satisfiability problem (MAX-SAT)	Given a Boolean formula in conjunctive normal form, this is the problem of determining the maximum number of clauses that can be made true by an assignment of truth values to the constituent variables. It is a generalization of the Boolean satisfiability problem.
Complexity theory, NP-complete	A key *classical* complexity class is "NP-complete" ("nondeterministic polynomial-time complete"). This class comprises "languages" that can be verified in polynomial time by a deterministic verifier. Here the validity of a solution can be rapidly verified (that is, in polynomial time), and a brute-force search algorithm can find solution(s) by evaluating all possible solution candidates. Notes: (i) "nondeterministic" refers to nondeterministic Turing machines (classical computers)—that is, mechanized tools/systems for formalizing mathematically the applicability of a brute-force search algorithm; (ii) "complete" infers the property of being able to simulate/address all cases in the same complexity class; (iii) "polynomial time" refers to an amount of search/solution time that is relatively "quick" for a deterministic algorithm to identify a solution to the problem on a Turing machine (classical computer).

Basic Glossary of Key Concepts in Chapter 5 *(cont.)*

Concept	Definition
Confidence interval of QC computation	A QC operates at a probabilistic level: measurements of algorithmic outputs provide a solution within a confidence interval; the computation is repeated until a satisfactory probabilistic certainty is achieved.
Diagonal matrix	A square matrix wherein the entries outside the main diagonal are all 0. One uses the process of diagonalizing the Hamiltonian matrix to find the energy eigenvalues and eigenvectors of a generic one-dimensional quantum system. In general, matrix methods are used for solving the time-independent Schrödinger equation.
	One approach basic for matrix diagonalization is to solve the characteristic polynomial for the eigenvalues, then re-insert the eigenvalues back into the eigenvalue equation and individually solve the resulting linear systems for the eigenvectors
Eigengap, Gap (of a system)	(Aka spectral gap.) The eigengap of a linear operator is the difference between two successive eigenvalues, wherein eigenvalues are sorted in ascending order; the difference between the smallest two eigenvalues. Specifically, in Quacomp: the difference between the smallest two eigenvalues of the Hamiltonian $\hat{H}$ of the energy; the difference between its ground state and its first excited state. A Hamiltonian $\hat{H}$ with a spectral gap is called a *gapped Hamiltonian,* and those that do not are called *gapless.* The spectral gap is a traditionally important quantity in the context of ground-state properties of any physical system: many important families of Hamiltonians in physics have the "gap property", meaning that the spectral gap in the limit of large system size $n \rightarrow \infty$ is lower-bounded by a constant [DES22]. In AQC, there is interest in the eigenvalue gap Δ of the Hamiltonian and the Hamiltonian's time-derivative.
Fabrication of a time-dependent non-unitary approximate unitary transformation circuit	Time-dependent non-unitary evolution is mapped to time-dependent unitary evolution through an approximation method to fabricate a circuit on which a circuit architecture of a current quantum computer may be placed.
Hybrid quantum applications (hQApps)	Current approach to computation (applications) that combines classical and QCs systems in order to undertake calculations directed at various (but specific) optimization or related use cases. Tasks such as preparing inputs for quantum algorithms, storing data, or interacting with the user are typically still performed using classical computers.
Hybrid quantum computing	A quantum computing architecture composed of different physical components and different quantum computing models [XIA13].
Intermediate representation (IR) of a computation	The IR of a computation is a representation of the algorism's quantum code that is "in between" the source language description and the target machine detailed instructions. Compilers may use several IRs during a process of translating and optimizing a program: the input is source code describing a quantum algorithm and compile time parameter(s); the output is a combined quantum/classical program expressed using a high-level IR [MAZ23].
Ising model	The Ising model (named after Ernst Ising and Wilhelm Lenz, also known as the Lenz-Ising model or Ising-Lenz model) is a mathematical model of ferromagnetism in statistical mechanics used to study the magnetic dipole moments of atomic spins. The model entails discrete variables $s_i \in \{-1, +1\}$ that represent magnetic dipole moments of atomic "spins"—these can be in one of two states (+1 or −1).

(continues)

Concept	Definition
Ising model *(cont.)*	The spins are displayed in a graph, usually a lattice, enabling each spin to interact (*only*) with its neighbors. Near-by spins that agree have a lower energy than those that do not agree. The model allows the identification of phase transitions as a simplified model of reality. The optimal configuration is given by [WIT14]: $$\operatorname*{argmin}_{s}(s^{\top}Js + h^{\top}s),$$ where s_i are the spins; the J operator describes the interactions between the spins; and h represents the impact of an external magnetic field. The Ising model transforms to a Quadratic Unconstrained Binary Optimization (QUBO) problem with a change of variables $s = 2w-1$; QUBO represents a key problem class for AQC, where it is solved through the physical process of quantum annealing. The Ising objective function is represented by the following Hamiltonian: $$H_I = \sum_{i,j} J_{i,j}\sigma_i^z\sigma_j^z + \sum_i h_i\sigma_i^z,$$ where σ^z_i is the Pauli Z operator acting on qubit i. The ground-state energy of this Hamiltonian equates to the optimum of the QUBO. One looks for the following ground state: $$\operatorname*{argmin}_{s}(s^{\top}H_I s).$$ Consider the base Hamiltonian of an adiabatic process as $$H_B = \sum_i \left(\frac{1 - \sigma_i^x}{2}\right).$$ Then an implementable formulation of the Ising problem on an adiabatic quantum processor can be expressed as [WIT14]: $$H(\lambda) = (1 - \lambda)H_B + \lambda H_I$$
k-local Hamiltonian problem	*k*-local Hamiltonian can be viewed as a set of local constraints on n qubits, each involving at most k of them. Given a k-local Hamiltonian $$H = \sum_{i=1}^{m} H_i$$ the *k*-local Hamiltonian problem deals with finding the smallest eigenvalue λ of H (λ is also called the ground state energy of the Hamiltonian). Another way of describing the k-local Hamiltonian problem is, given a k-local Hamiltonian and given values α, β where $\alpha > \beta$, establish if there exists a quantum eigenstate $|\psi\rangle$ of H with associated eigenvalue I such that $\lambda \geq \beta$ or determine if $\lambda \geq \alpha$. Yet another way of expressing the problem is as follows, as noted in [KEM06]: One wishes to determine whether there is a state of the n qubits such that the expected number of violated constraints is either below a certain threshold or above another, with a promise that one of the two cases holds and both thresholds are at least a constant apart—that is, determine whether the groundstate energy of a given k-local Hamiltonian is below one threshold or above another. The local Hamiltonian problem is the quantum analogue of

(continues)

Basic Glossary of Key Concepts in Chapter 5 *(cont.)*

Concept	Definition
k-local Hamiltonian problem *(cont.)*	the classical maximum satisfiability (MAX-SAT) problem. The 5-, 3-, 2-local Hamiltonian problem are Quantum Merlin Arthur (QMA)-complete (the *k*-local Hamiltonian problem is QMA complete for any $k \geq 2$).
	There are several complexity classes in Quacomp. Reference [DES22] has an extensive treatment of the topic.
Linear optical quantum computer	A quantum computing model based on the linear optical elements, such as beam splitter, phase shifter, single-photon detector [KOK07].
Logical qubits	Qubits connected through quantum entanglement that are used to reduce errors in QCs by storing the same data in different places. For example, a common encoding might use seven physical qubits to represent one logical qubit. Error-correction systems rely on data redundancy, where the same data object is stored in multiple places; logical qubits perform the same calculations across several physical qubits, reducing error rates in case one or more physical qubits fail.
Modular quantum computing	A modular approach for constructing large QCs.
Obtaining a ground state of a quantum system	Obtaining a (or the) most stable state of the quantum system. Has important applications in research of basic properties of quantum physics/chemistry systems, solving of combinatorial optimization problems, and pharmaceutical research. An important application scenario of a QC is to effectively solve or express the ground state of the quantum system.
One-way quantum computer	Computation decomposed into sequence of one-qubit measurements applied to a highly entangled initial state or cluster state [ISO19].
Quantum advantage	A QC can perform some particular computation significantly faster than a classical computer, or that no classical computer can perform it at all [BRA18].
Quantum annealer	A system/device that finds the solution of combinatorial optimization problems by applying an annealing method to quantum entangled system.
Quantum chip	Quantum circuits integrated on substrates to carry the functions of quantum information processing [ISO19].
Quantum circuit	A representation of a quantum general-purpose computer, representing a hardware implementation of a corresponding quantum algorithm/program under a quantum gate model. If the quantum circuit contains adjustable parameters that control a quantum gate, the quantum circuit is called a *Parameterized Quantum Circuit* (PQC) or a *Variational Quantum Circuit* (VQC), both of which are the same concept.
Quantum clock	A quantum system in a state of superposition of energy eigenstates can be used as a reference clock, according to which time evolution is defined [RUI17].
Quantum coding	The coding methods in quantum communication include quantum error correction code, quantum error averse code and quantum error correcting code [ISO19].
Quantum computation	A computation method based on quantum logic.
Quantum diagonal control	Controlling a quantum state evolution process by adding an adjustable group of diagonal Hamiltonian operators, and reducing observations by using the properties of a diagonal matrix.

Concept	Definition
Quantum Error Correction (QEC)	QCs can utilize one or more QEC codes to identify, using syndrome measurements, error-sensitive events that occur in the underlying quantum circuits, so that the proper corrections can be applied. QEC codes are topological stabilizer codes utilized to protect quantum information from errors caused by quantum noise, in order to prepare and/or preserve logical quantum states. Quantum decoder algorithms include but are not limited to Minimum-Weight Perfect-Matching (MWPM), union-find, and/or maximum likelihood.
Quantum gate	In quantum computation (especially in a computing model of a quantum circuit), a quantum gate (or quantum logic gate) is a basic quantum entity (circuit) that operates a small quantity of qubits.
Quantum gate array	Computation decomposed into a sequence of few-qubit quantum gates.
Quantum Merlin Arthur (QMA)	In computational complexity theory, QMA is the set of languages {L} for which, given a string, there is a polynomial-size quantum proof (a quantum state) that enables a polynomial time quantum verifier (running on a QC) to establish membership of the sting to L with high probability; when the string is not in the language, every polynomial-size quantum state is rejected by the verifier with high probability. The QMA class is the quantum analogue of NP in a probabilistic setting—namely, the class of all "languages" that can be verified probabilistically by a quantum verifier in polynomial time (the name is derived from the classical MA/Merlin Arthur class, which is the randomized analogue of NP) [KEM06]. (Arthur and Merlin are fictional agents performing certain tasks—for example, Arthur generates a random string, Merlin answers with a quantum certificate, and Arthur verifies it).
Quantum processor	A processor to handle quantum information—for example, superconducting qubits chip for quantum computing, integrated photonic chip for multiphoton entanglement and boson sampling [ARU19].
Quantum program	A computing program that identifies instructions that may be compiled to be executed on a QC in order to execute a given quantum algorithm. A general quantum program entails processes and abstractions involved in specifying a quantum algorithm, transforming the algorithm into executable form, running an experiment or simulation, and analyzing the results. Also see Chapter 8.
Qudit-based computation	Qudit is a multi-level computational unit alternative to the conventional 2-level qubit: qudit provides a larger state space to store and process information. As a consequence, qudit methods lead to a reduction of the circuit complexity, simplification of the experimental setup, and enhancement of the algorithm efficiency [WAN20]. Qudit methods apply both to the circuit model and to adiabatic quantum computing.
Semiconductor quantum computer	A QC based on the electric spin in semiconductors.
Stabilizer code	A class of quantum codes for performing quantum error correction to restore a noisy, decohered quantum state to a pure quantum state. To protect qubits, a stabilizer quantum error correction code appends ancilla qubits to qubits to be protected.

Basic Glossary of Key Concepts in Chapter 5 *(cont.)*

Concept	Definition
Superconducting quantum computer	A QC based on superconducting quantum circuits.
Syndrome	Measure of whether it is likely an error has occurred or not.
Time-dependent Schrödinger equation	$$i\hbar\frac{d|\psi(t)\rangle}{dt} = H|\psi(t)\rangle$$ where H is a Hamiltonian of a target quantum system, $|\psi\rangle$ represents a quantum state of the target quantum system at time t, and i are imaginary time units. Define t as $$\tau = \frac{i}{\hbar}t,$$ and the Schrödinger equation is obtained by rewriting: $$\frac{d|\psi(\tau)\rangle}{d\tau} = -H|\psi(\tau)\rangle$$ In this case, a solution of the Schrödinger equation is: $$|\psi(\tau)\rangle = e^{-H\tau}|\psi(0)\rangle$$ $$|\psi(\tau)\rangle = A(\tau)e^{-H\tau}|\psi(0)\rangle$$ $$A(\tau) = \frac{1}{\sqrt{\langle\psi(0)|e^{-2H\tau}|\psi(0)\rangle}}$$ An initial minimum eigenstate may be inferred thereby: $$\psi_0(0) = \lim_{\tau\to\infty}\frac{\psi(\tau)}{\langle\psi(\tau)\,|\,\psi(\tau)\rangle}$$ Time evolution is a powerful mechanism for finding a ground state of a quantum system and has many feasible solutions on quantum computers.
Universal set of gates	A set of gates is said to be universal for QC if any unitary operation may be approximated to arbitrary accuracy by a quantum circuit involving only those gates [NIE10].
Variational Quantum Eigensolver (VQE)	An approach for estimating ground state energy of a specific quantum system through parameterized/variational quantum circuits (PQCs or VQCs), it is a classical quantum hybrid computing paradigm and has been widely used in the field of quantum chemistry. Also see Glossary to Chapter 1.

Chapter 6

Quantum Communications and the Quantum Internet

6.1 Overview

Although Quantum Computers (QCs) can operate locally for some basic, discrete applications, their full value arises when a (large) number of such computers can be interconnected over a distance to form a synergistic array or web of processing engines, just as has been the case with classical computers and local area networks (LANs), wide area networks (WANs), and global area networks (GANs) such as the Internet. Networking has always been known as having the ability to enhance inherent computational power by linking multiple systems together.

Quantum Communication (Quacom) aims at supporting transmission of quantum-based communications between physically remote QCs, particularly for *sets* of QCs not just connected by discrete point-to-point links. Quacom handles the exchange of quantum information by utilizing unique features of Quantum Mechanics (QM) to perform communication tasks that are difficult or even impossible using classical mechanisms; Quacom makes extensive use of QM phenomena such as superposition, no-cloning, and entanglement. A (or the) Quantum Internet (Quaint) is a (large) collection of remotely distributed quantum-based (possibly standardized, interworking) (QC) nodes, wherein, using some quantum communication protocol, one can perform various distributed (quantum) computation or dispersed sensing. Quacom R&D is currently underway in many physics laboratories around the world; the follow-on step is network engineering and commercialization.

Transferring quantum information between distant parties is a basic, yet still challenging, requirement of Quacom. Nonetheless, quantum-encoded messages are already being transmitted at this juncture: to date, the most practical commercial application of Quacom is key distribution for encrypted communication—in the QM context, an unknown quantum state cannot be copied, and this no-cloning circumstance provides intrinsic ecosystem security; other applications are also quickly emerging (as discussed in Chapter 1).

As noted elsewhere in this text, entanglement is a property of QM systems that shows correlations between two physical systems that cannot be accounted for by classical physics—it is a phenomenon in which two or more qubits exist in a physically distributed state so that operations (measurements) on one of the qubits change the mutual state of the pair. In other words, entanglement is the phenomenon of quantum correlation between two qubits, it being impossible to describe the properties of each particle individually when they are entangled. Emerging quantum-based applications require the existence of a reliable network for the distribution of entanglement; thus, Quacom can be perceived as a set of mechanisms and supportive infrastructure for enabling the sharing of entangled quantum state between multiple parties at various physical distances.

Quacom allows information to be transmitted while encoded in QM quantum states. Entanglement and entanglement distribution are fundamental concepts in Quacom: entangled qubits have non-local properties wherein one can send one of the qubits to a remote device and, on the assumption that only limited noise has been introduced in the transmission, the two qubits remain in the entangled state until a measurement is undertaken—that is, outcomes of independent remote measurements on entangled objects show intrinsic correlations. In a Quantum Network (Quanet) entanglement is initially generated at a local (computing or transmission) node; thereafter, there is a transfer of one or both of the entangled qubits across a quantum channel. Currently, Quanets rely principally on (i) *teleportation* and (ii) *entanglement swapping*.

Clearly, there is a distinction between collection of QCs connected over a classical network (for example, to share computational results) and a Quanet where nodes are able to reliably transmit/deliver/distribute entangled qubits (entangled states) amongst themselves [WEH18], [KIM08], [BRI98], [CHI06], [CAL18]. Entanglement distribution encompasses methods for distributing quantum entanglement to distinct nodes. Thus, one cannot simply extrapolate the classical transmission methods to a quantum analogue to create a Quanet. Although the entangled pairs of qubits are the basic unit of communication in Quanets, it should be promptly noted that entangled (Bell) pairs are not directly equivalent to traditional data packets. The architecture of a Quanet is predictably different from that of a classical WAN or the classical Internet.

Various quantum-based building blocks, ranging from lower-level components to higher level systems, are the enabling elements of a Quanet. These elements include but are not limited to quantum memories, Quantum Repeaters (Quareps), quantum network end-nodes (for the data plane), and technologies that extend traditional network control technology (control plane) to support Quanet functionality; these elements also include FCAPS—fault, configuration, accounting, performance, and security—capabilities for network management (management plane).

Traditional networks are typically hierarchical—tiered—where local computers are connected via LANs; metropolitan and regional networks are connected via WANs; and international networks are connected via GANs, or Internet elements. The expectation is that Quanets will ultimately utilize a similar architecture, and the Quantum Internet (Quaint) will serve as the global vehicle for large-scale connectivity. Quacom and Quaint enable the linking of QCs at regional or global distances to support cluster and cloud computing, and applications such as cybersecurity—confidentiality, in particular—improved distributed sensing, improved clock synchronization, and other applications. Google and IBM are already offering cloud-based Quantum Computing (Quacomp) services. The Quaint will enable the synergistic deployment of integrated Quacomp technologies and protocols on a global scale; however, several technical challenges must be resolved before it becomes a reality. Some view the most crucial of these being the realization of a Quarep as an essential component in the long-distance transmission of quantum information [AZU23]. As the parallel of a classical repeater, the Quarep is designed to overcome loss and noise in the quantum channels constituting a Quanet.

6.1.1 Underpinning Precepts

The underpinning precepts defining Quacom are the *quantum measurement postulate* and the *no-cloning theorem* (e.g., see [BUZ96]); these precepts result in the impossibility of reading and copying quantum information without altering it. This predicament is in stark contrast with classical communication, in which in-path reading and duplicating of information are the fundamental mechanisms of the protocols at multiple layers of the protocol model (e.g., seven layers of the Open Systems Interconnection Reference Model (OSIRM) and five layers of the TCP/IP model). These Quacom characteristics drive the specific design of a Quanet or Quaint; nonetheless, the goal is to retain and institute as many of the modularity (that is, stack-driven functional separability) and scalability concepts that are ingrained in the classical Internet, although, as implied, a straightforward mapping between classical and Quaint protocol stacks is not achievable *a priori*.

Significant R&D effort has been expended of late by researchers and development labs into physically realizing and connecting quantum devices while endeavoring to improve their speed and error tolerance. Transferring quantum information and entanglement between remote parties is a basic requisite, but there are some fundamental and technical challenges at this time that need to be addressed in order to foster widespread adoption of the technology. Fundamental to Quacom's realization at the commercial level is the ability to generate and store entanglement in interconnected nodes that reside at non-local (non-trivial) distances. Important considerations relate to (i) the rate of entanglement generation, and (ii) the fidelity of the generated entanglement.

- *Entanglement generation* on a two-node link is not an expeditious process, and it may require a number of attempts to succeed. For example, the typical rate of success between Nitrogen-Vacancy (NV) nodes is in the range of 10 Hz [DAH19], [WEI22]. Other quantum technologies have higher entanglement rates but may not directly embody quantum memory, which NV center nodes support, and/or may have limited processing capabilities. The current, relatively low, entanglement generation rates are not adequate for practical Quanets beyond the lab, even though there have been important advances in generating entanglement with improved efficiency and fidelity over the course of the past decade (for example, in trapped-ion systems, the entanglement generation rate has improved to approximately 200 events per second in recent years). These advances are due to improvements in the efficiency of photon collection from atoms, the reduction of photon loss in the channels, and the use of single-photon detectors with higher detection efficiencies [AWS21].
- As noted elsewhere, *fidelity* characterizes the quality of a quantum state, describing how close a quantum state is to the state one, say the sender, tried to create; it is the probability that the state behaves identically to the desired state—in particular, if fidelity is less than 0.5 the

state is unusable. Fidelity can be used to quantify how much a given state has been degraded by noise, including gate errors and transmission channel losses.

Engineering challenges include limits dictated by channel losses, coexistence with background noise, and the dependance on imperfect detectors and single-photon sources. As discussed in previous chapters, quantum signals are weak, fragile, and impacted by noise. In addition, they cannot be copied or amplified; therefore, special approaches and techniques are needed. Both transmission rate and distance of Quacom are intrinsically limited by the channel loss, thus rate-loss tradeoffs have to be taken into consideration: noise and attenuation, even loss, limit the maximum distance for unrepeated communications. It is a given that Quanets need to address noise and loss; this can be accomplished, to various levels of success, using purification and quantum error correction.

As should be well understood at this juncture, before a photon (or more generally, a qubit) is measured, it exists in a superposition of all its possible quantum states, each state with an associated probability; a measurement effectively selects one state among these various states, but the fundamental predicament is that the photon's quantum state cannot be measured without triggering a "disturbance" that exposes the attempt. In spite of the fact that photons are hard to store, they are often used in Quacom because they exhibit low interaction with the environment and low transmission loss in certain media (e.g., singlemode fibers), resulting in relatively moderate decoherence; in addition, there is classical knowledge on how to control photons-based signals with off-the-shelf optical components.

An inherent advantage of Quacom is its ability to protect data being transmitted, frustrating eavesdropping; it may also offer novel functionality compared with classical communication. In fact, Quacom has already found applications in the area of "network security", as discussed more thoroughly in Chapter 7, where Quantum Key Distribution (QKD) has already been used for various financial transactions. However, this type of communication is currently reliably achievable only at metropolitan-scale distances. At a more general level, Quanets are currently limited to simple configurations and are typically focused on supporting specific experimental goals; the development of network components, such as transducers and quantum memories, is in its relative infancy [NIS24].

6.1.2 Relation to and Dependence on Classical Networks

It should be noted that in Quacom there is still a dependence on classical communication. There is currently a need for a degree of hybrid integration between Quacom and classical communications, given the need to associate quantum data stored in quantum devices with classical control information that is transmitted over an overlay communications link. See, for example, Figure 6.1. IETF's RFC 9340, *Architectural Principles for a Quantum Internet* [KOZ23], observes that the role of classical communication in Quanets is to (i) transact classical bits of information in the context of distributed protocols such as entanglement swapping and teleportation, and (ii) transact control information to set up end-to-end entanglement generation. For example, teleportation consumes an entangled pair and requires the transmission of two classical bits per qubit teleported.

This dependency may have some deleterious impacts; RFC 9340 notes that:

- Qubits do not have (traditional) headers, implying that Quanets need to transmit all control information required to complete the transaction of interest over a (separate) classical channel. As discussed elsewhere in this chapter, error management data utilized in entanglement purification/distillation is an illustration of such control information. Classical latency in the classical network, which depends on transmission time and distance, impacts various activities in the Quanet—for example, to create far-removed Bell pairs, the nodes typically store the qubits in quantum memories and await the reception of control information before undertaking the next operation in the process.

- In what some call "store and swap" Quanets, nodes store Bell pairs and then undertake entanglement swapping; there is no need to keep track of the order with which the Bell pairs are generated, but obviously, the correct entangled pairs need to be swapped and the related measurement outcomes need to be signaled to and correlated with the correct qubits at the far-end nodes.
- A quantum node must be able to make determinations about qubits that it receives. Given that qubits do not carry headers, the receipt of an entangled pair contains no control information that can be immediately used by a Quarep to make a networking or processing decision; the related control information will arrive separately over a classical channel. The repeater is, thus, required to store temporary state while awaiting the control information.

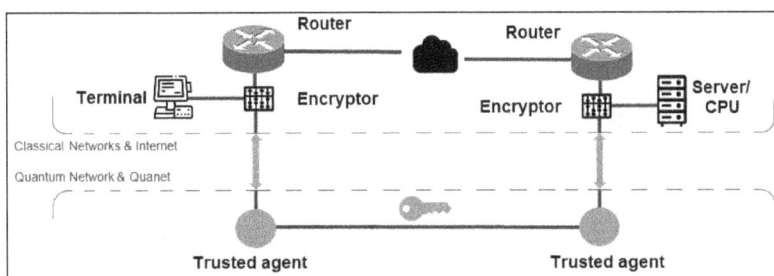

Figure 6.1 Example of hybrid network/arrangement

Table 6.1 identifies some key concepts that play important roles in Quacom, Quanets, and Quaint. The issues introduced thus far are discussed in some additional detail in the sections that follow. To that end, this chapter covers, among other topics, teleportation, swapping, how quantum communication channels are implemented, purification, quantum error correction, and repeaters. Some aspects of QKD are assessed in this chapter, but the topic is covered in more detail in Chapter 7 (QKD is entanglement based and relies on the correlations and inherent secretiveness of entangled Bell pairs—it does not require the transmission of quantum states information).

6.2 Basic Quacom/Quanet Concepts

Practical Quanets outside the lab require a combination of efficient photon coupling, long-lived heralded memory, and multi-qubit operations with existing telecom fiber infrastructure; this technology milieu is currently an outstanding challenge. Quanets, as currently conceived, rely on *entanglement, teleportation,* and *entanglement swapping.*

[Text continues on 238]

Table 6.1 Key Quacom Concepts

Quacom concept	Description	Exemplary references
Ancilla	Ancilla refers to an auxiliary subsystem. For example, photons can be generated from the vacuum by modulating an artificial atom that does not interact directly with the cavity, but instead is indirectly coupled to the field through some auxiliary subsystem—the ancilla.	[DEP23]
Cluster state(s)	An entangled state of multiple qubits. Entails a special type of entangled state: they refer to a family of quantum states of n-qubit two- or three-dimensional lattice in which each vertex corresponds to a qubit. This approach may be utilized in Quareps and in Quacomp. Also see Glossary in Chapter 5 (page 221).	[RAU01]
Eberhard bound	The required efficiency and transmission to perform a loophole-free experiment with photon polarization.	[EBE93], [LAR01]
Entangled photon source	The signal/light source of entangled photons	[KWI95]
Entanglement distribution	A method for distributing quantum entanglement to different nodes	[YIN17]
Entanglement swapping	The transfer of entanglement from a priori entangled systems to a priori separable systems. It is a useful mechanism for entanglement purification, teleportation, Quacomp and quantum cryptography. Entanglement swapping can be used to distribute Bell pairs across long distances by teleporting the state of one member of a Bell pair over progressively longer distances until the pair stretches from end to end [ABE23]. Bell pairs are "directionless (undirected)" network resources. Consider Bell's pairs. Suppose Alice has a particle that is entangled with a particle owned by Bob, and Bob teleports it to Christine, then afterward, Alice's particle is entangled with Christine's particle. More generally, suppose Alice has a particle that is entangled with a particle owned by Bob, and Christine has a particle that is entangled with a particle owned by Daniel. Bob teleports the particle to Christine, then afterward, Alice's particle is entangled with Daniel's particle.	[BEN93], [JI19], [JI22], [VED06], [ZAN23]
Protocol	In this context, a list of steps to be performed by Quacomp and/or Quacom entities to achieve a given goal. See Quantum Protocol.	
Purification and error correction	The process of enhancing or improving knowledge about the state, seeking an increase in the fidelity of the density matrix that represents knowledge about the state. It is the process of recovering a nearly pure copy of an unknown pure quantum state utilizing several noisy copies of the state. A number of techniques for handling quantum errors have been developed. For example, performing some types of testing on input states that are two imperfect Bell pairs, intending to produce an output Bell pair with higher fidelity.	[CHI23]

Quacom concept	Description	Exemplary references
Quantum channel	A physical connection that enables the transmission of qubits. Communication channel for transmitting quantum signals. Traditional optical fibers can be used by transmitting light pulses (photons) light. An engineering fact is that quantum channels, like other telecommunications links, are noisy and lossy (as discussed in Chapter 1 and Chapter 3)	
Quantum communication (Quacom)	A communication method using quantum entanglement mechanisms for information transmission. Quanets support quantum-secure communication and entanglement-assisted communication and enable the development of large-scale QC ensembles by interconnecting smaller dispersed systems utilizing quantum/QM means.	
Quantum concurrence C (ρ)	Metric used to assess the entanglement status of mixed state of two qubits. A pure state is completely specified by a point in phase space, a vector in a complex vector space. It turns out that in QM for any composite system, for example, $W = S_1 \otimes S_2$, there are pure states of the system in which the parts of the system do not have pure states of their own; such states are called *entangled*. Entanglement also exists for mixed quantum states: a mixed state is entangled if it cannot be represented as a mixture of unentangled pure states. For both pure and mixed quantum states, there are a number of measures to describe and/or assess the degree of entanglement; in the case of pure states of a bipartite system, there is a single widely accepted measure of entanglement—here entanglement can be described by a measure of entanglement called the *concurrence*—but for mixed states there are several measures that have been proposed and studied [WOO01].	[WOO01]
Quantum entanglement	Quantum entanglement enables qubits/particles to influence each other's measurement outcomes even when qubits/particles separated by considerable distances; it is the phenomenon, the process, by which two physical systems enjoy correlated properties, even though they may be widely separated. Quantum entanglement is an intrinsic process in support of Quacom, Quacomp, and Quantum metrology. In particular, entangled qubits can be used (i) to establish confidential (private) information, extending quantum cryptography to long distances, or (ii) to implement quantum logical gates.	[ABE23], [ZUK93]
Quantum error correction (QEC)	Methods to identify and address error-sensitive events that occur in the underlying quantum circuits; these methods (e.g., minimum-weight perfect matching, maximum likelihood, and so on) are employed to protect quantum information from errors stemming from quantum noise and decoherence, in order to prepare and/or preserve logical quantum states. *(continues)*	

Table 6.1 Key Quacom Concepts *(cont.)*

Quacom concept	Description	Exemplary references
Quantum error correction (QEC) *(cont.)*	This is typically accomplished by encoding some number of logical qubits into a larger number of physical qubits. Syndrome measurements, parity measurements between multiple qubits, are then used to determine the presence and location of an error, and a recovery operation may be applied to correct the error. One example (of many) of a QEC code is the Steane [[7,1,3]] code where one logical qubit is encoded into seven physical qubits; this encoding is robust against all single (physical) qubit errors.	[STE97], [WEI15], [NIE02]
Quantum memories	Devices that can store and retrieve quantum mechanical states.	
Quantum node	A node that can receive and emit quantum information.	
Quantum nondemolition measurements	Approaches for measuring a property of a quantum state without disturbing its other properties; for example, measuring the presence of a photon without disturbing its qubit state.	
Quantum photon source	Optical source generating photons that carry quantum information.	
Quantum protocols	A number of "protocols"[a] have been proposed and demonstrated for transporting quantum information between dispersed nodes, including the following, as discussed in [AWS21]: *Pitch-and-catch protocol.* Here a flying qubit (such as a photon) emitted by a stationary qubit (or reflected off a cavity holding a qubit) at the transmitting end carries the quantum state over the communication channel and transfers that state to another qubit at the receiving end. Demonstrations for this method have been documented using atomic qubits optical cavities; however, the loss in the photonic channel rapidly degrades the performance, making this protocol impractical at optical frequencies. On the other hand, it is possible to create strong coupling in superconducting circuits between the transmitting and receiving qubits with a microwave photon over a transmission line connecting the two modules; therefore, this protocol is more practical in these environments. *Heralded (i.e., confirmed) entanglement generation protocol.* Here a pair of entangled qubits is first generated probabilistically using photon emissions from the qubits and from the detection of emitted photons; then a deterministic teleportation of the qubit is accomplished using the generated entanglement as a resource. In this protocol, first the communication qubit on each module (such as a trapped ion module) emits a photon in such a manner that a selected feature of the photon (such as polarization, *(continues)*	

[a] In classical networking a "protocol" is a formal, well delineated, well defined, agreed upon (even officially standardized) procedure to carry out a computing/communications task among various homogenous or heterogeneous (multi-vendor/multi-technology) entities. In Quacom, a protocol is a described, general-class, technical or technology methodology; it may not necessarily be an (industry-) agreed-upon, standardized, or formal/canonical procedure. However, some efforts are underway to define a number of Quacom standards that are more akin to the concept of traditional communication standards, as described later on in this chapter.

Quacom concept	Description	Exemplary references
Quantum protocols (*cont.*)	frequency, phase or time bin) is entangled with the qubit. The emitted photons are collected, interfered on a 50:50 beam splitter, and detected at the outputs. The detection event signals a successful generation of entanglement between the two qubits that emitted the photons. Although the successful execution of the protocol occurs only probabilistically, success is heralded by detection of two photons at the output of the beam splitters, and reliable entanglement can be generated at low-to-moderate rates.	[AWS21]
Quantum repeaters (Quareps)	Quantum devices that can extend the range of quantum communication between a sender and a receiver. Quareps create links that entail multiple entangled pairs in an entangled chain, supporting entanglement swapping. Photons are used as carriers of entanglement over a distance, utilizing optical fiber transmission channels or free-space transmission. Quacomp nodes generate, process, and store quantum information, while Quacom utilize photons to transport entangled quantum states between endpoints. Signal attenuation occurs even in low-loss fiber, limiting transmission distance to a few hundred miles; this, as noted in Chapter 1, drives the need for Quareps. At this juncture researchers are focusing mostly on developing metropolitan-level networks, which are small enough to avoid the need for Quareps; some researchers have demonstrated larger networks that link multiple quantum devices, entangling over a dozen qubits, and using quantum teleportation over satellite links [JOH22], [YIN17]. Current-generation Quareps are based on linear optics and quantum memories. Using Quareps, the entanglement distribution time scales polynomially with the transmission distance, provided entanglement between quantum memories in remote locations can be heralded [USM12].	[BHA20], [BRI98], [DUA01], [JIA09], [MUN10], [MUN12], [MUR14], [YUA08], [USM12]
Quantum routers (Quarous)	Devices that deal with path computation between a source and a target destination node in a Quanet.	[CUB24]
Quantum storage	Qubit storage is a needed capability in Quacom, particularly for Quantum Repeaters (Quarep). Rare earth-doped crystals are possible candidates for storing qubits. Rare earth-doped crystals (for example, silicon-based crystal doped with neodymium) provide a large number of atoms trapped in a crystalline structure, which at cryogenic temperatures enjoy good coherence properties—here entangled photons can excite rare-earth atoms embedded in two crystals [SER19], [MAR17], [LAG21]. Memories based on rare earth-doped crystals have been utilized in a number of short-distance quantum teleportation experiments [LAG23].	[LAG21], [MAR17], [SER19]
Quantum switches	Devices that can route optical signals between different channels while maintaining quantum coherence and entanglement.	

Table 6.1 Key Quacom Concepts *(cont.)*

Quacom concept	Description	Exemplary references
Quantum teleportation	The transmission and reconstruction over arbitrary distances of the state of a quantum system. Quantum state teleportation consumes an unknown qubit state that one wants to transmit and regenerates it at the destination (this process does not violate the no-cloning theorem, given that the original state is destroyed in the process). During teleportation, an initial photon that carries the polarization to be transferred and one of a pair of entangled photons are subjected to a measurement such that the second photon of the entangled pair acquires the polarization of the initial photon; this latter photon can be arbitrarily far away from the initial one [BOU97], [KOZ23].	[BEN93], [BAU97], [BOS98], [FUR98], [HAO24], [LAG23], [REN17], [TAU24], [ZHA23]
Quantum transducers	Devices that convert a physical quantum state from one physical system to another. Consider, as an example, connecting two ion-based quantum nodes over a given distance: an ion excitation is converted to a photon using a quantum transducer, so that its carrier frequency is compatible with a telecom fiberoptic band. After propagation, the photon is converted back to an ion excitation, again using an appropriate quantum transducer.	
Repeaterless bound	Fundamentals in physics place limits on protocols for direct quantum communications. There are upper bounds for the maximum achievable rates for transmitting quantum information (two-way quantum capacity), distributing entanglement (two-way entanglement distribution capacity), and generating secret keys (secret key capacity K), over the most fundamental (basic) quantum channels.	[PIR17]
Syndrome measurements	A syndrome measurement provides information about an error that may have occurred, but it does not provide or retrieve the information that is stored in the qubit; this follows from the fact a measurement destroys any quantum superposition of the qubit, which, in turn, would prevent it from being used to convey quantum information. QEC codes are used to protect quantum information from errors.	
Telecom heralding	The process of detecting one of a pair of qubits (photons) to herald the other; one photon can be heralded by the detection of its twin. Heralded qubit amplification is emerging as a mechanism to address channel losses in Quacom as an alternative to the difficult design challenge of eliminating such losses.	

6.2.1 Entanglement

Entanglement, already discussed in this text in various places, particularly in Section 2.6 (starting on page 78), is an intrinsic feature of QM and Quacom; the most basic requirement to support Quacom is the ability to generate and to store entanglement on dispersed quantum nodes. Quantum entanglement is the process by which two physical systems enjoy correlated properties; observers of two entangled particles see correlations in their measurement results, even when they are widely separated by physical space. In a Quanet there is a need to support the *distribution entangled state(s)* with sufficient fidelity, while dealing with channel noise; the Quanet seeks to deliver quantum states in such a way that the requisite fidelity is maintained across the distance. This is a non-trivial task. A measurement of a qubit's state consumes the superposition and the entanglement linkage the qubit may have been part of; hence, one cannot utilize the same methods used in classical (bit) processing to support error detection and correction—appropriate quantum error detection and correction mechanisms and procedures are needed. The no-cloning theorem cited earlier makes it impossible to create an identical copy of an arbitrary, unknown quantum state, again making it impossible to use the same mechanisms that are employed for classical networks for signal amplification or retransmission—these rely on the ability to copy the stream of data.

Long-distance Quacom/Quanet systems need to make use of quantum entanglement. At this juncture, entanglement is required in nearly all Quacom and Quacomp protocols for applications such as quantum teleportation, dense coding, quantum key distribution, quantum secure direct communication, and other applications. Tasks such as dense coding, quantum key distribution over long distance and at high rate, quantum teleportation over long-distance free-space (see further below), and tests of local realism based on Bell's inequality have all been implemented in some practical Quacom settings using entanglement; the fact remains, however, that it is still challenging to actually create, on a routine commercialized basis, a generalized entangled system in which the two physical systems are separated by some non-trivial geography.

Entanglement is created through local interactions between two qubits or as a product of the way the qubits were created (e.g., entangled photon pairs). To create a distributed entangled state, one can then physically send one of the qubits to a remote node [KOS23]: consider preparing the state $(|00\rangle + |11\rangle)/\sqrt{2}$ with each node at some distance and retaining one of the two qubits; once either one of the two nodes performs a measurement, the state of the two qubits collapses to either $|00\rangle$ or $|11\rangle$—the outcome of the reading is random, and the final status does not exist prior to the measurement; however, the two nodes measure the same value.

A Quacom's basic supporting physical infrastructure aims at enabling the distribution of entanglement between distant locations, thus allowing the realization of applications that require collaboration between far-flung entities, such as quantum sensing and distributed quantum computation. At the hardware level, to connect two quantum nodes, distinct quantum devices such as quantum light sources, transducers, quantum channels, and detectors at each end, and elements supporting the channel itself must be compatible and must operate simultaneously.

In assessing the performance of Quanets, one is typically interested in metrics such as the channel throughput, the *quality* and *latency* of entanglement generation (for quantum performance, fidelity of entangled state; for classical performance, entanglement request latency), and the quality of the local gate execution[1] (how noisy quantum states and operations are). The quality of a quantum state is assessed by its fidelity, and the quality of executing a gate is assessed by its gate fidelity [VAR22]. Regarding entanglement generation, there are tradeoffs between the quality of the quantum execution and throughput or latency; for example, one can choose to exchange a

---

[1] For some applications noise-free transmission and quantum gate execution are not necessarily limiting because noise at the quantum level is addressed by using classical error correction after measuring the quantum state, as might be the case in QKD.

higher throughput of entanglement generation in favor of a lower quality of the resulting entanglement and vice versa. Regarding the quality of a quantum gate's execution, such nodal quality determines the quality of the quantum program under consideration; this limiting noise can result from intrinsic imperfections of the device, but it can also appear as a time-dependent factor associated with the waiting time before the quantum state can undergo further processing.

Entanglement between physically separate systems has been demonstrated with single trapped ions and atoms, atomic ensembles, photons, and solid-state systems. Light photons can encode qubits in their polarization; light photons have limited interaction with the environment (but are subject to attenuation and noise) and can travel rapidly through a medium—a photon can be characterized as possessing right or left circular polarization (or a superposition of the two), or as possessing horizontal or vertical linear polarization (or a superposition of the two). However, it is difficult to detect a single photon without destroying it, being that "off-the-shelf" detectors absorb the photon, neutralizing the quantum information. Validating a photon's presence is a critical function of any Quacom detector that employs light as its communication method, including the case of device-independent quantum cryptography, which relies on Bell tests for security.

It is well known in QM that the achievable distance for Quacom links is limited by the far-end ability to detect single photons (in addition, trust in imperfect devices used for cryptography may allow eavesdroppers to undertake side-channel attacks, exploiting unintended leakages of information or controlling the endpoint detectors). Classical photon amplification, as utilized in classical telecommunication networks by multiplying the number of photons, is not exploitable in QM for communication applications because the no-cloning theorem precludes copying (cloning) an unknown qubit with perfect fidelity. Given that predicament, quantum teleportation and entanglement swapping supported by Quareps aim at extending the usable length of quantum communication links.

Researchers have demonstrated remote entanglement in systems using neutral atoms, trapped ions, semiconductor quantum dots, and Nitrogen-Vacancy centers in diamond [HOF12], [DAI21], [RIT12], [VAN22], [KRU23], [STE20], [STO17], [BER12], [HUM18]. Also, recently, two atomic ensemble memories have been entangled through a metropolitan fiber network [LUO22], [YON20], [LIU24].

6.2.2 Teleportation

As noted in RFC 9340 [KOZ23], a basic way to distribute an entangled state is to transmit one of the qubits directly to the far end over a series of nodes, in conjunction with Quantum Error Correction (QEC) to manage path degradation. However, QEC makes relatively high demands on both resources (physical qubits needed) and their initial fidelity; at the practical level, QEC implementation is challenging, and QEC is not expected to be used until later generations of Quanets are possible. Until such time, Quanets rely on teleportation and entanglement swapping, as noted earlier. These approaches rely on the observation that one does not need to distribute *any arbitrary entangled quantum state*, but one only needs to be *able to distribute any one*[1] *of the entangled two qubits Bell pair*[2] *states*: $(|00\rangle + |11\rangle)/\sqrt{2}$; $(|00\rangle - |11\rangle)/\sqrt{2}$;

---

[1] Any of the four Bell pair states can be utilized because one can transform any Bell pair into another Bell pair utilizing local operations involving single-qubit gates applied to only one of the qubits.

[2] The motivation for using Bell pairs (as contrasted to any other two-qubit state) is that Bell pairs are the maximally entangled two-qubit set of basis states, as discussed elsewhere in this text; maximal

$(|01\rangle + |10\rangle)/\sqrt{2}$; and $(|01\rangle - |10\rangle)/\sqrt{2}$. The ability to distribute Bell pairs enables the distribution of any other arbitrary entangled state. Distributing a Bell pair between two nodes over a network is easier than transmitting a general quantum state; given that the state is known, handling errors becomes easier, and simpler error correction (such as entanglement purification/distillation) along with transfer reattempts is a viable approach.

Quantum teleportation is a critical process in support of Quanets. Teleportation is the transmission and reconstruction over arbitrary distances of the state of a quantum system. Teleportation allows the transfer of quantum states using previously shared entanglement in conjunction with classical (optical) communication: the qubit to be teleported is measured jointly with one part of the entangled state, projecting it onto a Bell state [BAU97], [LAG23]. Thus, at the QM level, quantum teleportation requires the ability to create entangled Bell pairs over the appropriate distance under consideration—teleportation is the mechanism by which quantum information is transmitted from one endpoint to another endpoint in conjunction with (i) the ability to transfer classical information over an appropriate network, and (ii) the establishment of the just-cited previously shared quantum entanglement between the two endpoints. Quantum teleportation functions as a basic primitive for several other quantum protocols. Teleportation[1] was first proposed in the early 1990s and demonstrated soon thereafter using light-based systems [BEN93]. Early quantum teleportation protocols were performed utilizing photonic states, but following the development of quantum network nodes with stationary qubits, remote qubit teleportation has been achieved between nodes that are based on trapped ions, trapped atoms, and diamond NV centers. Naturally, it is desirable to have access to the qubit after the teleportation occurs, given that this allows for downstream use of the quantum information.

Stated differently, quantum teleportation occurs when an arbitrary and unknown quantum state is teleported from one party to another by sharing entanglement between the parties and subsequently performing a Bell State Measurement (BSM) on the state to be teleported and one-half of the entangled pair [BAY23].[2] Teleportation requires the application of a unitary transformation to the stored qubit, based on the outcome of the remote BSM. To support the protocol over long distances, the qubit of interest has to be teleported to a physical qubit stored in a quantum memory that enjoys a *storage time that exceeds the two-way communication time between the two endpoints*.

---

entanglement implies that these states enjoy the strongest correlation of all other possible two-qubit states.

[1] Also see descriptions in Table 1.2 on page 34 and the Glossary of Chapter 1 on page 52.

[2] Some limitations of BSMs are described in [BAY23]: the basic approach to realizing an optical BSM is to allow two entangled photons to impinge on two inputs of a balanced beam splitter and measuring the resulting output patterns; the simplicity of this linear-optical approach comes at the cost of being able to identify only two of four possible Bell states: in 50% of all cases, the obtained results are ambiguous, thus limiting optical quantum technology that relies on successful projection onto the Bell basis (some approaches that offer ameliorations to this predicament have been developed of late). In atomic systems, complete BSMs have been performed; however, these require complex experimental setups that are challenging to scale up. Complete BSMs on spin qubits are possible in solid-state systems, but thermal effects typically prevent operations at room temperature. In general, non-optical approaches suffer from limited intrinsic clock rates of the order of megahertz; only the photonics platform offers, in principle, high processing clock rates at room temperature.

Consider the illustrative setup wherein the endpoints, Alice and Bob, each generate a Bell Einstein-Podolsky-Rosen (EPR) pair of two entangled qubits (each individual has one element of the Bell pair). Next, assume Alice wishes to deliver a qubit $|\psi\rangle$ of the pair to Bob—the state to be teleported is $|\psi\rangle = \alpha|0\rangle + \beta|1\rangle$, where α and β are unknown amplitudes (namely, Alice does not know the state of the qubit to be teleported). Alice proceeds along in the process and interacts the qubit $|\psi\rangle$ that corresponds to her portion of the EPR pair: she measures the two qubits in her possession, thereby obtaining one of four possible classical results, 00, 01, 10, and 11.

As stated, Alice does not know the state of the qubit to be teleported and can only forward classical information to Bob; upon performing the measurement, she sends this resulting information to Bob. Based on Alice's classical (classically delivered) message, Bob performs one of four operations on his portion of the EPR pair; by doing this he can recover the original state $|\psi\rangle$ (two classical bits of measurement results must be communicated to Bob for him to be able to recover the original state $|\psi\rangle$). Teleportation consumes precisely one Bell pair.[1] Figure 6.2 depicts the quantum circuit and infrastructure for implementing the teleportation of a qubit.

Teleportation of quantum states has been demonstrated by various researchers in recent years, with various physical systems, as noted earlier. NV nodes can generate entangled pairs and store the resulting qubits; entanglement between spatially separated quantum material systems has also been demonstrated utilizing entanglement swapping—recently reported experiments in this arena included a long-distance demonstration with two NV-based centers one mile apart and a demonstration with two single atoms separated by 20 miles with the utilization of optical fibers and telecom heralding (initially discussed in Chapter 4) [LAG23]. Also, make reference to the time progression called out in Figure 6.2.

[HER22] describes qubit teleportation *between non-neighboring network nodes*, which is a pragmatic requirement for more practical Quacom applications that have commercial value.

---

[1] An example of an experimental set-up for teleportation between two remote parties, Alice and Bob, is described by [LAG23], as follows directly here from the authors. At Alice there is a source of light-matter entanglement composed of an energy-time entangled photon pair source and a quantum memory, enabling the storage of qubits. The entanglement source is based on cavity-enhanced spontaneous parametric down conversion (cSPDC) designed to emit 1.8 MHz-wide photons. One photon of the pair is at 606 nm, and it is stored in a Pr^{3+}:Y2SiO5 crystal as a collective excitation of Pr^{3+} ions using the Atomic Frequency Comb (AFC) protocol. Stored photons are reemitted after a fixed time. The second photon of the pair is instead at telecom wavelength and is sent through an optical fiber towards Bob. At Bob there is a time-bin qubit source consisting of an amplitude and phase modulator that shapes light generated by an optical parametric oscillator as two weak coherent pulses $|e\rangle$ and $|l\rangle$; it is then possible to generate an arbitrary superposition state: $\alpha|e\rangle + e^{i\phi}\beta|l\rangle$ with $\alpha^2 + \beta^2 = 1$ and ϕ representing the phase relation between the two time bins. The second photon of the pair is instead at telecom wavelength and is sent through an optical fiber towards Bob. The Bell-state measurement/BSM for the teleportation protocol is implemented by sending the generated qubits to interfere with telecom idler photons from Alice at a beam splitter and by detecting the photons at the output modes with superconducting nanowire detectors. Provided that the modes at the beam splitter are indistinguishable, two consecutive detections at the same output of the beam splitter project the joint state of the telecom time bin qubit and the idler photon into the Bell-state $|\psi^+\rangle = (1/\sqrt{2})(|el\rangle + |le\rangle)$ that heralds the teleportation of the $e^{i\phi}\beta|e\rangle + \alpha|l\rangle$ qubit into the Pr^{3+} ions. On the other hand, two consecutive detections in the two time-bin modes in different beam splitter outputs project the joint state into the Bell state $|\psi^-\rangle = (1/\sqrt{2})(|el\rangle - |le\rangle)$, heralding instead the teleportation of the $e^{i\phi}|e\rangle - \alpha|l\rangle$ qubit. One sends the detection triggers back to Alice, where one uses an electronic logic control system in order to discriminate between these two scenarios.

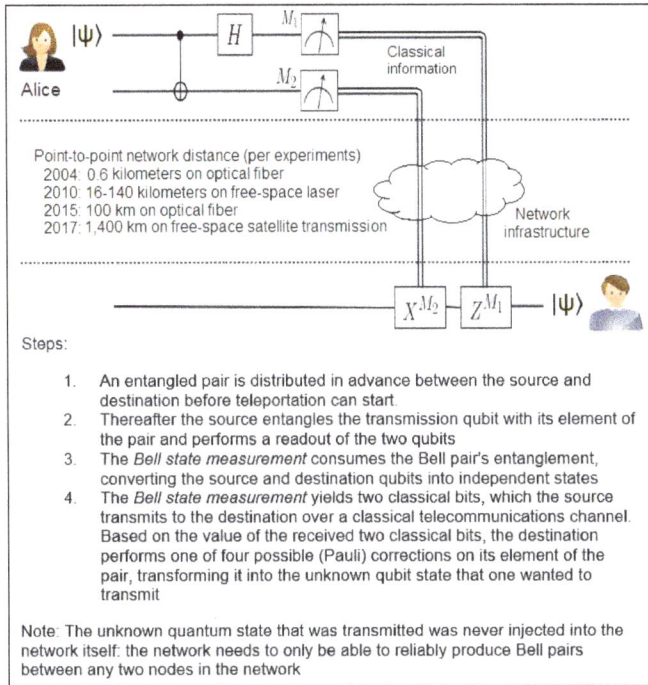

Figure 6.2 Mechanisms and Infrastructure for the Teleportation of a Qubit

The Quanet considered consists of three nodes in a line configuration—Alice, Bob, and Charlie. Each node contains a NV center in diamond. Using the NV electronic spin as the communication qubit, one can generate remote entanglement between each pair of neighboring nodes. In addition, Bob and Charlie each use a nearby ^{13}C nuclear spin as a memory qubit. The steps of the teleportation protocol are as follows. To prepare the teleporter, one can use an entanglement-swapping protocol mediated by Bob (similar to a quantum repeater protocol) to establish entanglement between Alice and Charlie. Once successful preparation of the teleporter is heralded, the input qubit state is prepared on Charlie and finally teleported to Alice. Thus, in summary, the steps are [HER22]:

(1) Prepare the teleporter by establishing entanglement between Alice and Charlie using an entanglement swapping protocol on Bob, followed by swapping the state at Charlie to the memory qubit.

(2) Prepare the qubit state to be teleported on the communication qubit on Charlie.

(3) Perform a BSM on Charlie's qubits. The outcome is communicated to Alice over a classical channel (a BSM is a projection onto a maximally entangled basis, this being the Bell basis). Dependent on this outcome, Alice applies a quantum gate to obtain the teleported qubit state.

A key parameter for quantum teleportation is the fidelity[1] of the pre-shared entangled state between Alice and Charlie. Given the generation of this state by entanglement swapping, its

---

[1] Using a trapped-ion platform consisting of up to 30 trapped ions, in late 2024 the Quantinuum H2 quantum processor demonstrated fault-tolerant teleportation of quantum states. Quantum entanglement

fidelity can be increased by mitigating errors on the individual links. The network in [HER22] generates entanglement between neighboring nodes using a single-photon protocol in an optical-phase-stabilized architecture. Refer to [HER22] for a lengthy discussion of how multi-hop end-to-end teleportation can be achieved in a reliable manner.

6.2.3 Entanglement Swapping

The effectiveness of direct management of entangled pairs across a Quacom link decreases with channel length; furthermore, due to the no-cloning theorem one is unable to amplify the signal. The pragmatic solution is *entanglement swapping*, which is a mechanism for transferring and distributing entanglement correlation between distant systems that have not interacted previously. Entanglement swapping is a mechanism for distributing the entanglement correlation between distant systems; it is a protocol by which quantum systems that have never interacted in the past can become entangled, this being a core mechanism for long-distance Quacom and Quanets [VED06], [ZAN23], [KOK00], [BRA10], [ADH10], [YUR92], [KHA14], [ROA14] (among others). Entanglement swapping is useful for entanglement purification, teleportation, communication, and quantum cryptography. Notice that Quareps are quantum nodes that create long-distance entangled pairs utilizing entanglement swapping.

Consider two entangled particles/qubits (A, B) that are shared between Alice and Bob; similarly, Christine and Daniel also share a different entangled pair of particles (C, D). Initially there is no entanglement between Alice's and Daniel's particles (A, D). But if Bob and Christine, who are located at the same location, make measurement in a selected basis on the pair (B, C) and classically communicate the outcome with distant partners, then Alice's and Daniel's particles, who are at a significant distance (by definition), thus become entangled, achieving entanglement swapping. See Figure 6.3, partially inspired by reference [ZAN23] (refer to the reference for a full explanation of the swapping process).

A measure of interest is *quantum concurrence*. Quantum concurrence C (ρ) is a manner of measuring entanglement in the mixed state of two qubits. C (ρ) = max (0, $\lambda_1, -\lambda_2, -\lambda_3, -\lambda_4$) where the λs are the non-negative eigenvalue of the (Hermitian) matrix R = $\sqrt{((\sqrt{\rho}) \rho^\sim (\sqrt{\rho}))}$ with $\rho^\sim = (\sigma_y \otimes \sigma_y) \rho^*(\sigma_y \otimes \sigma_y)$ ($\rho^\sim$ being the spin-flipped state of ρ and Pauli's σ_y—ρ^* is the complex conjugate of ρ). Concurrence provides an effective method to quantify entanglement. Interested readers may consult [WOO01], [WOO98], [ZHO15] for a thorough discussion of this topic.

---

and the teleportation of a quantum state across the processor are fundamental elements of Quacomp, and quantum state teleportation is commonly used in designs for large-scale QCs. The fragility of the quantum states, however, requires QEC to ensure their faithful processing. The implemented error-correcting color code effectively stabilizes the qubits, allowing quantum teleportation to be carried out in a fault-tolerant manner. The Quantinuum's H2 trapped-ion quantum processor was able to demonstrate fault-tolerant state teleportation circuits for the Steane QEC code. The circuits used up to 30 qubits at the physical level and employed real-time QEC. Over several variations of logical teleportation circuits using both transversal gates and lattice surgery, the researchers measured the logical process fidelity to be 0.975 ± 0.002 for the transversal teleportation implementation and 0.851 ± 0.009 for the lattice surgery teleportation implementation, as well as 0.989 ± 0.002 for an implementation of Knill-style QEC [RYA24], [GEN24].

Alice   Daniel

Alice - Bob entanglement

Daniel - Christine entanglement

Bob   Christine

Entangled qubits A and B comprise the first pair of entangled qubits; C and D comprise the second pair of entangled qubits

$$|\phi\rangle_{AB} = \sqrt{p_0}|00\rangle_{AB} + \sqrt{p_1}|11\rangle_{AB}$$

$$|\phi\rangle_{CD} = \sqrt{p_0'}|00\rangle_{CD} + \sqrt{p_1'}|11\rangle_{CD}$$

The initial state of our four-qubit system is

$$|\phi\rangle = |\phi\rangle_{AB} \otimes |\phi\rangle_{CD}$$

$$= \sqrt{p_0 p_0'}|00\rangle_{AD}|00\rangle_{BC} + \sqrt{p_0 p_1'}|01\rangle_{AD}|01\rangle_{BC} +$$
$$\sqrt{p_1 p_0'}|10\rangle_{AD}|10\rangle_{BC} + \sqrt{p_1 p_1'}|11\rangle_{AD}|11\rangle_{BC}$$

In support of measurements over BC qubits the following orthonormal basis is used

$$|\tilde{\Phi}_+\rangle_{BC} = \alpha_0|00\rangle_{BC} + \beta_0|11\rangle_{BC}$$

$$|\tilde{\Phi}_-\rangle_{BC} = \beta_0^*|00\rangle_{BC} - \alpha_0^*|11\rangle_{BC}$$

$$|\tilde{\Psi}_+\rangle_{BC} = \alpha_1|01\rangle_{BC} + \beta_1|10\rangle_{BC}$$

$$|\tilde{\Psi}_-\rangle_{BC} = \beta_1^*|01\rangle_{BC} - \alpha_1^*|10\rangle_{BC}$$

where α_i and β_i are unknown coefficients and for normalization $|\alpha_i|^2 + |\beta_i|^2 = 1$.

Notice that the state of qubits A and D is similar to the state of the basis BC. After measurements in the basis BC, Alice and Daniel's qubits A, D which are initially separable, become entangled in one of the four possible forms:

$$|\Phi\rangle = \sqrt{P_{\tilde{\Phi}_+}}|\tilde{\Phi}_+\rangle_{AD}|\tilde{\Phi}_+\rangle_{BC} + \sqrt{P_{\tilde{\Phi}_-}}|\tilde{\Phi}_-\rangle_{AD}|\tilde{\Phi}_-\rangle_{BC}$$
$$+ \sqrt{P_{\tilde{\Psi}_+}}|\tilde{\Psi}_+\rangle_{AD}|\tilde{\Psi}_+\rangle_{BC} + \sqrt{P_{\tilde{\Psi}_-}}|\tilde{\Psi}_-\rangle_{AD}|\tilde{\Psi}_-\rangle_{BC}$$

where $|\tilde{\Phi}_+\rangle_{AD} = (\sqrt{p_0 p_0'}\alpha_0^*|0\rangle_A|0\rangle_D + \sqrt{p_1 p_1'}\beta_0^*|1\rangle_A|1\rangle_D)/\sqrt{P_{\tilde{\Phi}_+}}$

with probability $P_{\tilde{\Phi}_+} = p_0 p_0'|\alpha_0|^2 + p_1 p_1'|\beta_0|^2$ (similarly for $|\tilde{\Phi}_-\rangle_{AD}, |\tilde{\Psi}_+\rangle_{AD}, |\tilde{\Psi}_-\rangle_{AD}$)

(One can also extend entanglement swapping between two pairs of qubits to three pairs of qubits)

Alice   Alice - Daniel entanglement   Daniel

Bob   Christine

$|\psi\rangle$ —— qubit —— M —— classical bit

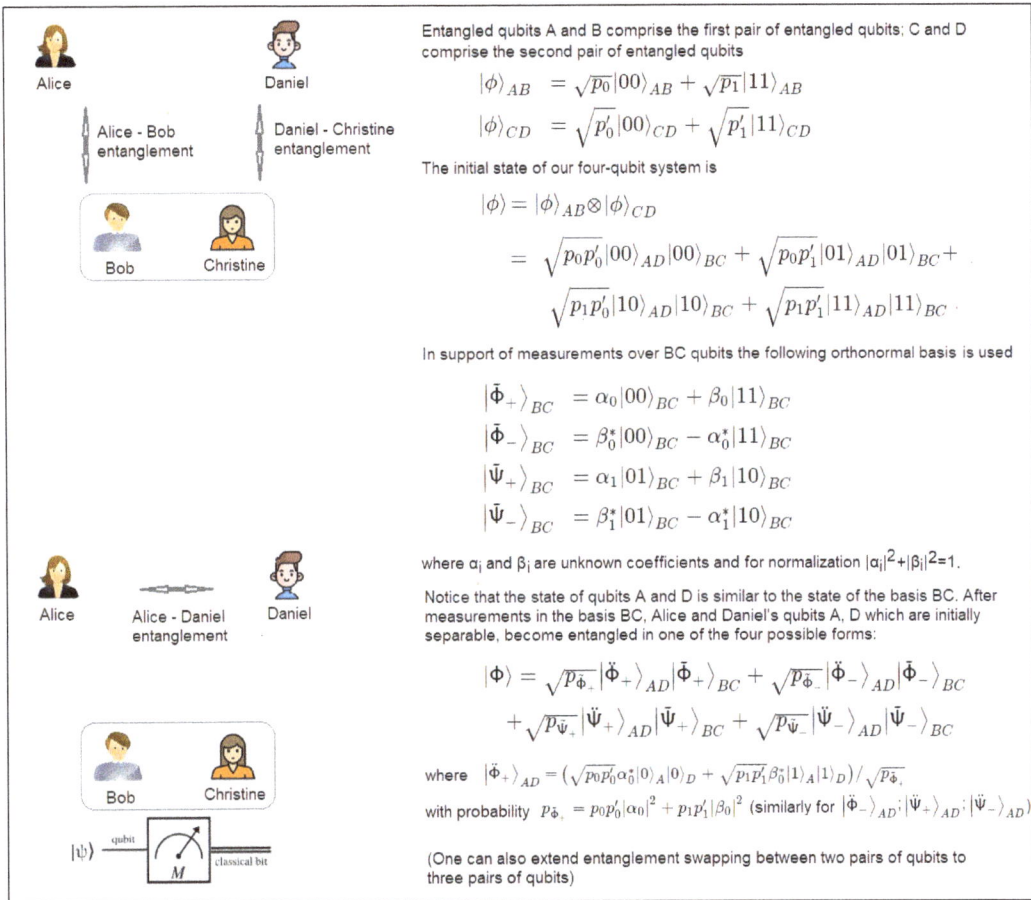

Figure 6.3   Entanglement swapping (partially based on reference [ZAN23]). Top: initial status; Bottom: post-measurement status.

6.2.4 Purification

As noted in previous chapters, fidelity is a number ranging from 0 to 1 that measures the closeness of the state to a sought target implementation; the higher the fidelity, the higher the quality of the quantum state (similarly, it measures the closeness of the quantum gate to a sought target implementation where, again, the higher the fidelity, the higher the quality of the quantum gate).

Quantum states are known to be susceptible to the effects of noise-induced decoherence. The generation of Bell pairs and/or the entanglement swapping process can typically introduce noise: the fidelity of the state degrades with each link and each swap. One can create higher-fidelity Bell pair states from two (or more) lower-fidelity pairs through quantum state purification. Quantum state purification (also known as distillation) is the process of enhancing knowledge about the state of a system, this increasing the fidelity in the density matrix that represents knowledge about the state of a system; it is the process of recovering a nearly pure copy of an unknown pure quantum state utilizing several noisy copies of the state. Quantum state purification has applications to Quacom over noisy channels and Quacomp with imperfect devices.

Quantum state purification is the process of recovering a nearly pure copy of an unknown pure quantum state using multiple noisy copies of the state. This process has obvious applications to Quacom over noisy channels and quantum computation with imperfect devices. It follows that a basic challenge in *Quantum Information Science and Technology* (QIST) is thus to identify and utilize practical mechanisms for protecting quantum systems from noise and/ or for removing the effects of noise that have already occurred in order to reverse (or at least partially reverse) the effect of decoherence and produce less noisy states out of mixed ones. The task of state purification comprises these processes. Given that noise renders quantum states less distinguishable, it is not possible to purify a single copy of a noisy state; however, given multiple copies of a noise-impacted state, one can endeavor to reconstruct a single copy that is closer to the original pure state. In general, one expects the quality of the reconstructed state to be higher if one could make use of several copies of the noise-impacted state [CHI23]. One issue of interest relates to the (minimum) number of samples needed to produce a high-fidelity copy of the original state; another question deals with establishing efficient procedures for undertaking such purification.

Typically, the input states are two imperfect Bell pairs where one has the goal of producing an output Bell pair of higher fidelity; see Figure 6.4. The output fidelity can be higher than the input fidelity in some circumstances. In the example of Figure 6.4, the final fidelity and states are respectively [ABE23], [VAN14]:

$$F_{\text{fin}} = P^2/(P^2 + (1-P)^2)$$

$$\rho_{\text{fin}} = F_{\text{fin}} \, |\phi^+\rangle\langle\phi^+| + (1 - F_{\text{fin}})\, |\psi^+\rangle\langle\psi^+|$$

$$= \frac{P^2}{P^2 + (1-P)^2} |\phi^+\rangle\langle\phi^+| + (1 - \frac{P^2}{P^2 + (1-P)^2}) |\psi^+\rangle\langle\psi^+|$$

with $1 - P$ the error/flip probability of one of the Bell pairs—when P is low, the flip probability is high; when P is high, the flip probability is low. (Here P is just a scalar, not an operator.)

- At the low value of $P = 0$, with high flip probability, $\rho_{\text{fin}} = |\psi^+\rangle\langle\psi^+|$ and $F_{\text{fin}} = 0$;
- At a high value of $P = 1$, with low flip probability, $\rho_{\text{fin}} = |\phi^+\rangle\langle\phi^+|$ and $F_{\text{fin}} = 1$.

(for $P > 0.5$, $F_{\text{fin}} > F$, the input/initial fidelity).

Compared to QEC methods (see below), purification makes relatively low demands on fidelity and resources, but protocols implemented over considerable distances have to deal with round-trip delays incurred in the classical communication portion of the process.

6.2.5 Quantum Error Correction

A basic way to distribute an entangled state over an appropriate network, a Quanet, is to transmit one of the qubits directly to the far end, possibly over a series of nodes, in conjunction with the application of QEC, whose goal is to manage path degradation (including but not limited to decoherence). QEC is the process used to protect quantum information against noise: QEC methods aim at identifying and addressing error-sensitive events that occur in the underlying quantum circuits in order to prepare and/or preserve logical quantum states. Also see Chapter 5, Section 5.6, starting on page 200.

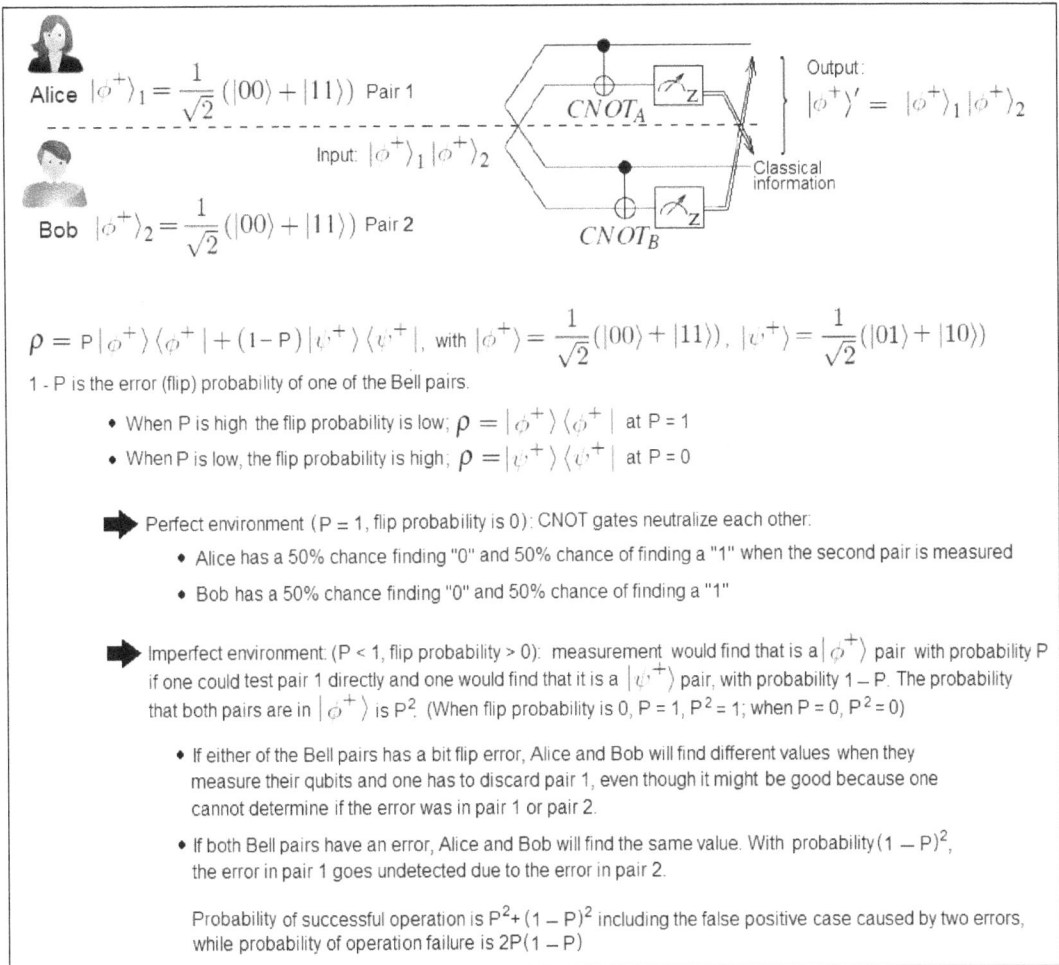

Figure 6.4 Basic purification (some observations in this figure based on [ABE23])

Classical error correction utilizes redundancy in conjunction with a syndrome measurement to establish which error impacted an encoded state; a simple (e.g., one- or few-bit) error can be reversed by applying a corrective operation implied by the syndrome. While copying quantum information is not possible due to the no-cloning theorem, one can disperse the information embodied in one qubit over a highly entangled state of a number of physical qubits. For example, Shor defined a QEC that entailed storing the information of *one qubit* onto a highly entangled state of *nine qubits*. More generally, error management is typically accomplished by encoding some number of logical qubits into a larger number of physical qubits (e.g., see, among others: [CAI24], [DEV13], [ERH21], [GOU21], [LI24], [RYA 22], [SHO95], [STE97], [WEI15]).

The QEC process entails encoding input information into entangled quantum states known as the *code space*; in addition, to establish whether the encoded information is corrupted or not, the step of *syndrome detection* is undertaken. Thus, QEC encodes logical qubits using several physical (raw) qubits to protect them from errors. Parity measurements between multiple qubits are then used to determine the presence and location of an error, and a recovery operation may be applied to correct the error.

QEC performs a multi-qubit syndrome measurement that obtains information about the error yet does not disturb the quantum information in the encoded state (that is, it does not obtain the information actually stored in the logical qubit). The syndrome measurement is able to determine the existence of error(s), the location of error(s), and the type of error(s) (errors such as bit flip, a sign/phase flip, or possibly both). To address the identified error, the Pauli operator corresponding to the type of error is applied to the corrupted qubit in order to counteract the impact of the error.

Thus, the basic QEC concept, as noted in [JOU23], is that the quantum information content of some logical qubits can be encoded in the Hilbert space of a larger number of physical qubits. When errors affect the state of these physical qubits, the original quantum information content of the logical qubits can be recovered with some probability. In the stabilizer formalism,[1] a quantum code space $|\psi\rangle$ is defined by the quantum states (existing in the space of physical qubits) stabilized by a set of operators $\{G_i\}$—that is, $G_i|\psi\rangle = (+1)|\psi\rangle$ for all i. When quantum information is stored in the code space and a correctable error occurs, measurements of some of the stabilizers will return the value -1 (the syndrome), indicating the type of error that has corrupted the information. The error can then be removed by applying a recovery operation to the quantum state of the physical qubits. One example (among many) of a QEC code is the Steane [[7,1,3]] code where one logical qubit is encoded into seven physical qubits; this encoding is robust against all single (physical) qubit errors.

The challenge is that QEC makes relatively high demands on both resources (physical qubits needed) and their initial fidelity. Hence, as noted earlier, QEC techniques are not expected to be utilized in the short term but in the near future; until such time, Quanets rely on *entanglement swapping* and *teleportation*. In fact, some researchers characterize Quanets as belonging to three classes (or categories) based on the error management schemes they employ for loss management (dealing with lost qubits[2]) and for error management (dealing with quantum state errors). These classes are defined by the use and directions of the classical signaling required in their distributed protocols—classical signaling transmit the classical bits; this transmission incurs round-trip delays, thus affecting the Quanet's performance. See Table 6.2 (synthetized from [KOZ23]).

6.3 Underlying Technologies

6.3.1 Quantum Optics

Quantum Optics is a subdiscipline of Quacom that has received significant R&D focus in the past few decades. Networks and/or links using Quantum Optics mechanisms are able to

---

[1] The stabilizer theory allows one to extend some classical binary error correcting codes for use in Quacomp/Quacom. Stabilizer codes are the quantum generalization of linear codes. The stabilizer formalism utilizes elements of the Pauli group (briefly described in Chapter 5); a stabilizer code is a class of quantum codes for performing quantum error correction to restore a noisy, decohered quantum state to a pure quantum state. To protect qubits, a stabilizer quantum error correction code appends ancilla qubits to qubits to be protected [GOT97], [PES23]. A unitary encoding circuit rotates the global state into a subspace of a larger Hilbert space, and this highly entangled, encoded state corrects for local noisy errors.

[2] Photon loss rate from transmission to measurement is generally expected to be less than 50% in order for the Quanet to operate within acceptable bounds.

Table 6.2 Quanet Classes and Signaling Methods Used[a]

	First class Quanet	Second class Quanet	Third class Quanet
Loss tolerance	Heralded entanglement generation (bidirectional classical signaling) Heralded entanglement generation confirms the receipt of an entangled qubit using a heralding signal. A pair of directly connected quantum nodes repeatedly attempt to generate an entangled pair until the heralding signal is received	Heralded entanglement generation (bidirectional classical signaling)	QEC (no classical signaling) QEC can be applied to complement lost qubits, eliminating the need for reattempts; and since the correction procedure is composed of local operations, it does not require a heralding signal
Error tolerance	Entanglement purification (bidirectional classical signaling) Entanglement purification is simple, but it incurs round-trip delays due to the requirement for bidirectional classical signaling	Entanglement purification (unidirectional classical signaling) or QEC (no classical signaling)	QEC (no classical signaling) -- QEC is able to correct state errors locally so that it does not need any classical signaling between the quantum nodes
Description	Implementable with a limited set of quantum gates	Heralding is still used to compensate for transmission losses. QEC used initially for entanglement purification only (only needs unidirectional classical signaling). Further along QEC codes used to create logical Bell pairs that (no classical signaling needed)	Directly transmit QEC-encoded qubits between adjacent nodes. Elementary link Bell pairs can be created without heralding or any other classical signaling. Qubits themselves are forwarded end to rather than relying on Bell pairs and entanglement swapping: networks support a "store and forward" architecture in addition to "store and swap" arrangement.

[a](*Source:* Content synthetized from [KOZ23])

transport, manipulate, and store qubits; key applications of Quantum Optics include QKD, the best well-known quantum application, and superdense coding (see Figure 6.5).[1] Quantum

---

[1] As alluded to in Chapter 1, in superdense coding the transmission of a semi-nibble (two bits) by using a single qubit is facilitated by the fact that Alice, at one end of a transmission link, is able choose among four quantum gate operations to apply on her portion of the entangled state – the selected operation depends the on the value of the semi-nibble Alice wants to transmit; Alice then transmits to Bob the qubit state evolved through the chosen gate: this qubit encodes information about the semi-nibble bits Alice utilized to select the operation; the state information is retrieved by Bob given the pre-shared entanglement between the pair, and upon receiving Alice's qubit, Bob obtains two classical bits of information, the semi-nibble, by operating on the entangled pair and measuring both—with A as a control qubit (previously received during the entanglement preparation stage) and B as target (received) qubit. Notice that an inimical third party

(Step 1): Preparation of an entangled state

$$|\Phi^+\rangle = \frac{1}{\sqrt{2}}(|0\rangle_A \otimes |0\rangle_B + |1\rangle_A \otimes |1\rangle_B) = \frac{1}{\sqrt{2}}(|0_A 0_B\rangle + |1_A 1_B\rangle)$$

(Step 2): The entangled state is shared between Alice and Bob.

Qubit noted by subscript A is sent to Alice and the qubit noted by subscript B is sent to Bob.

(Step 3): Encoding (transforming): by applying a quantum gate to her qubit locally, Alice can transform the entangled state $|\Phi^+\rangle$ into any of the four Bell states. Two classical bits are mapped (packed) into one qubit "B":

00: When Alice desires to send the classical two-bit string 00 to Bob, she applies the identity quantum gate $\mathbb{I} = \begin{bmatrix} 1 & 0 \\ 0 & 1 \end{bmatrix}$ to her qubit. The resultant entangled quantum state remains unchanged:
$$|B_{00}\rangle = \frac{1}{\sqrt{2}}(|0_A 0_B\rangle + |1_A 1_B\rangle)$$

01: When Alice desires to send the classical two-bit string 01 to Bob, she applies the quantum NOT (or bit-flip) gate $X = \begin{bmatrix} 0 & 1 \\ 1 & 0 \end{bmatrix}$ to her qubit. The resultant entangled quantum state becomes:
$$|B_{01}\rangle = \frac{1}{\sqrt{2}}(|1_A 0_B\rangle + |0_A 1_B\rangle)$$

10: When Alice desires to send the classical two-bit string 10 to Bob, she applies the quantum phase-flip gate $Z = \begin{bmatrix} 1 & 0 \\ 0 & -1 \end{bmatrix}$ to her qubit. The resultant entangled state becomes :
$$|B_{10}\rangle = \frac{1}{\sqrt{2}}(|0_A 0_B\rangle - |1_A 1_B\rangle)$$

11: When Alice desires to send the classical two-bit string 11 to Bob, she applies the quantum gate $iY = i\begin{bmatrix} 0 & -i \\ i & 0 \end{bmatrix} = Z * X$ to her qubit. The resultant entangled state becomes:
$$|B_{11}\rangle = \frac{1}{\sqrt{2}}(|0_A 1_B\rangle - |1_A 0_B\rangle)$$

(Step 4): Alice transmits her entangled qubit to Bob using a quantum network using some conventional physical medium (e.g., telecom fiberoptic network)

(Step 5): Decoding

Bob endeavors to establish which classical bits Alice sent by performing the CNOT unitary operation, with A as control qubit and B as target qubit. Next he perform the unitary $H \otimes I$ operation on the entangled qubit A (the Hadamard quantum gate H is only applied to B).

Resultant entangled state	Entangled state after unitary operations
B_{00}	$|00\rangle$
B_{01}	$|01\rangle$
B_{10}	$|10\rangle$
B_{11}	$|11\rangle$

(measurements projects the entangled state onto one of the four two-qubit basis vectors $|00\rangle, |01\rangle, |10\rangle$ or $|11\rangle$)

Example: if Bob derives the basis state $|01\rangle$ he then infers that Alice intended to send the semi-nibble 01

Preparation and sharing of a Bell Pair ("certificate entity")

Transmission

Alice: Encoded semi-nibble

Transmission (unlimited distance)

B

Transmission (unlimited distance)

A

Bob: decoding semi-nibble

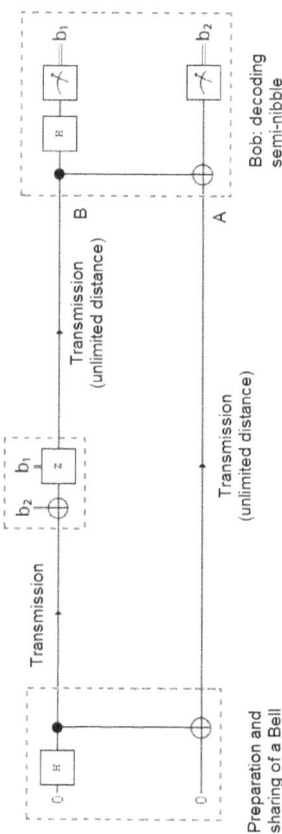

Figure 6.5 Superdense coding

optical communications is based on using photons as *flying qubits*, namely, transporting qubits from a local physical quantum emitter through the network infrastructure in order to convey quantum information—to a remote physical quantum receiver. Photons, as noted elsewhere, have weak interaction with the environment, thus reducing the risk of decoherence, and have the ability to be carried, with appropriate arrangements, over (some) existing telecom infrastructure.

Unlike classical optical communications where light intensity from a laser source is measured at the other end of an optical fiber transmission link (which may also possibly include various network elements along the way such as, but not limited to add-drop multiplexers), Quantum Optics is a technique to transport and control optical signals utilizing QM phenomena such as no-cloning, entanglement, and teleportation; these techniques are used, for example, by Quareps to extend Quacom distances. Quantum Optics deals with QM phenomena (e.g., entanglement and teleportation) related to the interactions between quanta of the EM field (i.e., photons of light) with matter (i.e., with atoms and molecules) in the context of nonlinear optics.

In the medium-to-long term, support of advanced quantum applications—such as quantum switching and routing—likely will require that correlated complex states be transmitted by utilizing the multimode (vectorial) nature of light. Quantum Optics' realms from a quantum states perspective include the following, as noted in [MAN20]:

- *Basic non-classical states of light*, such as single photons, squeezed states, twin optical beams and EPR states, characterized by just a few modes of the electromagnetic field.
- *Quantum states of light with multiple quantum degrees of freedom* (i.e., either spatial, temporal, frequency, or polarization modes), thus facilitating a more advanced utilization of the supporting technology. Here, each mode of the EM field can be considered as an individual quantum degree of freedom; depending on the transmission mode, the axis of oscillation in EM transmission may have different orientations to the direction of transmission. For example, in a Transverse Electric (TE) mode, the electric field is transverse to the direction of propagation while the magnetic field is normal to it; in a Transverse Magnetic (TM) mode, the magnetic field is transverse to the direction of propagation while the electric field is normal to it; in a Transverse Electric and Magnetic (TEM) mode, both the electric field and the magnetic field (always perpendicular to one another in free space) are transverse to a direction of travel. Different sets of modes allow one to consider the same quantum state from different perspectives: a given state can be entangled on one basis and factorized on another one. Thus, it is possible to utilize quantum fields not only in choosing the modes that participate, but also optimizing their spatial-temporal shapes. That is to say, the qubits can be encoded into a photon's degrees of freedom (e.g., the photon's polarization, which is the direction of the electric field oscillations); in addition, the Orbital Angular Momentum (OAM) of light may be considered a promising further degree of freedom for multiplexing data in free space and optical fibers and at the nanoscale. OAM represents the component of angular momentum of a light beam that is dependent on the field spatial distribution, and not on the polarization.

---

is unable to eavesdrop on information being communicated through superdense coding and any effort to measure either qubit will collapse the state of that qubit, thus alerting Bob and Alice. If the inimical eavesdropper Eve, intercepts Alice's qubit to Bob along the transmission path, she only has part of an entangled state: without access to Bob's qubit, Eve is unable to extract any usable information from Alice's qubit.

6.3.2 Heralded photons

As alluded to earlier, in Quacom there is a need for heralding schemes. Quanets require the existence in the network of quantum nodes that support multiple long-lived qubits and that can collect, store and process information communicated by photonic channels preferably using telecommunication-grade fiber links (or satellite-based links); in addition, the ability to herald successful photon arrival events is critical to the ability to deploy scalable implementations, as is the need to detect quantum-gate and communications errors. Heralded entanglement generation schemes are mechanisms that are used to announce successful entanglement generation. Entanglement between qubits in two remote quantum memories is facilitated when photons are generated in a heralded manner.

For commercially practical Quacom applications, there is (i) a need to deploy appropriately placed/spaced repeater nodes performing operations with high fidelity, and (ii) a need to create entangled states consisting of a large number of photons (preferably beyond the dozen or so realizable at press time). In the context of item (i), one way to create long-distance entanglement is to create short-distance entangled links and then chain them to form longer-distance entanglements using entanglement swapping; this enables one to iteratively to create entanglement along a series of linear links—note that to store the entanglement both nodes need to store one qubit per entangled link.

In the context of item (ii), producing heralded entanglement allows long-distance Quacom *without the need to create entanglement consisting of a large number of qubits*; here, the heralding signal provides a confirmation that an entanglement generation attempt was successful, as illustrated in Figure 6.6 [DAH19]. In the heralded generation of entanglement, a 'flag' signal is provided once an entangled pair is successfully prepared; heralding is the process of detecting one of the pair of qubits (photons) to herald the other qubit; one photon can be heralded by the detection of its twin.

Heralded entanglement includes path heralding, source heralding, and bilateral heralding, as noted in [CAC20]:

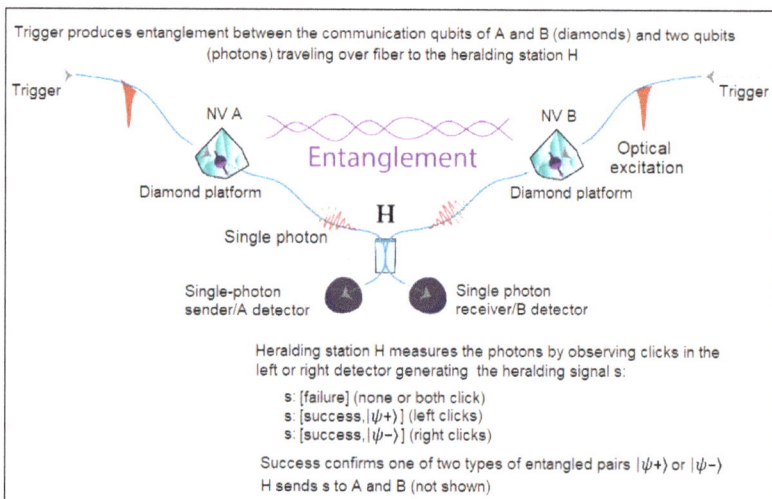

Figure 6.6   Heralded entanglement generation in a NV platform (illustration synthetized from [DAH19])

- *Path heralding*: An entangled photon source-pair located somewhere in-path between two link endnodes transmits an entangled photon (a flying qubit) over a quantum channel to each of the link endnodes, where transducers then map the entanglement from the flying qubits to matter qubits. Here the transducers are aware if the transfers were successful; thereupon, the transducers herald successful entanglement generation utilizing a message exchange over a classical channel.

- *Source heralding*: One of the two link endnodes transmits a flying qubit that is entangled with one of its matter qubits. A transducer at the far end of the link maps the entanglement from the flying qubit to one of its matter qubits. The transducer is aware if its transfer was successful and is able to herald successful entanglement generation utilizing a classical message that is transmitted to the other node.

- *Bilateral heralding*: Both link endnodes transmit a flying qubit that is entangled with one of their matter qubits. A pertinent in-path detector undertakes a joint measurement on the flying qubits; the measurement transforms (cascades) the remote matter qubits into an entangled quantum state. The detector is aware if the entanglement succeeded and is able to herald successful entanglement generation by transmitting a message to each node over the classical channel.

Until a decade ago (unheralded) entanglement between separated physical quantum objects had been achieved for various systems for *relatively local distances* [BER12]. But to support long-distance Quacom, the heralded generation of entanglement between distant physical quantum systems is a practical necessity. In recent years, heralded entanglement has been achieved terrestrially over regional distances (say, 100 km or less), and it has been demonstrated with single trapped ions, diamond crystals, and cold atomic ensembles; for the realization of heralded entanglement over long distances, single neutral atoms are promising candidates [HOF12]. Because of signal impairments and degradation in communication channels, the distance of entanglement distribution is limited to about 100 km on the ground. Quareps are utilized to address this limitation by using quantum memory and entanglement swapping.

The detection of an idler photon heralds the presence of a single signal photon. For example, in Figure 6.7, the detection of an idler photon at 1338 nm (using a low-noise superconducting Single Photon Detector [SPD]) heralds the presence of a signal photon at 883 nm [USM12]. Rare-Earth-Ion-Doped Crystals (REICs) can absorb one photon and then emit two photons that have opposite polarization states: the atoms re-emit photons that are indistinguishable from a photon they have absorbed, implying that they can act as repeaters. The specific state of each of these photons is undetermined until measured; but because the photons are entangled—that is, correlated—measuring the polarization of one photon instantly reveals the state

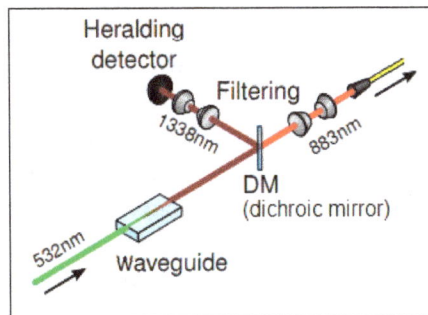

Figure 6.7   Heralded quantum entanglement (simplified from [USM12])

of the second, regardless of how far they are separated in space. Both the photon from the absorption-re-emission process and a different photon passing by without interaction may trigger a detector; thus, to ensure that only the right photons are counted in the Quarep, one needs to adequately prepare the initial signal so that each "proper" photon is partnered with a second photon of a different wavelength; the second photon is known as a "herald".

A Quarep may be built where light is directed from a laser onto a Periodically Poled Potassium Titanyl Phosphate (PPKTP) crystal [FRA12]. The crystal responds by emitting a photon of the exact wavelength needed to excite the neodymium ions, along with a second photon with a much higher energy. Given that this particular pair of photons is not emitted every time the crystal is excited, the higher-energy partner acts as a herald to the photon that is needed. Detecting the herald photon implies that the rest of the system can work as desired, allowing entanglement to occur. Because of the presence of herald photons, the problem of false negatives (due to non-entangled photons showing up in detectors) is mitigated, as noted in [FRA12]: (1) laser light is used to excite the PPKTP crystal; if the right reaction occurs, a herald photon shows up in the detector, while the photon needed for the rest of the experiment travels on; (2) that photon is split into two entangled photons, which are sent down separate paths until they reach the crystals doped with neodymium—the neodymium ions absorb the light and re-emit it; and (3) the new photons are then recombined at a beamsplitter. Depending on the polarization states of the entangled photons, they will either pass through or reflect, landing in one of two detectors. This confirms that entanglement between the neodymium-doped crystals actually occurred.

A number of protocols for generating remote entanglement make use of interference between photons produced by the different network nodes; this requires frequency conversion to match the photons from the different sources. For example, a two-photon interference platform dealing with photons generated from a rubidium atomic ensemble and a trapped barium ion has been demonstrated—this requires closely matching their center frequencies via difference frequency generation [CAD19].

Sources of photons that utilize various technologies can be compared using a number of parameters, as elaborated by NIST and highlighted in Table 6.3 [NIS24].

6.3.3 Quantum Channels in the Telecommunication Wavelength Range

As hinted at earlier in this chapter, typical Quacoms' engineering desiderata include, but are not limited to, the following: (i) entanglement transfer over quantum communication channels of some geographic scope while *enjoying compatibility with the classical telecommunication infrastructure*; and (ii) support of *multiplexed operations* to enable system scaling. Although various remote physical systems have been successfully entangled, so far none of these realizations capture all of the requirements for Quacom, including direct compatibility with telecommunication fiberoptic systems operating at well-known telecom wavelengths, in order to utilize existing optical transmission technologies. The efficient interface between a non-classical state of light and various material systems (including fiberoptic waveguides) is an important requirement for Quacom.

When qubit processing nodes, say QCs, are separated by metropolitan or regional distances, Quareps can be utilized to address the issue of rapid transmission loss of single photons and allow one to extend the distance of entanglement distribution utilizing quantum channels that transport curated qubits in "telecom-range" photons. Quareps support the establishment of entanglement between these distant nodes; and as noted earlier, Quanets do not utilize

Table 6.3 Photon Source Characterizations (Source: NIST [NIS24])

Characterization	Description
Technical parameters	Metrics related to wavelength, bandwidth, temporal, and spatial characteristics.
Single-photon emission probability	The probability of exactly one photon in a well-defined mode at a given time. It is measured using our well-characterized, high-efficiency single-photon detectors.
Spectral/Spatial mode structure	The mode structure of the single photon output determines the possible use of the source; in the context of networking, a single spatial mode (i.e., confinement to a single mode fiber) is required.
Photon-number statistics	Sources are characterized using their photon statistics and purity of photon number (this can be done using transition edge sensors and multiplexed click detectors); for various applications the output of a single photon source should not emit two-photon or multiphoton states, or as few as possible.
Indistinguishability	Some network protocols require single photon interference; the degree of such interference determines how indistinguishable the states are. When two photon states are fully indistinguishable, the interference leads to a full coalescence of single photon states to a state of two photons, whereas partially indistinguishable states do not fully coalesce.
Detector's first-order effects (detection efficiency, dark counts)	Detection efficiency is measured using the detector substitution and correlated method. Dark counts are measured in the absence of background radiation, including blackbody radiation. Both detection efficiency and dark counts are defined for a detector that is fully armed.
Detector's second-order effects (afterpulsing/recovery characterization)	The transient process that occurs after a photon detection includes a recovery (dead) time and afterpulsing.
Detector's latency and timing jitter	A measurable classical signal appears with a delay with respect to the absorption of a photon at the detector.

traditional signal amplifiers because detecting and then reinjecting qubits disrupts their entanglement, frustrating the process of transmission.

To facilitate deployment and keep network costs low, the quantum channels should preferably be based on existing telecommunication optical facilities operating at well-known telecom wavelength ranges. Some telecom-range quantum memories are based on erbium doped solids, but the *efficiency and storage time are very limited*. Most of the platforms that can be used as quantum nodes operate with photons at visible to near-infrared wavelength, which is away from the telecom fiberoptic window of operation [HUA19]. Quantum Frequency Conversion (QFC) and the use of nondegenerate photon-pair source can be employed to realize a nonclassical correlation between a stationary quantum node and a flying telecom photon. QFC has been demonstrated in various platforms including atomic gas, NV defect in diamond, trap ions, and rare-earth-ion-doped crystals (for example, $Nd^{3+}:Y_2SiO_5$ or $Ti:Tm:LiNbO_3$, and $Pr^{3+}:Y_2SiO_5$; of these, the latter exhibits better storage efficiency and storage time).

Furthermore, as regional-scale Quanets are envisioned, the need arises, for practical and economic reasons, to be able to *multiplex* quantum optical channels with classical channels in a single fiber. Given that quantum links may require exchange of classical data in addition to exchanging qubits (as seen for example in Figure 6.2), methodologies and technologies for co-delivery and coexistence of quantum and classical signals is desirable. One challenge in that context is the crosstalk between channels, where classical channels may degrade quantum channels with background noise; classical channels entail the transmission of thousands of photons to transmit a bit, while quantum channels use just one photon to transmit a qubit. Classical/quantum network coexistence can be achieved with classical communications with ultra-faint light [POL19], [BUR22]; these solutions utilize very weak laser pulses with faint optical states entailing just a handful (less than a dozen) photons per bit at the receiver, for the classical channel.

6.3.4 Detection-Related Matters

Determining if and when a photon has arrived at a receiver without destroying it is a relatively difficult task; this determination is obviously needed for Quacom, including quantum communication protocols such as device-independent secure communication. Determining if and when a photon is received is challenging in the presence of even relatively low channel losses, but this determination can be achieved with techniques that bypass the effects of loss by certifying that the quantum system is ready after transmission; this certification is a practical alternative to the arduous task of greatly mitigating or eliminating intrinsic transmission loss. Approaches for determining if and when a photon has arrived at a receiver include the following, as noted and elaborated in [MEY16]:

- Traditional destructive photon detection such as *entanglement swapping*. Entanglement swapping certifies that stationary quantum systems (e.g., atoms or crystal defect centers, such as Nitrogen-Vacancy centers) are ready in the desired quantum states. However, the synchronization and indistinguishability of photons from disparate sources required for entanglement swapping is difficult to achieve.
- *Heralded qubit amplification* is a nondestructive optical method of certifying a photon's arrival, which uses ancilla photons interfering with the signal photon and specific detection patterns to herald the signal's arrival. Heralded qubit amplification, rather than detecting photons, increases the relative one-photon component $|1\rangle$ over the vacuum component $|0\rangle$ in a heralded manner. Given that deterministic noiseless amplification is not possible in QM, these amplifiers operate non-deterministically, emitting a ready signal when amplification was successful. The idea of heralded qubit amplification is to teleport the input state in the vacuum/one-photon basis using linear optics, with an ancilla photon detection heralding successful teleportation to the output. The teleportation is biased with a tuneable beam splitter towards the one-photon state, amplifying this over the vacuum. Two of these amplifiers operated coherently in parallel, one for each polarization, allow qubits to be amplified.
- In contrast to the usage of linear optics for heralded qubit amplification, *photonic qubit precertification* employs nonlinear optics to certify the presence of a photonic qubit. Qubit precertification is similar in effect to heralded qubit amplification, in the sense that it removes the vacuum component of a photonic qubit state at the cost of small success probability. In precertification, the sender, Alice, transmits a photonic qubit encoded in polarization to the receiver, Bob, over a lossy channel. Bob splits the incoming photon into two photons, labeled *flag* and *signal,* through polarization-preserving single-photon parametric down-conversion. Since the flag photon is never produced without the corresponding signal photon, the detection of the flag pre-certifies the presence of the signal. The configuration is such that the signal photon bears the same quantum information initially encoded by Alice. No timing synchronization is required between Alice and Bob, but the wavelength of Alice's photon

must match the acceptance band of Bob's down-conversion crystal. This approach offers greater utility than entanglement swapping and qubit amplification and can be employed in a plethora of Quacom implementations.

6.3.5 Device Independence

Device independence in a Quanet ensemble is a design desideratum. This is because by violating a Bell inequality, the legitimate parties in device-independent communication can exclude inimical eavesdroppers, even without detailed knowledge or characterization of their devices. The challenge in implementing device-independent Quacom is that it requires a Bell inequality violation without loopholes[1]; this has only recently become possible experimentally. The establishment of non-local correlations, achieved through the violation of a Bell inequality, is the basis for device-independent QIST applications. Several non-locality tests have been conducted in the past decade, but many of them suffered from either locality or detection loopholes[2]; among the proposals for overcoming these problems are (i) the use of atom-photon entanglement, and (ii) hybrid photonic measurements leading to Bell inequality violations with manageable transmission and detection efficiencies [TEO13].

Device-independent communication is limited by the challenges of the loophole-free Bell test—namely, closing the locality and detector loopholes. In particular, the detector loophole is difficult to address because every source of loss between the production of photon pairs and final detection accumulates against the efficiency of the complete link (for example, note that commercial optical fiber has attenuation of around 0.16 dB/km; this allows only about six miles of fiber between the source and receiver before falling below the Eberhard bound of 66.7% efficiency, even before including the component losses and detector efficiency) [MEY16]. To achieve full device-independent Quacom and perform loophole-free Bell tests over distances of more than a few miles, one needs to develop strategies for dealing with photon loss due to phenomena such as scattering, absorption, and diffraction.

6.4 Constituent Elements

Obviously, a number of elements are required to build out a Quanet, and there are distinct physical hardware platforms for implementing Quanets' nodal hardware. These platforms differ in how they operate on qubits, store qubits, and generate entanglement across a channel. For

---

[1] Tests of Bell's inequalities: tests of local realism based on Bell's inequality; "loophole-free Bell test" means that the Bell test is performed under only two assumptions: that local realism is true, where locality is justified by space and time measurements; and that superdeterminism does not hold. Violating a Bell inequality in a loophole-free scenario rejects one or both of those assumptions [MEY16].

[2] Non-local correlations can be obtained by measuring entangled quantum systems in appropriately chosen local observables. This a Bell test: the non-local nature of the measurement outcomes can be certified by the violation of certain constraints, these being the Bell inequalities briefly discussed at the end of Chapter 3. Several Bell tests have been conducted in the last few decades, but no non-local correlations have been routinely established, because all of the performed experiments suffered either from the locality loophole or from the detection loophole (experiments using entangled photons have reported Bell inequality violations closing the locality and detection loopholes separately); the main technological challenge to closing both loopholes simultaneously is having both efficient detection and long-distance entanglement [TEO13].

example, hardware based on trapped ion platforms or NV-based platforms enjoy good operational control over the qubits but have a challenge in generating entanglement at high rates. By contrast, hardware based on atomic ensembles and optical elements is efficient at generating entanglement at workably high rates, but once the entanglement is generated, such hardware has limited processing capabilities.

6.4.1 Quantum Memory

Quantum memories, needed for Quareps and other Quacomp applications, were already discussed in Chapter 4 (starting on page 153); additional information is provided herewith. Storing a qubit in memory is a technically demanding task given that it is relatively difficult to isolate a quantum system from the environment while mitigating noise—noise negatively impacts the fidelity of the state. The memory lifetime depends on the specific physical system utilized and the hardware implementation; storage time ranging from seconds to a few minutes have been achieved in the recent past.[1]

There are limitations and disadvantages of existing memory technologies (e.g., see Table 6.4, summarized from [BUS23], on next page). For example, high-efficiency memory operation is a prerequisite for many processes, but some protocols are limited by QM theory to low efficiency; in addition, optically controlled memories utilize bright coherent pulses of light, which can introduce spurious noise photons through unwanted processes (such as fluorescence and four-wave mixing); these processes degrade the quality of the memory operation. Hence, there is a need for efficient quantum memories able to store a quantum state of light, such as a photon, for a period of time in a fiber-integrated optical cavity, and then, at a later time, recall the quantum state of light and quantum information with a high probability of success.

[BUS23] offers one such system (also see Figure 6.8).

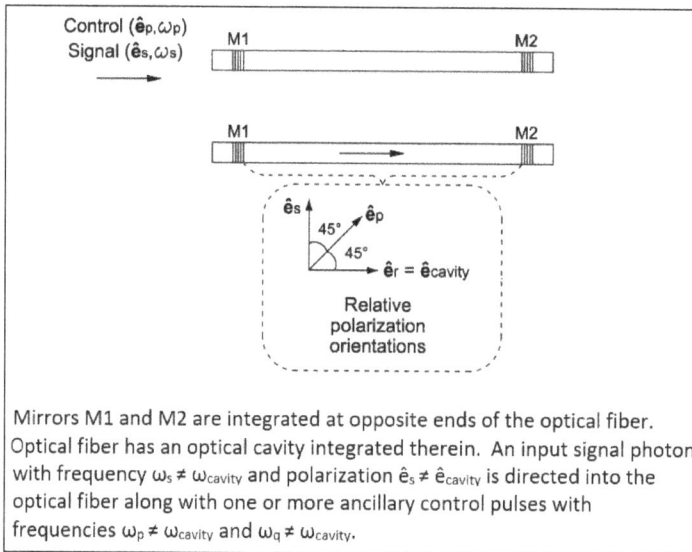

Control ($\hat{e}_p, \omega_p$)
Signal ($\hat{e}_s, \omega_s$)

M1   M2

M1   M2

$\hat{e}_s$
45° $\hat{e}_p$
45°
$\hat{e}_r = \hat{e}_{cavity}$

Relative polarization orientations

Mirrors M1 and M2 are integrated at opposite ends of the optical fiber. Optical fiber has an optical cavity integrated therein. An input signal photon with frequency $\omega_s \neq \omega_{cavity}$ and polarization $\hat{e}_s \neq \hat{e}_{cavity}$ is directed into the optical fiber along with one or more ancillary control pulses with frequencies $\omega_p \neq \omega_{cavity}$ and $\omega_q \neq \omega_{cavity}$.

Figure 6.8   Example of quantum memory (according to [BUS23])

---

[1] For example, a diamond-based quantum memory that holds 10 qubits and stores information for 75 seconds; both of the parameters being records for such a device has been demonstrated in the recent past [NAT19].

Table 6.4 Partial Survey of Existing Memory Technologies[a]

Technique/ technology	Description	Short set of basic references
Single emitter approach	A single entity such as an atom or quantum dot is driven to an excited state and then subsequently emits a single photon when it relaxes.	[NIS23]
Photon pair source	For example, spontaneous parametric down-conversion or spontaneous four-wave mixing. These sources, however, have drawbacks because they only operate in a stochastic (probabilistic) mode.	[COH09] [COU18]
Multiplexing photon pair sources	Combining multiple independent probabilistic photon pair sources together such that their overall success probability as a single photon source is higher.	[EIS11] [NUN13]
Four-wave mixing (FWM)	Creating light within a fiber Bragg grating defect which constitutes a cavity between two fiber Bragg gratings, one at each end; this provides a way of trapping light in a fiber Bragg grating defect, but the light stored in the defect is not the light that is input to the fiber. This technique is used where an input "idler" is amplified by a pump pulse; the amplification of the idler also creates a weak signal pulse that is trapped in the defect; however, the signal pulse created in the defect will be degraded by noise, and thus it is not possible to use the disclosed method as a quantum memory.	[MAR02]
Trapping a quantum state in an optical fiber	Trapping a quantum state of light in an optical fiber consisting of a fiber that is optically coupled to two narrow-linewidth cavities; the narrow-linewidth cavities reflect or transmit a given frequency depending on the optical distance between the reflective elements that are comprised by each narrow-linewidth cavity. Entails the use of mechanical switching of narrow-linewidth cavities, placing restrictions on the speed with which the device can operate.	[LEU06]
Storage ring resonator	Photon source to deliver single photons that includes a storage ring resonator to receive pump photons and generate a signal photon and an idler photon. An idler resonator is coupled to the storage resonator to couple the idler photon out of the storage resonator and into a detector. The photon source can be fabricated into a Photonic Integrated Circuit (PIC). However, integration with fiber optic networks is a challenge because of imperfect mode-matching between the PIC and optical fiber.	[MIH19]
Storage excitations	These include: (i) "spin-waves" on the energy levels of atomic ensembles and rare-earth-ion-doped crystals; (ii) acoustic phonons in an optical fiber; (iii) optical phonons in a diamond crystal; (iv) optical phonons in an ensemble of hydrogen molecules; (v) microwaves in a cavity; (vi) an electronic excitation in a single atom in an optical cavity; and (vii) buffering the signal photon in a free-space cavity or an optical fiber cavity. In such systems, protocols for mapping the signal photon into the memory include: (a) optically controlled memories; (b) electromagnetically induced transparency; (c) Raman scattering protocol; (d) Autler-Townes protocol; (e) engineered absorption; (f) controlled reversible inhomogeneous broadening; (g) atomic frequency combs; (h) gradient echo memory; and (i) retrieval of silenced echo (hybrids of these protocols also exist).	[BUS23]

[a] (Summarized from [BUS23])

In this system a nonlinear optical switching mechanism can be used to modify at least one property of the quantum light, or cavity, to trap the quantum light in the optical cavity. Subsequent application of the nonlinear optical switching mechanism switches at least one property of the stored quantum light, or cavity, to release the quantum light from the optical cavity, providing a photon source integrated within the quantum memory. This fiber optic quantum memory for storing a quantum state of light is composed of: (i) an optical fiber having an optical cavity integrated into the optical fiber; (ii) reflective elements integrated within the optical fiber at opposing ends of the optical fiber, the respective elements forming a part of the optical fiber; and (iii) a nonlinear optical switching mechanism for selectively storing quantum light from an external source within the cavity.

The first application of the switching mechanism modifies at least one property of the quantum light such that the reflective elements trap the quantum light within the cavity; a subsequent application of the switching mechanism modifies at least one property of the quantum light, such that the quantum light can be extracted from the cavity; quantum information encoded in the quantum light is preserved during storage in the cavity. Most sources of light emit a large number of photons in an uncontrolled or stochastic way; the system in [BUS23] seeks to address the creation of true single photons (a "Fock state" with photon number equal to one) "on demand". In some realizations, it is possible to incorporate a photon pair source within the quantum memory such that the memory can be used both as a quantum memory and as a single photon source that operates by temporal multiplexing. The proposed system also proposes to use switching of the photon properties as a switching mechanism to control its storage and release. Refer to [BUS23] for a detailed description.

Among other *physical systems,* quantum memories can be implemented as solid-state quantum emitters, including the nitrogen and silicon vacancy centers in diamond, vacancies in silicon carbide, vacancies in silicon, and quantum dots [BER23]. Rare-earth doped crystals have also been utilized for quantum memories; these crystals provide a system with a large number of atoms trapped in a solid-state matrix and with desirable coherence properties. Rare-earth based systems have demonstrated high-efficiency, large multiplexing capability, and long-lived quantum storage; however, direct compatibility with the telecom facilities has been a work-in-progress challenge. To that end, there have been reported demonstrations of heralded entanglement between two spatially separated quantum nodes, wherein the entanglement is stored in multimode solid-state quantum memories; at each node a doped crystal stores a photon of a correlated pair, with the second photon at telecom wavelengths—entanglement between quantum memories located at different locations (laboratories) is heralded by the detection of a telecom photon at a rate up to 1.4 kilohertz, and the entanglement is stored in the crystals for a pre-determined storage time up to 25 microseconds, as described in [LAG21]. A high entanglement rate with heralded operation, compatibility with the telecom facilities, and long storage times are desired; however, the ability to achieve all three simultaneously has not yet become an off-the-shelf type of implementation.

6.4.2 Quantum Repeaters

As discussed earlier in this text, Quareps are critical Network Elements (NEs) in a Quanet because Quacom is impaired by signal loss (attenuation, absorption) and decoherence. Quantum information is typically stored and manipulated in stationary, matter-implemented qubits, but because quantum information is more fragile than classical information owing to interactions with the environment, the network performance of quantum protocols degrades, often rapidly, with nodal distance (channel length). In particular, such degradation affects any point-to-point quantum communication links; thus, as might be expected, there are maximum rates that are achievable by repeaterless Quacom—for example, there are maximum achievable rates for transmitting quantum information by distributing entanglement [PIR17]; the remedy, clearly, is to utilize Quareps. Quareps are used to extend the range of fiber-based entanglement distribution beyond a few hundred miles; the technology is under development.

Quantum teleportation and entanglement swapping are utilized by Quareps to extend the usable length of Quacom links. Quareps typically (i) generate entanglement between neighboring nodes; (ii) extend entanglement to long-range pairs employing entanglement swapping; and (iii) undertake purification (distillation) to maintain the fidelity of the produced pairs.

As is the case in classical communication, repeaters are needed at various distances along the way—for example, signal attenuation in telecommunication fiber is around 0.1 to 0.2 db/km, necessitating a repeater every 15–20 miles (some experimental fibers have slightly better performance). Recall that qubits cannot be copied, thus Quareps must work in a quantum-specific manner; in particular, they must support entanglement preservation/distribution—the fundamental goal is to create entangled quantum states between nodes that are at a distance (see Figures 6.9 and 6.10). A Quarep's architecture may use heralding techniques to distribute entanglement between remote quantum nodes. There is an expectation that at the current pace of advance, by the end of the decade one should be able to establish entanglement using entanglement swapping,

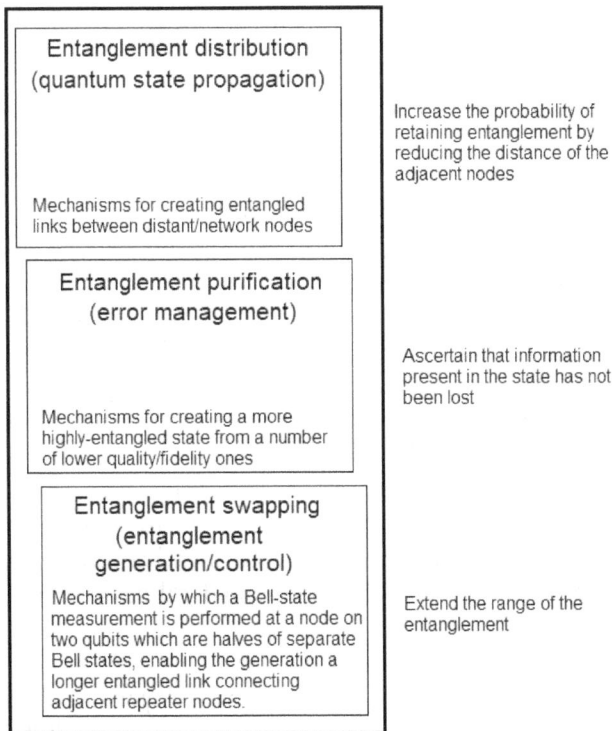

Entanglement distribution (quantum state propagation)

Mechanisms for creating entangled links between distant/network nodes

Increase the probability of retaining entanglement by reducing the distance of the adjacent nodes

Entanglement purification (error management)

Mechanisms for creating a more highly-entangled state from a number of lower quality/fidelity ones

Ascertain that information present in the state has not been lost

Entanglement swapping (entanglement generation/control)

Mechanisms by which a Bell-state measurement is performed at a node on two qubits which are halves of separate Bell states, enabling the generation a longer entangled link connecting adjacent repeater nodes.

Extend the range of the entanglement

Figure 6.9 Typical, basic Quarep functions

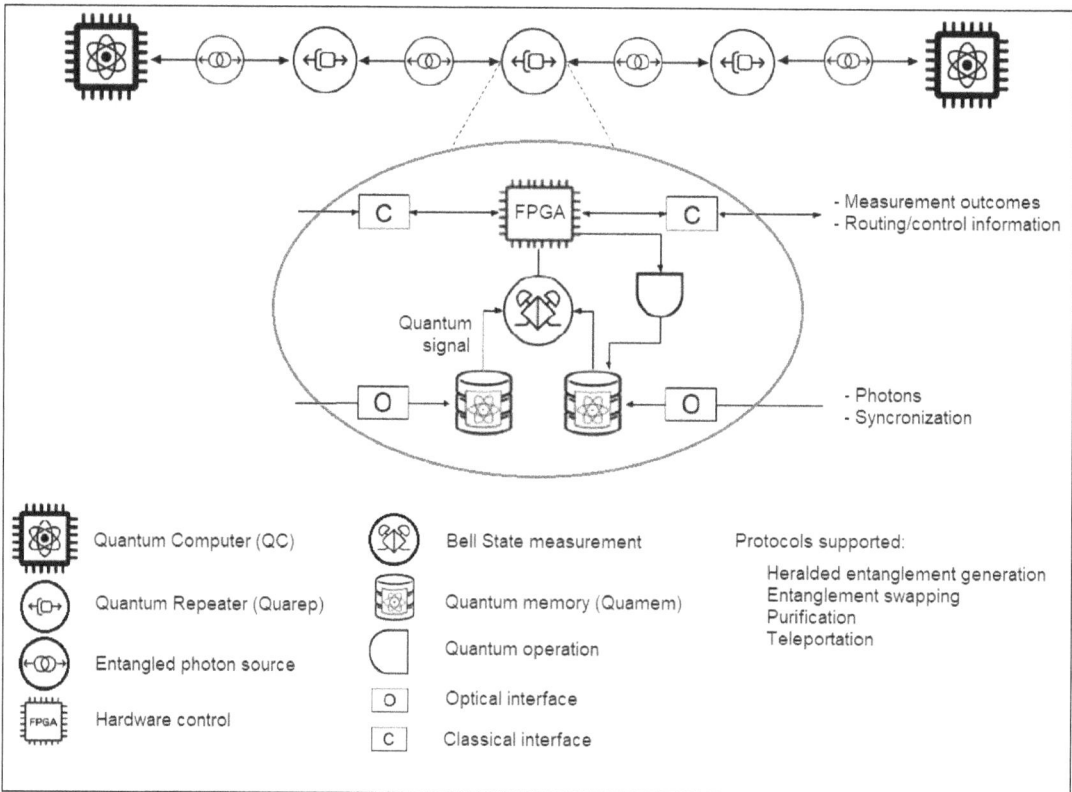

Figure 6.10   Quarep ecosystem (*Source:* [CUB24])

over 1,000 kilometers of optical fibers using perhaps ten Quareps (initially supporting perhaps one entanglement per second, making such a link a low datarate, but the generation rate will improve over time) [CAS24]. A related NE is a *Quantum Router (Quarou),* which is a Quarep that also has a management plane to support the management of the network.

Sustained progress in Quarep architectures has been experienced in the past decade both from a research perspective as well as from an engineering perspective; in fact, there have been three generations of Quareps, with an evolving fourth generation, as depicted in Table 6.5 [AWS21].

Early Quareps architectures utilized entanglement swapping to extend entanglement and purification to detect errors, as noted in [ABE23]:

- In these early Quareps, entanglement swapping is used for teleportation to adjoin two Bell pairs spanning adjacent short distances into one pair over the requisite longer distance (entanglement swapping is independent of the distances between endpoint A and midpoint B, and between midpoint B and endpoint C; only local quantum operations are required, facilitated by classical communication).

- In these early Quareps purification is employed to address the errors introduced. Local quantum operations are performed at both nodes on two Bell pairs, then one of the Bell pairs is measured. The measurement results are exchanged and compared: if they agree, the pair's fidelity has improved and the pair is kept for reuse; if the measurement results disagree, the pair is discarded.

Table 6.5 Generations of Quareps (Synthetized from [AWS21])

Quarep generation	Description	Exemplary reference
1G Quareps	Quareps that utilize heralded entanglement generation and heralded entanglement purification. These Quareps can tolerate a higher rate of errors. These Quareps require two-way classical signaling over the entire string of repeaters along the path; in turn this requires that the quantum memory lifetimes and coherence times must be longer than the round-trip communication delay.	[BRI98], [DUA01]
2G Quareps	Quareps that utilize quantum encoding and classical error correction to substitute the entanglement purification with classical error correction, handling all operational errors. This approach is more demanding in physical resources but requires classical signaling only two-way between neighboring repeaters.	[JIA09], [MUN10]
3G Quareps	Quareps that utilize quantum encoding to deterministically correct photon losses as well as operation errors. By eliminating two-way classical signaling, these Quareps can support high entanglement distribution rates (limited only by the speed of local operations, such as photon source rates, detector saturation rates, and so on).	[MUN12], [MUR14]
Near-term 4G Quareps	Integrated Quarep systems that utilize memory-enhanced quantum communication surpassing a repeaterless bound—for example, utilizing solid-state spin memories associated with Silicon-Vacancy (SiV) color center integrated in a diamond nanophotonic resonator (a series of repeaterless bounds impose fundamental limits the direct quantum communication protocols[a]).	[BHA20]

[a] Repeaterless bounds impose fundamental limits on the direct quantum communication protocols. The interested reader should consult [PIR17] for an extensive treatment of the topic, which notes that there are upper bounds based on the Relative Entropy of Entanglement (REE). [PIR17] notes that there are maximum achievable rates for transmitting quantum information (two-way quantum capacity), distributing entanglement (two-way entanglement distribution capacity), and generating secret keys (secret key capacity) through the most fundamental quantum channels. The two-way assisted capacities are benchmarks for quantum repeaters because they are derived by removing any restriction from the point-to-point protocols between the remote parties, who may perform the most general strategies allowed by quantum mechanics in the absence of pre-shared entanglement. These upper bounds on maximum achievable capacities have been derived for many fundamental channels—namely, bosonic lossy channels, quantum-limited amplifiers, dephasing and erasure channels in arbitrary dimension (including the fundamental rate-loss trade-off affecting any protocol of quantum key distribution).

As the intrinsic element of a Quarep, the heralded distribution of two-party entanglement between two distant nodes has only been realized with built-in-type quantum memories; this approach, however, suffers from the trade-off between multiplexing capacity and deterministic properties; this, in turn, hinders the development of high-efficiency Quareps. On the other hand, Quareps based on absorptive quantum memories can overcome such limitations, given that they separate the quantum memories and the quantum light sources; moreover, wideband absorptive quantum memories used in the Quarep nodes are compatible with deterministic entanglement sources and can simultaneously support multiplexing, enabling the (possible near-term) development of practical solid-state Quareps and high-speed Quanets [LIU21].

As illustrative example of a Quarep, reference [BER23] describes a Spectrally Multiplexed Quantum Repeater (SMuQR) based on spatially arrayed nodes of frequency-multiplexed multi-qubit registers; SMuQR uses the natural inhomogeneous distribution of optical transition frequencies in solid state defect centers. This distribution enables spectrally selective, individual addressing of large numbers of defect centers—artificial atom qubits—within an optical diffraction-limited spot along a long cavity or waveguide. The spectral selection relies on frequency shifting an incident optical field at a rate as fast as once-per-defect-center lifetime. The defect centers are resonant at visible frequencies and emit visible single photons, which are then down-converted to a wavelength compatible with long-distance transmission via conventional optical fiber. The down-converted photons are all at the same telecommunications wavelength, with the different spectral bins mapped to different temporal bins to preserve the multiplexing in the time domain, for distribution to other nodes in the quantum network. See Figure 6.11. By

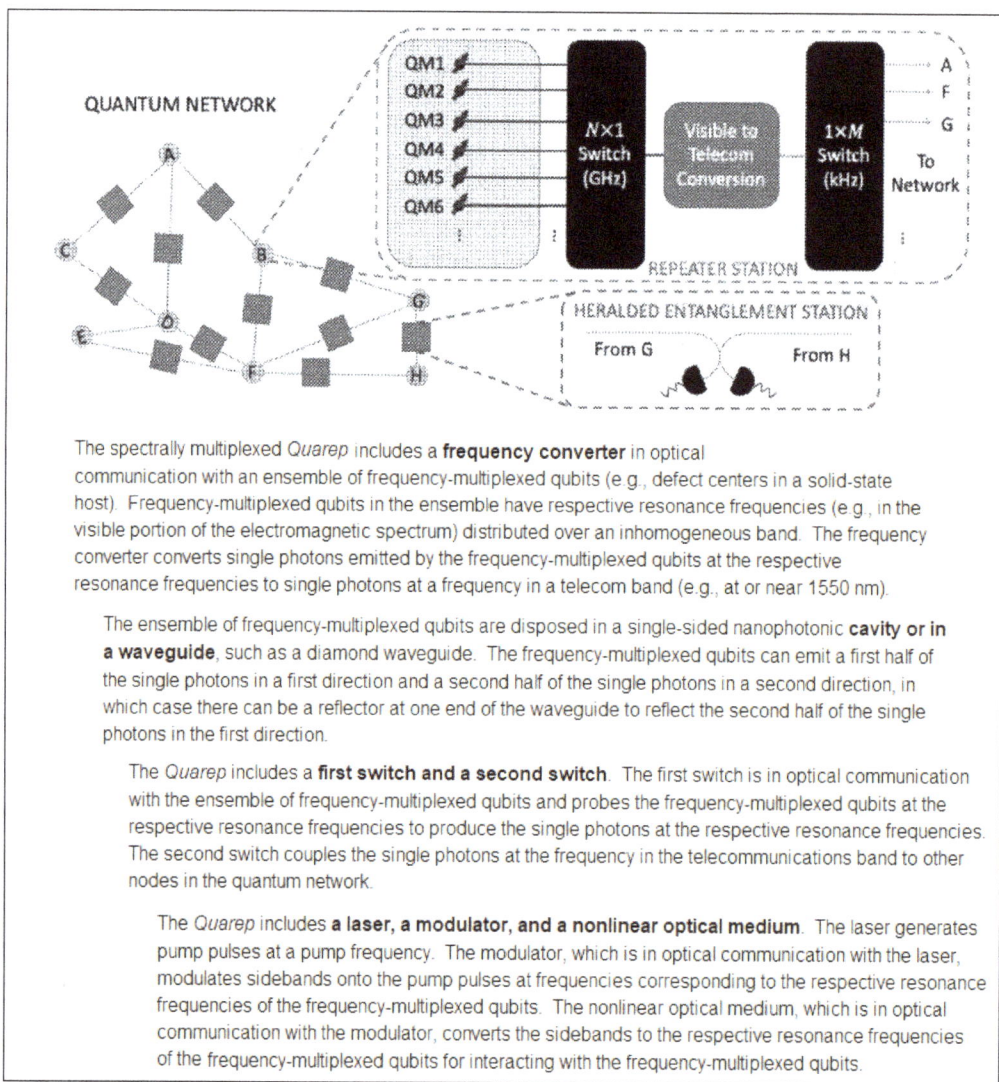

The spectrally multiplexed *Quarep* includes a **frequency converter** in optical communication with an ensemble of frequency-multiplexed qubits (e.g., defect centers in a solid-state host). Frequency-multiplexed qubits in the ensemble have respective resonance frequencies (e.g., in the visible portion of the electromagnetic spectrum) distributed over an inhomogeneous band. The frequency converter converts single photons emitted by the frequency-multiplexed qubits at the respective resonance frequencies to single photons at a frequency in a telecom band (e.g., at or near 1550 nm).

The ensemble of frequency-multiplexed qubits are disposed in a single-sided nanophotonic **cavity or in a waveguide**, such as a diamond waveguide. The frequency-multiplexed qubits can emit a first half of the single photons in a first direction and a second half of the single photons in a second direction, in which case there can be a reflector at one end of the waveguide to reflect the second half of the single photons in the first direction.

The *Quarep* includes a **first switch and a second switch**. The first switch is in optical communication with the ensemble of frequency-multiplexed qubits and probes the frequency-multiplexed qubits at the respective resonance frequencies to produce the single photons at the respective resonance frequencies. The second switch couples the single photons at the frequency in the telecommunications band to other nodes in the quantum network.

The *Quarep* includes **a laser, a modulator, and a nonlinear optical medium**. The laser generates pump pulses at a pump frequency. The modulator, which is in optical communication with the laser, modulates sidebands onto the pump pulses at frequencies corresponding to the respective resonance frequencies of the frequency-multiplexed qubits. The nonlinear optical medium, which is in optical communication with the modulator, converts the sidebands to the respective resonance frequencies of the frequency-multiplexed qubits for interacting with the frequency-multiplexed qubits.

Figure 6.11   Quanet with spectrally multiplexed Quareps—each Quarep contains a bank of quantum memories used to distribute entanglement (synthetized from [BER23])

electro-optic phase modulation, it is possible to turn on and off the desired spectral bands at a rate as fast as once-per-emitter lifetime (i.e., the reciprocal of the emitter lifetime), in principle allowing on the order of 10^9 individual defect centers to be addressed per second when scanning in frequency and spatial dimensions across a diamond waveguide containing a high density of artificial atoms. The qubits can be spectrally multiplexed at arbitrary frequencies. Single photons emitted by the qubits at these arbitrary frequencies are converted to temporally multiplexed photons at the same frequency (e.g., the telecom output frequency of choice, the so-called *quantum ITU* [International Telecommunication Union] frequency).

6.4.3 Quantum Switches

Since the dawn of electronic communications (starting with telephony and then with data communications), switches have played a critical interconnectivity role. The same applies to Quacom: there

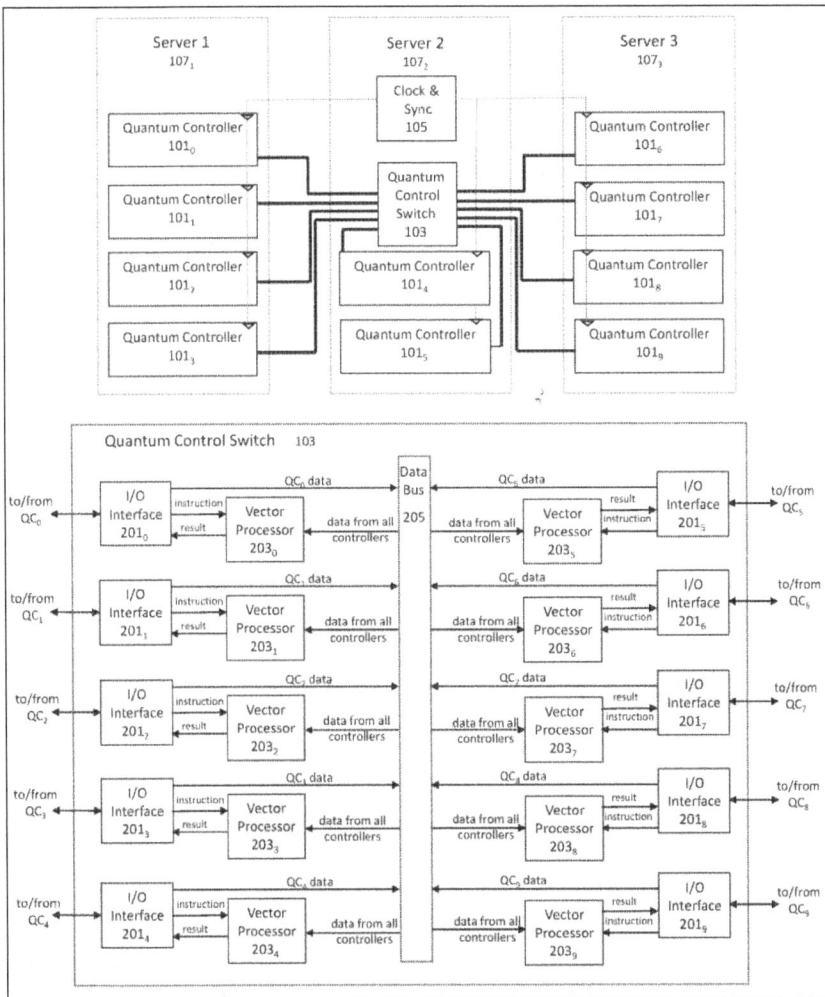

Figure 6.12   Example of Quantum Control Switch (*Source:* [SIV22]) *[Described on next page]*

is a need for (optical) quantum switches capable of routing qubits to the intended and specified remote QC. In particular, quantum switches must preserve quantum coherence. Existing Layer 1 optical switches are unsuitable for routing entangled photons; while electromechanical-based systems retain entangled states status, they are not sufficiently fast in their switching functionality; furthermore, optoelectronic switches annihilate quantum information and/or inject so much noise that single photons cannot be reliably detected. R&D work is underway to develop usable quantum switches. Figure 6.12 [SIV22} depicts an example of a quantum switch; other configurations are naturally possible.

6.4.4 Quantum Routers

Quarous are devices that deal with path computation between a source and a target destination node, say, a pair (s,t), and route installation. The goal of a router in a Quanet is to maximize the total number of end-to-end entanglements in support of all requests, say for (s_1,t_1), (s_2,t_2), and so on, while simultaneously maximizing fidelity support. Path computation can be distributed or centralized. Routes can be static or dynamic; the latter can in turn be reactive, opportunistic, proactive, or virtual. Route establishment algorisms can be based on path searching, linear programming, or AI methodologies. The result is a set of paths, schedules, and path installations. See Figure 6.13, modeled after [CUB24]. Some of the parameters of the quantum routing algorithms include the end-to-end entanglement generation rate goal and the minimum fidelity threshold required by the sources.

As seen in the figure, some of the intrinsic factors to be taken into consideration in designing and/or selecting a routing process are: (i) the probability $P_{u,v}$ of generating elementary entanglement on a path (channel) connecting two (internal) nodes u, v in the network (which in turn

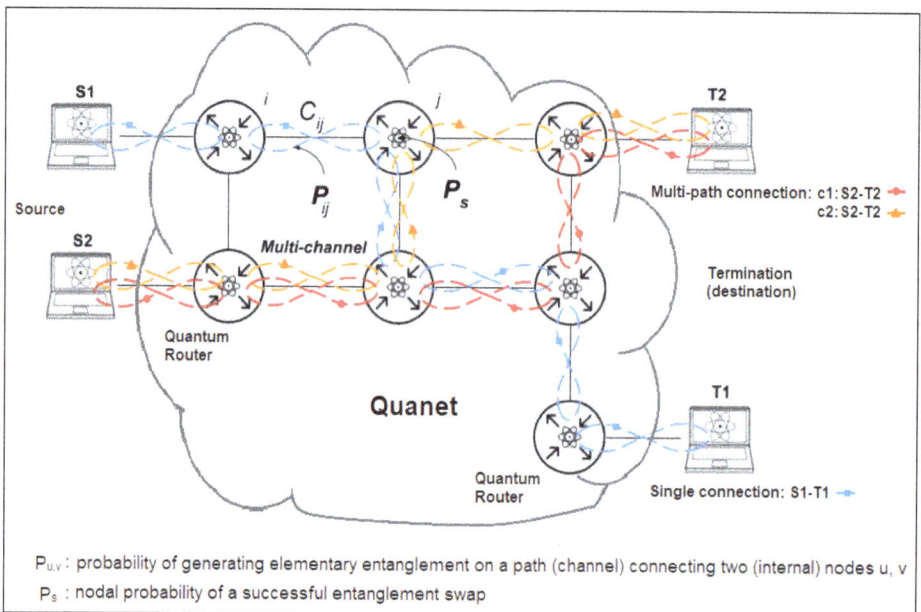

$P_{u,v}$: probability of generating elementary entanglement on a path (channel) connecting two (internal) nodes u, v
P_s : nodal probability of a successful entanglement swap

Figure 6.13 Quantum Router example (*Source:* Inspired by and based on [CUB24])

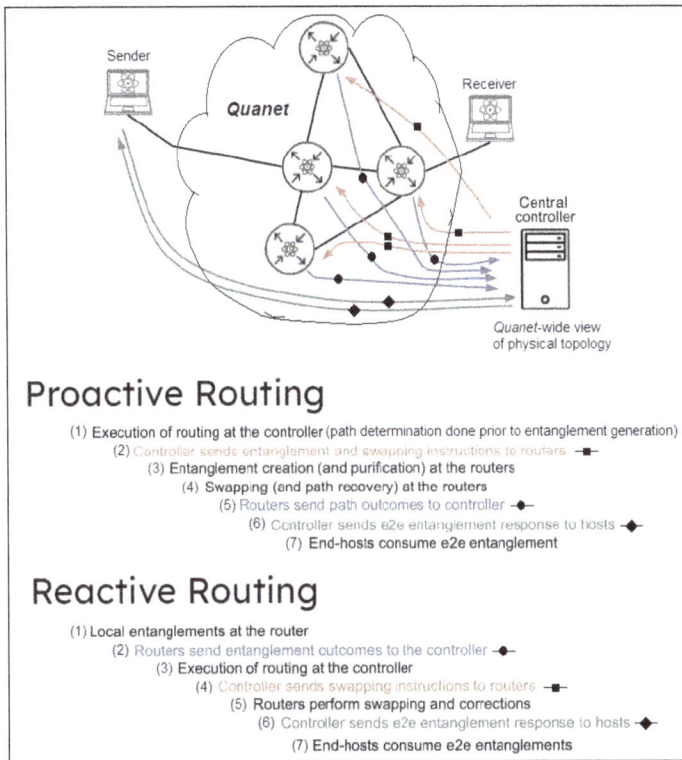

Figure within:

Quanet-wide view
of physical topology

Proactive Routing

(1) Execution of routing at the controller (path determination done prior to entanglement generation)
(2) Controller sends entanglement and swapping instructions to routers ▪
(3) Entanglement creation (and purification) at the routers
(4) Swapping (and path recovery) at the routers
(5) Routers send path outcomes to controller ◆
(6) Controller sends e2e entanglement response to hosts ◆
(7) End-hosts consume e2e entanglement

Reactive Routing

(1) Local entanglements at the router
(2) Routers send entanglement outcomes to the controller ◆
(3) Execution of routing at the controller
(4) Controller sends swapping instructions to routers ▪
(5) Routers perform swapping and corrections
(6) Controller sends e2e entanglement response to hosts ◆
(7) End-hosts consume e2e entanglements

Figure 6.14   Examples of quantum routing methodologies (*Source:* Inspired by and based on [CUB24])

depends on source and detector efficiencies, channel parameters—attenuation, distance—and number of attempts per time slot); (ii) the nodal probability P_s of a successful entanglement swap (which is typically less than 0.5 for swapping based on linear optics); and, (iii) the edge capacity $C_{s,d}$ that defines the maximum number of entanglements that can be generated in a single time slot (this capacity depends on the multiplexing modes—time, space, wavelength—and the qubit capabilities at nodes u and v).

As noted above, two basic routing schemes are *proactive* schemes or *reactive* schemes. Some of the basic steps for both are shown in Figure 6.14, also modeled after [CUB24].

6.5 Topological Network Considerations

As noted, fundamentally, in Quacom there is a need to create entangled states across a distance. The underlying technology and the network infrastructure affects how quickly one can create these entangled states, how long entanglement persists (can be maintained), and how much time it takes to run a given application on this network. The transportation of qubits from one node to another requires the presence of physical quantum communication channels. These physical channels may be implemented using telecommunications-grade fibers or free space optics; in either case, they support the transmission of electromagnetic (EM) states—single photons or other quantum optical states—over the physical medium. As is the case in classical networks, communication entails a single transmitter and receiver, or it can entail multiple receivers that can individually be enabled or disabled in a shared bus arrangement. Hierarchical networks in which nodes are grouped into tiers, which are in turn grouped into higher-level tiers, may in fact be optimal for certain applications [BAP18], [ELD20]. In addition to the quantum channel, a practical QC-to-QC link utilizes a classical channel to share some supportive information between two or more nodes associated with Quacom.

6.5.1 Network Layering

Modularity is a common approach used in (classical) computing and communications. For both local area applications and wide area communication, it was determined in the early days of data communications that layering the communications protocols was desirable and useful. Modularity mitigates overall complexity by aggregating a relatively small number of specialized, standard, off-the-shelf modules into a larger architecture, in which each module deals with a specific, well-defined function. This gave rise to the development of the OSIRM in the early 1980s, which encompasses hierarchically based architectural tiering. More specifically, network layering can be perceived at the logical as well at the physical level: the OSIRM is a logical protocol layering, while having technology-based tiers in a network (such as in local area networks, wide area networks, and global area networks) is an example of physical layering. Thus, in classical communication, the architecture—the underlying protocols—and the physical infrastructure are supported by functional/implementation tiering.

It has been recognized that layering in Quacom and Quanets could also prove useful, as already alluded to in Chapter 1 (e.g., [JIA09] [MET11], [LEE20], [ALS21], [WEH18], [ABE23] among others). The classical Internet entails a *"separation of concern"* principle, in which the network functionalities are organized as a stack of layers, each providing some specific functionality through specific network protocols that are embodied in well-known abstract models for packet-switching systems, particularly the OSIRM model and the TCP/IP model [ILL22]. This approach has enabled the classical Internet to evolve rapidly since the 1970s, from a skinny network linking a few static nodes in a handful of university labs to a ubiquitous system interconnecting billions of devices and accessible to more than half of the world's population.

Thus, there has been extensive advocacy for adopting and adapting similar separation of concern principles, or even the classical Internet protocol stack itself, to the Quaint, while at the same time dealing with the "peculiarities" of quantum entanglement. Analysis and assessments have shown that Quacom/Quanets (and Quacomp) also benefit from a quantum modular architecture, wherein separate quantum subsystems are integrated into a more extensive quantum system (utilizing quantum interconnects). At the same time, keep in mind that Quanets also may require some type of classical communication/messaging to operate in various applications.

A (classical) network usually has *three (hierarchical) planes*: (i) a *control* plane; (ii) a *data* plane; and (iii) a *management* plane. Often—not always but increasingly so—the control plane and the data plane are separate and distinct. The control plane focuses on issues such as topology management and, as such, may perform its core function(s) even in a Quanet environment by exchanging classical messages—at this juncture there still are ongoing assessments to establish whether a separate quantum control plane is advantageous or not. In Quanets, the data plane actually entails two planes: a *quantum* data plane and a *classical* data plane. The former deals with processing and swapping entangled pairs; the latter processes and forwards classical packets.

Table 6.6 Example of Quacom Layering[a]

Layer	Proposed functionality
Physical Entanglement layer	Responsible for the physical interaction that creates Bell pairs between two different stations
Entanglement Control layer	Responsible for managing the single-hop physical entanglement process, selecting qubits to attempt entanglement at each end of the link, and utilizing classical messages to report the results
Error Management layer	Responsible for choosing two Bell pairs and electing one pair to have its fidelity boosted and the other to be sacrificed, assuring that both stations make the same decisions
Quantum State Propagation layer	Responsible for administering the Bell pairs, such as purification, swapping, and so on (for example, purify or swap first or vice versa)
Application layer	Responsible for end-to-end entanglement management when required

[a] As suggested by Abelém, Vardoyan, and Towsley [ABE23]

Table 6.6 depicts a proposed layering model.

Simpler (somewhat more traditional) layered models are proposed by others—for example, in RFC 9340, which is extended herewith as depicted in Figure 6.15. In this model, the application could be a traditional application—for example, a financial application, a complex bio-engineering application, or a physics-phenomenon simulation application. The application may rely on a Quacomp process. To distribute the information remotely, or to invoke supportive remote computing capabilities (akin to cloud computing), a Quacom function is invoked. The Quacom function will generate end-to-end entangled pairs. This may require negotiation between the end nodes utilizing classical links to convey some signaling or seek control information (e.g., information related to routing paths to be used). Following the distribution of control information, the Quanet performs the necessary quantum operations, such as generating entanglement over individual quantum links, performing entanglement swaps at Quarep(s), and undertaking additional signaling to transmit the swap outcomes. The entangled pair is delivered to the remote

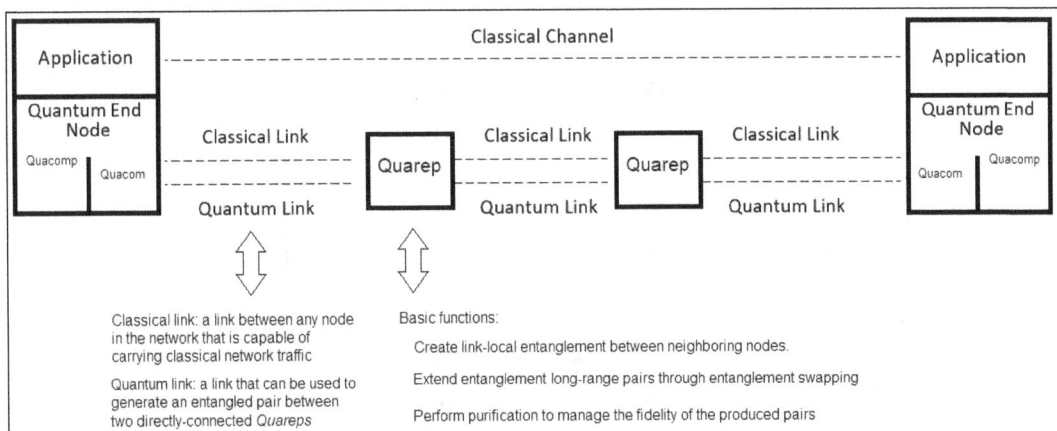

Figure 6.15 Basic layered model (extended from RFC 9340)

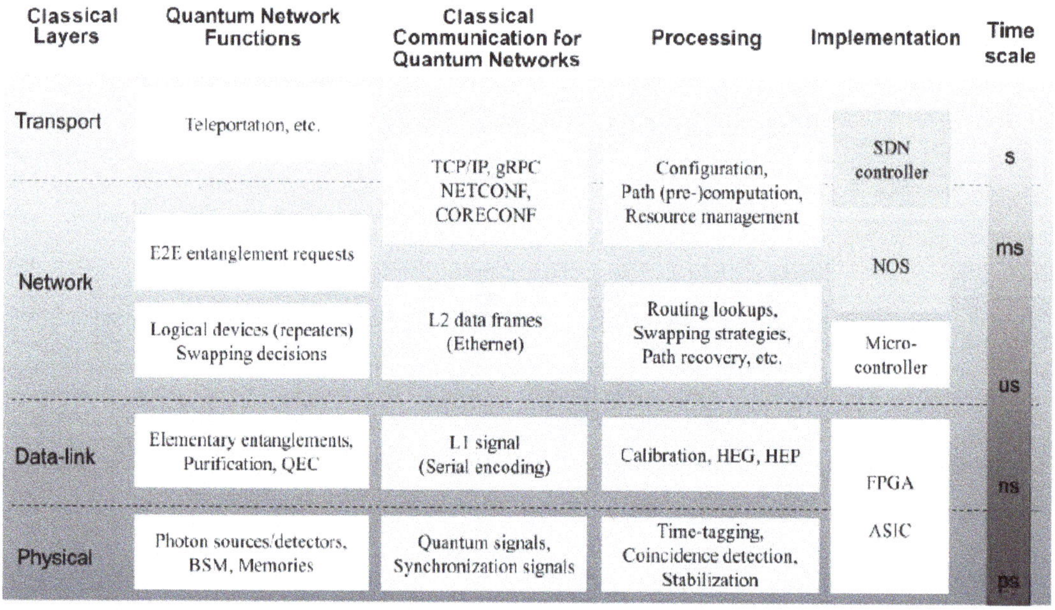

Classical Layers	Quantum Network Functions	Classical Communication for Quantum Networks	Processing	Implementation	Time scale
Transport	Teleportation, etc.			SDN controller	s
		TCP/IP, gRPC NETCONF, CORECONF	Configuration, Path (pre-)computation, Resource management		
Network	E2E entanglement requests			NOS	ms
	Logical devices (repeaters) Swapping decisions	L2 data frames (Ethernet)	Routing lookups, Swapping strategies, Path recovery, etc.	Micro-controller	us
Data-link	Elementary entanglements, Purification, QEC	L1 signal (Serial encoding)	Calibration, HEG, HEP	FPGA	ns
Physical	Photon sources/detectors, BSM, Memories	Quantum signals, Synchronization signals	Time-tagging, Coincidence detection, Stabilization	ASIC	ps

Figure 6.16   Other example of Quanet layering (*Source:* [CUB24])

application (the Quacomp process in particular), once it is ready, along with the relevant pair identifier. Figure 6.16 provides another example of Quanet layering, as suggested in [CUB24].

Further, in the context of layering, Figure 6.17 illustrates a seven-layer software-oriented model advanced by FGQT of CEN-CENELEC [OVA22], which, to a large extent, is similar

Layer 7	Applications/Services/Use Cases
Layer 6	Programming tools
Layer 5	Assembly programming
Layer 4	Hardware Abstraction Layer
Layer 3	Software drivers
Layer 2	OS communication primitives
Layer 1	Control software

Figure 6.17   CEN-CENELEC proposed (software) stack

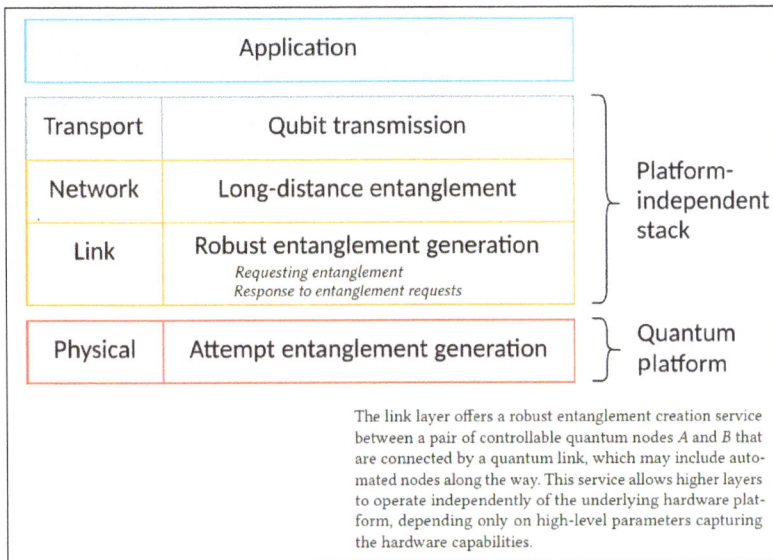

Application		
Transport	Qubit transmission	
Network	Long-distance entanglement	Platform-independent stack
Link	Robust entanglement generation *Requesting entanglement* *Response to entanglement requests*	
Physical	Attempt entanglement generation	Quantum platform

The link layer offers a robust entanglement creation service between a pair of controllable quantum nodes A and B that are connected by a quantum link, which may include automated nodes along the way. This service allows higher layers to operate independently of the underlying hardware platform, depending only on high-level parameters capturing the hardware capabilities.

Figure 6.18   Other proposed quantum stack (as advanced in [DAH19])

to the stack one would find in a classical setting, but with quantum-specific particularities. Yet another proposed stack is shown in Figure 6.18, as advanced in [DAH19], also with support in [POM21], [POM22], [KOZ20]. Another hierarchical model that is specific to QKD is defined in the ITU-T Y.3802 recommendation and is discussed in Chapter 7; this ITU model entails a service layer, a control layer, a key management layer, and a quantum layer.

Although a final reference model has not been agreed to by the Quacom stakeholders, each of these stacks, and the research efforts associated with them, point to the desirability of following the classical, and successful, paradigm of protocol segmentation and layering.

6.5.2 Endpoint Considerations

A number of endpoint arrangements are available. Architecturally, a quantum-based system may comprise (i) an integrated computing/communication device that can be used in Quacomp/Quacom settings where the same device is used to perform both local quantum computation and the network function, or (ii) two distinct devices, one for computing and one for communication. Figure 6.19 depicts two possible nodal architectures for the quantum endpoint in a Quacom environment.

In the former architectural arrangement, two sets of qubits can be perceived: (i) storage qubits used for local processing, and (ii) communication-focused qubits for remote entanglement generation that are coupled with an optical interface to the actual communication medium (e.g., fiber). In this case, the generation of the processing qubits and the communications qubits must be done sequentially (simultaneous activation of the local qubit gates and the entangling function is infeasible at this juncture). The time necessary for local gate execution depends solely on the local processing speed, while the time required for entanglement generation depends on the physical distance to the remote network node. Thus, in a situation in which the remote node is at a non-trivial distance, local processing may need to be suspended for a significant amount of time while entanglement generation is in progress [VAR22]. At this juncture, the same device is often used to perform both local quantum computation and the network function.

Networked QC

COMPUTING COMPONENT / DEVICE — State transfer requests — NETWORKING COMPONENT / DEVICE — Entanglement with remote node

Computational requests
Entanglement requests

⟨⟩ **Single-Device (SD) architecture**: processor and the network device are the same device; device may have an internal logical or physical partitioning into a computing or networking component

- example of a physical division is the use of a subset of its qubits for networking and others purely for computing
- example of a logical division is a scheduler switching between both functions but networking and computation are performed using the same qubits

⟨⟩ **Double Device (DD) architecture**: two separate devices are utilized

Networked QC

COMPUTING COMPONENT / DEVICE — State transfer requests — NETWORKING COMPONENT / DEVICE — Entanglement with remote node

Computational requests
Entanglement requests

Application

Application invokes the system via three types of requests:

(i) local quantum computations (on the computing component/device),

(ii) network operations (entanglement generation), and

(iii) movement (state transfer) of generated entanglement into the processor for further processing – requires cooperation from both processing and network devices/components

For SD, a move could be achieved simply by transferring the state to another set of qubits on the same device.

For DD, a move can be realized e.g., using entanglement generation between the processor and the network devices, followed by teleportation

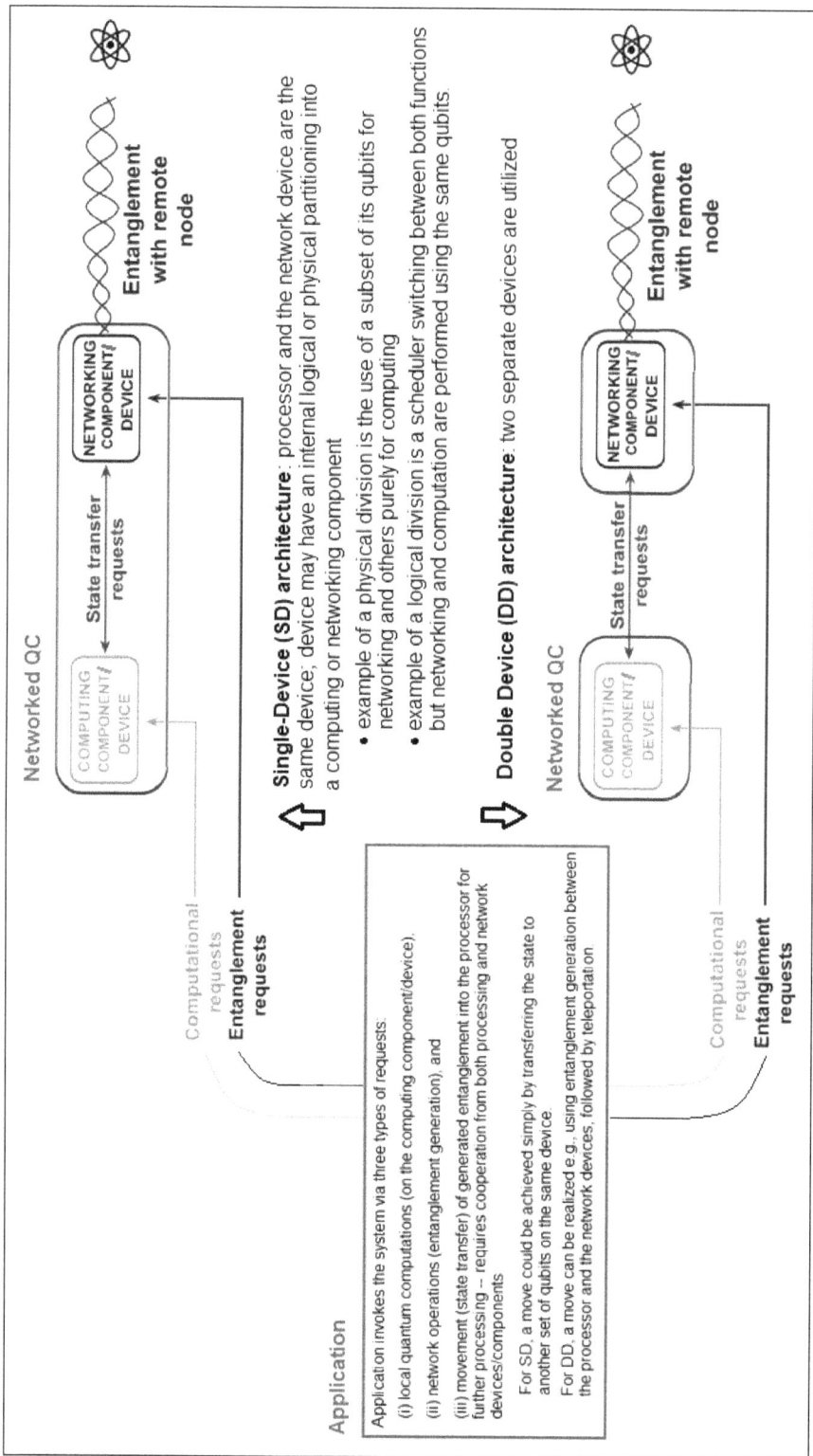

Figure 6.19 Two possible endpoint architectures for a quantum processor interfaced to a quantum network (inspired by [VAR22])

In the latter architectural arrangement, a function-specific network interface device is used for entanglement generation with remote network nodes. This separation of computing functions from communication functions has a long precedent, starting in the 1980s when the IBM® mainframe had an associated Front End Processor (FEP) to manage communication, and later with PCs and the use of a Network Interface Card (NIC) to attach to the LAN. This approach requires an additional step of producing entanglement between the network interface device and the processor to perform the teleportation transfer—this entanglement needs to be produced only at short distances, implying that the remote entanglement generation via the external network interface device and computations on the processor are only suspended for a short amount of time when the entanglement is transferred from the computing element to the nearby network interface device.

6.5.3 Point-to-Point Links

The simplest form of Quacom is represented by point-to-point links that span a geographical distance; these point-to-point links are a first step in advancing Quacom. Link connectivity allows the engineers to quantify parameter thresholds for the components and to validate communication protocols aimed at preserving quantum information in the presence of noise, attenuation, and loss.

Just as was the case with classical networks of the 1970s and 1980s, where implementations were homogeneous at the hardware level (typically using equipment from the same vendor at all network nodes), current Quanets are also homogeneous, given the fact that it is challenging to interface and aggregate different quantum systems. A near-term desideratum—even a requirement—is the ability to interface the different technologies in an efficient manner, say at the two ends of the point-to-point link. For example, QCs are typically based on different hardware platforms than are Quareps, thus requiring appropriate transducers between the two (e.g., QCs based on superconducting technology require transducers to connect to optical transmission media).

One example from [NIS23], [KYL22] entails optical networking of microwave entanglement nodes using appropriate transduction devices, as illustrated in Figure 6.20. The testbed contained a source of two-mode squeezed states and microwave-to-optical transducers; the squeezed optical states provide the continuous variable quantum resource that is provided to the network nodes by optical fibers in order to establish remote entanglement. At each of the endpoint nodes, the optical states are converted to the microwave domain with resonator devices.

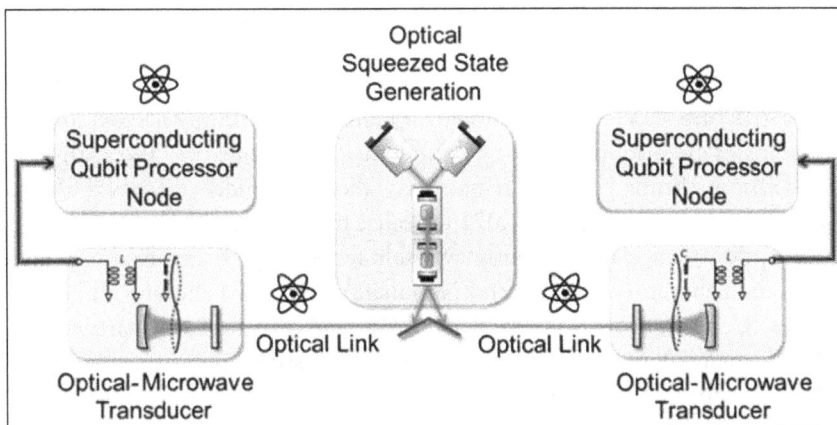

Figure 6.20   Networked microwave entanglement (simplified from [NIS23])

6.5.4 Network Management

Beyond being able to physically connect a few (or preferably many) quantum devices and transfer information using a quantum stack, there is a need for network management capabilities, including FCAPS (Fault, Configuration, Accounting, Performance, and Security) functionality, in order to run and operate Quanets; typically, the management plane and stack is an adjunct capability complementing the transmission stack.

6.6 Quanet Examples

A few illustrative Quanet examples are included in this section; this list is not exhaustive by any means. Quanet testbeds are being deployed in a variety of locations worldwide, including, but not limited to, the US (e.g., Illinois, New York, Chicago), Japan, The Netherlands, and the UK. Table 6.7 depicts a possible evolution of Quanets in the next few years, according to some observers [ROZ23]. Each phase progressively enables more advanced networking, computational, and cryptographic functionality.

6.6.1 Local Area Networks

Researchers recently demonstrated a quantum LAN design that mitigates deployment limitations in timing and security, in a scalable fashion and using commercial off-the-shelf components [ALS22]. In such a demonstration, White Rabbit switches were utilized to synchronize three remote nodes with ultra-low timing jitter, significantly increasing the fidelities of the distributed entangled states over designs using Global Positioning System (GPS) clocks. A parallel quantum key distribution channel was used to secure the classical communications needed for instrument control and data management. With this approach, the conventional network that manages the entanglement network is secured using keys generated by an underlying QKD layer.

6.6.2 Repeaterless Metropolitan Area Network Quanet: Illinois Express Quantum Network (IEQNET)

While several technical aspects required to support Quacom have been demonstrated in the laboratory in the recent past, including the production of entangled photons using lasers in direct line of sight, the emancipation from the lab to a city or regional environment utilizing fiberoptic infrastructure is much more complex. Quareps are certainly needed to address transmission losses; losses are also impacted by temperature variations over the span of the day or seasons. Furthermore, timing precision remains a challenge in widely dispersed Quacom nodes.

The IEQNET program was active in 2022 to realize metropolitan-scale quantum networking over deployed optical fiber using currently available technology. It involved two national labs (Argonne National Laboratory/ANL, Fermi National Accelerator Laboratory/FNAL/Fermilab), two universities (Caltech, Northwestern University), and two industry partners (Hyperlight, NuCrypt). While many Quanet recent demonstrations focused on point-to-point or linear topologies, IEQNET sought to develop architectures supporting multi-user, multi-node capabilities in metropolitan geographies that go beyond linear topologies, with coexistence between quantum and classical information in the same optical fibers. See Figure 6.21 [CHU22], [CHU23].

Table 6.7 Possible Evolution of Quanets in the Next Few Years (Modeled after [ROZ23])

Phase	Capabilities
Phase 1: Establishment of reliable repeater networks; nodes prepare and broadcast quantum states to neighboring nodes	Quantum nodes can prepare and broadcast quantum states to their adjacent nodes. In particular, this functionality supports protocols for the distribution of prepared and measurable quantum keys between adjacent nodes—for example, to establish a network of individual QKD (Quantum Key Distribution) links that exchange secure keys (quantum information, however, cannot be transmitted to non-adjacent nodes).
Phase 2: Development of preparation and measurement networks, introducing the capability of qubit transmission to any other node	Quantum nodes can prepare and transmit individual qubits to any other node in the network. Prepare-and-measure Quanets can be utilized for a number of applications, including safe identification in two-party cryptography with noisy quantum memory and key distribution—entanglement is utilized to ensure security, however nodes do share an entangled state. Alternatively, it is sufficient for nodes to confirm if entanglement can be communicated if the end nodes have performed a coherent version of a prepared and measured protocol.
Phase 3: Building of entanglement distribution networks, enabling end-to-end quantum entanglement	At this stage, two quantum nodes can obtain end-to-end quantum entanglement in either a deterministic or an announced method. End nodes do not require quantum memories at this stage. In a low-loss environment this added functionality allows device-independent QKD.
Phase 4: Construction of quantum memory networks; end users can store and transport quantum information	At this stage, quantum nodes can store quantum information in their memories and transport quantum information to one another. The transit time between the two end nodes determines the minimum storage time. This functionality enables, for example, blind quantum calculations and also supports cryptographic activities such as anonymous quantum communication, secret sharing, and clock synchronization protocols.
Phase 5: Creation of few-qubit fault-tolerant networks, enabling local quantum operations on logical qubits	This stage entails the construction of few-qubit fault-tolerant networks. Here, the end nodes can perform local quantum operations on a few logical qubits fault tolerantly. Here end nodes are capable of fault-tolerant execution of a universal gate set on multiple logical qubits, and by connecting the end nodes, this functionality enables the creation of a distributed Quacomp.
Phase 6: Establishment of quantum computing networks; large-scale fault-tolerant quantum processing	In this phase, full-fledged Quanets for Quacomp are constructed, enabling large-scale fault-tolerant quantum processing. Each end node is capable of undertaking large-scale quantum computations that cannot be efficiently simulated on a conventional computer.

IEQNET made use of a layered architecture that decouples control and data planes. Centralized control and management functions addressed routing and quantum channel assignment, dynamic optical path establishment, and calibration optimization functions; software-defined networking (SDN) mechanisms were used to orchestrate optical switches and multi-wavelength entangled photon sources. Specifically, control and management functions included (i) time synchronization of remote locations for distribution of entanglement and their use in subsequent applications; (ii) path Routing and Wavelength Assignment (RWA), this being a means of selecting lightpaths through the network; and (iii) quantum channel calibration and optimization, given that the single-photon nature of quantum communication signals makes them extremely sensitive to noise on the quantum channels. See Figure 6.22 [CHU22], [CHU23].

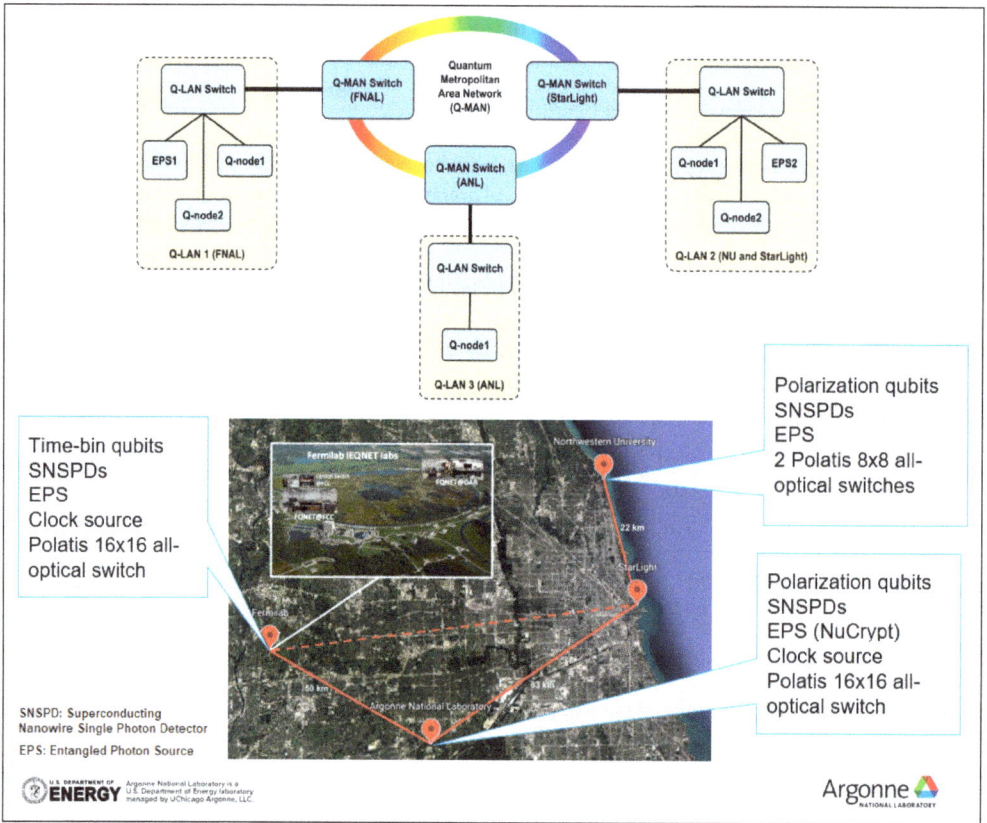

Figure 6.21 Illinois Express Quantum Network (IEQNET) [based on CHA23]

In a related activity [KAP23], the Q-LAN1 nodes deployed in two separate locations at Fermilab (one at the D0 Assembly Building and the other at the Fermilab Computing Center) are utilized to support a demonstration of a quantum teleportation protocol in which a photonic qubit (provided by Alice) is interfered with one member of an entangled photon-pair (from Bob) and projected (by Charlie) onto a Bell state, whereby the state of Alice's qubit can be transferred to the remaining member of Bob's entangled photon pair.

6.6.3 Recent Metropolitan Quanet Demonstrations

A major challenge in realizing practical regional or global Quanets relates to the generation of robust entanglement between quantum (memory) nodes connected by fiberoptic infrastructure. At press time three separate research groups based in the United States, China, and the Netherlands announced successful demonstrations of quantum entanglement distribution over several kilometers of existing optical fibers in urban areas, a step towards the creation of the Quaint [KNA24], [LIU24], [STO24]. Together, the experiments are seen as "major milestones" and "the most advanced demonstrations so far" of the technology needed for the Quaint [CAS24]. Each of the research teams was able to connect various portions of a network using photons in the optical-fiber-friendly infrared portion of the electromagnetic spectrum. Each of the demonstrations used different kinds of quantum memory devices to store the qubit(s).

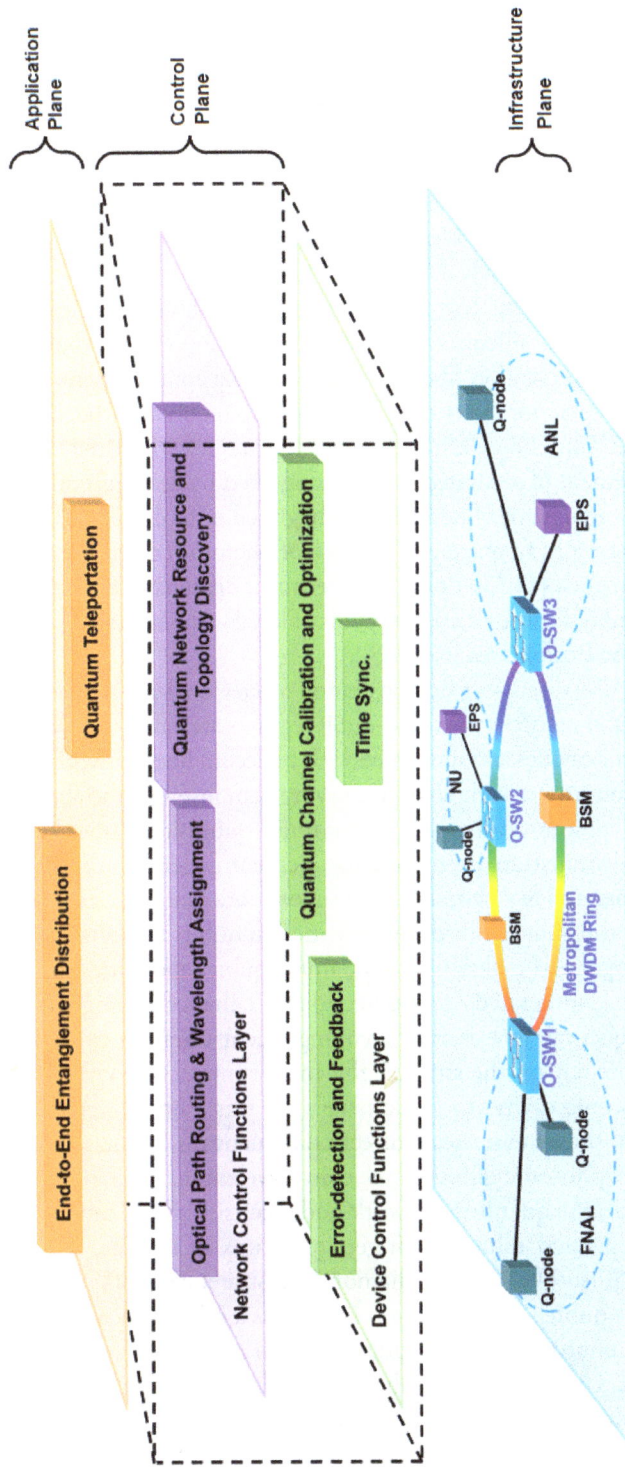

Figure 6.22 IEQNET made use of a layered architecture (*Source: [CHU23]*)

- At the University of Science and Technology of China, qubits were encoded in the collective states of clouds of rubidium atoms. The quantum memories were set up in three separate labs in the Hefei area. Each lab was connected by optical fibers to a central "photonic server" around 10 kilometers away. Any two of these nodes could be put in an entangled state if the photons from the two atom clouds arrived at the server at exactly the same time [CAS24], [LIU24].
- At Delft University in The Hague, the experiment established a link between individual nitrogen atoms embedded in small diamond crystals with qubits encoded in the electron states of the nitrogen and in the nuclear states of nearby carbon atoms. The optical fiber ran from the University in Delft over a 25-kilometer link across the suburbs of The Hague to reach a second laboratory in the city [CAS24], [STO24].
- A Boston-area urban environment test demonstrated a two-node Quanet composed of multi-qubit registers based on silicon vacancy (SiV) in nanophotonic diamond cavities integrated with a telecommunication fiber network. Remote entanglement was generated by the cavity-enhanced interactions between the electron spin qubits of the SiVs and optical photons. Serial, heralded spin-photon entangling gate operations with time-bin qubits were used for robust entanglement of separated nodes. Long-lived nuclear spin qubits were used to provide second-long entanglement storage and integrated error detection. By integrating efficient bidirectional quantum frequency conversion of photonic communication qubits to telecommunication frequencies (1,350 nm), the experiment demonstrated the entanglement of two nuclear spin memories through 40 km spools of low-loss fiber and a 35-km-long fiber loop deployed around the Boston area [KNA24].

As highlighted above, [KNA24] discusses a demonstration of a regional (35 kilometer) two-node Quanet composed of multi-qubit registers; the multi-qubit registers are based on SiV centers in diamond coupled to nanophotonic cavities and integrated with a telecom fiber network. SiVs coupled to cavities are a promising quantum network platform, having demonstrated the ability to support memory-enhanced quantum communication and robust multi-qubit single-node operation.

The system of [KNA24] demonstrated remote entanglement generation between two electron spins in two spatially separated SiV centers with a success rate of up to 1 Hz. The approach uses serial, heralded spin-photon gate operations with time-bin qubits for robust entanglement of separated nodes. The arrangement makes use of the multi-qubit capabilities to entangle two long-lived nuclear spins, using integrated error detection to enhance entanglement fidelities and dynamic decoupling sequences to extend the entanglement duration to 1 s. Both entanglement generation techniques rely on the strong light–matter interaction enabled by the coupling of SiV to the nanophotonic cavity. To demonstrate the feasibility of deployed quantum networks using this platform, the system uses bidirectional quantum frequency conversion to convert the wavelength of the photonic qubits to telecom wavelengths.

As shown in Figure 6.23 (a), the quantum network nodes consist of SiV centers in diamond that reside in individually operated dilution refrigerator setups in separate laboratories. By selectively implanting the ^{29}Si isotope into the diamond substrate, each SiV deterministically contains two addressable spin qubits: one electron spin used as a communication qubit, which couples strongly to itinerant photons, and one long-lived ^{29}Si nuclear spin used as a memory qubit to store entanglement. Under an externally applied magnetic field, Zeeman sublevels define the electronic spin qubit states ($|\downarrow_e\rangle$, $|\uparrow_e\rangle$) and the nuclear spin qubit states ($|\downarrow_n\rangle$, $|\uparrow_n\rangle$) (Figure 6.23 (b), left). Microwave pulses are used to drive the electronic spin-flipping transitions, whereas radio-frequency pulses drive the nuclear spin-flipping transitions.

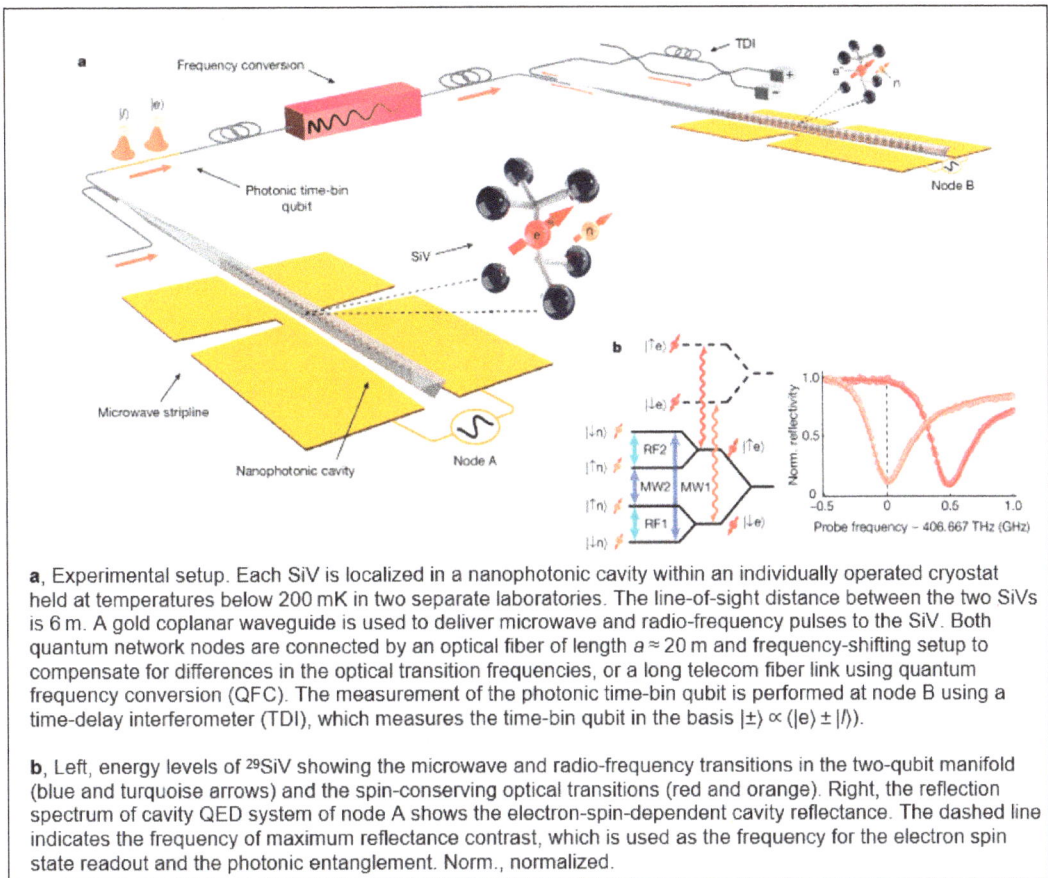

a, Experimental setup. Each SiV is localized in a nanophotonic cavity within an individually operated cryostat held at temperatures below 200 mK in two separate laboratories. The line-of-sight distance between the two SiVs is 6 m. A gold coplanar waveguide is used to deliver microwave and radio-frequency pulses to the SiV. Both quantum network nodes are connected by an optical fiber of length $a \approx 20$ m and frequency-shifting setup to compensate for differences in the optical transition frequencies, or a long telecom fiber link using quantum frequency conversion (QFC). The measurement of the photonic time-bin qubit is performed at node B using a time-delay interferometer (TDI), which measures the time-bin qubit in the basis $|\pm\rangle \propto (|e\rangle \pm |l\rangle)$.

b, Left, energy levels of ^{29}SiV showing the microwave and radio-frequency transitions in the two-qubit manifold (blue and turquoise arrows) and the spin-conserving optical transitions (red and orange). Right, the reflection spectrum of cavity QED system of node A shows the electron-spin-dependent cavity reflectance. The dashed line indicates the frequency of maximum reflectance contrast, which is used as the frequency for the electron spin state readout and the photonic entanglement. Norm., normalized.

Figure 6.23   A two-node Quantum Network of cavity coupled solid-state emitters. (Courtesy: C. M. Knaut, A. Suleymanzade, et al. [KNA24]; Open Source; see notes in reference list)

The SiV centers are embedded into nanophotonic diamond cavities, which enhance interactions between light and the electron spin. The strong emitter–cavity coupling results in an electron-spin-dependent cavity reflectance (Figure 6.23 (b), right). This can be used to construct a reflection-based spin-photon gate that contains a sequence of rapid microwave gates generating entanglement between the electron spin of the SiV and the photonic qubits. The two nodes are connected either directly by an optical fiber of length $a \approx 20$ m or by a considerably longer telecom fiber link, as shown in Figure 6.24. Figure 6.25 (on page 281) shows the entanglement generation sequence.

6.6.4 Low Earth Orbit Satellite Quanets

Classical communication networks consist of multiple nodes connected by various types of channels, including, but not limited to, optical fibers, ground-to-satellite links, wireless radio frequency (RF) spectrum, and free-space optical links. Such a heterogeneous architecture is equally important for Quanets, since quantum nodes may access a Quanet by different kinds of channels. Indeed, Quacom has been individually validated in free-space, optical fibers, and

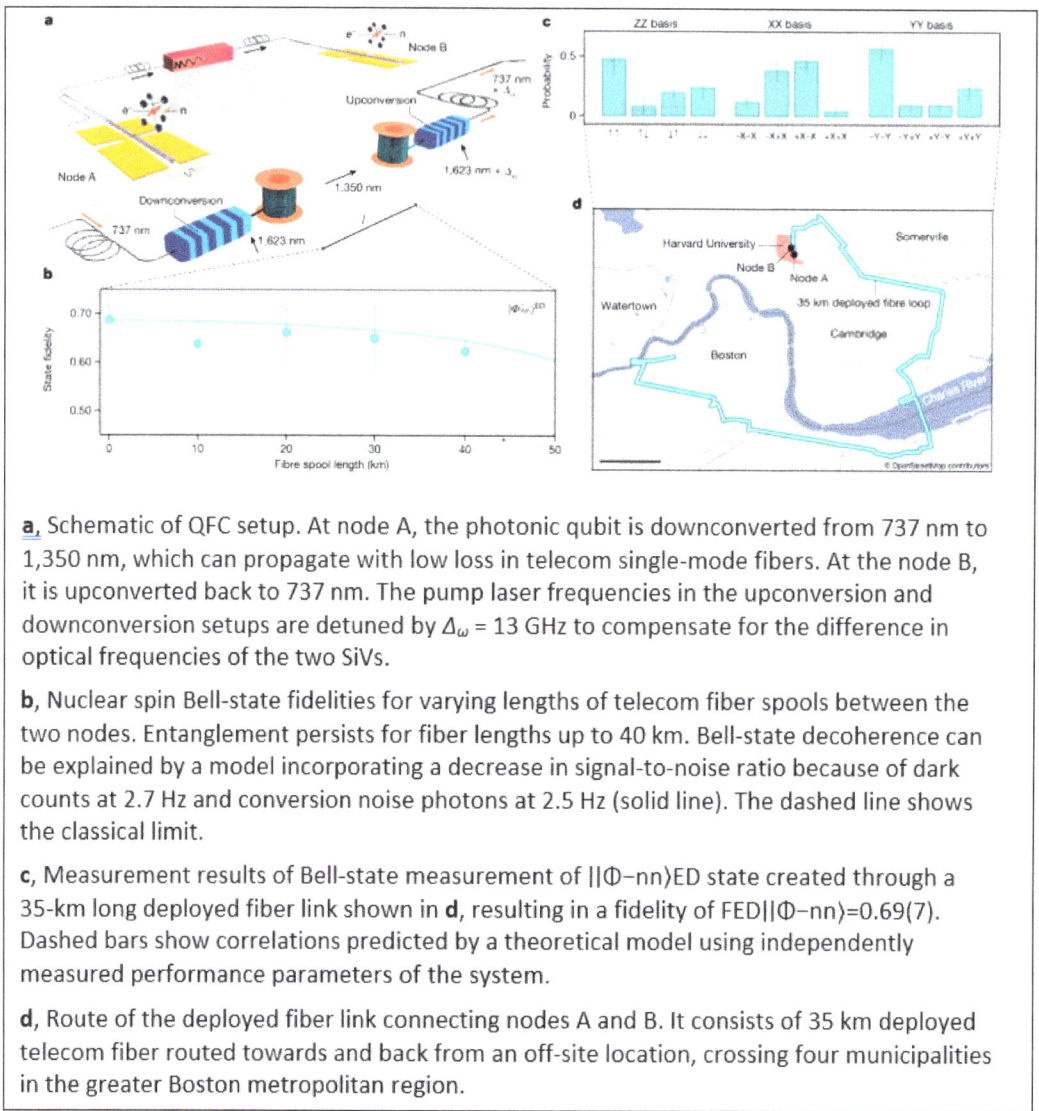

a, Schematic of QFC setup. At node A, the photonic qubit is downconverted from 737 nm to 1,350 nm, which can propagate with low loss in telecom single-mode fibers. At the node B, it is upconverted back to 737 nm. The pump laser frequencies in the upconversion and downconversion setups are detuned by $\Delta_\omega = 13$ GHz to compensate for the difference in optical frequencies of the two SiVs.

b, Nuclear spin Bell-state fidelities for varying lengths of telecom fiber spools between the two nodes. Entanglement persists for fiber lengths up to 40 km. Bell-state decoherence can be explained by a model incorporating a decrease in signal-to-noise ratio because of dark counts at 2.7 Hz and conversion noise photons at 2.5 Hz (solid line). The dashed line shows the classical limit.

c, Measurement results of Bell-state measurement of $||\Phi-nn\rangle$ED state created through a 35-km long deployed fiber link shown in **d**, resulting in a fidelity of FED$||\Phi-nn\rangle$=0.69(7). Dashed bars show correlations predicted by a theoretical model using independently measured performance parameters of the system.

d, Route of the deployed fiber link connecting nodes A and B. It consists of 35 km deployed telecom fiber routed towards and back from an off-site location, crossing four municipalities in the greater Boston metropolitan region.

Figure 6.24   Nuclear spin entanglement distribution through 35 km of deployed fiber. (Courtesy: C. M. Knaut, A. Suleymanzade, et al. [KNA24]; Open Source; see notes in reference list)

between a satellite and a ground station, but a combined heterogeneous Quanet employing multiple types of channels has *remained a work-in-progress*. To enable the next generation of Quacom, disconnected terrestrial cluster states–based Quanets are coupled through satellite (cluster state) quantum networks, thus providing global coverage. A goal is, thus, to be able to entangle disconnected terrestrial Quanets coupled through Low Earth Orbit (LEO) satellite Quanets to form a heterogeneous satellite-terrestrial Quanet. Such a Quanet needs to (i) be robust against channel impairments, (ii) operate over heterogeneous links, (iii) provide security for 5G/6G (and beyond) wireless and/or optical networks, and (iv) support, among other applications, Internet-of-Things (IoT) sensing applications. An engineering challenge is that, unlike in the point-to-point communication case, the fundamental Quacom rate limits are not well understood even for the simplest broadcast and multiple-access channel settings.

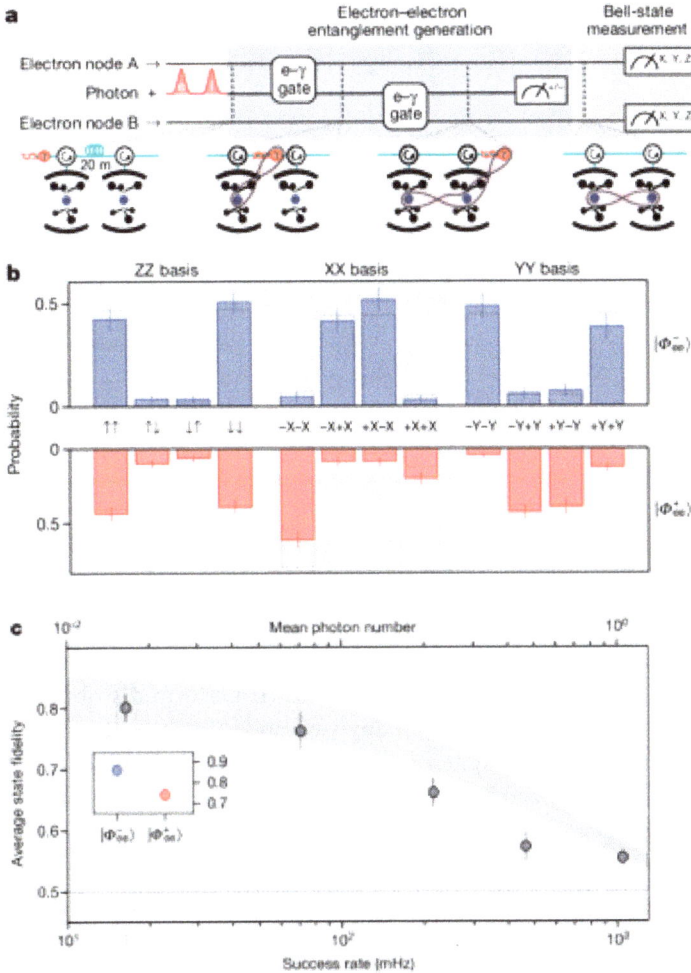

a, Entanglement generation sequence. A photonic qubit is entangled with the electron spin in node A using the e–γ gate. A second e–γ gate entangles the photonic qubit with node B, generating a GHZ state among the two electronic qubits and the photonic qubit. A measurement of the photonic qubit in the $|\pm\rangle$ basis heralds the generation of an electronic Bell state $||\Phi\pm ee\rangle$.

b, Measurement results of Bell-state measurement. Measured correlations in the ZZ, XX and YY bases of the electronic spin corresponding to a Bell-state fidelity of $F||\Phi-ee\rangle=0.86(3)$ (blue) and $F||\Phi+ee\rangle=0.74(3)$ (red). Dashed bars show correlations predicted by a theoretical model using independently measured performance parameters of the system.

c, Sweep of mean photon number of the photonic qubit showing that the success rates can be increased by sending photonic qubits with a higher mean photon number. The average fidelity of the generated $||\Phi+ee\rangle$ and $||\Phi-ee\rangle$ states is plotted.

Figure 6.25   Remote entanglement between two electronic spins. Courtesy: C. M. Knaut, A. Suleymanzade, et al. [KNA24] (Open Source; see notes in reference list.)

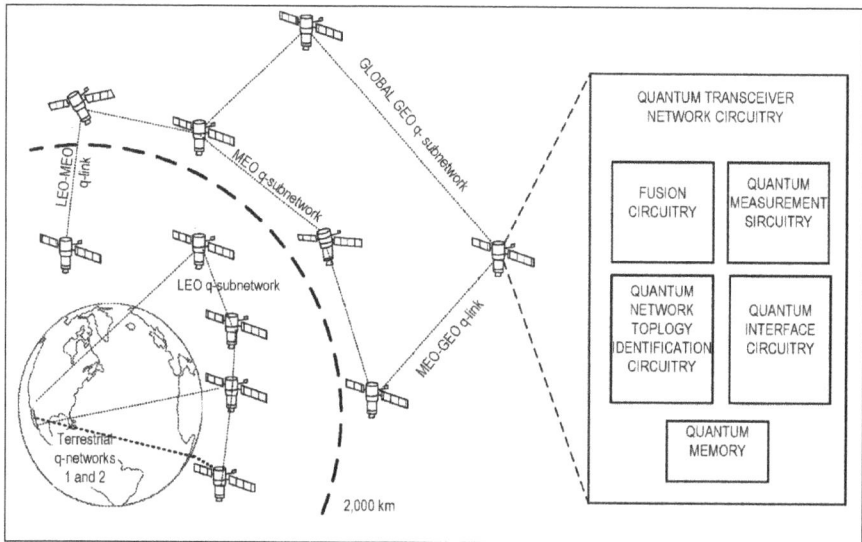

Figure 6.26   Integrated Quanet (*Source:* [DJO24])

[DJO24] proposed a system that implements the multipartite Quanet by employing the cluster state–based concept[1]; such system is able to (i) perform distributed quantum computing, (ii) teleport quantum states between any two nodes in the network, and (iii) enable the next generation of cyber security systems. The cluster states can be described using the stabilizer formalism, and as such they can be certified by simple syndrome measurements; in this formalism, the cluster states can be interpreted as codewords of corresponding quantum error correction code, and corresponding errors can be corrected by simple syndrome decoding, among other methods. By performing X, Y, and/or Z measurements (by using X-basis and Z-basis, respectively) on properly selected nodes, EPR pairs (Bell states) may be established between any two nodes in the network; moreover, multiple EPR pairs can be established simultaneously.

The systems of [DJO24], depicted in Figure 6.26, include a cluster state-based Quanet of satellites that enables global coverage. As seen in the figure, the quantum satellite network may be composed of quantum subnetworks comprising LEO satellites; some of these LEO satellite-based quantum subnetworks may be connected to a subnetwork of Medium-Earth Orbit (MEO)/Geosynchronous Earth Orbit (GEO) satellites (e.g., [MIN09], [MIN15]). The LEO satellites may also be used to interconnect terrestrial cluster state-based quantum networks.

Quantum links between ground-based nodes may be established, for example, using fiber-based connections or free-space optical connections. This quantum global network may also be used to distribute the entangled states for quantum sensing applications and enable distributed

---

[1]   Raussendorf and Briegel proposed a model based on a special type of entangled states, called *cluster states* [RAU01]. A cluster state refers to a family of quantum states of n-qubit two- or three-dimensional lattice, or even a more general graph, in which each vertex corresponds to a qubit. A specific preparation procedure is applied (e.g., vertices are connected using controlled-Z two-qubit gates). Thus, the cluster state is a highly entangled state that has applications in measurement-based quantum computing and one-way quantum computing applications; by utilizing the cluster state and properly measuring the qubits in specific order and basis, one can undertake arbitrary quantum computations [MAR12], [DJO21].

quantum computing on a global scale. In particular, QKD Quacom applications utilize quantum information theory concepts to realize the distribution of keys with verifiable security, where security is ensured by the fundamental laws of physics, as opposed to mathematical approaches employed in computational security-based cryptography. To overcome the rate-distance limit of discrete variable (DV)-QKD protocols, two approaches have been pursued recently: (i) development of quantum relays, and (ii) the employment of the trusted relays. The quantum relays require the use of long-duration quantum memories and high-fidelity entanglement purification (distillation). On the other hand, the trusted-relay methodology assumes that the relay between two users can be trusted; unfortunately, this assumption is difficult to verify in practice. See [DJO24] for additional information on this global Quanet.

6.6.5 Packet Switched Quantum Network

Quacom systems typically operate on optical networks by distributing entangled states for teleporting quantum states between endpoints. This circuit-switching approach is based on (complex) quantum repeaters with robust quantum memories and high rates of entanglement generation, which may not scale up easily. Transporting quantum information through a packet-switched optical network presents additional challenges due to the inability to amplify or copy data signals containing quantum information without destroying the quantum properties of the data. [SHA24] offers to modify the typical link layer and node architecture to accommodate packet-switched routing of quantum information. This enables packet switching in a hybrid classical/quantum network by defining a data frame structure capable of handling a quantum payload. The hybrid frame structure includes a classical header, a quantum payload, and an optional classical trailer. The classical header includes information for routing the quantum payload, as well as information for error mitigation and correction. The classical trailer may indicate the end of the quantum payload signal. The quantum payload encodes one or more qubits in an optical format (e.g., polarization encoding or time-bin encoding) that may be measured by the destination endpoint. Quacom systems avoid direct measurement of the quantum payload, allowing the quantum information encoded at the source endpoint to reach the destination endpoint without collapsing into a measured state. The classical header/trailer and the quantum payload may be generated using different photonic sources and be multiplexed into a hybrid data frame by using different degrees of freedom of light (e.g., time, wavelength, polarization, spatial mode). Refer to [SHA24] for additional information on this proposed approach.

6.6.6 Other Examples/Initiatives

A (grossly) incomplete list of various Quanet press time activities and testbeds follows.

U.S. Department of Energy Quanet efforts

In the US, the National Quantum Initiative Act was a government attempt in the late 2010s to speed up the study and growth of quantum physics [COM18]; in 2023 the Department of Energy (DOE) earmarked $24 million for three collaborative Quanet projects, and more recently, the US and Germany agreed to collaborate on QC projects. Achieving scalable infrastructures for quantum information flows requires advancements in devices, error mitigation techniques, and new Quanet architectures and protocols; thus, the Quanet projects included [ENE23]:

- A collaborative research effort led by Argonne National Laboratory, partnering with Northwestern University, the University of Chicago, the University of Illinois-Urbana-Champaign, and Fermi National Accelerator Laboratory following a heterogeneous, full-stack approach in codesigning scalable Quanets.
- A collaborative research effort led by Oak Ridge National Laboratory, partnering with the University of Massachusetts-Amherst, the University of Arizona, and Arizona State University developing the architecture and protocols for a performance-integrated scalable Quaint.
- A collaborative research effort led by Fermi National Accelerator Laboratory, partnering with the California Institute of Technology, the University of Illinois-Urbana-Champaign, Northwestern University, and Argonne National Laboratory developing hyper-entanglement-based networking and error noise-robust correction techniques for developing advanced Quanets for science discovery.

There have been efforts to establish a Quantum Computing Center of Excellence for the US Army [SWA24].

Testbed plans in Japan

Japan's Moonshot Goal 6 aims at realizing a fault-tolerant universal QC that would revolutionize the economy, industry, and security by 2050 [NAG23]. To that end, various elements are being studied in Japan. For example, at a laboratory at Shinkawasaki-city, (Tokyo-area) researchers have constructed a four-node star network spanning several miles that supports routing. (Nodes include: Shinkawasaki-city, Keio University Shin-Kawasaki Campus; Keio University Yagami Campus; and Yokohama National University).

Quantum Network Developments in the Netherlands

The Netherlands is geographically positioned to be a major node in evolving Quanets: in the EU the digital infrastructure is concentrated around Frankfurt, London, Amsterdam, and Paris. Goals of R&D work included (i) establishing quantum connectivity between different locations over a NL National Quantum Network, and (ii) connecting the industry on the testbed to test hardware, software, and applications. Three specific activities that were underway at press time included [ROB23]:
- NL Quantum R&D Network: Development and integration of key components of Quantum networks:
  - Network elements (nodes, hardware, frequency conversion, switches, and repeaters)
  - Operating systems (OSs), stack, backplane, middleware
  - Protocols and standardization (network and routing)
  - Development of an integrated plan for quantum and classical networks
  - Cost efficient components and tools
- NL Quantum Staging Network
  - Setup and simulation of quantum ecosystems
  - Stable and accessible operational network
  - (Online) open network access for the development, compilation, and testing of software application, related hardware, and additional (network) components
  - Connect to EU Space Quantum Communication platform
  - Exchange of data/qubits between relevant NL nodes

ITU-T
- SG 17 focuses on Quantum security
- SG 13 focuses on QKD
- FG-QIT4N Quantum information technology for network
- JCA-QKDN

IEC
- WP on Quantum Information Technologies
- IEC SEG 14 Quantum technologies

ISO/IEC JTC1
- SC 6-ISO/IEC 9594-11, key distribution
- SC 27 -focuses on security and privacy in ICT systems
- SC 38 - ISO/IEC TS 7339 PaaS

QED-C
- QED-C – Quantum Economic Development Consortium

ETSI
- ISG QKD – Quantum Key Distribution
- TC Cyber WG QSC – Quantum-Safe Cryptography

CEN CENELEC
- FGQT
- JTC 22 Quantum technology

IEEE
- P7130 Standard for QC Definitions
- P1913 for Software Quantum Communications
- P7131 for QC performance Metrics & Performance Benchmarking
- P2995 Quantum algorithm design and development
- P3120 Standard for Quantum Computing Architecture
- P3155 Standard for Programmable Quantum Simulator
- P3185 Standard for Hybrid Quantum-Classical Computing
- P3172 Recommended Practice for Post-Quantum Cryptography Migration
- P1943 Standard for Post-Quantum Network Security

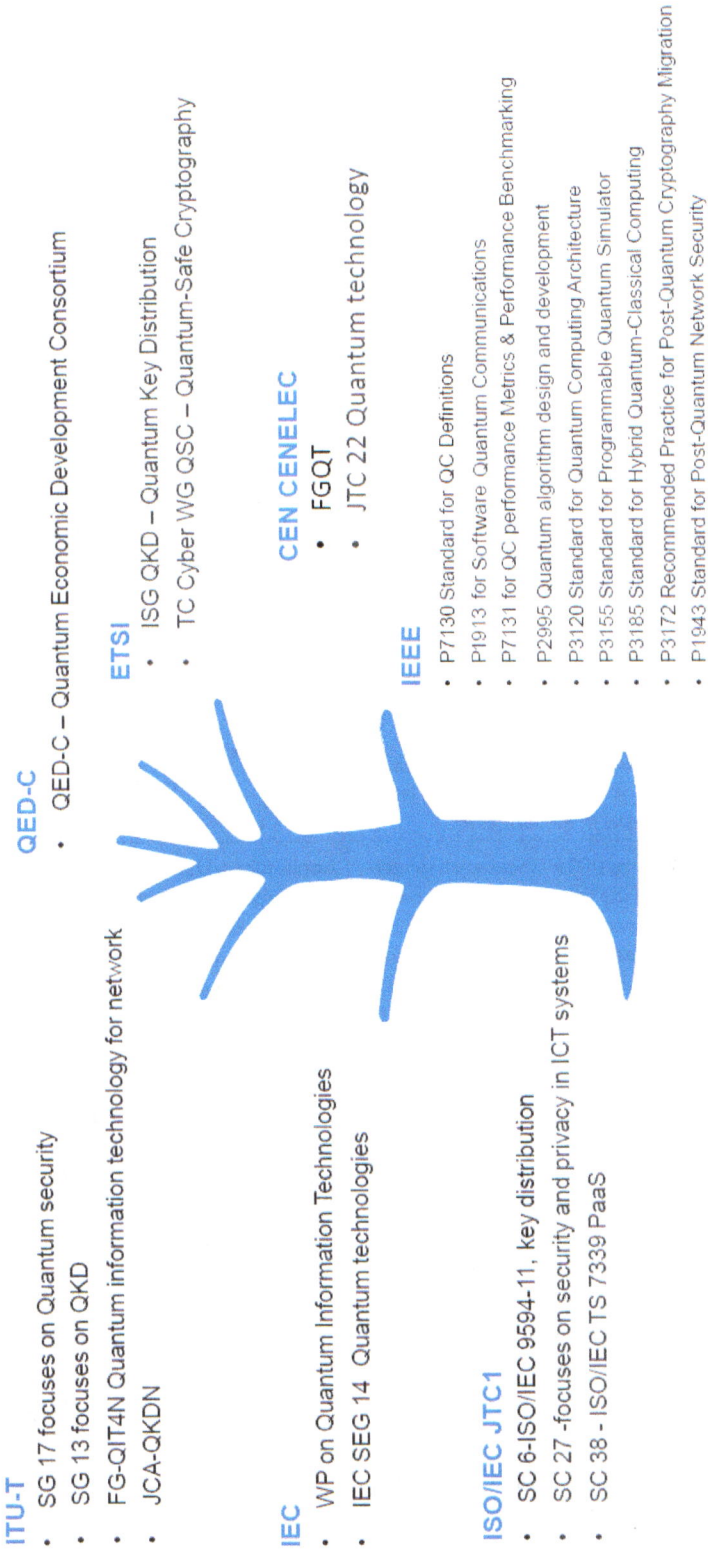

Figure 6.27 Multi-agency quantum standardization initiatives and/or standards (*Source:* ISO/IEC JTC 1 [YAN23])

Table 6.8 Snapshot of Current Quantum Standardization Efforts

Description of quantum-related activity	Selected deliverable topics	Type of output[a]
The European Telecommunications Standards Institute (ETSI) is the EU's recognized regional standards body for telecommunications, broadcasting and other electronic communications networks and services. The ETSI Industry Specification Group on Quantum Key Distribution (ISG-QKD) was the first forum aiming at standardization of Quacom technologies; it was created in 2008 by a group of European QKD manufacturers, telecommunications providers, universities, and research institutions. At this juncture, relevant work takes place in the Technical Committee on Cyber Security (CYBER) and the Industry Specification Group (ISG) on QKD.	QKD: Authentication, Components and internal Interfaces, Architectures and Frameworks, Vocabulary, Case Studies, Optical Characterization Quantum Computing Impact of QIST Systems Quantum-safe Cryptography: Security, Schema, Assurance	Informative: Group Reports, Technical Reports, White Papers, ETSI Guides Normative: Technical Specifications; Group Specifications; security certification of QKD
The Institute of Electrical and Electronics Engineers (IEEE) is a US-based professional association that has developed thousands of standards for consumer electronics, computers, and telecommunications. IEEE Quantum is an IEEE Future Directions initiative launched in 2019 that serves as IEEE's leading community for all projects and activities on quantum technologies; it has developed a project plan to address the current landscape of quantum technologies, identify challenges and opportunities, leverage and collaborate with existing initiatives, engage the quantum community at large, and sustain the Quantum Initiative in the long term.	Software-Defined Quantum Communication Quantum Technologies Definitions Quantum Computing Performance Metrics & Performance Benchmarking	Normative: Standards
The Internet Research Task Force (IRTF) focuses on longer-term research issues related to the Internet, while the parallel organization, the Internet Engineering Task Force (IETF), focuses on the shorter-term issues of engineering and standards making. The Quantum Internet Research Group (QIRG) is addressing the design and build of quantum networks. Issues to be explored include routing, resource allocation, connection establishment, interoperability, and security. This group will also perform coordination with other SDOs.	Applications, Use Cases and Architectural Principles for Quantum Internet Transition from Classical to Post Quantum Cryptography	Informative: Informational Documents Proposed Standards
The International Organization for Standardization (ISO) / International Electrotechnical Commission (IEC) /Joint Technical Committee (JTC) 1 has 2 entities developing quantum technology standards: most efforts will be from Working Group (WG) 14 Quantum computing, while sub-committee (SC) 27 Information security is specifically addressing QKD security.	Terminology Security requirements, test and evaluation methods for quantum key distribution	Normative: International Standards
The Study Groups (SG) of the International Telecommunications Union's Telecommunication Standardization Sector (ITU-T) assembles global experts to develop international standards (ITU-T Recommendations). SG11 (Signaling Requirements), SG13 (Future Networks), SG15 (Transport, Access and Home), and SG17 (Security) are in the process of developing documents of interest to quantum technologies.	Key efforts: QKD networks—Security, Management, Architecture	Recommendations Normative: International Standards

Description of quantum-related activity	Selected deliverable topics	Type of output[a]
The European CEN and CENELEC Joint Technical Committee 22 (CEN/CLC/JTC 22) is undertaking standardization efforts in the quantum field, building upon two deliverables from the CEN and CENELEC Focus Group on Quantum Technologies (FGQT, 2020-2023). Founded in 2022, the CEN/CLC/JTC 22 on Quantum Technologies has already developed several technical documents.	Quantum metrology, sensing and enhanced imaging; quantum computing and simulation; Quacom; and quantum cryptography.	Informative: Informational Documents; quantum technologies use cases Proposed Standards

[a] (e.g., report, interoperability standard, test protocol, procurement specification, etc.)

(Source: ITU-T Focus Group on Quantum Information Technology for Networks [ITU21] and CEN/CENELEC Focus Group on Quantum Technologies (FGQT) [OVA22])

- Quantum Network Construction Support
  - Support and construction of other Quanets
  - Roll-out of Quanets in different uses cases, driven by and in cooperation with the industry and government agencies
  - Translate feedback and experience from industry and government agencies back to the R&D hubs for further applied research
  - Creation of start-up companies
  - International support (e.g., EuroQCI)

EC strategy on Quantum Networks in EU27

All 27 EU member states signed a declaration agreeing to work together to explore and build a quantum communications Infrastructure (QCI) across Europe (EuroQCI) to boost their capabilities in quantum technologies, cybersecurity, and industrial competitiveness.

Quantum Internet Alliance (QIA)

QIA's objective is to build by 2029 two metropolitan networks containing quantum processors, connected by a long-distance fibre backbone using quantum repeaters. This network will be fully programmable to allow the realization of any application supported by the hardware using platform-independent software. Large EU industry players are involved in this effort: Cellnex Telecom, RHEA, SAP, Telecom Italia, Telefonica, Thales.

China's Quantum Satellite Micius

A project using China's quantum satellite Micius demonstrated in 2017 the first quantum-enabled communications in space by beaming photons between the satellite and two distant ground stations; particles can remain in a linked quantum state (quantum entanglement) at a record-breaking distance of more than 1,200 kilometers [YIN17], [LU22].

6.7 Standardization Efforts

Standardization always drives expedited deployment of a technology by enabling a large number of vendors to provide interoperable devices. The same applies to Quanets/Quaint. The term *"protocol"* typically refers to a list of steps to be performed by participating entities to achieve a defined goal. Quanet protocols, including protocols for QKD, are a relatively new initiative for Standards Developing Organizations (SDOs); fortunately, progress has been made in the recent past. The Internet Engineering Task Force (IETF), the ITU (International Telecommunication Union), the ETSI (European Telecommunications Standards Institute—for example, GR QKD) and the ISO (International Organization for Standardization—for example, ISO/IEC JTC1 AG4) are all pursuing Quanet standardization efforts. See Table 6.9 based on [ITU21] (and [ALL21]); additional activities have taken place more recently (e.g., see Figure 6.27 [YAN23]) on page 285, but these activities still fit the broad descriptions of this table.

6.7.1 Quantum Internet Research Group (QIRG)

QIRG operates under the IETF (which operates under the Internet Research Task Force). The goal of the QIRG is to address some key questions on how to design Quanets. Some of the areas under study by QIRG that need to be addressed include [QIR20]:
- Routing: Finding an optimal path in a Quanet considering the requirement of achieving a certain fidelity threshold and the low coherence time of quantum memories. There are a number of proposals, and which routing schemes are appropriate for which circumstances needs to be assessed.
- Resource allocation: Quanets entail new resource considerations, such as the coherence time of quantum memories. Some of the routing proposals already include a notion of dynamic traffic in the network.
- Connection establishment: Quanets deliver entangled states instead of packets, thus the connection semantics may be different; these semantics need to be defined.
- Interoperability: Different Quanets based on different hardware (ion traps, atomic ensembles, Nitrogen-Vacancy centers) and using different protocols are currently being designed and built. There is a need for internetworking.
- Security: Quanets aim at offering enhanced security for applications. The question focuses on the security of quantum network elements such as Quareps.
- Application Programming Interface (API) design: Develop API for entangled states, given new considerations such as fidelity and the low coherence time of quantum memories.

As discussed earlier in this chapter, a Quanet and a Quanet stack must take into account the fundamentally new properties of quantum entanglement. To that end, the 2023 IETF RFC 9340 describes the framework of a Quanet and introduces some basic goals and architectural principles for a Quaint. Goals discussed include the following (refer to the RFC for additional details):
- Support distributed quantum applications
- Support future distributed quantum applications
- Support heterogeneity
- Ensure security at the network level
- Make Quanets easy to monitor
- Ensure availability and resilience

Principles discussed in RFC 9340 include the following:
- Entanglement is the fundamental service
- Bell pairs are indistinguishable
- Fidelity is part of the service
- Time is an expensive resource

[Text continues on page 291]

Table 6.9 Standardization Activities on Quanets by the ITU[a]

Recommendation or activity	Description
Recommendation ITU-T Y.3800	"Overview on networks supporting quantum key distribution"
Recommendation ITU-T Y.3801	"Functional requirements for quantum key distribution networks"
Recommendation ITU-T Y.3802	"Quantum key distribution networks—Functional architecture"
Recommendation ITU-T Y.3803	"Quantum key distribution networks—Key management"
Recommendation ITU-T Y.3804	"Quantum key distribution networks—Control and management"
Recommendation ITU-T Y.3805	"Quantum key distribution networks—Software defined networking control"
Recommendation ITU-T Y.3806	"Quantum key distribution networks—Requirements for quality of service assurance"
Recommendation ITU-T Y.3807	"Quantum key distribution networks—QoS parameters"
Recommendation ITU-T Y.3808	"Framework for integration of quantum key distribution network and secure storage network"
Recommendation ITU-T Y.3809	"A role-based model in quantum key distribution networks deployment"
Recommendation ITU-T Y.3810	"Quantum key distribution network interworking—Framework". For QKDNs, Recommendation ITU-T Y.QKDN-iwfr specifies framework of QKDN interworking (QKDNi). This recommendation describes the overview of interworking QKDNs, the reference models, and the functional models of Gateway Functions (GWFs) and Interworking Functions (IWFs).
Recommendation ITU-T Y.3811	"Quantum key distribution networks—Functional architecture for quality of service assurance". This recommendation specifies a functional architecture of QoS assurance for QKDNs. This recommendation first provides an overview of the functional architecture of QoS assurance for the QKDN. It then describes the functional architecture of QoS assurance, which includes functional entities such as QoS data collection, data processing, data storage, data analytics, QoS anomaly detection and prediction, QoS policy decision-making, and enforcement and reporting. Based on the functional entities described in the functional architecture, this recommendation specifies a basic operational procedure of QoS assurance for the QKDN.
Recommendation ITU-T Y.3812	"Quantum key distribution networks—Requirements for machine learning based quality of service assurance". This recommendation specifies high-level and functional requirements of machine learning (ML) based QoS assurance for QKDNs. This recommendation first provides an overview of requirements of ML based QoS assurance for the QKDN. It describes a functional model of ML based QoS assurance and followed by associated high level and functional requirements of ML based QoS assurance. Some use cases are described.

[a] At the time of this writing

Table 6.9 Standardization Activities on Quanets by the ITU *(cont.)*

Recommendation or activity	Description
Supplement ITU-T Y.Sup70	"Quantum key distribution networks—Applications of machine learning".
Draft Recommendation ITU-T Y.3813 (ex Y.QKDN-iwrq)	"Quantum key distribution networks interworking—Functional requirements". For QKDN, Recommendation ITU-T Y.QKDN_iwrq specifies functional requirements for QKDN interworking (QKDNi). This recommendation describes the functional requirements for key management layer, QKDN control layer, and QKDN management layer, for interworking using Gateway Nodes (GWNs) and/or Interworking Nodes (IWNs).
Draft Recommendation ITU-T Y.3814 (ex Y.QKDN-ml-fra)	"Quantum key distribution networks—Functional requirements and architecture for machine learning enablement". QKDN is expected to maintain stable operations and meet the requirements of various cryptographic applications efficiently. Due to the advantages of machine learning (ML) related to autonomous learning, ML can help to overcome the challenges of QKDN in terms of quantum layer performances, key management layer performances, and QKDN control and management efficiency. Based on the functional requirements and architecture of QKDN in [ITU-T Y.3801] and [ITU-T Y.3802], this recommendation is to specify one possible set of functional requirements and a possible architecture for ML-enabled QKDN (QKDNml), including the overview, the functional requirements, architecture and operational procedures of QKDNml.
Draft Recommendation ITU-T Y.QKDN-iwac	"Quantum key distribution network interworking—Architecture". For QKDN, Recommendation ITU-T Y.QKDN_iwac specifies functional architecture for QKDNi.
Draft Recommendation ITU-T Y.QKDN-rsfr	"Framework of quantum key distribution network resilience". For QKDNs, this recommendation describes a framework of QKDN resilience. This recommendation describes the overview of QKDN resilience, models, and requirements of QKDN protection and recovery. It also includes different use cases of QKDN resilience in the appendix.
Draft Technical Report ITU-T TR-QEFN	"ITU-T's views for quantum-enabled future networks". The scope of this technical report is to describe ITU-T's views for Quantum-Enabled Future Networks (QEFN) for the future networks study to act as a document to help SG13 to study the future network evolution towards Quantum era.
Draft Recommendation ITU-T Y.QKDNi_SDNC	"Quantum key distribution network interworking—Software-defined networking control". This draft recommendation specifies the software-defined network control for the interworking including the overview of the role of SDN control for the interworking between QKDN providers, the functional entities of SDN control for the interworking, the interfaces of SDN control for the interworking, the functional requirements of SDN control for the interworking, and the security considerations.
Draft Recommendation ITU-T Y.QKDNf_fr	"Framework of Quantum Key Distribution Network Federation". This draft recommendation specifies the framework of Quantum Key Distribution Network Federation (QKDNf), including the overview of QKDNf, reference architecture for enabling QKDNf, functional entities of QKDNf, reference points for the QKDNf, functional requirements of the QKDNf, overall operational procedures of QKDNf, and security considerations.
Draft Supplement ITU-T Y.supp. QKDN-roadmap	"Standardization roadmap on quantum key distribution networks". Supplement Y.supp. QKDN-roadmap to ITU-T Y-series recommendations provides the standardization roadmap on quantum key distribution networks. It describes the landscape with related technical areas of trust technologies from an ITU-T perspective and identifies related standards and publications developed in Standards Development Organizations (SDOs).

Recommendation or activity	Description
Draft Technical Report ITU-T TR-QN-UC	"Use cases of quantum networks beyond QKDN". This technical report presents the use cases of quantum networks beyond QKDN under three categories as follows: (i) use cases based on Quantum Information Networks (QINs): use cases that depend on QIN to realize their function as, for example, but not exclusive to, distributed quantum computing, distributed quantum sensing, quantum clock network, and so on; (ii) use cases beneficial for classic networks: use cases that can provide additional functionality, new characteristics, or improved performance for classic ICT networks, as for example, but not exclusive to, Quantum Random Number Generator (QRNG), Quantum Time Synchronization (QTS), quantum cryptography beyond QKD, and so on; (iii) use cases where the network plays an intrinsic role for the QIT application: use cases in which the QIT application is significantly defined or enhanced by the functionality provided by a QIN and/or a classical network and is beyond simple remote access of a QIT application via a classical network. Some examples include synchronization of quantum clocks, distributed QRNG beacons for smart contracting, etc.
Draft Supplement ITU-T Y.supp. QKDN-UC	"Use cases of quantum key distribution networks". This supplement presents use cases of QKD networks. In particular, the scope of this supplement includes: (i) competitive advantage brought by QKDN; (ii) overview of QKDN use cases; and (iii) analysis of collected QKDN use cases including categorization.
Draft Recommendation ITU-T Y.QKDN-AMC	"Quantum key distribution networks—Requirements and architectural model for autonomic management and control". To support autonomic networking, cognitive networking, and self-management for improving the QoS in a quantum key distribution network (QKDN), this draft recommendation specifies the requirements and architectural model for Autonomic Management and Control (AMC) in QKDNs. In particular, the scope of this recommendation includes: (i) overview of AMC in QKDN; (ii) requirements for AMC in QKDN; (iii) consideration for cognition process of AMC in QKDN, and (iv) architectural model for AMC in QKDN.
Draft Recommendation ITU-T Y.QKDN_SSNreq	"Functional requirements for integration of quantum key distribution network and secure storage network". This recommendation specifies functional requirements for integration of quantum key distribution network and secure storage network. It includes detailed description of the following: (i) functional requirements for SSN user plane; (ii) functional requirements for SSN control plane; (iii) functional requirements for SSN storage plane; (iv) functional requirements for SSN management plane.
Draft Recommendation ITU-T Y.QKDN_SSNarch	"Functional architecture for integration of quantum key distribution network and secure storage network". This draft recommendation studies functional architecture for integration of quantum key distribution network and secure storage network. It includes detailed description of the following: (i) functional architecture model; (ii) functional elements; (iii) reference points; (iv) share format and metadata; (v) storage configuration; and (vi) operational procedures.

- Be flexible with regard to capabilities and limitations
  - The architecture should be able to function under the physical constraints imposed by the current-generation hardware. Near-future hardware will likely have low entanglement generation rates, quantum memories able to hold a handful of qubits at best, and decoherence rates that will render many generated pairs unusable.
  - The architecture should not make it difficult to run the network over any hardware that may come along in the near future. The physical capabilities of repeaters will improve, and redeploying a technology is extremely challenging.

Figure 6.15 (on page 269) depicted a basic layered model presented in RFC 9340 (with some extensions).

6.7.2 ITU (International Telecommunication Union)

There has been a *substantive amount* of standardization activities on Quanets by the ITU. The work on QKD networks and related security considerations is undertaken in ITU-T Study Group 13 (Future Networks and Cloud [45]) and in ITU-T Study Group 17 (SG17 Security). In particular, as of late 2022, ITU SG13 (QKD Network Architecture Aspects) published numerous Recommendations and Supplements on Quantum Key Distribution Networks (QKDN), as depicted in Table 6.8 [ITU22]; while these standards and drafts may be superseded over time, this list illustrates the level of interest and the standardization focus. A QKDN is a cryptographic infrastructure aimed at providing secure symmetric keys to cryptographic applications in institutional networks. ITU standards for QKDNs target the enablement of the integration of QKD technology into large-scale networks and also address its own (QKDN) security.

SG13 also assessed use cases of Quanets beyond QKDNs, such as quantum time synchronization, quantum computing, quantum random number generator use cases, and Quacom (e.g., quantum digital signatures, quantum anonymous transmission, and quantum currency).

The ITU-T Focus Group on Quantum Information Technology for Networks (FG-QIT4N) was established in September 2019 to provide a collaborative platform for pre-standardization aspects of Quanets focusing on matters that were not in the scope of SG13 and SG17 (Security Aspects of QKD Network and Applications of Quantum Random Number Generator [QRNG] for Security), Its main objectives were [ITU24]:

- To study the evolution and applications of Quanets (e.g., quantum computing, quantum communication) and develop technical report(s) on these matters.
- To focus on terminology and use cases for Quanets and develop technical report(s) on these matters.
- To provide necessary technical background information and collaborative conditions to effectively support Quanet-related standardization work in ITU-T Study Groups such as ETSI ISG-QKD, ETSI TC Cyber, IEEE, ISO/IEC JTC 1/SC 27/WG3, ISO/IEC JTC1 AG4, IETF, IRTF.
- To develop technical report(s) on telecom/network aspects of QKD networks that are identified in close coordination with ITU-T SG13 and SG17 as not within the scope of SG13 (QKD network architecture aspects) and SG17 (security aspects of QKD network and applications of QRNG for security), focused on terminologies, new use cases, protocols and transport technologies.

FG QIT4N produced the following reports:

- D1.1: Quantum information technology for networks terminology: Network aspects of quantum information technologies
- D1.2: Quantum information technology for networks use cases: Network aspects of quantum information technologies
- D1.4: Standardization outlook and technology maturity: Network aspects of quantum information technologies
- D2.1: Quantum information technology for networks terminology: Quantum Key Distribution Network (QKDN)
- D2.2: Quantum information technology for networks use cases: Quantum Key Distribution Network (QKDN)

- D2.3: Quantum key distribution network protocols: Quantum layer
- D2.3: Quantum key distribution network protocols: Key management layer, Quantum Key Distribution Network (QKDN) control layer and QKDN management layer
- D2.4: Quantum key distribution network transport technologies
- D2.5: Standardization outlook and technology maturity: Quantum Key Distribution Network (QKDN)

6.7.3 Other Standardization Efforts

Several other SDOs have Quacom/Quanet standardization activities, including (but not limited to) the following:

- The ISO and the IEC Joint Technical Committee JTC1 (ISO/IEC JTC1) has been pursuing quantum technologies standardization activities in two working groups: (i) WG14 on Quantum Computing, and (ii) SubCommittee SC27 WG3 for security certification of QKD systems.
- The European CEN and CENELEC Joint Technical Committee 22 (CEN/CLC/JTC 22) has addressed standardization in the quantum field and had developed several technical documents by the time of this writing. Emphasis has been on quantum metrology, sensing, and enhanced imaging; quantum computing and simulation; and quantum cryptography.
- IEEE had four active quantum standards efforts at press time: (i) P7130, which defines terminology for quantum technologies; (ii) P1913, Software-Defined Quantum Communication (SDQC), an application-layer protocol that uses TCP/IP and manages configuration of quantum endpoints to dynamically create, modify, or remove quantum applications and/ or protocols; (iii) P7131, which specifies Quacomp performance metrics to standardize performance benchmarking of quantum computing hardware and software; and (iv) P3120, which describes architectures for QCs.

In addition, there are many commercial and industrial activities to develop "de facto" agreements to facilitate deployment of the quantum technology (e.g., see [OVA22]).

References

[ABE23] A. J. G. Abelém, G. Vardoyan, D. Towsley, "Quantum Internet: The Future of Internetworking". April 2023, arXiv:2305.00598v1.

[ADH10] S. Adhikari, A. Majumdar, et al., "Swapping Path-Spin Intraparticle Entanglement onto Spin-Spin Interparticle Entanglement". EPL (Europhys. Lett.) 2010;89:10005. doi: 10.1209/0295-5075/89/10005.

[ALL21] C. Allocca, "Quantum Computing Standardization". Innovation, and Business". Inside Quantum Technology (IQT) Conference, November 3, 2021, New York. https://www.intriq.org/events/conference-dinside-quantum-technology-new-york

[ALS21] M. Alshowkan, B. P. Williams, et al., "Reconfigurable Quantum Local Area Network Over Deployed Fiber". PRX Quantum 2, 040304; 2021. doi: 10.1103/PRXQuantum. 2.040304.

[ALS22] M. Alshowkan, P. G. Evans, et al., "Advanced Architectures for High-Performance Quantum Networking" J. Opt. Commun. Netw. **14**, 493-499. May 2022. https://opg.optica.org/jocn/abstract.cfm?URI=jocn-14-6-493

[AWS21] D. Awschalom, K.K. Berggren, et al., "Development of Quantum Interconnects (QuICs) for Next-Generation Information Technologies". PRX Quantum 2, 017002

– Published 24 February 2021. Available online on 11/12/2023 at https://journals.aps.org/prxquantum/abstract/10.1103/PRXQuantum.2.017002.

[AZU23] K. Azuma, S.E. Economou, et al. "Quantum Repeaters: From Quantum Networks To The Quantum Internet". Rev. Mod. Phys. 95, 045006, 2023. https://doi.org/10.1103/RevModPhys.95.045006.

[BAP18] A. Bapat, Z. Eldredge, et al., "Unitary Entanglement Construction in Hierarchical Networks". Physical Review A, 98, 062328, 2018. https://doi.org/10.1103/PhysRevA.98.062328.

[BAY23] M. J. Bayerbach, S. E. D'aurelio, et al., "Bell-State Measurement Exceeding 50% Success Probability with Linear Optics". Science Advances, 9 Aug 2023, Vol 9, Issue 32. doi: 10.1126/sciadv.adf4080.

[BEN93] C. H. Bennett, G Brassard, et al., "Teleporting an Unknown Quantum State Via Dual Classical and Einstein-Podolsky-Rosen Channels". Phys. Rev. Lett. 70, 1895; 1993. PMID: 10053414. doi: 10.1103/PhysRevLett.70.1895

[BER12] H. Bernien, B. Hensen, et al., "Heralded Entanglement Between Solid-State Qubits Separated By 3 Meters". 26 Dec 2012 arXiv:1212.6136. https://doi.org/10.48550/arXiv.1212.6136. Also, Nature 497, 86-90, 2013. https://doi.org/10.1038/nature12016.

[BER23] E. A. Bersin, C. Errando-Herranz, D. R. Englund, Spectrally Multiplexed Solid State Quantum Emitters and Memories for Quantum Repeaters. U.S. Patent Application 20230344516, 2023-10-26. Uncopyrighted material.

[BHA20] M. K. Bhaskar, R. Riedinger, et al., "Experimental Demonstration of Memory-Enhanced Quantum Communication". Nature 580, 60, March 23, 2020. https://doi.org/10.1038/s41586-020-2103-5.

[BOS98] D. Boschi, S. Branca, et al., "Experimental Realization of Teleporting an Unknown Pure Quantum State via Dual Classical and Einstein-Podolsky-Rosen Channels". Phys. Rev. Lett. 80, 1121–1125;1998. doi: https://doi.org/10.1103/PhysRevLett.80.1121.

[BOU97] D. Bouwmeester, J.W. Pan, et al., "Experimental Quantum Teleportation". Nature 390, 575–579 (1997).

[BRA10] C. Branciard, N. Gisin, S. Pironio, "Characterizing the Nonlocal Correlations Created via Entanglement Swapping". *Phys. Rev. Lett.* 2010; 104:170401. doi: 10.1103/PhysRevLett.104.170401.

[BRI98] H.-J. Briegel, W. Dür, et al., "Quantum Repeaters: The Role of Imperfect Local Operations in Quantum Communication". Phys. Rev. Lett. 81, 5932–5935, 1998. https://doi.org/10.1103/PhysRevLett.81.5932.

[BUR22] I.A. Burenkov, N.F.R. Annafianto, et al., Experimental Shot-by-Shot Estimation of Quantum Measurement Confidence, Physical Review Letters, 128, 040404. 2022. https://doi.org/10.1103/PhysRevLett.128.040404

[BUS23] P. J. Bustard, D.G., England, et al., A Fibre Optic Integrated Quantum Memory for Light. U.S. Patent Application 20230420046, 2023-12-28. Uncopyrighted material.

[BUZ96] V. Bužek, M. Hillery, "Quantum Copying: Beyond the No-Cloning Theorem". Phys. Rev. A 54, 1844. 1 September 1996. https://doi.org/10.1103/PhysRevA.54.1844.

[CAC20] A.S. Cacciapuoti, M. Caleffi, et al., "When Entanglement Meets Classical Communications: Quantum Teleportation for the Quantum Internet". IEEE Transactions on Communications Vol. 68, Issue. 6, pp. 3808-3833, June 2020. https://ieeexplore.ieee.org/document/9023997. doi: 10.1109/TCOMM.2020.2978071.

[CAD19] A. Craddock, J. Hannegan, et al., "Quantum Interference Between Photons from

An Atomic Ensemble and A Remote Atomic Ion". Physical Review Letters, 123, 213601, 2019, https://doi.org/10.1103/PhysRevLett.123.213601.

[CAI24] W. Cai, X. MuLuyan, et al., "Protecting Entanglement Between Logical Qubits Via Quantum Error Correction". Nature Physics, March 2024, https://doi.org/10.1038/s41567-024-02446-8.

[CAL18] M. Caleffi, A.S. Cacciapuoti, G. Bianchi, "Quantum Internet: From Communication to Distributed Computing". In: Proceedings of the 5th ACM International Conference on Nanoscale Computing and Communication, pp. 1–4, 2018. https://doi.org/10.1145/3233188.3233224.

[CAS24] D. Castelvecchi, "'Quantum Internet' Demonstration in Cities Is Most Advanced Yet". Nature, May 15, 2024. Available online on May 16 at https://www.nature.com/articles/d41586-024-01445-2. doi: https://doi.org/10.1038/d41586-024-01445-2.

[CHI06] L. Childress, J. M. Taylor, et al., "Fault-Tolerant Quantum Communication Based on Solid-State Photon Emitters". Phys. Rev. Lett. 96, 070504, 2006. doi: 10.1103/PhysRev Lett.96.070504.

[CHI23] A. M. Childs, H. Fu, et al., "Streaming Quantum State Purification". arXiv:2309.16387. 28 September 2023. https://doi.org/10.48550/arXiv.2309.16387.

[CHU22] J. Chung, E. M. Eastman, et al., "Design and Implementation of the Illinois Express Quantum Metropolitan Area Network". IEEE Transactions on Quantum Engineering, 11 November 2022. doi: 10.1109/TQE.2022.3221029

[CHU23] J. Chung, "Design and Implementation of the Illinois Express Quantum Metropolitan Area Network". IETF 116, QIRG Meeting - March 27, 2023.

[COH09] O. Cohen, J. S. Lundeen, et al., "Tailored Photon-Pair Generation in Optical Fibers". Physical Review Letters 102, 123603; 2009. https://doi.org/10.1103/PhysRevLett.102.123603.

[COM18] Committee on Science, Space, & Technology, Lamar Smith Chairman. https://science.house.gov/2018/6/smith-announces-intent-introduce-national-quantum-initiative-act-hosts-first.

[COU18] C. Couteau "Spontaneous Parametric Down-Conversion". arXiv:1809.00127. https://doi.org/10.48550/arXiv.1809.00127. Also Contemporary Physics 59, 291-304; 2018. https://doi.org/10.1080/00107514.2018.1488463.

[CUB24] M. Cubeddu, (in collaboration with A. Abane, V. S. Mai), "Routing in Quantum Networks". March 19, 2024. IRTF QIRG. Available online on April 14, 2024 at https://datatracker.ietf.org/meeting/119/session/qirg

[DAH19] A. Dahlberg, M. Skrzypczyk, et al., "A Link Layer Protocol for Quantum Networks". arXiv:1903.09778, https://doi.org/10.48550/arXiv.1903.09778. SIGCOMM '19 Proceedings of the ACM Special Interest Group on Data Communication, August 2019, pp. 159-173. https://dl.acm.org/doi/10.1145/3341302.3342070.

[DAI21] S. Daiss, S. Langenfeld, et al., "A Quantum-Logic Gate Between Distant Quantum-Network Modules". Science. 2021;371:614–617. doi:10.1126/science.abe3150.

[DEP23] M.V.S. de Paula, W.W.T. Sinesio, A. V. Dodonov, "Ancilla-Assisted Generation of Photons from Vacuum via Time-Modulation of Extracavity Qubit". Entropy 2023, 25, 901. https://doi.org/10.3390/e25060901.

[DEV13] S. J. Devitt, W. J. Munro, K. Nemoto, "Quantum Error Correction for Beginners". Reports on Progress in Physics, Volume 76, Number 7. 20 June 2013. doi: 10.1088/0034 -4885/76/7/076001.

[DJO21] I. B. Djordjevic, "Cluster State-based Quantum Computing". *in Quantum Information Processing, Quantum Computing, and Quantum Error Correction* (Second Edition), Elsevier Inc, 2021. ISBN 978-0-12-821982-9. doi: 10.1016/C2019-0-04873-X.

[DJO24] I. Djordjevic, Global Quantum Communication Networks. U.S. Patent 11876565, 2024-01-16. Uncopyrighted material.

[DUA01] L. M. Duan, M. D. Lukin, et al., "Long Distance Quantum Communication with Atomic Ensembles and Linear Optics". Nature 414, 413, 2001.

[EBE93] P. H. Eberhard, "Background Level and Counter Efficiencies Required for A Loophole-Free Einstein-Podolsky-Rosen Experiment". Phys. Rev. A 47, 747–750, 1993.

[EIS11] M. D. Eisaman, J. Fan, et al., "Invited Review Article: Single-Photon Sources and Detectors". Rev. Sci. Instrum. 82, 071101; 2011. https://doi.org/10.1063/1.3610677.

[ELD20] Z. Eldredge, L. Zhou, et al., "Entanglement Bounds on the Performance of Quantum Computing Architectures". Physical Review Research, 2, 033316 (2020) https://doi.org/10.1103/PhysRevResearch.2.033316

[ENE23] Department of Energy Announces $24 Million for Research on Quantum Networks", Energy.gov. August 29, 2023. Available online on June 21, 2024 at https://www.energy.gov/science/articles/department-energy-announces-24-million-research-quantum-networks.

[ERH21] A. Erhard, H. Poulsen Nautrup, et al., "Entangling Logical Qubits with Lattice Surgery". Nature 589, 220–224 (2021). https://doi.org/10.1038/s41586-020-03079-6

[FRA12] M. Francis, "Photon Heralds Entanglement in New Quantum Repeater". 3/6/2012. Available online on January 23, 2024 at https://arstechnica.com/science/2012/03/herald-photons-announce-crystal-entanglement-and-there-was-much-rejoicing/

[FUR98] A. Furusawa, J. L. Sørensen, et al., "Unconditional Quantum Teleportation". Science, 23 Oct 1998. Vol 282, Issue 5389, pp. 706-709. doi: 10.1126/science.282.5389.706.

[GEN24] E. Gent, "Quantinuum Successfully Teleports A Logical Qubit - The Feat Could Enable Large-Scale, Fault-Tolerant Quantum Computers". IEEE Spectrum. 9/30/2024. Available online on 10/7/2024 at https://spectrum.ieee.org/logical-qubit.

[GOT97] D. Gottesman, "Stabilizer Codes and Quantum Error Correction". Thesis in partial fulfillment of the requirements for the Degree of Doctor of Philosophy. California Institute of Technology Pasadena, California 2004 (Submitted May 21, 1997). Available online on May 3, 2024 at https://thesis.library.caltech.edu/2900/2/THESIS.pdf.

[GOU21] Q. Guo, Y.-Y. Zhao, et al., "Testing A Quantum Error-Correcting Code on Various Platforms". Science Bulletin. 66 (1): 29–35, 2021. arXiv:2001.07998. doi:10.1016/j.scib.2020.07.033.

[HAO24] H. Hao Q.-Y Zhao, et al., "A Compact Multi-Pixel Superconducting Nanowire Single-Photon Detector Array Supporting Gigabit Space-To-Ground Communications". Light Sci Appl 13, 25; 2024. https://doi.org/10.1038/s41377-023-01374-1

[HER22] S. L. N. Hermans, M. Pompili, et al., "Qubit Teleportation Between Non-Neighbouring Nodes in A Quantum Network". Nature, Vol 605, 26 May 2022, pages 663ff. https://doi.org/10.1038/s41586-022-04697-y.

[HOF12] J. Hofmann, M. Krug, et al., "Heralded Entanglement Between Widely Separated Atoms". Science 337(6090):72-5. July 2012. doi:10.1126/science.1221856.

[HUA19] Y.-L. Hua, T.-S. Yang, et al., "Storage of Telecom-C-Band Heralded Single Photons with Orbital-Angular-Momentum Encoding in A Crystal". Science Bulletin, Volume 64, Issue 21, 2019, Pages 1577-1583, ISSN 2095-9273. https://doi.org/10.1016/j.scib.2019.09.006.

[HUM18] P. C. Humphreys, K. Norbert, et al., "Deterministic Delivery of Remote Entanglement on a Quantum Network". Nature. 2018;558:268–273. doi:10.1038/s41586- 018-0200-5.

[ILL22] J. Illiano, M. Caleffi, et al., "Quantum Internet Protocol Stack: A Comprehensive

Survey". Computer Networks, Volume 213, August 2022, 109092, ISSN 1389-1286. https://doi.org/10.1016/j.comnet.2022.109092.

[ITU21] ITU-T Focus Group on Quantum Information Technology for Networks (FG-QIT4N). "FG QIT4N D1.4 Standardization Outlook and Technology Maturity: Network Aspects of Quantum Information Technologies". Available online on April 15, 2024 at https://www.itu.int/dms_pub/itu-t/opb/fg/T-FG-QIT4N-2021-D1.4-PDF-E.pdf.

[ITU22] International Telecommunication Union—Study Period 2022-2024. "LS on Work Progress On Quantum Key Distribution (QKD) Network in SG13 (as of November 2022)". ITU-T Working Party 3/13.

[ITU24] ITU-T Focus Group on Quantum Information Technology for Networks (FG-QIT4N). Available online on April 15, 2024 at https://www.itu.int/en/ITU-T/focusgroups/qit4n/Pages/default.aspx

[JI19] Z. Ji, H. Zhang, et al., "Quantum Protocols for Secure Multi-Party Summation". *Quantum Inf. Process.* 2019;18:168. doi: 10.1007/s11128-018-2141-1.

[JI22] Z. Ji, P. Fan P., H. Zhang, "Entanglement Swapping for Bell States and Greenberger–Horne–Zeilinger States in Qubit Systems". *Phys. A Stat. Mech. Its Appl.* 2022; 585:126400. doi: 10.1016/j.physa.2021.126400.

[JIA09] L. Jiang, J. M. Taylor, et al., "Quantum Repeater with Encoding". Physical Review A, 79(3), 2009. https://doi.org/10.1103/PhysRevA.79.032325.

[JOH22] M. Johnson-Groh, "What Is a Quantum Network?" Symmetry Magazine, 01/26/22. Fermilab/SLAC publication. Available at https://www.symmetrymagazine.org/article/what-is-a-quantum-network?language_content_entity=und

[JOU23] P. Jouzdani, H. Arslan Hashim, E. R. Mucciolo, "Quantum Algorithms for State Preparation and Data Classification based on Stabilizer Codes". 18 September 2023. arXiv:2309.10087. https://doi.org/10.48550/arXiv.2309.10087.

[KAP23] K. Kapoor, S. Xie, et al., "Picosecond Synchronization System for The Distribution of Photon Pairs Through a Fiber Link Between Fermilab and Argonne National Laboratories," IEEE Journal of Quantum Electronics, Volume: 59, Issue: 4, August 2023. doi: 10.1109/JQE.2023.3240756.

[KHA14] A. Khalique, B. C. Sanders, "Long-Distance Quantum Communication Through Any Number of Entanglement-Swapping Operations". *Phys. Rev. A.* 2014;90:032304. doi: 10.1103/PhysRevA.90.032304.

[KIM08] H. J. Kimble, "The Quantum Internet". Nature 453, 1023–1030 (2008). doi:10.1038/nature07127.

[KNA24] C. M. Knaut, A. Suleymanzade, et al.," Entanglement of Nanophotonic Quantum Memory Nodes in A Telecom Network". Nature 629, 573–578 (2024). 2024 May;629(8012):573-578. doi: 10.1038/s41586-024-07252-z. Open Access: This article is licensed under a Creative Commons Attribution 4.0 International License, which permits use, sharing, adaptation, distribution and reproduction in any medium or format, as long as you give appropriate credit to the original author(s) and the source, provide a link to the Creative Commons license, and indicate if changes were made. To view a copy of this license, visit http://creativecommons.org/licenses/by/4.0/.

[KOK00] P. Kok, S. L. Braunstein, "Entanglement Swapping as Event-Ready Entanglement Preparation". *Fortschr. Phys. Prog. Phys.* 2000;48:553–557. doi: 10.1002/(SICI)1521-3978(200005)48:5/7<553::AID-PROP553>3.0.CO;2-K.

[KOZ20] W. Kozlowski, A. Dahlberg, S. Wehner, "Designing a Quantum Network Protocol".

CoNEXT '20: Proceedings of the 16th International Conference on Emerging Networking EXperiments and Technologies. November 2020. Pages 1–16. https://doi.org/10.1145/3386367.3431293.

[KOZ23] W. Kozlowski, S. Wehner, et al., RFC 9340 - Architectural Principles for a Quantum Internet, March 2023. https://doi.org/10.17487/RFC9340.

[KRU23] V. Krutyanskiy, M. Galli, et al., "Entanglement of Trapped-Ion Qubits Separated By 230 Meters". Phys. Rev. Lett. 2023;130:050803. doi:10.1103/PhysRevLett.130.050803.

[KWI95] P. G. Kwiat, K. Mattle, et al., "New High-Intensity Source of Polarization Entangled Photon Pairs". Physical Review Letters Vol. 75, No. 24, 1995, pp. 4337-4341.

[KYL22] A. Kyle, C. Rau, et al., "Entanglement Thresholds of Double-Parametric Quantum Transducers". Physical Review Applied, 17, 044057, 2022, https://doi.org/10.1103/PhysRevApplied.17.044057.

[LAG21] D. Lago-Rivera, S. Grandi, et al., "Telecom-Heralded Entanglement Between Multimode Solid-State Quantum Memories". Nature 594,37–40, 2021. https://doi.org/10.1038/s41586-021-03481-8.

[LAG23] D. Lago-Rivera, J. V. Rakonjac, et al., "Long Distance Multiplexed Quantum Teleportation from A Telecom Photon to A Solid-State Qubit". Nat Commun. 14, 1889; 2023. https://doi.org/10.1038/s41467-023-37518-5.

[LAR01] J.-Å. Larsson, J. Semitecolos, "Strict Detector-Efficiency Bounds For N-Site Clauser-Horne Inequalities". Phys. Rev. A 63, 022117, 2001.

[LEE20] Y. Lee, E. Bersin, et al., "A Quantum Router Architecture for High-Fidelity Entanglement Flows in Multi-User Quantum Networks". 2020/05/04, https://doi.org/10.48550/arXiv.2005.01852.

[LEU06] M. Leung, T. C. Ralph, "Quantum Memory Scheme Based on Optical Fibers and Cavities". Phys. Rev. A 74, 022311 – Published 10 August 2006. https://doi.org/10.1103/PhysRevA.74.022311.

[LI24] J. Li, Z. Xie, et al., "Heralded Entanglement Between Error-Protected Logical Qubits for Fault-Tolerant Distributed Quantum Computing". Sci. China Phys. Mech. Astron. 67, 220311, 2024. https://doi.org/10.1007/s11433-023-2245-9.

[LIU21] X. Liu, J. Hu, et al., "Heralded Entanglement Distribution Between Two Absorptive Quantum Memories". Nature, 2021; 594(7861):41-45. doi: 10.1038/s41586-021-03505-3.

[LIU24] J. L. Liu, X. Y. Luo, et al., "Creation of Memory–Memory Entanglement in A Metropolitan Quantum Network". Nature 629, 579–585, 2024. https://doi.org/10.1038/s41586-024-07308-0.

[LU22] C.-Y. Lu, Yuan Cao, et al., "Micius Quantum Experiments in Space". *Rev. Mod. Phys.* 94(3), 035001, 2022. https://doi.org/10.1103/revmodphys.94.035001

[LUO22] X-Y. Luo, Y. Yu, et al., "Postselected Entanglement Between Two Atomic Ensembles Separated By 12.5 Km". Phys. Rev. Lett. 2022;129:050503. doi:10.1103/PhysRevLett.129.050503.

[MAR02] C. Martijn de Sterke, E. N. Tsoy, and J. E. Sipe, "Light Trapping in A Fiber Grating Defect by Four-Wave Mixing". Vol. 27, Issue 7, pp. 485-487; 2002. https://doi.org/10.1364/OL.27.000485

[MAR12] D. C. Marinescu, G. M. Marinescu, "Chapter 1—Preliminaries". in *Classical and Quantum Information*, Elsevier Inc., 2012. ISBN 978-0-12-383874-2. doi: https://doi.org/10.1016/C2009-0-64195-7.

[MAR17] N. Maring, P. Ferrara, et al., "Photonic Quantum State Transfer Between a Cold Atomic Gas and a Crystal". Nature 551, 485–488, 2017. https://doi.org/10.1038/nature24468.

[MET11] R.V. Meter, J. Touch, "Recursive Quantum Repeater Networks". Progress in Informatics, 1(8):65–79, 2011. DOI:10.2201/NiiPi.2011.8.8. Corpus ID: 18165387.

[MEY16] E. Meyer-Scott, *Heralding Photonic Qubits for Quantum Communication*. Thesis presented to the University of Waterloo in fulfillment of the thesis requirement for the degree of Doctor of Philosophy in Physics (Quantum Information) Waterloo, Ontario, Canada, 2016.

[MIH19] P. Mihir, D.R. Englund, M. Heuck, Apparatus and Methods for Single Photon Sources. U.S. Patent 10,429,718. 2019-10-01. Uncopyrighted material.

[MIN09] D. Minoli, *Satellite Systems Engineering in an IPv6 Environment* (Francis and Taylor 2009).

[MIN15] D. Minoli, *Innovations in Satellite Communications and Satellite Technology: The Industry Implications of DVB-S2X, High Throughput Satellites, Ultra HD, M2M, and IP* (Wiley 2015).

[MUN10] W. J. Munro, K. A. Harrison, et al., "From Quantum Multiplexing to High Performance Quantum Networking". Nat. Photonics 4, 792, 2010.

[MUN12] W. J. Munro, A. M. Stephens, et al., "Quantum Communication Without the Necessity of Quantum Memories". Nat. Photonics 6, 777, 2012.

[MUR14] S. Muralidharan, J. Kim, et al., "Ultrafast and Fault-Tolerant Quantum Communication Across Long Distances". Phys. Rev. Lett. 112, 250501, 2014.

[NAG23] S. Nagayama, "Quantum Internet Testbed Efforts in Japan (overview)". Slides-116-qirg. Yokohama. March 27, 2023. https://datatracker.ietf.org/meeting/116/session/qirg.

[NAT19] Nature Staff, "A Diamond's Quantum Memory Sets a Glittering Record". Nature 573, 466-467, 2019. doi: https://doi.org/10.1038/d41586-019-02766-3.

[NIE02] M.A. Nielsen, I. Chuang, "Quantum computation and quantum information". American Journal of Physics Vol. 70, 2002, pp. 558-559.

[NIS23] NIST, "Quantum Networks at NIST". Available on Nov. 11, 2023 at https://www.nist.gov/pml/productsservices/quantum-networks-nist.

[NIS24] NIST Staff, "Quantum Network Metrology". Available online on 3/11/2024 at https://www.nist.gov/pml/quantum-networks-nist/quantum-network-metrology

[NUN13] J. Nunn, N. K. Langford, et al., "Enhancing Multiphoton Rates with Quantum Memories". Physical Review Letters 110, 133601; 2013. https://doi.org/10.1103/PhysRevLett.110.133601.

[OVA22] O. van Deventer, N. Spethmann, et al., "Towards European Standards for Quantum Technologies". 3 Oct 2022, arXiv:2203.01622. https://doi.org/10.48550/arXiv.2203.01622. EPJ Quantum Technol. 9 (2022) 33. https://doi.org/10.1140/epjqt/s40507-022-00150-1.

[PES23] A. Pesah, "The Stabilizer Trilogy I — Stabilizer codes". Quantum Computing, 31 Jan 2023. Available online on May 3, 2024 at https://arthurpesah.me/blog/2023-01-31-stabilizer-formalism-1/

[PIR17] S. Pirandola, R. Laurenza, et al., "Fundamental Limits of Repeaterless Quantum Communications". arXiv:1510.08863. 24 Jan 2017. https://doi.org/10.48550/arXiv.1510.08863. Also, Nat. Commun. 8, 15043, 2017. https://doi.org/10.1038/ncomms15043.

[POL19] S.V. Polyakov, I.A. Burenkov, Communication Linker for Communication Linking. U.S. Patent 10,382,141. 2019-08-13. Uncopyrighted Material.

[POM21] M. Pompili, S. L. N. Hermans, et al., "Realization of a Multinode Quantum Network of Remote Solid-State Qubits". Science. 2021 Apr 16;372(6539):259-264. doi: 10.1126/science.abg1919.

[POM22] M. Pompili, C. Delle Donne, et al., "Experimental Demonstration of Entanglement

Delivery Using a Quantum Network Stack". npj Quantum Information 8.1, 2022. doi: 10.1038/s41534-022-00631-2.

[QIR20] Quantum Internet Research Group (QIRG), https://datatracker.ietf.org/group/qirg/about/, 2020.

[RAU01] R. Raussendorf, H. J. Briegel, "A One-Way Quantum Computer". PhysRev Lett.86.5188. Volume 86, Issue 22, May 2001. doi: 10.1103/PhysRevLett.86.5188.

[REN17] J.-G. Ren, P. Xu, et al., "Ground-to-Satellite Quantum Teleportation". Nature, 549(7670):70–73, 2017. https://doi.org/10.1038/nature23675

[RIT12] S. Ritter, C. Nölleke, et al., "An Elementary Quantum Network of Single Atoms in Optical Cavities". Nature. 2012;484:195–200. doi:10.1038/nature11023.

[ROA14] L. Roa, A. Muñoz, G. Grüning, "Entanglement Swapping for X States Demands Threshold Values". *Phys. Rev. A.* 2014;89:064301. doi: 10.1103/PhysRevA.89.064301.

[ROB23] J. Robbers, "Quantum Network Testbed Developments in NL". IETF 116 Meeting March 27, 2023, Yokohama. https://datatracker.ietf.org/meeting/116/session/qirg.

[ROZ23] G. G. Rozenman, N. K. Kundu, et al., "The Quantum Internet: A Synergy of Quantum Information Technologies And 6G Networks". IET Quantum Communication, Volume 4, Issue 4, December 2023. Pages 147-166. This article also appeared in: Quantum Industry: Applications in Quantum Communication (Quantum. Tech Europe 2022). https://doi.org/10.1049/qtc2.12069.

[RYA22] C. Ryan-Anderson, N. C. Brown, et al., "Implementing Fault-tolerant Entangling Gates on the Five-qubit Code and the Color Code". 3 August 2022, arXiv:2208.01863. https://doi.org/10.48550/arXiv.2208.01863.

[RYA24] C. Ryan-Anderson, N. C. Brown, et al, "High-Fidelity Teleportation Of A Logical Qubit Using Transversal Gates And Lattice Surgery". Science, 19 Sep 2024, Vol 385, Issue 6715, pp. 1327-1331. doi: 10.1126/science.adp601.

[SER19] A. Seri, D. Lago-Rivera, et al., "Quantum Storage of Frequency-Multiplexed Heralded Single Photons". Phys. Rev. Lett. 123, 080502, 2019. https://doi.org/10.1103/PhysRevLett.123.080502.

[SHA24] A. Shabani, B. Qi, et al., Packet Switched Quantum Network. U.S. Patent Application US 2024/0147101. May 2, 2024. Uncopyrighted material.

[SHO95] P. W. Shor, "Scheme for Reducing Decoherence in Quantum Computer Memory". Physical Review A. 52 (4): R2493–R2496. October 1995. doi: https://doi.org/10.1103/PhysRevA.52.R2493.

[SIV22] I. Sivan, Y. Cohen, et al., System and Method for Processing Between a Plurality of Quantum Controllers, WO 2022/238759, Nov. 17, 2022. International Application published under the Patent Cooperation Treaty (PCT). Uncopyrighted material.

[STE20] L. J. Stephenson, D. P. Nadlinger, et al., "High-Rate, High-Fidelity Entanglement of Qubits Across an Elementary Quantum Network". Phys. Rev. Lett. 2020;124:110501. doi:10.1103/PhysRevLett.124.110501.

[STE97] A. Steane, "Active Stabilization, Quantum Computation, and Quantum State Synthesis". Phys. Rev. Lett. 78, 2252 (1997). https://doi.org/10.1103/PhysRevLett.78.2252.

[STO17] R. Stockill, M. J. Stanley, et al., "Phase-Tuned Entangled State Generation Between Distant Spin Qubits". Phys. Rev. Lett. 2017;119:010503. doi: 10.1103/PhysRevLett.119.010503.

[STO24] A. J. Stolk, K. L. van der Enden, et al., "Metropolitan-Scale Heralded Entanglement of Solid-State Qubit". Preprint at arXiv. https://doi.org/10.48550/arXiv.2404.03723. April 4, 2024.

[SWA24] M. Swayne, "$20 Million Investment Proposed for DoD-Led Quantum Center of Excellence". Quantum Computing Business. June 24, 2024. Available online on June 24, 2024 at https://thequantuminsider. com/2024/06/24/20-million-investment-proposed-for-dod-led-quantum-center-of-excellence/.

[TAU24] M. Taufiqi, A. Purwanto, et al., "Cyclic Quantum Teleportation with Multi-Level of Control". Int J Theor Phys 63, 9; 2024. https://doi.org/10.1007/s10773-023-05513-7

[TEO13] C. Teo, Araújo, M., Quintino, et al., "Realistic Loophole-Free Bell Test with Atom–Photon Entanglement". Nat. Commun. 4, 2104, 2013. https://doi.org/10.1038/ncomms3104

[USM12] I. Usmani, C. Clausen, "Heralded Quantum Entanglement Between Two Crystals". arXiv:1109.0440. 2 Sept 2011. https://doi.org/10.48550/arXiv.1109.0440. Also, Nature Photonics 6, 234-237, 2012. https://doi.org/10.1038/NPHOTON.2012.34

[VAN14] R. Van Meter, *Quantum Networking*, Wiley 2014. ISBN: 9781848215375, ISBN: 9781118648919. doi:10.1002/9781118648919.

[VAN22] T. van Leent, M. Bock, et al., "Entangling Single Atoms Over 33 Km Telecom fibre". Nature. 2022;607:69–73. doi:10.1038/s41586-022-04764-4.

[VAR22] G. Vardoyan, M. Skrzypczyk, S. Wehner, "On the Quantum Performance Evaluation Of Two Distributed Quantum Architectures". Performance Evaluation 153, 2022. https://doi.org/10.1016/j.peva.2021.102242.

[VED06] V. Vedral, *Introduction to Quantum Information Science*. Oxford University Press on Demand; Oxford, UK: 2006. (2013 Reprint Edition, ISBN-1: 9780199673483, ISBN-13: 978-0199673483.

[WEH18] S. Wehner, D. Elkouss, R. Hanson, "Quantum Internet: A Vision for The Road Ahead". Science 362, 303, 19 October 2018. Available online at https://www.science.org on November 01, 2023. doi: 10.1126/science.aam9288.

[WEI15] Y. S. Weinstein, "Syndrome Measurement Strategies for the [[7,1,3]] Code". Quantum Information Processing; Issue 6/2015.

[WEI22] S. H. Wei, B. Jing, et al., "Towards Real-World Quantum Networks: A Review". Laser and Photonics Reviews, January 2022, Vol. 16, 2100219. https://onlinelibrary. wiley.com/doi/10.1002/lpor.202100219. doi: 10.1002/lpor.202100219.

[WOO01] W. K. Wootters, "Entanglement of Formation and Concurrence". Quantum Information and Computation, Vol. 1, No. 1 (2001) 27-44.

[WOO82] W. Wootters, W. Zurek, "A Single Quantum Cannot Be Cloned". *Nature* 299, 802–803 (1982). https://doi.org/10.1038/299802a0.

[WOO98] W. K. Wootters, "Entanglement of Formation of An Arbitrary State of Two Qubits". Phys. Rev. Lett. 1998, 80, 2245–2248.

[YAN23] H. Yang, "Quantum Information Technology Standardization Updates". ISO/IEC JTC 1 Workshop on Quantum Information Technology Standardization, ISO/IEC JTC 1/WG 14 Convenor. April 2023.

[YIN17] J. Yin, Y. Cao, et al., "Satellite-Based Entanglement Distribution Over 1200 Kilometers". Science Vol. 356, No. 6343, June 2017, pp. 1140-1144. doi: 10.1126/science.aan3211.

[YON20] Y. Yong, M. Fei, et al., "Entanglement of Two Quantum Memories Via Fibres Over Dozens of Kilometres". Nature. 2020;578:240–245. doi: 10.1038/s41586-020-1976-7.

[YUA08] Z.S. Yuan, Y. A. Chen, et al., "Experimental Demonstration of a BDCZ Quantum Repeater Node". Nature, Vol. 454, No. 7208, 2008, pp. 1098-1101.

[YUR92] B. Yurke, D. Stoler, "Bell's-Inequality Experiments Using Independent-Particle Sources". *Phys. Rev. A.* 1992; 46:2229–2234. doi: 10.1103/PhysRevA.46.2229.

[ZAN23] S. M. Zangi, C. Shukla, et al., "Entanglement Swapping and Swapped Entanglement". Entropy (Basel). 2023 Feb 25;25(3):415. doi: 10.3390/e25030415. PMID: 36981304; PMCID: PMC10047960.

[ZHA23] J. Zhao, H. Jeng, et al., "Enhancing Quantum Teleportation Efficacy with Noiseless Linear Amplification". Nat. Commun. 14, 4745; 2023. https://doi.org/10.1038/s41467-023-40438-z.

[ZHO15] L. Zhou, Y.-B. Sheng, "Concurrence Measurement for the Two-Qubit Optical and Atomic States". Entropy 2015, 17, 4293-4322; 19 June 2015. doi:10.3390/e17064293.

[ZUK93] M Zukowski, A Zeilinger, et al., "'Event-ready-detectors' Bell Experiment Via Entanglement Swapping". Phys Rev Lett., 1993 Dec 27; 71(26):4287-4290. PMID: 10055208. doi: 10.1103/PhysRevLett.71.4287.

Chapter 7

Quantum-Based Security

This chapter elaborates on some aspects of quantum-based security, in particular Quantum Key Distribution (QKD) concepts. More generally, Quantum Cryptography refers to cryptosystems based on the properties of Quantum Mechanics (QM). This chapter only briefly covers Quantum Cryptography per se—a burgeoning R&D field which would require an extensive treatment by itself and/or in a separate text.

7.1 Introduction and Overview

7.1.1 Introduction

QKD is a method—a protocol—that utilizes QM to distribute keys between parties, as briefly discussed in Chapter 1. QKD systems and networks allow one to establish provably *secure classical cryptographic keys* between two or more legitimate users. QKD is an important subdiscipline of Quantum Information Science and Technology (QIST) that is reaching a technical level that makes it suited for commercialization. It is a methodology for securely sharing an encryption key by utilizing a single photon continuously transmitted between a transmission node and a reception node connected by an optical fiber (network).

QKD is used to create and distribute a pair of symmetric encryption keys in such a manner that the security of the process relies on the principles of QM rather than on the complexity of various mathematical problems. QKD systems are able to exchange keys between two parties in a secure way that cannot be guessed. Any cryptographic technique becomes ineffective if the key distribution mechanism is fragile; thus, QKD offers unconditionally secure communication based on QM principles.

A QKD protocol is a set of steps to establish a key: QKD enables two remote parties, a sender (typically known as Alice) and a receiver (typically known as Bob), to establish a secret key over a quantum channel—Alice and Bob can thus share unconditional secure keys (e.g., see [LIU22], [CAO22], [STR20], [MEH20], [TSA21], [BED17]). See Figures 7.1 and 7.2 for some simplified examples. Under a number of assumptions, the QKD key-establishment process is guaranteed to be secure against all inimical agents. Perfect cloning of an arbitrary unknown quantum state is forbidden in QM by the no-cloning theorem and is the intrinsic concept of QKD-based security.

Initially, in the 2000s, only point-to-point (PTP) links supporting QKD were demonstrated; more recently, medium-size terrestrial networks have been deployed over regional distances (e.g., 75–100 miles); one has also seen demonstrations of satellite-based systems. Fiber-based terrestrial QKD networks have reached a level of maturity, and a number of testbeds have already been deployed in various countries. Fiber-based QKD systems have also been integrated with classical optical networks and/or wireless/satellite (including CubeSats) systems and/or atmospheric free-space lasers systems [SHA21], [BOA18], [PHA19], [LIA17], [LU22], [VIL20], [TRI18]. Higher, improved data rates have also been achieved in the recent past. QKD can also be integrated with other cryptographic algorithms to provide enhanced security: the quantum keys obtained from QKD can be used for implementing post-quantum cryptographic algorithms [YAV23].

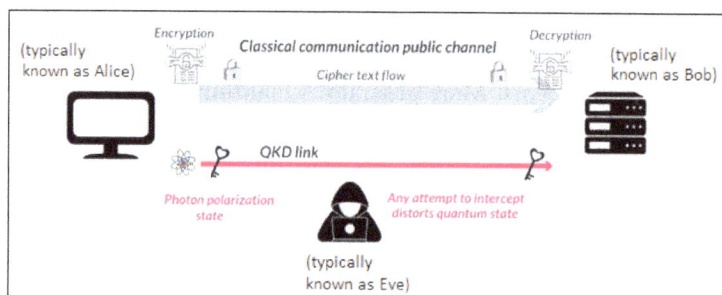

Figure 7.1   Logical, simplified view of QKD-based security

7.1.2 Overview

Cryptographic algorithms include a variety of protocols, procedures, schemes, and infrastructures, but they rely on a relatively small set of cryptographic mechanisms: mathematical functions that transform data, generally using a parameter, the key, to support confidentiality and protect the information under consideration. The protection of the keys themselves is critical to the unfaltering security of the protected data. Public-key cryptographic algorithms (also known as *asymmetric algorithms*) require the originator to utilize one key and the recipient to use a different but related key. One of these asymmetric keys, specifically the private key, must be kept secret, while the other key, the public key, can be made public without degrading the security of the cryptographic process. Asymmetric (public-key) algorithms are routinely employed for data integrity applications (e.g., digital signatures) and for the protected exchange of shared keys used by symmetric algorithms [NIS23]. Symmetric algorithms necessitate a secret key to be shared by sender and receiver; symmetric algorithms are more efficient for the protection of large volumes of information, but the secure exchange and establishment of shared keys requires protection of its own—currently by asymmetric cryptography, but in the future by QKD techniques.

A Quantum Network (Quanet) protocol is a kind of communication protocol wherein a set of rules allows two or more entities of a communication ecosystem to transmit quantum information; this communication is accomplished via some kind of variation of a physical quantity. In classical networking, many symmetric key and public key cryptographic protocols have been standardized, and a set of comparable interoperability supporting efforts in the quantum realm

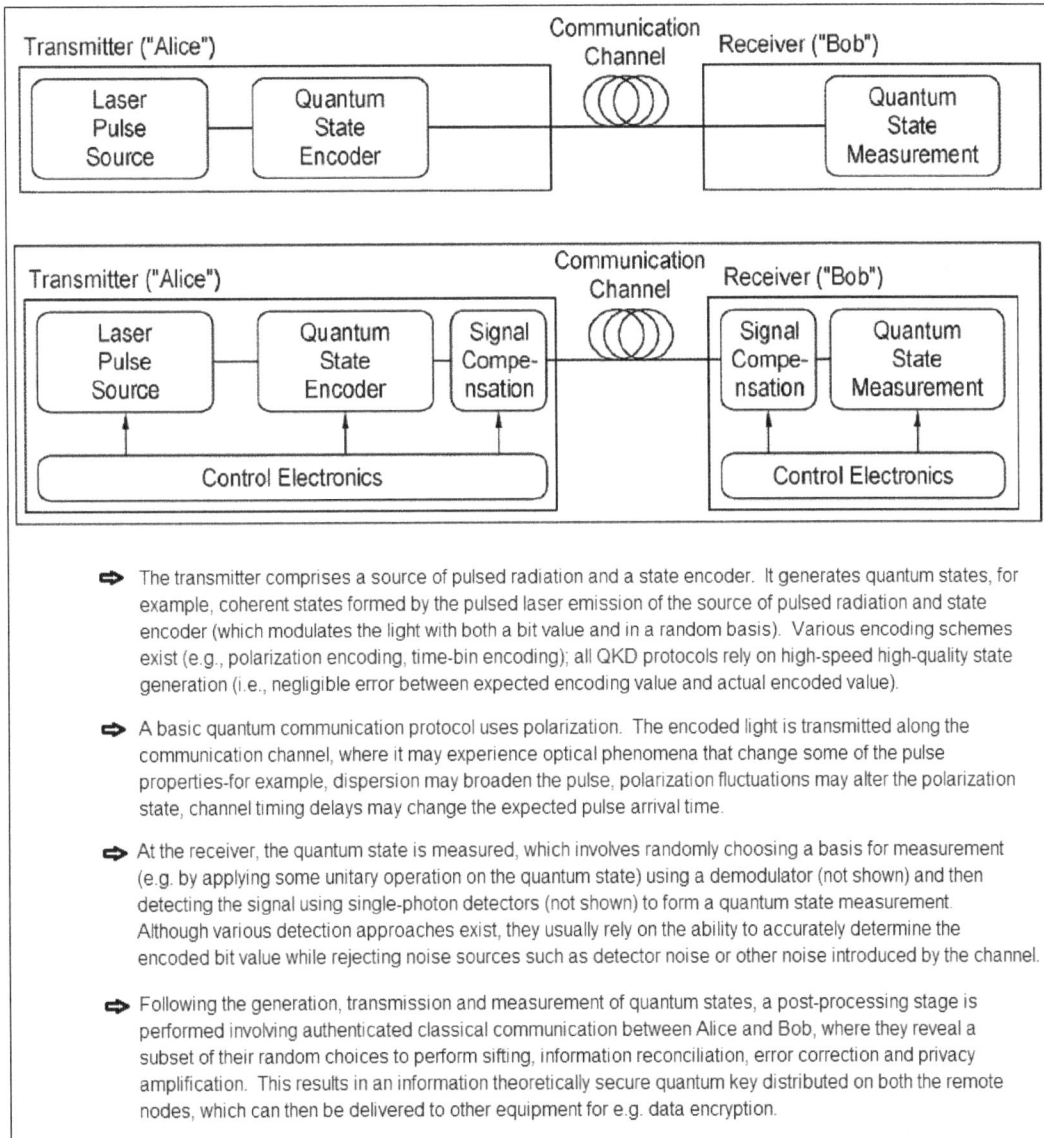

Top schematic: PTP QKD system
Bottom schematic: PTP QKD system with control electronics

Figure 7.2   Physical QKD system example ([WOO23])

is also desirable. A standardized protocol is a set of rules that are or have been developed by various standards developing organizations (SDOs) or industry groups that support (heterogeneous) system interoperability (including component- and subsystem-level interoperability).

Quantum Key Distribution Networks (QKDNs) have been defined in several ITU standards, as alluded to in Chapter 6. A QKDN is a cryptographic infrastructure aimed at providing secure symmetric keys to cryptographic applications in institutional networks. A QKDN comprises several hierarchical layers: the service layer, the control layer, the key management layer, and the quantum layer, as depicted in Figure 7.3 which illustrates the ITU-T Y.3802 model.

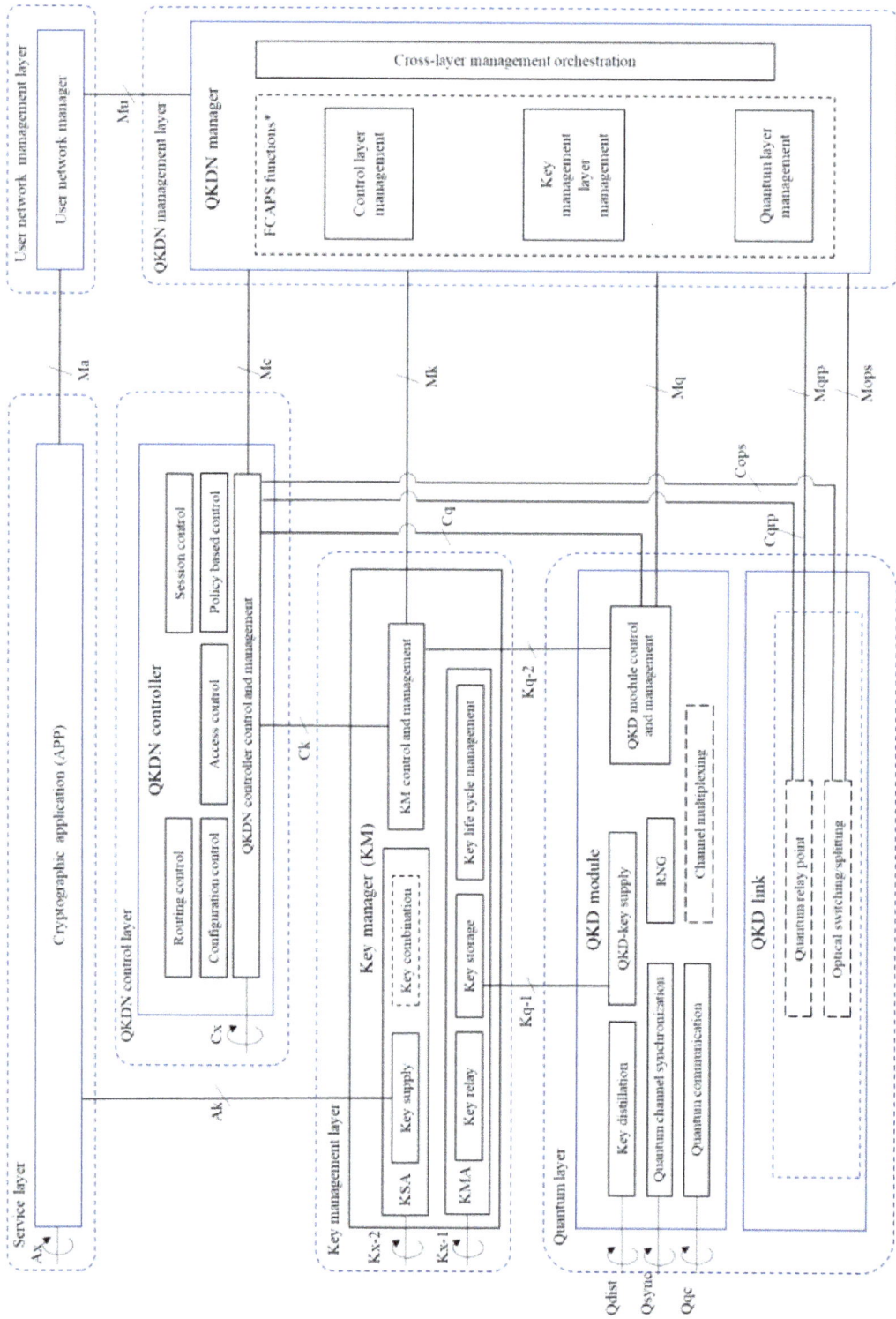

Figure 7.3 Functional architecture model of QKDN [ITU-T Y.3802] (with permission from the ITU)

The FG QIT4N D2.3-Part 1 document [ITU21], published by the ITU-T Focus Group on Quantum Information Technology for Networks (FG QIT4N), provides a detailed review of protocols in the quantum layer of a QKDN—in particular, different types of QKD protocols, the protocol workflows, features and parameters, and the commercialization status of QKD systems. Some early portions of this chapter are loosely based on [ITU21].

QKD can be implemented using a number of different protocols. As implied, QKD protocol is a quantum layer key establishment protocol, a step-by-step procedure, in which two remote parties negotiate a secret symmetric key. QKD protocols are implemented using dedicated hardware to transmit quantum states through physical channels and software to postprocess classical information to output random bits as keys. Some useful references on QKD mechanisms and technologies, among a large body of literature, include [BEN84], [DIA16], [EKE91], [FXU20], [GIS02], [GIS04], [GIS06], [LO14], [SCA09], [PIR15], and [PIR20]. The Glossary at the end of this chapter defines some basic quantum cryptography concepts and terms.

As discussed in Chapter 5, in 1994 Shor created a quantum algorithm that achieves an exponential speed-up for integer factorization and discrete logarithm. As a consequence, QCs will catastrophically shatter public key cryptographic schemes; the security of currently used public key cryptographic schemes (e.g., Elliptic-Curve Cryptography [ECC]; RSA, named after Ron Rivest, Adi Shamir, and Leonard Adleman), is based on the hardness of either integer factorization or discrete algorithm problems, and thus they are not secure in a QC context. To identify alternatives, the field of Quantum Resistant Cryptography, also known as Post Quantum Cryptography (PQC) or Post Quantum Encryption (PQE), has seen extensive research in the past decade.[1] As noted, QKD is a method that uses QM to distribute keys between different parties; thus, it does not rely on computing complexity assumptions, classical or quantum, and is not impacted by evolving QC capabilities.

The BB84 QKD protocol introduced by Bennet and Brassard in 1984 was the first QKD protocol to be invented, and it has been extensively studied and implemented in various commercial QKD systems [BEN84], [BEN92], [BEN92a] (also see [CHE22], [DUE06], [GIS02], [GIS07], [KRA05], [SHO00]). Since then, several dozen other QKD protocols have been developed; some of these more recent protocols have also progressed to the point where they have been demonstrated experimentally and/or implemented in commercial products and deployed in QKDNs. The motivations for developing new QKD protocols have been to: (i) improve QKD performance in terms of key rate and distance; (ii) reduce the implementation complexities, the device requirements, and the costs; and (iii) enhance the *implementation* security. Despite these efforts and the progress made in QKD protocol development, none of these protocols have yet been properly standardized; the majority of the existing QKD protocols are currently available in various reported versions, but the procedures for implementing these protocols can only be found in the research literature. Efforts by the various SDOs mentioned in Chapter 6 aim at addressing this lacuna.

7.2 Quantum Layer Functionality of a QKDN

QKD protocols that establish symmetric keys between two trusted nodes are implemented in QKD modules in the quantum layer of the QKDN/Y.3802 model, and symmetric keys are established

---

[1] Some actively researched PQC methods include lattice-based methods, code-based methods, multivariate-based methods, hash/symmetric key-based signatures methods, and elliptic curve isogeny-based methods. These particular topics are not further discussed in this text.

through point-to-point QKD links. In conventional QKD (BB84) implementations, Alice encodes a photon with a randomly chosen value of 0 or 1; Alice then transmits the encoded photon to Bob in one of two randomly chosen basis states. In the ITU-T Y.3802 model of Figure 7.3, a QKD protocol is realized inside the QKD module; the Random Number Generation (RNG) function is the source of randomness for quantum communication and for key purification (distillation). More specifically, the quantum channel synchronization and the channel multiplexing in the quantum layer's *QKD module*, as well as the optical switching/splitting function and the quantum relay link in the quantum layer's *QKD link,* constitute the physical layer transport mechanisms in which the implementations of QKD protocols occur, as follows (per [ITU21]):

- The quantum channel synchronization function module provides clock and timing synchronization for the quantum channel with sufficient precision to support quantum signal transmission and measurement.
- The channel multiplexing function module enables wavelength division multiplexing of quantum and classical channels.
- The optical switching/splitting function enables the switching or splitting of quantum channel traffic and synchronization signals.
- The quantum relay point function module serves as an untrusted intermediate point in the QKD link, as required by the QKD protocol to extend the QKD distance, such as with measurement-device-independent (MDI) QKD protocols.

The QKD module control and management function is responsible for the overall control and management of the functional elements in the QKD modules.

Supporting large, scalable, secure Quanets requires connections beyond simple QKD PTP links. Two topology structures to expand QKD beyond PTP inks have been proposed: (i) utilize a trusted relay (a Quantum Repeater [Quarep]) architecture; and (ii) implement a transparent optical network architecture [CHE09], [CHE10]. By employing optical technologies (including passive quantum networks that use passive optical components with passive optical splitting, Wavelength-Division Multiplexing [WDM] links, or systems with active optical switching), one can achieve transparent connections over considerable distances using low complexity networks; however, the application of optical technologies by themselves cannot increase communication distance and key generation rate for QKD, thus proposed solutions have opted to combine additional trusted relays to construct a hybrid scalable network.

7.2.1 Categories of QKD Protocols

Among other methods, the availability of a set of research-defined QKD protocols can be classified based on the different encoding and decoding methods utilized—specifically, herewith, the *discrete-variable* (DV)-QKD schemes and *continuous-variable* (CV)-QKD schemes (refer to [ITU-21] for a more detailed description of all these various schemes and protocols):

i.   In DV-QKD schemes, Alice, the sender, encodes information with discrete variables such as phase, polarization, or time bin of single photons, and the receiver, Bob, uses Single Photon Detectors (SPDs) to decode information. Examples of DV-QKD schemes include the following: the BB84 protocol, the E91 protocol, the B92 protocol, the six-state protocol, the BBM92 protocol, the SARG04 protocol, the coherent-one way protocol, the DPS protocol, the RRDPS protocol, the Twin-Field protocol QKD, the DV MDI protocol, and the DI QKD protocol.

ii. In CV-QKD schemes, the sender, Alice, encodes information utilizing the position and momentum quadrature of a quantized electromagnetic field in an (infinite dimensional) Hilbert space; the receiver, Bob, utilizes coherent detection—such as homodyne or heterodyne detection—to decode the information. Some examples of CV-QKD schemes include the Gaussian-modulation-based CV protocol, the discrete-modulation-based CV protocol, and the CV-MDI protocol.

Pursuant to this classification, a *short list of QKD protocols* includes the following [ITU21]: (i) the *BB84 protocol* is the original, most well-researched, and widely implemented QKD protocol. It was initially designed to emit optical pulses with a perfect single photon source; an alternative approach is using an attenuated laser. A drawback is that BB84 is vulnerable to the photon number splitting (PNS) attack, which reduces the key rate of the protocol. The decoy state method can be used to defeat the PNS attack [LO05], [LI22]. (ii) *The E91 protocol* was the first QKD protocol to utilize quantum entanglement; it detects information leakage by monitoring the violation of Bell inequality of data obtained by measuring the bipartite quantum states shared between the legitimate communication parties [EKE91]. A drawback is that it is challenging to design an entangled-photon source that outputs high-fidelity entangled photon pairs with high repetition rate, as discussed in earlier chapters; consequently, the entangled-photon source poses a limitation in implementing a high-throughput E91-based systems. (iii) The *B92 protocol* is a simplified version of the BB84 protocol that transmits two non-orthogonal quantum states instead of four. (iv) The *Six-state protocol* can be perceived as a revised BB84 protocol that utilizes six quantum states on three orthogonal bases instead of four states on two bases. (v) The *BBM92 protocol* can be seen as an entanglement-based version of the BB84 protocol. (Again, the reader may refer to [ITU-21] and the numerous references listed there for a more detailed description of these protocols).

7.2.2 QKD Operation

In the basic BB84 protocol, Alice can transmit a secret (random) key to Bob by sending a string of photons with the private key encoded in their polarization. The no-cloning theorem specifies that Eve (an eavesdropper) cannot measure these photons and transmit them to Bob without disturbing the photon's state in a detectable way [ASI21]. To work properly, there should be no error-inducing degradation in the quantum channel; otherwise, if the channel is subject to transmission errors, Alice and Bob cannot detect Eve's presence all the time.

The operation of a QKD protocol entails two main phases (or steps) [ITU21]: (i) the *raw key exchange* phase (also known as *quantum communication* stage), and (ii) the *classical post-processing* phase. The raw key exchange is carried via the quantum channel, and the classical post-processing is carried via the classical channel. This general operation is present in all QKD protocols, while the specific steps differ in various QKD protocols. The raw key exchange process can be realized using a prepare-and-measure (P&M) scheme, an entanglement-based scheme, or a measurement device independent scheme; here techniques such as entangled photons generation and Bell state measurements are utilized. In the classical post-processing phase of a QKD protocol, Alice and Bob process their raw key (partially correlated and partially secret bit strings) by transacting information over a classical channel. See Figure 7.4 based on the description in [ITU21].

Raw key exchange

Prepare And Measure (P&M) scheme

1: Alice encodes classical information on the quantum states. Alice encodes a classical random variable a on a set of non-orthogonal quantum states

2: Alice transmits these quantum states using a (quantum) communication channel to Bob

3: At the output of the quantum channel, Bob measures the received quantum states to obtain a classical random variable b, it being partially correlated with the random variable a of Alice

4: Alice and Bob repeat this process and exchange a number of quantum states thus generating two sets of partially correlated data on each side. These two sets of data are the *raw key*

Classical post-processing

1: Sifting: Alice and Bob exchange a classical message to signal which orthogonal subsets (basis) of a have been used in preparation for the encoding and/or the measurement in the raw key exchange phase – in the Prepare-and-Measure (P&M) phase. Alice and Bob then discard the part of the *raw key* for which the encoding and measurement basis are inconsistent. The retained portion is known as the *sifted key*

2: Parameter estimation: Alice and Bob compare a random subset of their sifted key and estimate statistics aimed at establishing various parameters of the quantum channel (e.g., quantum bit error rate). Pursuant to such parameter estimation, Alice and Bob proceed to assess the mutual correlation between their *sifted key* and then compute an upper bound of information that is accessible to Eve, the eavesdropper. If the upper bound of Eve's information is higher than Alice and Bob's mutual information (a measure of the correlation), then no secret key can be generated and the QKD process is aborted; otherwise, the process proceeds to the next step

3: Error correction (aka, information reconciliation): Alice and Bob agree on an identical bit string utilizing classical error correction techniques. Information reconciliation is achieved by Bob sharing a key identical to Alice's data (direct reconciliation) or by Alice sharing a key identical to Bob's data (reverse reconciliation). After error correction, the partially *correlated key* of Alice and Bob becomes *fully correlated* – however, some information may be leaked to Eve in the preceding steps or during transmission over the quantum channel

4: Privacy amplification: Alice and Bob process the *correlated key* from the error correction step to eliminate the information of the key that Eve may have acquired. The portion of the key that needs to be discarded is based on the upper bound information of Eve as computed in the parameter estimation for direct or reverse reconciliation. After removing the corresponding fraction of the key, Alice and Bob have an *identical secret key* which is practically unknown by Eve (but up to a negligible failure probability ε)

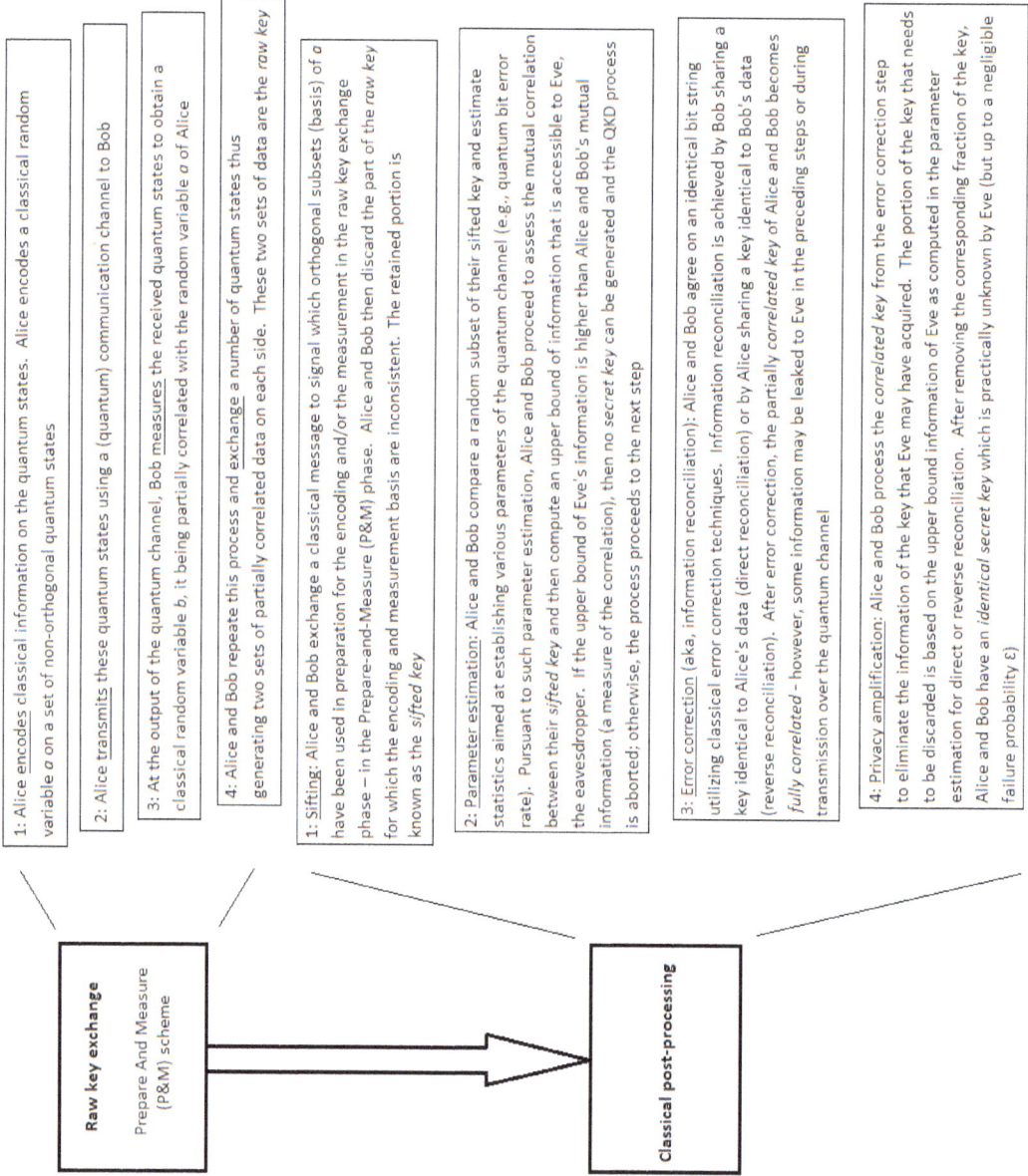

Figure 7.4 Phases of a QKD protocol (as described in [ITU21])

Figure 7.5 [DOI23] (on next page) is a diagram illustrating an example of a *functional structure* of a quantum cryptographic communication system. Figure 7.6 is a diagram illustrating an example of a basic *configuration of a quantum cryptography communication* system along with a sequence diagram illustrating an example of a global key supply process [TAK23]. The generation unit is configured to generate a response message including at least one of a random number and an encryption key shared by QKD via a communication network. The supply unit is configured to supply the response message to the application.

Figure 7.7 is another QKD arrangement [OZA23]. The top portion of Figure 7.7 is a diagram schematically illustrating an encryption communication system; the bottom portion is a sequence chart representing a processing procedure of an encryption communication method executed by the encryption communication system.

7.2.3 QKD Network Topologies

As hinted at earlier, current designs for a QKD network usually fall into three designs [YEO24]:

- A mesh of terrestrial QKD links connected with trusted nodes acting as Quareps, with some form of routing between the trusted nodes. There are technical limitations on the locations of nodes, given that current terrestrial QKD technologies have a distance limitation; the quantum channel may be, for example, an optical fiber or an optical free space channel.
- A mesh of terrestrial QKD links, in which the quantum signals are switched between end points. This can be supported with optical switches (or beam-splitters) in order to randomly route the quantum signals across the network; the quantum channel may be, for example, an optical fiber or an optical free space channel. In addition to distance limitations, it is difficult to scale this type of switched network to thousands of end points.
- Satellite QKD system with (potentially) global coverage without the need for any intermediate trusted nodes; however, optical ground receiving (OGR) stations are needed, and the complexity of the QKD transmitter/receiver technology, as well as the cost of satellite links, are relatively high, possibly making this design unaffordable for mass-market deployment.

In consideration of these limitations, there is a perceived need for an improved terrestrial QKD communications system that enables a QKD network to be created between multiple endpoint devices and massively scales in a secure manner for use—for example, as a "last-mile" solution to connect offices and homes into a QKD network whilst keeping connection costs relatively low. To that end, [YEO24] proposes methods and systems (i) for enabling QKD communication links to form a QKD communication network with a plurality of endpoints and (ii) for controlling data flow over the QKD communication links to one or more of the plurality of endpoints for use between multiple endpoints.

A QKD linking element acts as a hub and provides or assigns a QKD link to each endpoint of the plurality of endpoints and is configured such that control/switch/route data flowing via the QKD links to/from the endpoints to allow any endpoint to exchange quantum-encoded signals and classical signals with the QKD linking apparatus and/or any other endpoint via their quantum and classical channels of their corresponding QKD link. The QKD linking element and one or more of the plurality of endpoints may perform any QKD protocol for securely exchanging keys via the quantum and classical channels provided by the corresponding QKD links of said one or more endpoints. See Figure 7.8 for an example of a linking element.

[Test continues on page 314]

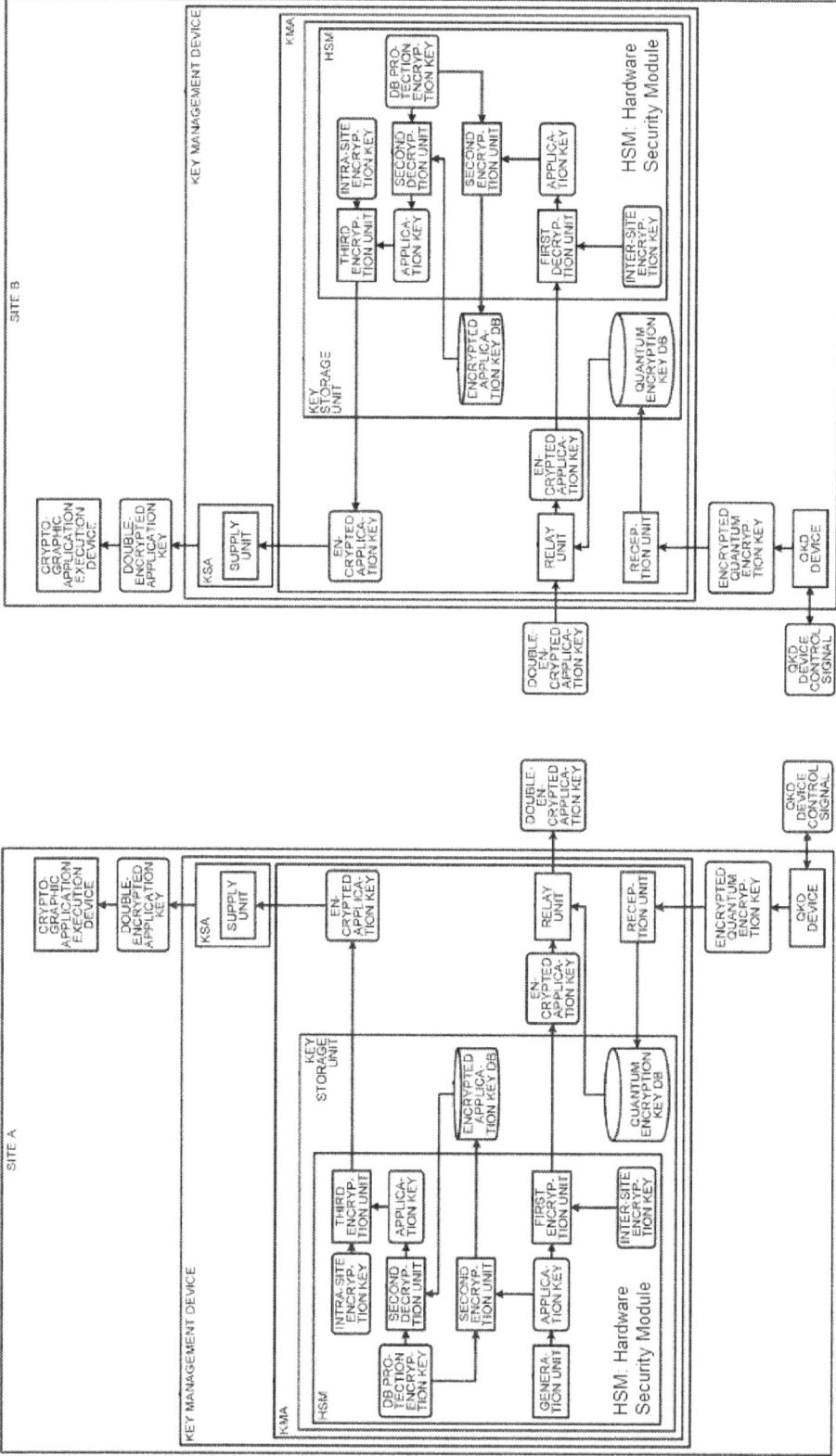

Figure 7.5 Example of a functional structure of a quantum cryptographic communication system (*Source:* [DOI23])

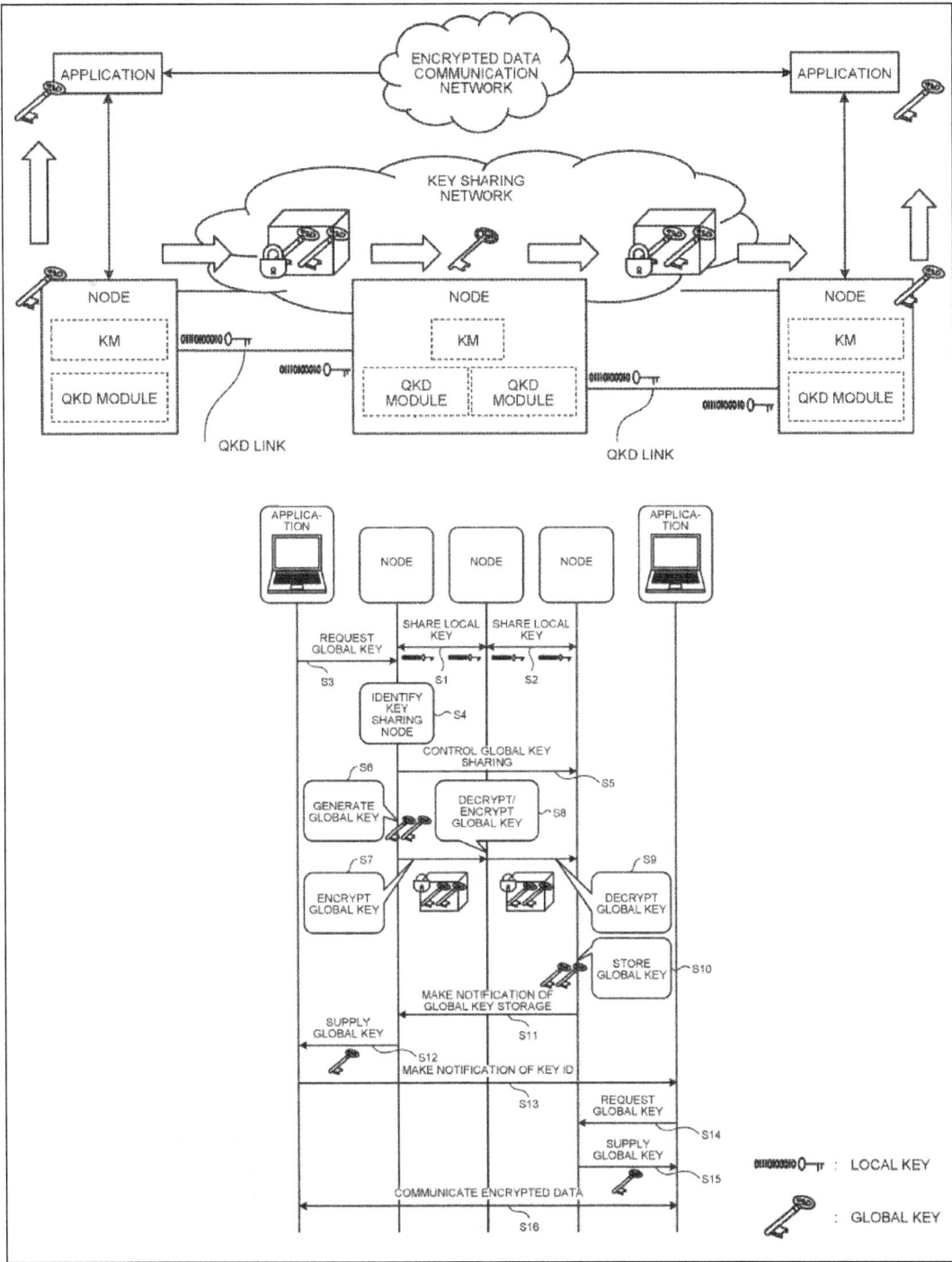

Figure 7.6 Quantum cryptography communication system and of a global key supply process (*Source:* [TAK23])

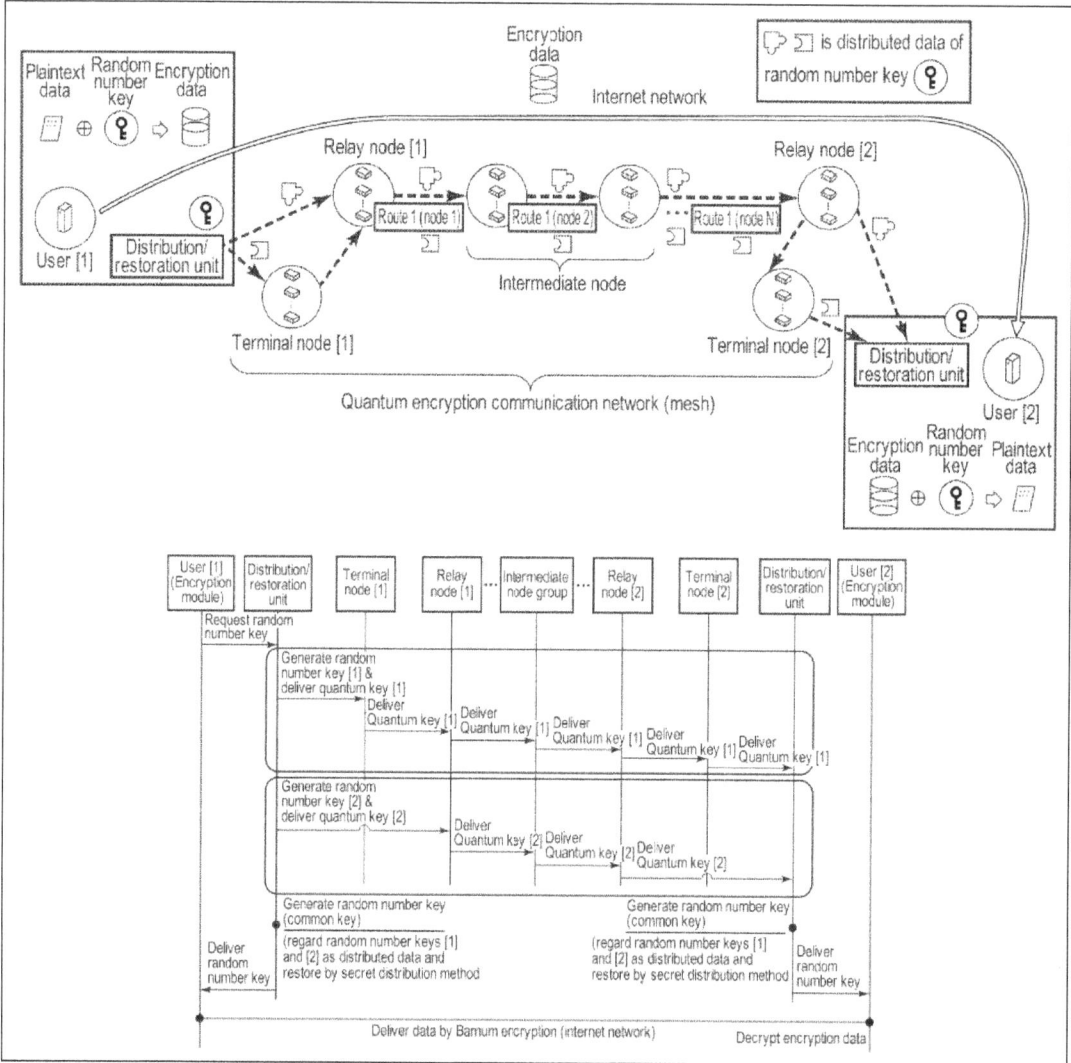

Figure 7.7   Yet another example of a QKD arrangement (*Source:* [OZA23])

Each QKD link may include a communications channel including at least a quantum channel and a classical channel that may share the same communication medium. One or more QKD links may share the same communication medium as one or more other QKD links. Two or more QKD links may also be configured to share the same communication medium. The QKD linking apparatus may include a controller configured to perform any QKD protocol for exchanging keys between endpoints and switch/route quantum and classical data flows in the appropriate sequence, between the QKD linking apparatus and/or one or more endpoints via the quantum and classical channels of the QKD links.

Thereafter, the controller may be configured to enable data flow between endpoints via the classical channels of the QKD links of the endpoints that allows secure communications between the endpoints. Figures 7.9 and 7.10 depict some topological variants. Refer to [YEO24] for additional specifics related to this proposed approach.

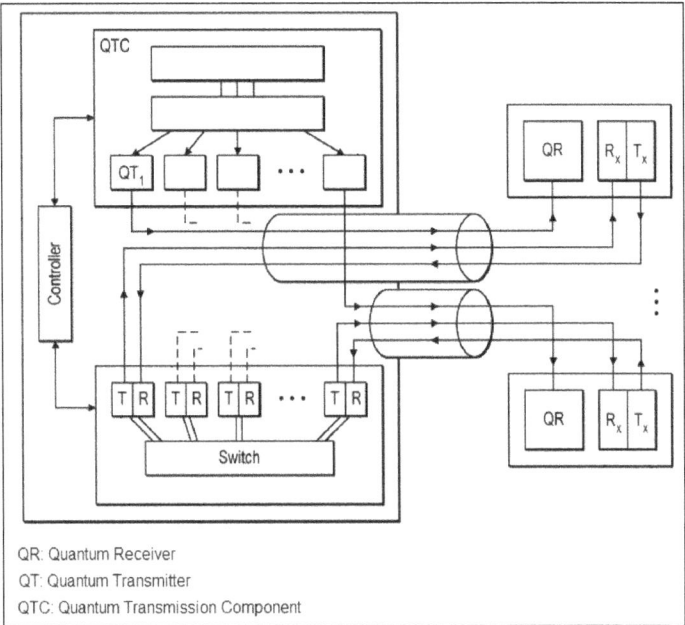

Figure 7.8 QKD linking apparatus configured for enabling secure communications between two or more endpoint devices (*Source:* [YEO24])

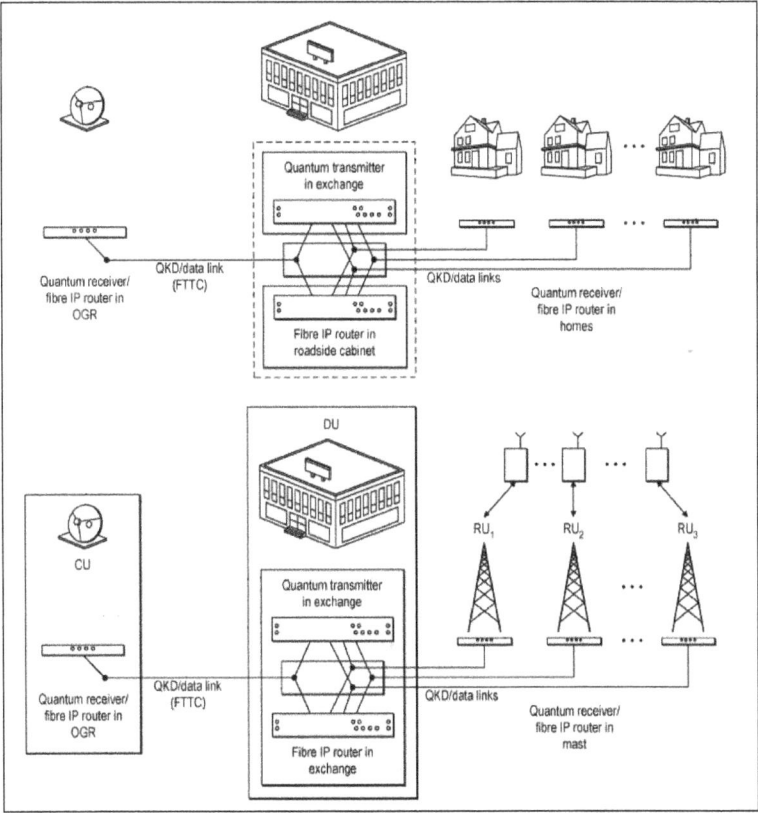

Figure 7.9 Example of a QKD system with quantum transmitter at an exchange (central office) location (*Source:* [YEO24])

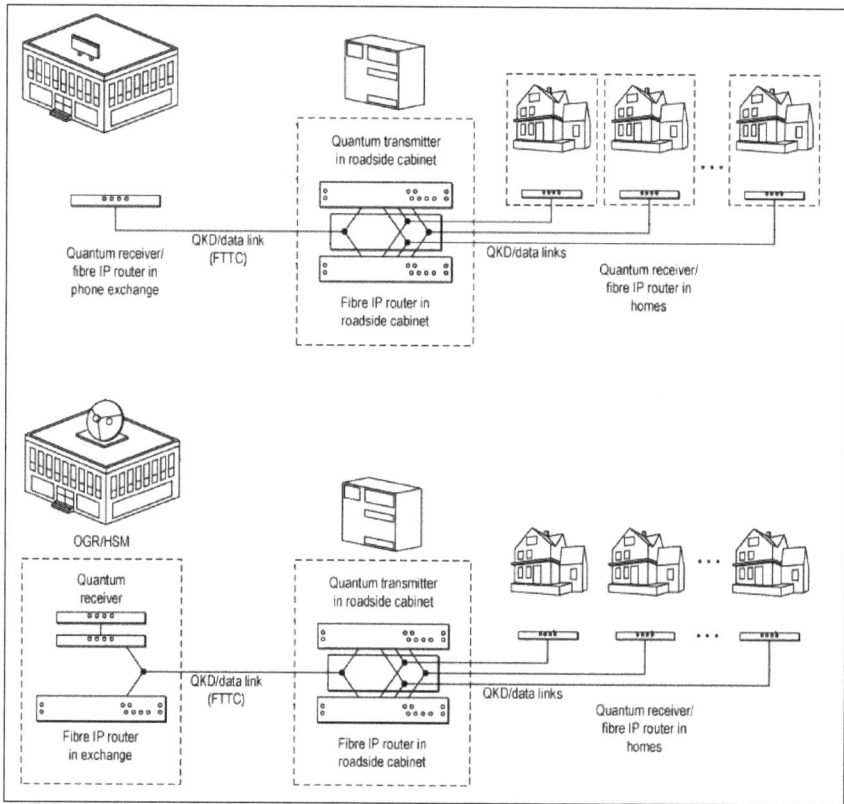

Figure 7.10   Example of a QKD system with quantum transmitter at a "last mile" hub (central office) (*Source:* [YEO24])

While QKD networks have been built upon existing fiber optic network infrastructure, the cost and complexity of QKD network development has led others to pursue QKD as a Service (QaaS) as a potential solution [RAH23]. QaaS allows multiple users to share the same QKD network infrastructure to obtain their required secret-key rates and opt in to its use as a service instead of developing and deploying their own QKD network. At this time QaaS implementations rely on software-defined networking (SDN) to obtain a Software-Defined Networking for Quantum Key Distribution as a Service (SDQaaS). SDQaaS uses SDN as the control plane to provide QaaS based upon an existing QKD network. The QaaS functions are developed in an SDN controller.

SDQaaS also describes the protocol extension, intercommunication workflow, and routing and SKR assignment strategy for a QaaS implementation in the SDQaaS framework. See Figure 7.11 for an example and refer to [RAH23] for additional details. Other implementations and/or applications are clearly possible. Figure 7.12 illustrates an example of a QKD-based wireless communication service.

7.2.4 QKD Challenges

Real-world QKD implementations may typically have vulnerabilities: theoretical QKD protocol security does not systematically imply security of a QKD implementation. These issues are

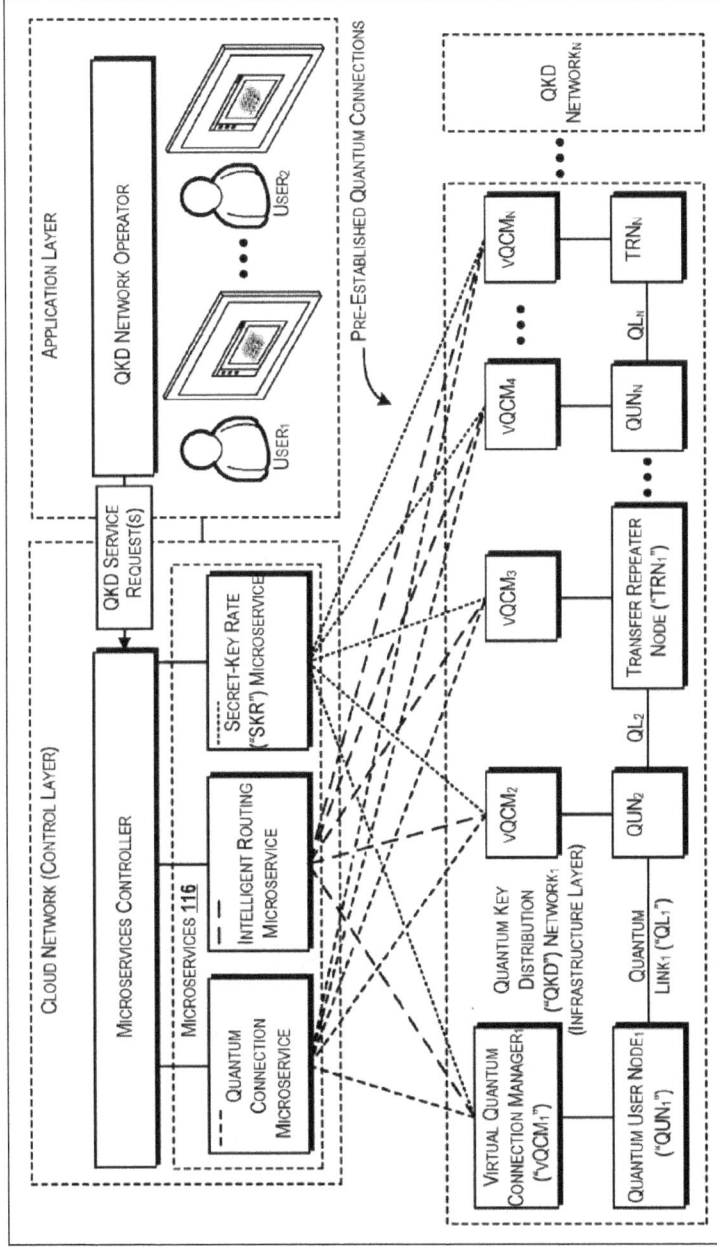

Figure 7.11 Quantum Key Distribution Networking as a Service (*Source:* [RAH23])

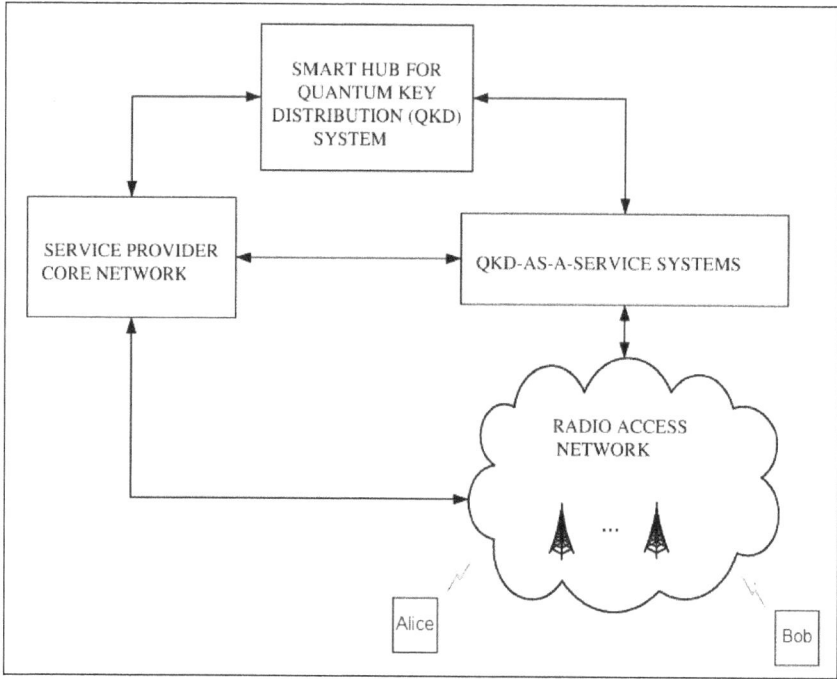

Figure 7.12   QKD in a wireless environment (*Source:* [RAM23])

the subject of sustained R&D studies at the present time. In theory, QKD provides a secure mechanism to distribute secret keys between two distant parties; in practice, this is not precisely the case because real devices do not conform to the requirements imposed by the security proofs. In fact, various types of quantum hacking attacks have surfaced and experimentally demonstrated that exploit device imperfections in practical QKD settings [WAN21], [LO14].

High channel loss engenders a severe threat to the BB84 protocol—an exemplar of early QKD protocols—due to possible PNS attacks; for example, satellite channels suffer high loss, and so do unrepeated long fiber-based optical links. Classical reference [HWA03], on which this explanation is based, describes the PNS attack as follows:

Say that Alice (a legitimate participant) utilizes the following photon sources in BB84 protocol: generate a pulse that contains a single photon with, for example, 90% probability and emits a pulse that contains multi-photons with 10% probability. The issue here is that multi-photons are inadvertently generated, and thus one does not know when they have been emitted. Also assume that the channel loss ℓ is, for example, 90%, or it has 10% yield y. One has $y = 1 - \ell$. Here we assume that Bob (the other legitimate participant) uses more practical detectors that are insensitive to photon numbers.

Eve's attacking method is the following: First, Eve (an eavesdropper) measures the number of photons of each pulse; when it is one, she just blocks it; when it is more than one, she splits the photons. Then Eve preserves one and sends the other photons via an ideal lossless channel to Bob. Then what Bob observes is that only 10% of photon pulses arrive at him, as expected. However, Eve can get full information about the key by measuring each of the preserved photons in a proper basis that is publicly announced later by Alice. The assumption for Eve is that all multi-photon pulses were used for the PNS attacks.

Herewith one can see that if the yield y is less than the probability of multi-photon generations, then the scheme is totally insecure due to the PNS attacks. In other words, yield y must be greater than the probability of multiphoton generations in order that the scheme be secure. Thus, the probability of multi-photon generations is a parameter for quality of (imperfect) single-photon sources. The smaller it is, the higher the quality is. For a single-photon source with a given quality, the loss that can be tolerated is when y exceeds probability of multi-photon generations. Therefore, when yield y is very low, almost perfect single-photon generation is required. However, the problem is that the sources cannot be perfect single photons in practice. In the case of surface-to-surface free-space BB84 protocol, this condition is barely satisfied; however, the loss in satellite channels is higher than that in the surface-to-surface scheme; (in addition, the security of surface-to-satellite protocol where a satellite plays a role of a legitimate user is based on physical security of the satellite).

The decoy-pulse (decoy-state) method overcomes the PNS attack for BB84 in the presence of high loss. The basic concept of the decoy-state method is the following, as elaborated in [HWA03]: In PNS attack, Eve selectively transports subsets of multi-photons to Bob. Thus, the yield of multi-photon pulses must be abnormally higher than that of single-photon pulses. Assume that Alice had intentionally and randomly replaced photon pulses from signal sources by multi-photon pulses (the decoy states). Because Eve cannot distinguish multi-photon pulses of signal source from those of decoy source, the yields of the two pulses must be similar. Thus, Alice and Bob can detect the PNS attack by checking the yield of decoy source. In other words, a legitimate user intentionally and randomly replaces signal pulses by multiphoton pulses (decoy pulses [LO05]). Then, the stakeholders (Alice and Bob) check the loss of the decoy pulses. If the loss of the decoy pulses is significantly less than that of signal pulses, the process is aborted; otherwise, to continue the process, they estimate the loss of signal for the multiphoton pulses based on that of decoy pulses; this estimation can be undertaken with the assumption that the two losses have similar values [HWA03].

To address these implementation security "loopholes", a number of approaches in addition to the decoy method have been proposed, including the device-independent (DI) QKD approach and the measurement-device-independent (MDI)[1] QKD approach [STE09], [LO12], [BAR05], [ACI07], [VAZ14]. The security of DI-QKD relies on the violation of a Bell inequality; no knowledge related to the inner working of the quantum apparatus is needed, given that they follow the prescriptions of the protocol and not those of Eve [BEL64], [BRU14], [CUR19], [GIS10], [ZAP19]. DI-QKD is, however, difficult if not impractical to implement at this time, particularly for systems at a distance (DI-QKD needs close to 100% detection efficiency). By contrast, MDI-QKD has been widely demonstrated experimentally in recent years.

7.2.5 Measurement-Device-Independent Quantum Key Distribution (MDI-QKD)

MDI-QKD closes all side channels in the detection unit, thus facilitating the implementation of assured security in QKD, given that now one only needs to secure the source. MDI-QKD can remove all detection side channels from quantum communication systems and *allows the Quanet* to scale up to a large set of users [WAN21], [RUB13], [SIL13], [LIU13], [TAN14], [YIN16]. MDI-QKD, however, requires that certain specific assumptions on the sources are

---

[1] Also known as the *time-inversion* version of EPR (Einstein–Podolsky–Rosen) protocol

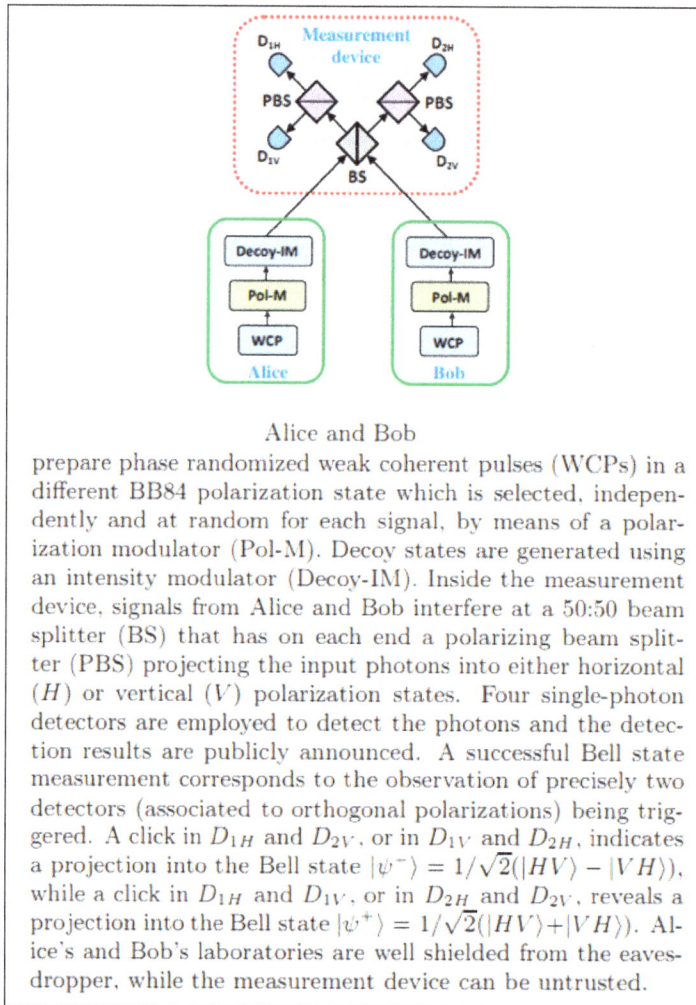

Alice and Bob

prepare phase randomized weak coherent pulses (WCPs) in a different BB84 polarization state which is selected, independently and at random for each signal, by means of a polarization modulator (Pol-M). Decoy states are generated using an intensity modulator (Decoy-IM). Inside the measurement device, signals from Alice and Bob interfere at a 50:50 beam splitter (BS) that has on each end a polarizing beam splitter (PBS) projecting the input photons into either horizontal (H) or vertical (V) polarization states. Four single-photon detectors are employed to detect the photons and the detection results are publicly announced. A successful Bell state measurement corresponds to the observation of precisely two detectors (associated to orthogonal polarizations) being triggered. A click in D_{1H} and D_{2V}, or in D_{1V} and D_{2H}, indicates a projection into the Bell state $|\psi^-\rangle = 1/\sqrt{2}(|HV\rangle - |VH\rangle)$, while a click in D_{1H} and D_{1V}, or in D_{2H} and D_{2V}, reveals a projection into the Bell state $|\psi^+\rangle = 1/\sqrt{2}(|HV\rangle + |VH\rangle)$. Alice's and Bob's laboratories are well shielded from the eavesdropper, while the measurement device can be untrusted.

Figure 7.13   Basic setup for MDI-QKD. (*Source:* H.-K. Lo, M. Curty, B. Qi [LO12])

satisfied. A common such assumption is that Alice's and Bob's transmitters do not leak information out of their security zones. Figure 7.13 from [LO12] depicts a basic setup for MDI-QKD.

As noted, BB84-QKD requires a perfect single-photon source and equally selective detectors. However, there is a lacuna between ideal environments and practical implementations. Due to the imperfection of system's equipment, the implementation of the QKD suffers attacks from the detection side and/or the source side. At the time of this writing, perfect single-photon sources are not available, and thus, weak coherent photon sources (WCPs) with phase randomization (polarization encoding [TAN14a]) are often utilized to replace the single-photon sources; however, given that the photon number of the pulses emitted by WCPs could be more than one, an eavesdropper (Eve) can launch a PNS attack (on either or both senders and receivers[1]),

---

[1]   On the receiver side, due to the low (detection) efficiency of the detectors, Eve can launch various attacks against the detectors themselves (e.g., [MAK06]); these attacks include detector blinding attack, dead time attack, faked state attack, and time shift attack (e.g., [GER11], [HAN04], [MAK05], [QI07]).

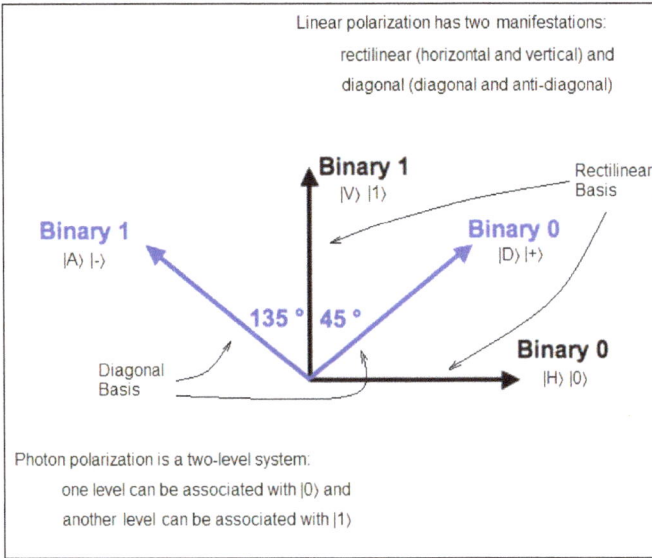

Figure 7.14 Photon polarizations

as has been known for over two decades [BRA00], [LUT00], [HWA03], [LIU11], [LIU13a], [CHE22]. (See Figure 7.14 for a view of the basic photon polarization process). Because of the limitations of technology, most MDI-QKD protocols use WCPs, but these arrangements can still suffer from a PNS attack from eavesdroppers; therefore, the existing MDI-QKD protocols also need the decoy-state method; the decoy state can resist PNS attacks reasonably well.

In MDI-QKD, Alice and Bob do not need to perform measurement operations, thus the protocol can be intrinsically immune to all detection attacks [LO12]. MDI-QKD with the decoy-state method can resist both source attacks and detection attacks; thus, decoy-state MDI-QKD is a promising QKD protocol that can provide secure keys in practical commercial applications. However, one of the design challenges is that the secure key rate of the existing decoy-state MDI-QKD is not high; [CH22], which the reader may wish to consult, endeavors to offer improvements.

7.3 Some Observations on Post-Quantum Encryption

Public-key encryption methods currently in widespread use rely on computational security that existing computers and algorithms cannot decrypt within a finite time. In the near future, when QCs become more prevalent, computational security is not guaranteed, and secure transmission confidentiality cannot be guaranteed by the conventional encryption methods [MUM21], [BRO16]. The Internet Engineering Task Force (IETF) Internet-Draft [DRI24] notes that the mathematical problems of integer factorization, discrete logarithms over finite fields, or elliptic curves are currently at the base of most of the asymmetric algorithms used for key establishment and digital signatures on the Internet. These mathematical "problems", and the algorithms based on them, will be vulnerable to attacks using Shor's Algorithm on a sufficiently large general-purpose QC, specifically on what is called a *Cryptographically Relevant Quantum Computer* (CRQC). Preparing for the near-term development of a CRQC requires

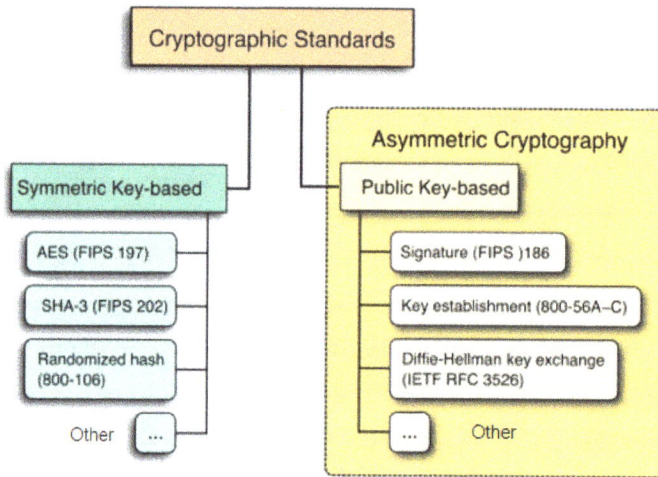

Figure 7.15   Common cryptographic standards

modifying established (standardized) protocols to use asymmetric algorithms that are provably secure against both QCs and today's classical computers.

Cryptography is used extensively, currently being embedded in systems and components as operating systems (OSs), communications systems, and Internet of Things (IoT) devices in environments such household appliances, embedded medical devices, and automobiles; see Figure 7.15. Protection of data is needed for data at rest, for data in transit, and for data in use. Examples of (asymmetric) cryptography uses include, but are not limited to [NIS23]:

- Digital signatures used to provide source authentication and integrity authentication as well as support the non-repudiation of messages, documents, or stored data
- Identity authentication processes used to establish an authenticated communication session or authorization to perform a particular action
- Key transport of symmetric keys (e.g., key-wrapping, data encryption, message authentication keys) and other keying material (e.g., initialization vectors)

Recent developments in Quacomp have prompted efforts supporting the development of encryption methods that can withstand attacks. As noted elsewhere in this text, it is possible for QCs to factorize large primes utilizing Shor's Algorithm and (soon) solve other resistant one-way functions significantly more quickly than with conventional computers. PQE approaches are being advanced and assessed; PQE endeavors to deliver encryption methods providing robust security without depending on the computational security of current mathematical algorithms.

When practical QCs become broadly available, they will be able to defeat many of the public-key cryptosystems currently in use, as well algorithms based on secret keys. The algorithms listed in Table 7.3 [NIS23] are considered vulnerable; this will compromise the confidentiality and integrity of all sorts of digital communications.

7.3.1 Vulnerabilities of Existing Methods

Encryption and digital signature schemes are widely used in Internet protocols such as (D)TLS, IKEv2/IPsec, and SSH. These protocols all include a key exchange phase and an authentication

Table 7.1 QC-Vulnerable Encryption Algorithms ([NIS23])

Algorithm	Function	Specification
Elliptic Curve Diffie Hellman (ECDH) Key Exchange	Asymmetric algorithm for digital signatures/key exchange	NIST SP 800-56A/B/C
Menezes Qu Vanstone (MQV) Key Exchange	Asymmetric algorithm for key exchange	NIST SP 800-56A/B/C
Elliptic Curve Digital Signature Algorithm (ECDSA)	Asymmetric algorithms for digital signatures/key exchange	FIPS PUB 186-5
Diffie Hellman (DH) Key Exchange	Asymmetric algorithms for digital signatures	IETF RFC 3526
RSA Encryption Algorithm	Asymmetric algorithms for digital signatures/key establishment	SP 800-56B Rev. 2
RSA Signature Algorithm	Asymmetric algorithms for digital signatures/key exchange	FIPS PUB 186-5
Digital Signature Algorithm	Asymmetric algorithms for digital signatures/key exchange	FIPS PUB 186-5
Edwards-curve Digital Signature Algorithm (EdDSA)	Asymmetric algorithms for digital signatures	FIPS PUB 186-5

phase. In the key exchange phase, the peers exchange asymmetric keys that enable them to establish a shared secret utilizing Elliptic-Curve Diffie-Hellman (EC-DH). These protocols then proceed to derive a symmetric key, which is used to symmetrically encrypt data exchanged between the peers. The authentication phase includes the capability of providing an asymmetric signature of a transcript of the exchanged data; this establishes that the peer signed this data with its private key. The corresponding public key is usually included with the identity of the peer and is authenticated by using public key infrastructure (PKI) (or other methods); in this manner the peers can verify they are talking to the peer with the expected identity who holds the expected public key [NIS23].

PKI (i) controls the issuance of digital certificates to protect sensitive data over end-to-end communications, and (ii) provides unique digital identities for users, devices, and applications. In 2016, the National Institute of Standards and Technology (NIST) began developing standards for PQC/PQE (also called *Quantum Resistant [QR] cryptography*) to enable the establishment of cryptographic systems that are secure against both QCs and classical computers, and, by using standards, can interoperate with existing communications protocols and networks. Initiatives to update or replace installed cryptographic technologies have taken many years. There is an extensive body of R&D and literature on PQC/PQE; some of these references are included in the Reference section at the end of this chapter.[1]

While currently some networks use EC-DH together with a static key, EC-DH is not QR; even if EC-DH were to be replaced with a QR-DH or QR-KEM (key encapsulation mechanism), the outcome would still not be secure, given that QR-KEMs, especially QR-KEMs that rely on noise, are not safe when used with a static key. A QC could frustrate the asymmetric schemes for

---

[1] Particularly [ACC22], [ALA22], [BAR18], [BAR19], [BAR21], [BAR21a], [BIN19], [CYB23], [CYB23b], [DRI24], [ETS20], [GSM23], [LEE21], [MA21], [MIT23], [MOS23], [NAT22], [NIS23a], [NIS23b], [NIS23c], [NIS23d], [OFF22], [POS23], [QUA23], [SOU22], [SOU22a], and [VER22], among a large corpus of references.

key exchange and signing. As noted elsewhere in this text, there are two quantum algorithms that can be utilized in attacks against extant data protection methods, also as described in [NIS23]:

1. *Shor's algorithm* provides an efficient method for computing the discrete-logarithm problem and the elliptic curve discrete-logarithm problem, as well as the problem of factoring large integers, thus breaking current key exchange, digital signature, and public key encryption methods that are based on asymmetric cryptography. Because of Shor's algorithm, new cryptographic algorithms are needed that are resistant to attacks that can be launched from both classical and QCs. Shor's algorithm could break EC-DH, implying that a quantum-capable threat actor could recover the symmetric key used to encrypt data.

2. *Grover's algorithm* can be used to speed up the identification of a secret key in a key address space. To neutralize this type of attack, strong encryption algorithms are needed with keys whose key address space is large enough to be considered not vulnerable.

New NIST standards that specify key establishment and digital signature schemes that are designed to resist attacks by QCs were expected to be published in 2024 or 2025. The replacement of algorithms requires changing or replacing (i) hardware that implements or accelerates algorithm performance, (ii) application code, (iii) communications devices and protocols, and (iv) user and administrative procedures including best practices.

Encryption techniques can be taxonomized into two broad groups, depending upon the frequency of exchange of private key information used to encrypt plaintext and decrypt ciphertext back to the original plaintext[1]:

- The first group of encryption methods involves the exchange of private key information[2] just once, in the form of one-time-pads (OTPs) containing apparently random information (these are also known as Stream Ciphers or VERNAM CIPHERS).

- The second group of encryption methods involves the exchange of private keys over a session, over a time interval of interest; here exchanging private keys (rather than references to private keys) occurs more than once.

Given that one-way functions are critical to the frequent exchange of private keys, QCs threaten to render the present encryption standards ineffective, enabling the interception of private information on public networks (a *one-way function* is easy to compute given an input,

---

[1] The discussion in the rest of this chapter is based in large measure on [ROU24], to whom we are indebted.

[2] As noted in [ROU24], many modern encryption standards use deterministic processes to produce and exchange small private keys which could be deciphered QCs; alternatively, modern encryption can use a stochastic process to produce a large private key, physically exchanged only once before use, where the amount of use is closely related to and limited by the size of the private key. The OTPs are mutually agreed texts (e.g., text produced from unpredictable physical processes, such as quantum noise). OTPs can be based on random sequences produced by either deterministic or stochastic processes or both. The best type of random information for OTPs is obtained when using True Random Number Generators (TRNGs), which derive randomness from one or more of various completely unpredictable quantum processes, such as radioactive decay, avalanche effect, shot noise or thermal noise, or properties of photons; these are *Quantum Random Number Generators* (QRNGs). One can use entropy as one measure of the quality of the OTP randomness; the higher the entropy of the ciphertext, the less information is revealed for analysis. Deterministic OTPs are vulnerable to cryptanalysis (decipherment techniques). Stochastic OTPs are not vulnerable to cryptanalysis; however, the OTP's random sequence cannot be reused without becoming deterministic, and so the random sequence is used at the same rate as the plaintext, so the OTPs become impractical by being exhausted when used with very large volumes of information transmitted over the (classical) Internet.

but computationally hard to invert given the image of a random input). To counter the new vulnerabilities, new algorithms such as *lattice encryption* and *ring learning with errors* have been proposed for private key exchange. However, given that frequent private key exchanges still rely on deterministic mathematical functions, the emergence of CRQCs requires that any future resistant one-way mathematical function be selected with caution—even though such function may still become susceptible to advances in quantum algorithms and mathematical theory.

Encryption such as *Advanced Encryption Standard* (AES) utilizes deterministic private key exchanges; QKD uses stochastic private key exchanges. AES and QKD both subsequently use deterministic algorithms in a *Pseudo Random Number Generator* (PRNG) to ostensibly mix random information with the private plaintext in order to encrypt or decrypt publicly secure ciphertext. Both the existing AES and the newer QKD rely on the frequency of key exchanges and the strength of their deterministic processes for the security of their encryption to protect against attack by adversarial entities and/or cryptanalysts. The bottom line is that both AES and QKD are susceptible to cryptanalysis since they are or may soon become vulnerable to CRQCs (which are exponentially more powerful than classical computers) and quantum algorithms.

7.3.2 Classes of PQE

There is a need for an approach that addresses the technical issues preventing a conventional OTP encryption from being truly unbreakable by QCs (i.e., the lack of truly random numbers, limited lifetime of encrypted data, and so on). Considering the fact that current methods of encryption may become ineffectual in the near future, two classes of PQE have emerged from the research community:

- *Quantum Computer Resistant Encryption* (QRE): An encryption scheme expected to resist decryption using CRQCs. The strength of the resistance depends on vulnerabilities (i) within *resistant one-way functions* (ROWFs), protecting publicly exchanged encryption keys, and/or (ii) within *pseudo-random functions* (PRF) (a form of ROWF[1]) producing very long cyclical, but yet predictable, random number sequences protecting plaintext/input information messages. QRE uses ROWFs or PRFs, thus it may never be completely safe.

---

[1] As noted in [ROU24], PRFs may be considered as a type of ROWF function. PRFs provide a long, disordered sequence of apparently random yet predictable numbers used to protect plaintext/input information and other types of information. PRFs are deterministic; by using a specific index into the sequence, called a key, the same random number is always returned. PRFs are finite and cyclical and will eventually repeat sequences of numbers; the underlaying encryption will never be completely safe. The PRF function is computationally simple, with a specific index, and easily provides a short random number sequence, whereas the PRF's reverse ROWF function to find a short random number sequence in an extremely long sequence is an impractically difficult polynomial time computation. When some known, or even guessed, plaintext/input information is present within a portion of cipher text, a portion of the PRF's random sequence may more easily be discovered. This known plaintext attack is called a *crib-drag*. Searching within the very long predictable sequence for the discovered portion of random number is lengthy and difficult; however, if found, it provides an index and a possible encryption key. If the encryption key is revealed, all previous and subsequent cipher text can easily be decrypted until the key is changed. In the future, QCs and quantum search algorithms may be capable of rapidly checking entire portions of cyclical/predictable random number sequences, rendering the exchange of private encryption keys with ROWFs ineffective.

- *Quantum Computer Safe Encryption* (QSE): An encryption scheme that requires a *secure one-way function* (SOWF) to exchange shared secrets/private keys and/or to encrypt plaintext/input information and decrypt ciphertext. No amount of computational power applied to either the exchange of private keys or the encryption of plaintext/input information can decrypt the ciphertext. QSE is completely immune from advances in QCs and does not contain ROWFs or PRFs, both of which introduce potential insecurities. A provably impenetrable QSE method cannot use predictable/deterministic processes to produce private keys, which would give rise to the possibility of decipherment. To prevent any possibility of decipherment, irrespective of advances in computational speed or algorithmic methods, all encryption must only use truly random/stochastic processes to produce a private key which is exchanged only once in private, before one public use, where the private key is sufficiently large to encrypt any practical volume of plaintext with minimal computational power.

As just noted, ROWF or SOWFs are used to privately exchange information and/or encryption keys:

- *ROWFs* are a computationally simple mathematical functions often requiring two or more parameters/values, whose reverse function produces one correct solution and (i) an impractically large number of incorrect solutions to check, and/or (ii) an impractically difficult, polynomial-time computation. Thus, ROWFs require impractically large computational power to check all permutations of possible two-key values to discover a valid result that reveals an encryption key. AES utilizes a ROWF function, previously believed to be highly resistant, which simply multiplies two large prime numbers; the reverse AES ROWF requires factoring the product of the two large primes, which has been considered to be impractical; at this juncture, however, AES is vulnerable to factoring with CRQCs using Shor's algorithm. Other ROWFs have been proposed, yet the practicality of their use and future resistance remains uncertain. As computing advances, the further development of CRQCs and the associated rapid increase in QC size, improvements in computer speed, and advances in mathematical theory all increase the vulnerability of ROWF functions. ROWF functions are deterministic and can never be completely safe. ROWFs are used in technologies such as quantum resistant block chains and lattice algorithms, which have been thought to be invulnerable to attack by QCs but in actuality may also not be completely secure.

- *SOWFs* are physical properties or functions that require only one parameter/value and where no reverse function exists; examples include entanglement within a quantum system, a private table of properties derived from a quantum effect, or a QRNG—no amount of computational power applied to a SOWF is sufficient to create a reverse function; in particular, Heisenberg's Uncertainty Principle dictates that it is impossible to compute random physical properties derived from quantum effects; any attempt to create a reverse function for a photon's quantum properties through observation only destroys any SOWF result, making recovery of the original value impossible. SOWF functions are completely safe.

Make note that QKD is a hybrid system, using both SOWFs to exchange encryption keys securely and then *using the encryption key with AES or other PRFs (a type of ROWF) to encrypt plaintext/input information*. As discussed earlier, QKD utilizes quantum effects, such as photon polarization, to transmit an encryption key in secret over a public network or medium. Typically, to improve performance, a PRF takes the private encryption key as an index into a predictable deterministic sequence utilized to protect the plaintext messages/input information, but if the PRFs are rendered insecure by advances in quantum algorithms then *initial QKD private key exchanges* become pointless when they are no longer effective at protecting the ciphertext from decryption.

Quantum photon polarization is inherently stochastic; however, the reliability and frequency of key exchanges based on polarization degrades as a function of the length of the fiber optic cables used. The farthest achieved distance at the time of this writing was around 400 km, with practical private exchanges typically being around 30 to 100 km. QKD that uses photon polarization in directional line-of-sight laser or satellite communication is degraded by the accuracy of the optical systems and by known atmospheric effects such as refraction and reflections from the likes of rain, water vapor, and particulates or other pollution. *After exchanging keys, the private key is typically used to create a seed value for a PRF.* The PRF produces what is an apparently random (but predetermined) sequence of numbers that is devoid of the limitation of degraded speed of QKD over longer distances, and it is thus used to mix with the plaintext to create the ciphertext. Since the PRF algorithm is typically publicly known, any encryption utilizing deterministic sequences requires (i) the PRF algorithm to have a very long period, and (ii) the frequency of key exchanges be high enough to prevent cryptanalysts from trying all possible combinations of PRF sequences by brute force. Consequently, QKD using ROWFs for plaintext encryption can never be considered completely safe.

[ROU24] offers techniques for implementing what is asserted to be an impenetrable, inherently quantum-safe, fully stochastic encryption method—a QSE method (see Figure 7.16). The method entails eight steps that can be implemented in hardware, software, or in combinations of hardware and software, as follows: (1) the identification of OTP data to be exchanged

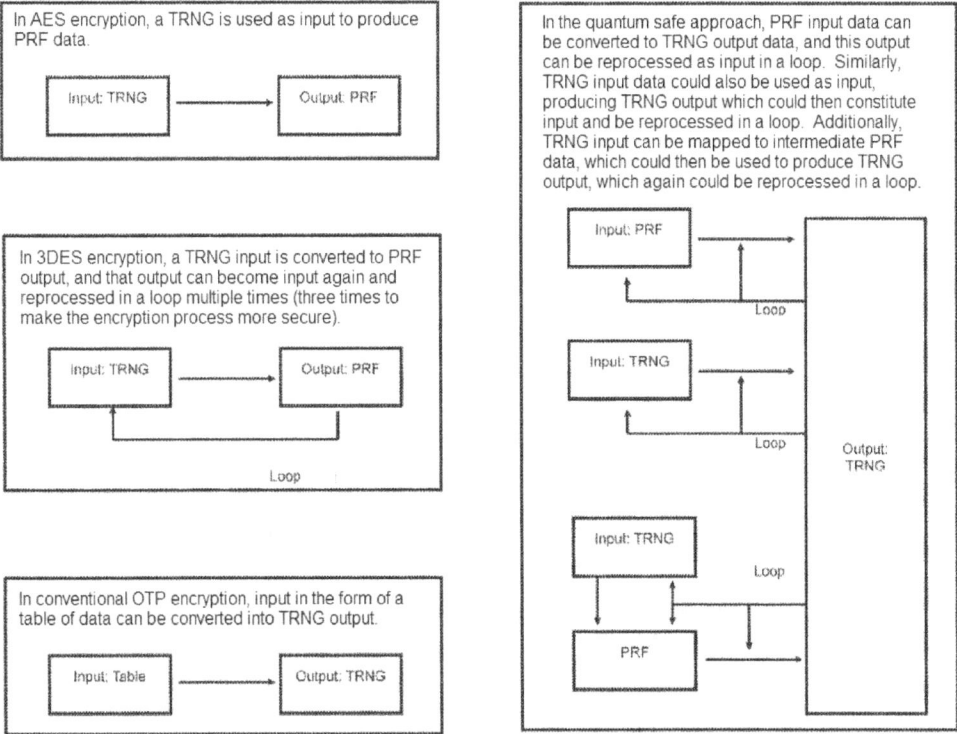

In AES encryption, a TRNG is used as input to produce PRF data.

Input: TRNG → Output: PRF

In 3DES encryption, a TRNG input is converted to PRF output, and that output can become input again and reprocessed in a loop multiple times (three times to make the encryption process more secure).

Input: TRNG → Output: PRF
Loop

In conventional OTP encryption, input in the form of a table of data can be converted into TRNG output.

Input: Table → Output: TRNG

In the quantum safe approach, PRF input data can be converted to TRNG output data, and this output can be reprocessed as input in a loop. Similarly, TRNG input data could also be used as input, producing TRNG output which could then constitute input and be reprocessed in a loop. Additionally, TRNG input can be mapped to intermediate PRF data, which could then be used to produce TRNG output; which again could be reprocessed in a loop.

Input: PRF
Loop
Input: TRNG
Loop
Output: TRNG
Input: TRNG
Loop
PRF

TRNG: True Random Number Generator
PRF: Pseudo-Random Functions

Figure 7.16 Conventional encryption schemes are contrasted with the Quantum-Safe Encryption (QSE) approach (*Source:* [ROU24])

between multiple devices (which can be real, physical devices or virtual devices) or stored on a given device; (2) partitioning data into segments for creating a partitioned OTP; (3) the creation of a virtual OTP by projecting OTP data into a virtualized complex space (a process in which random information corresponding to the original OTP data is generated and assigned to a hyper-dimensional/complex coordinate system, which expands the size of the data involved in the encryption/decryption process and is a ROWF); (4) entangling the OTP data from one or more devices (entangling involves mixing/merging the data from one or more devices); (5) physical separation of the devices (when more than one device is being used); (6) encryption and decryption of information using OTP data; when required (7) authentication (a process in which the connection between multiple devices is verified); and (8) certification (a process in which the communication between multiple devices is registered as valid).

[ROU24] also notes that OTPs can be utilized for a form of post-quantum encryption; however, there are some issues with this approach: (i) it does not use truly random numbers or produce truly random results, making it vulnerable to cryptanalysis; (ii) OTPs could be reused in this approach, making them potentially breakable—for instance, approaches such as frequency analysis and natural language processing can be used to brute-force attack reused OTPs; (iii) the lifetime of the OTP is limited, meaning that data cannot be secured for prolonged time periods; and (iv) the size of communication is limited by the storage capacity available for the corresponding OTP. To avoid the OTP storage capacity issues, symmetric key encryption based on the generation of a large dynamic OTP with a fractal convolution function has been proposed; however, this approach also does not use truly random numbers (timestamps) or produce truly random results (given that a deterministic convolution function is used to expand the timestamps). The lack of randomness in the key and generated OTP is a significant issue, as is the encryption which relies on a ROWF, leaving the system vulnerable to cryptanalysis.

Considerable R&D is expected in the next few years in the PQC/PQE field; interested readers should track ensuing developments.

References

[ACC22] Accredited Standards Committee X9, Quantum Computing Risks to the Financial Services Industry. X9 Informative Report—X9 IR F01-2022. Available on June 1, 2024 at https://x9.org/download-qc-ir/.

[ACI07] A. Acín, N Brunner, et al., "Device-Independent Security of Quantum Cryptography Against Collective Attacks". Phys. Rev. Lett. 98, 230501 (2007).

[ALA22] G. Alagic, D. Apon D, et al., Status Report on the Third Round of the NIST Post-Quantum Cryptography Standardization Process. (National Institute of Standards and Technology, Gaithersburg, MD), NIST Interagency or Internal Report (IR) NIST IR 8413-upd1, September 26, 2022. https://doi.org/10.6028/NIST.IR.8413-upd1.

[ASI21] M. R. Asif, "Quantum Key Distribution and BB84 Protocol". Quantum Untangled, June 24, 2021. Available online on May 21, 2024 at https://medium.com/quantum-untangled/quantum-key-distribution-and-bb84-protocol-6f03cc6263c5

[BAR05] J. Barrett, L. Hardy, A. Kent, "No Signaling and Quantum Key Distribution". Phys. Rev. Lett. 95, 010503 (2005). doi: 10.1103/PhysRevLett.95.010503

[BAR18] E.B. Barker, L. Chen L, et al., Recommendation for Pair-Wise Key Establishment Schemes Using Discrete Logarithm Cryptography. (National Institute of Standards and

Technology, Gaithersburg, MD), NIST Special Publication (SP) 800-56A, Rev. 3. 2018. https://doi.org/10.6028/NIST.SP.800-56Ar3.

[BAR19] E.B. Barker, L. Chen L, et al., Recommendation for Pair1299 Wise Key-Establishment Using Integer Factorization Cryptography. (National Institute of Standards and Technology, Gaithersburg, MD), NIST Special Publication (SP) 800-56B, Rev. 2. 2019. https://doi.org/10.6028/NIST.SP.800-56Br2

[BAR21] W. C. Barker, W. T. Polk, M. P. Souppaya, Getting Ready for Post-Quantum Cryptography: Exploring Challenges Associated with Adopting and Using Post-Quantum Cryptographic Algorithms. (National Institute of Standards and Technology, Gaithersburg, MD), NIST Cybersecurity White Paper (CSWP) NIST CSWP 15. 2021. https://doi.org/10.6028/NIST.CSWP.15.

[BAR21a] W. C. Barker, M. P. Souppaya, W. Newhouse, Migration to Post-Quantum Cryptography Project Description. (National Institute of Standards and Technology, Gaithersburg, MD). 2021. Available on June 1, 2024 at https://csrc.nist.gov/pubs/pd/2021/08/04/migration-to-postquantum-cryptography/final.

[BED17] R. Bedington, J. M. Arrazola, A. Ling, "Progress in Satellite Quantum Key Distribution". NPJ Quan. Inf. 3(1), 1–13, 2017. https://doi.org/10.1038/s41534-017-0031-5.

[BEL64] J. S. Bell, "On the Einstein Podolsky Rosen Paradox". Physics Physique Физика. 1 (3): 195–200. November 1964. doi: 10.1103/PhysicsPhysiqueFizika.1.195.

[BEN84] C. H. Bennett, G. Brassard, "Quantum Cryptography: Public-Key Distribution and Coin Tossing". in Proceedings of IEEE International Conference on Computers, Systems and Signal Processing, Bangalore, India, 1984, pp. 175-179. Also, Theor. Comput. Sci. 2014;560:7. doi: 10.1016/j.tcs.2014.05.025.

[BEN92] C. H. Bennett, "Quantum Cryptography Using Any Two Nonorthogonal States". Physical Review Letters Vol. 68, No. 21, 1992, pp. 3121-3124.

[BEN92a] C. H. Bennett, G. Brassard, N. Mermin, "Quantum Cryptography Without Bell's Theorem". Physical Review Letters Vol. 68, No. 5, 1992, pp. 557-559.

[BIN19] N. Bindel, J. Brendel, et al., "Hybrid Key Encapsulation Mechanisms and Authenticated Key Exchange". Post-Quantum Cryptography, pp. 206-226, DOI 10.1007/978-3-030-25510-7_12, July 2019, https://doi.org/10.1007/978-3-030-25510-7_12.

[BOA18] A. Boaron, G. Boso, et al., "Secure Quantum Key Distribution Over 421 Km of Optical Fiber". Phys. Rev. Lett. 121(19), 190502, 2018. https://doi.org/10.1103/physrevlett.121.190502.

[BRA00] G. Brassard, N. Lütkenhaus, et al., "Limitations on Practical Quantum Cryptography". Phys. Rev. Lett. 2000;85:1330. doi: 10.1103/PhysRevLett.85.1330.

[BRO16] A. Broadbent, C. Schaffner, "Quantum Cryptography Beyond Quantum Key Distribution". Des. Codes Cryptogr. 78(1), 351–382, 2016. https://doi.org/10.1007/s10623-015-0157-4.

[BRU14] N. Brunner, D. Cavalcanti, D., et al., "Bell nonlocality". Rev. Mod. Phys. 86, 419 (2014).

[CAO22] Y. Cao, Y. Zhao, et al., "The Evolution of Quantum Key Distribution Networks: On the Road to The Qinternet". IEEE Commun. Surv. Tutorials 24(2), 839–894, 2022. https://doi.org/10.1109/comst.2022.3144219.

[CHE09] T.-Y. Chen, H. Liang, et al., "Field Test of a Practical Secure Communication Network with Decoy-State Quantum Cryptography," Opt. Express 17, 6540–6549, 2009.

[CHE10] T.-Y. Chen, J. Wang, et al., "Metropolitan All-Pass and Inter-City Quantum Communication Network". Opt. Express 18, 27217. 2010. Available online on May 16, 2024 at https://opg.optica.org/oe/fulltext.cfm?uri=oe-18-26-27217&id=208760. https://doi.org/10.1364/OE.18.027217.

[CHE22] X. Chen, L. Chen, Y. Yan, "Detecting a Photon-Number Splitting Attack in Decoy-State Measurement-Device-Independent Quantum Key Distribution via Statistical Hypothesis Testing". Entropy (Basel). 2022 Sep 2;24(9):1232. doi: 10.3390/e24091232.

[CUR19] M. Curty, H.-K. Lo, "Foiling Covert Channels and Malicious Classical Post-Processing Units in Quantum Key Distribution". NPJ Quantum Inf. 5, 14, 2019.

[CYB23] Cybersecurity & Infrastructure Security Agency (and National Security Agency and National Institute of Standards and Technology), Quantum-Readiness: Migration to Post-Quantum Cryptography. (CISA, Arlington, Virginia), August 21, 2023. Available June 1, 2024 at https://www.cisa.gov/resources-tools/resources/quantum-readiness-migration-post-quantum1254 cryptography.

[CYB23b] Cybersecurity & Infrastructure Security Agency (and National Security Agency and National Institute of Standards and Technology), Quantum-Readiness: Migration to Post-Quantum Cryptography. (CISA, Arlington, Virginia), August 21, 2023. Available on June 1, 2024 at https://www.cisa.gov/resources-tools/resources/quantum-readiness-migration-post-quantum1325 cryptography.

[DIA16] E. Diamanti, H.-K. Lo, et al., "Practical Challenges in Quantum Key Distribution". npj Quantum Information, Vol. 2, 2016, No. 16025.

[DOI23] K. Doi, T. Nakashima, et al., Quantum Cryptographic Communication System, Key Management Device, And Key Management Method. U.S. Patent Application US 2023/0299953, Sep. 21, 2023. Uncopyrighted material.

[DRI24] F. Driscoll, Terminology for Post-Quantum Traditional Hybrid Schemes. (Internet Engineering Task Force (IETF)), Internet-Draft draft-ietf-pquip-pqt-hybrid-term inology. May 2024. Available on June 1, 2024 at https://datatracker.ietf.org/doc/draft-ietf-pquip-pqt-hybrid-terminology/00/

[DUE06] M. Dušek, N. Lütkenhaus, M. Hendrych, "Quantum Cryptography". Prog. Opt. 2006;49:381.

[EKE91] Ekert, A. K. "Quantum Cryptography Based on Bell's Theorem". Phys. Rev. Lett. 67, 661–663, 1991. doi: 10.1103/PhysRevLett.67.661.

[ETS20] ETSI TS 103 744 V1.1.1, "CYBER; Quantum-safe Hybrid Key Exchanges". December 2020, https://www.etsi.org/deliver/etsi_ts/103700_103799/103744/01.01.01_60/ts_103744v010101p.pdf.

[FXU] F. Xu, X. Ma, et al., "Secure quantum key distribution with realistic devices". Reviews of Modern Physics Vol. 92, No. 2, 2020, pp. 025002.

[GER11] I. Gerhardt, Q. Liu, et al., "Full-Field Implementation of a Perfect Eavesdropper on A Quantum Cryptography System". Nat. Commun. 2011;2:349. doi: 10.1038/ncomms1348.

[GIS02] N. Gisin, G. Ribordy, et al., "Quantum Cryptography". Rev. Mod. Phys. 74, 145–195, 2002. https://doi.org/10.1103/RevModPhys.74.145

[GIS04] N. Gisin, G. Ribordy, et al., "Towards Practical and Fast Quantum Cryptography". 2004. arXiv:quant-ph/0411022.

[GIS06] N. Gisin, S. Fasel, et al., "Trojan-Horse Attacks on Quantum-Key-Distribution Systems". Physical Review A, Vol. 73, 2006, No. 2, p. 022320.

[GIS07] N. Gisin, R. Thew, "Quantum Communication". Nat. Photon. 2007;1:165. doi: 10.1038/nphoton.2007.22.

[GIS10] N. Gisin, S. Pironio, N. Sangouard, "Proposal for Implementing Device-Independent Quantum Key Distribution Based on A Heralded Qubit Amplifier". Phys. Rev. Lett. 105, 070501, 2010.

[GSM23] GSM Association, Guidelines for Quantum Risk Management for Telco, Version 1.0. GSM Association. 2023. Available on June 1, 2024 at https://www.gsma.com/get-involved/working-groups/wp-content/uploads/2023/09/Guidelines-for-Quantum-Risk-Management-for-Telco-v1.0.pdf.

[HEN04] W. Henning, K. Harald, et al., "Quantum Eavesdropping Without Interception: An Attack Exploiting the Dead Time of Single-Photon Detectors." New J. Phys. 2004;13:073024.

[HER17] M. Herrero-Collantes, J. C. Garcia-Escartin, "Quantum Random Number Generators". Reviews of Modern Physics Vol. 89, No. 1, 2017, pp. 015004.

[HIL99] M. Hillery, V. Bužek, A. Berthiaume, "Quantum Secret Sharing". Physical Review A Vol. 59, No. 3, 1999, pp. 1829-1834.

[HWA03] W.-Y. Hwang, "Quantum Key Distribution with High Loss: Toward Global Secure Communication". Phys Rev Lett. 2003 Aug 1;91(5):057901. doi: 10.1103/PhysRevLett.91.057901.

[ITU17] ITU-T, "Quantum Key Distribution Networks - Interworking Framework". ITU-T work program 2017-2020 for SG13 Q16/13. Available online on May 14, 2024 at https://www.itu.int/ITU-T/workprog/wp_item.aspx?isn=17186.

[ITU21] ITU-T Focus Group on Quantum Information Technology for Networks (FG QIT4N), "Quantum Key Distribution Network Protocols: Quantum Layer". FG QIT4N D2.3-part 1. 24 November 2021.

[KRA05] B. Kraus, N. Gisin, R. Renner, "Lower and Upper Bounds on The Secret-Key Rate for Quantum Key Distribution Protocols Using One-Way Classical Communication". Phys. Rev. Lett. 2005;95:080501. doi: 10.1103/PhysRevLett.95.080501.

[LEE21] C. C. Lee, T.G. Tan, et al., Quantum Computing Threat Modelling on a Generic CPS Setup. 2021. Available on June 1, 2024 at https://link.springer.com/chapter/10.1007/978-3-030-81645-2_11.

[LI22] Z. Li, K. Wei, "Improving Parameter Optimization in Decoy-State Quantum Key Distribution". Hindawi Quantum Engineering, Volume 2022, Article ID 9717591, https://doi.org/10.1155/2022/9717591.

[LIA17] S.-K. Liao, W.Q. Cai, et al "Satellite-To-Ground Quantum Key Distribution". Nature 549, 43–47, 2017. https://doi.org/10.1038/nature23655.

[LIU11] W. T. Liu, S. H. Sun, et al., "Proof-Of-Principle Experiment of a Modified Photon-Number-Splitting Attack Against Quantum Key Distribution". Phys. Rev. A. 2011;83:042326. doi: 10.1103/PhysRevA.83.042326.

[LIU13] Y. Liu, TY Chen, et al., "Experimental Measurement-Device-Independent Quantum Key Distribution". Phys. Rev. Lett. 111, 130502, 2013.

[LIU13a] D. Liu, S. Wang, et al., "The Security of Decoy State Protocol in The Partial Photon Number Splitting Attack". Chin. Sci. Bull. 2013;58:3859. doi: 10.1007/s11434-013-6037-2.

[LIU22] R. Liu, G. G. Rozenman, et al., "Towards the Industrialisation of Quantum Key Distribution in Communication Networks: A Short Survey". IET Quan. Commun. 3(3), 151–163, 2022. https://doi.org/10.1049/qtc2.12044.

[LO05] H. K. Lo, X. F. Ma, K. Chen, "Decoy State Quantum Key Distribution". Phys. Rev. Lett. 2005;94:230504. doi: 10.1103/PhysRevLett.94.230504.

[LO12] H.-K. Lo, M. Curty, B. Qi, "Measurement-Device-Independent Quantum Key Distribution".

arXiv:1109.1473. 28 May 2012. Phys. Rev. Lett. 108, 130503 (2012). doi: https://doi.org/10.1103/PhysRevLett.108.130503.

[LO14] H.-K. Lo, M. Curty, K. Tamaki, "Secure Quantum Key Distribution". Nat. Photon. 8, 595–604 (2014).

[LU22] C.-Y. Lu, Yuan Cao, et al., "Micius Quantum Experiments in Space". Rev. Mod. Phys. 94(3), 035001, 2022. https://doi.org/10.1103/revmodphys.94.035001.

[LUT00] N. Lütkenhaus, "Security Against Individual Attacks for Realistic Quantum Key Distribution". Phys. Rev. A. 2000;61:052304. doi: 10.1103/PhysRevA.61.052304.

[MA21] C. Ma, L. Colon, et al., "CARAF: Crypto Agility Risk Assessment Framework". Journal of Cybersecurity, Volume 7, Issue 1, 2021, pp. 1–11. https://doi.org/10.1093/cybsec/tyab013

[MAK05] V. Makarov, D. R. Hjelme, "Faked States Attack on Quantum Cryptosystems". J. Mod. Opt. 2005;52:5. doi: 10.1080/09500340410001730986.

[MAK06] V. Makarov, A. Anisimov, J. Skaar, "Effects of Detector Efficiency Mismatch on Security of Quantum Cryptosystems". Phys. Rev. A. 2006;74:022313. doi: 10.1103/PhysRevA.74.022313.

[MEH20] M. Mehic, M. Niemiec, et al., "Quantum Key Distribution: A Networking Perspective". ACM Comput. Surv. (CSUR) 53(5), 1–41, 2020. https://doi.org/10.1145/3402192.

[MIT23] MITRE, Common Weakness Enumeration (CWE). 2023. Available on June 1, 2024 at https://cwe.mitre.org/

[MOS23] M. Mosca, J. Mulholland, A Methodology for Quantum Risk Assessment. (Global Risk Institute). 2023. Available on June 1, 2024 at https://globalriskinstitute.org/mp-files/a-methodology-for-quantum-riskassessment-pdf.pdf/

[MUM21] A. Kumar, S. Garhwal, "State-Of-The-Art Survey of Quantum Cryptography". Arch. Comput. Methods Eng. 28(5), 3831–3868, 2021. https://doi.org/10.1007/s11831-021-09561-2.

[NAT22] National Security Memorandum 10 (NSM-10), National Security Memorandum on Promoting United States Leadership in Quantum Computing While Mitigating Risks to Vulnerable Cryptographic Systems. 2022. (The White House, Washington, DC). Available on June 1, 2024 at https://www.whitehouse.gov/briefing-room/statements-releases/2022/05/04/national-security-memorandum-on-promoting-united-states-leadership-in-quantum-computing-while-mitigating-risks-to-vulnerable-cryptographic-systems/.

[NIS23] NIST SPECIAL PUBLICATION 1800-38B Migration to Post-Quantum Cryptography Quantum Readiness: Cryptographic Discovery—Volume B: Approach, Architecture, and Security Characteristics of Public Key Application Discovery Tools. December 2023. National Cybersecurity Center of Excellence, National Institute of Standards and Technology, 100 Bureau Drive 28 Mailstop 2002, Gaithersburg, MD 20899. This publication is available on May 25, 2023 at https://www.nccoe.nist.gov/crypto-agility-considerations-migrating-post-quantum-cryptographic-algorithms

[NIS23a] National Institute of Standards and Technology, Module-Lattice-based Key-Encapsulation Mechanism Standard. (U.S. Department of Commerce, Washington, DC), Federal Information Processing Standards Publication (FIPS) 203 (Draft). 2023. https://doi.org/10.6028/NIST.FIPS.203.ipd.

[NIS23b] National Institute of Standards and Technology, Module-Lattice-Based Digital Signature Standard. (U.S. Department of Commerce, Washington, DC), Federal Information Processing Standards Publication (FIPS) 204 (Draft). 2023. https://doi.org/10.6028/NIST.FIPS.204.ipd.

[NIS23c] National Institute of Standards and Technology, Stateless Hash-Based Digital Signature Standard. (U.S. Department of Commerce, Washington, DC), Federal Information Processing Standards Publication (FIPS) 205 (Draft). 2023. https://doi.org/10.6028/NIST.FIPS.205.ipd.

[NIS23d] National Institute of Standards and Technology, Digital Signature Standard (DSS). (U.S. Department of Commerce, Washington, DC), Federal Information Processing Standards Publication (FIPS) 186-5. 2023. https://doi.org/10.6028/NIST.FIPS.186-5.

[OFF22] Office of Management and Budget, Migrating to Post-Quantum Cryptography. (The White House, Washington, DC), OMB Memorandum M-23-02, November 18, 2022. Available on June 1, 2024 at https://www.whitehouse.gov/wp-content/uploads/2022/11/M-23-02-M-Memo-on-Migrating1258 to-Post-Quantum-Cryptography.pdf.

[OZA23] K. Ozaki, M. Matsuo, Encryption Communication System, Encryption Communication Apparatus, And Encryption Communication Method. U.S. Patent Application 2023/0336330, October 19, 2023. Uncopyrighted material.

[PHA19] T.A. Pham, N. T. Nguyen, N. T. Dang, "Quantum Key Distribution Over Hybrid Fiber-Wireless System for Mobile Networks". In: Proceedings of the 10th International Symposium on Information and Communication Technology, pp. 236–241, 2019. https://doi.org/10.1145/3368926.3369670.

[PIR15] S. Pirandola, C. Ottaviani, et al., "High-Rate Measurement-Device-Independent Quantum Cryptography". Nature Photonics Vol. 9, 2015, pp. 397-402.

[PIR20] S. Pirandola, U. L. Andersen, et al., "Advances in Quantum Cryptography". Advances in Optics and Photonics Vol. 12, No. 4, 2020, pp. 1012-1236.

[POS23] Post-Quantum Cryptography Working Group, Risk Model Technical Paper. FS-ISAC Post Quantum Cryptography Working Group. 2023. Available on June 1, 2024 at https://www.fsisac.com/hubfs/Knowledge/PQC/RiskModel.pdf?hsLang=en

[QI07] B. Qi, C.H.F. Fung, et al., "Time-Shift Attack in Practical Quantum Cryptosystems". Quantum Inf. Comput. 2007;7:73. doi: 10.26421/QIC7.1-2-3.

[QUA23] Quantum-Readiness Working Group of the Canadian Forum for Digital Infrastructure Resilience, Canadian National Quantum-Readiness Best Practices and Guidelines, Version 03. 2023. Available on June 1, 2024 at https://ised-isde.canada.ca/site/spectrum-management-telecommunications/sites/default/files/attachments/2023/cfdir-quantum-readiness-best-practices-v03.pdf

[RAH23] M. Rahman, W. Trost, et al., Quantum Key Distribution Networking as A Service. U.S. Patent Application US 2023/0344622, Oct. 26, 2023. Uncopyrighted material.

[RAM23] K. Raman, Smart Hub Quantum Key Distribution and Security Management in Advanced Networks. U.S. Patent Application 2023/0328064, Oct. 12, 2023. Uncopyrighted material.

[ROU24] O. F. Roussy Newton, A. Deonarine, et al., System and Method for Quantum-Safe Authentication, Encryption and Decryption of Information. U.S. Patent 11,991,275, May 21, 2024. Uncopyrighted material.

[RUB13] A. Rubenok, J. A. Slater, et al., "Real-World Two-Photon Interference and Proof-Of-Principle Quantum Key Distribution Immune to Detector Attacks". Phys. Rev. Lett. 111, 130501, 2013.

[SCA09] V. Scarani, H. Bechmann-Pasquinucci, et al., "The Security of Practical Quantum Key Distribution". Reviews of Modern Physics, Vol. 81, No. 3, 2009, pp. 1301-1350.

[SHA21] P. Sharma, A. Agrawal, et al., "Quantum Key Distribution Secured Optical Networks: A Survey". IEEE Open J. Commun. Soc. 2, 2049–2083, 2021. https://doi.org/10.1109/ojcoms.2021.3106659.

[SHO00] P.W. Shor, J. Preskill, "Simple Proof of Security of The BB84 Quantum Key Distribution Protocol". Phys. Rev. Lett. 2000;85:441. doi: 10.1103/PhysRevLett.85.441.

[SIL13] TF Silva, D Vitoreti, et al., "Proof-Of-Principle Demonstration of Measurement-Device-Independent Quantum Key Distribution Using Polarization Qubits". Phys. Rev. A 88, 052303 (2013).

[SOU22] M. P. Souppaya, M. Ogata, et al., Software Supply Chain and DevOps Security Practices: Implementing a Risk-Based Approach to DevSecOps Project Description. (National Institute of Standards and Technology, Gaithersburg, MD). 2022. Available on June 1, 2024 at https://csrc.nist.gov/News/2022/implementing-a-risk-based-approach-to-devsecops-fi and https://csrc.nist.gov/pubs/pd/2022/11/09/implementing-a-riskbased-approach-to-devsecops/final

[SOU22a] M. P. Souppaya, K.A. Scarfone, D. F. Dodson, Secure Software Development Framework (SSDF) Version 1.1: Recommendations for Mitigating the Risk of Software Vulnerabilities. (National Institute of Standards and Technology, Gaithersburg, MD), NIST Special Publication (SP) 800-218. 2022. https://doi.org/10.6028/NIST.SP.800-218

[STE09] P. Stefano, A. Antonio, et al., "Device-Independent Quantum Key Distribution Secure Against Collective Attacks". New J. Phys. 2009;11:045021.

[STR20] E. C. Strinati, S. Barbarossa, et al., "6G In the Sky: On-Demand Intelligence at The Edge Of 3D Networks". ETRI J. 42(5), 643–657, 2020. https://doi.org/10.4218/etrij.2020-0205.

[TAK23] R. Takahashi, Y. Tanizawa, Key Management Device, Quantum Cryptography Communication System, And Computer Program Product. U.S. Patent Application 2023/0308264, Sep. 28, 2023. Uncopyrighted material.

[TAN14] Y. L. Tang, H. L. Yin, et al., "Measurement-Device-Independent Quantum Key Distribution Over 200 Km". Phys. Rev. Lett. 113, 190501, 2014. https://doi.org/10.1103/PhysRevLett.113.190501.

[TAN14a] Z. Tang, Z. Liao, et al "Experimental Demonstration of Polarization Encoding Measurement-Device-Independent Quantum Key Distribution". Phys. Rev. Lett. 2014;112: 190503. doi: 10.1103/PhysRevLett.112.190503.

[TRI18] P. V. Trinh, T V. Pham, et al., "Design and Security Analysis of Quantum Key Distribution Protocol Over Free-Space Optics Using Dual-Threshold Direct-Detection Receiver". IEEE Access 6, 4159–4175, 2018. https://doi.org/10.1109/access.2018.2800291.

[TSA21] C.-W. Tsai, C.-W. Yang, et al., "Quantum Key Distribution Networks: Challenges and Future Research Issues in Security". Appl. Sci. 11(9), 3767, 2021. https://doi.org/10.3390/app11093767.

[VAZ14] U. Vazirani, T Vidick, "Fully Device-Independent Quantum Key Distribution". Phys. Rev. Lett. 113, 140501 (2014).

[VER22] M.J.D. Vermeer, E. Parker, A. Kochhar, Preparing for Post-Quantum Critical Infrastructure: Assessments of Quantum Computing Vulnerabilities of National Critical Functions. Homeland Security Operational Analysis Center operated by the Rand Corporation. 2022. Available on June 1, 2024 at https://www.rand.org/pubs/research_reports/RRA1367-6.html

[VIL20] A. Villar, A. Lohrmann, et al., "Entanglement Demonstration on Board A Nano-Satellite". Optica 7(7), 734–737, 2020. https://doi.org/10.1364/optica.387306.

[WAN21] W. Wang, K. Tamaki, M Curty, "Measurement-Device-Independent Quantum Key Distribution with Leaky Sources". Sci Rep 11, 1678 (2021). https://doi.org/10.1038/s41598-021-81003-2.

[WOO23] R. I. Woodward, Y. San, A. J. Shields, Transmitter for A Quantum Communication System, A Receiver for A Quantum Communication System and A Method of Controlling a Quantum Communication System. U.S. Patent Application 2023/0283370, September 7, 2023. Uncopyrighted material.

[YAV23] A. A. Yavuz; S. E. Nouma, et al., "Distributed Cyber-Infrastructures and Artificial Intelligence in Hybrid Post-Quantum Era". In: 2022 IEEE 4th International Conference on Trust, Privacy and Security in Intelligent Systems, and Applications (TPS-ISA), pp. 29–38. IEEE. 2022. doi: 10.1109/TPS-ISA56441.2022.00014.

[YEO24] A. Yeomans, D. Burns, W. David, QKD Switching System. U.S. Patent Application 20240097893. 2024-03-21. Uncopyrighted material.

[YIN16] H. L. Yin, T. Y. Chen, et al., "Measurement-Device-Independent Quantum Key Distribution Over A 404 Km Optical Fiber". Phys. Rev. Lett. 117, 190501, 2016.

[ZAP19] V. Zapatero, M. Curty, "Long-Distance Device-Independent Quantum Key Distribution". Sci. Rep. 9, 17749, 2019.

Basic Glossary of Key Concepts in Chapter 7

This Glossary is based on various industry sources, including [DOI23], [DRI24], [EKE91], [HER17], [HIL99], [ITU17], [ITU21], [NIS23], [ROU24], [WOO23], [YEO24] and others. This Glossary should be used not only as a repository of terms and terminology, but as a tool that further elaborates, explains, and expands on various fundamental quantum concepts.

Basic Glossary of Key Concepts in Chapter 7

Term	Description
Actors in QKD	(Classical) cryptographic jargon of the actors: Alice refers to the QKD transmitter, Bob refers to the QKD receiver, and Eve refers to the attacker.
Adversary-in-the-middle attacks	Cyberattacks where adversaries attempt to position themselves between two or more networked devices using network sniffing or transmitted data manipulation. By abusing features of common networking protocols that can determine the flow of network traffic (e.g., ARP, DNS), adversaries may force a device to communicate through an adversary-controlled system so they can collect information or perform additional actions [NIS23].
Automated data collection attacks	Cyberattacks where, once established within a system or network, an adversary may use automated techniques for collecting internal data (e.g., using a command and scripting interpreter to search for and copy information fitting set criteria such as file type, location, or name at specific time intervals; or using cloud APIs, command line interfaces, or extract, transform, and load services to automatically collect data) [NIS23].
BB84 QKD protocol	QKD protocol introduced by Charles Bennet and Gilles Brassard in 1984; the first QKD protocol to be invented. The protocol has been widely studied and implemented in various commercial key distribution systems. It employs photon polarization bases, time-bin encoding, or other modulation techniques to transmit the information. While several dozen QKD protocols have been proposed since then, only a handful have been thoroughly studied and implemented (at least progressing from proposals to demonstration in experiments).
	The modulation technique is used to form a set of bases in which each basis in the set is orthogonal and/or unique or distinguishable. When using photon polarization, the protocol may be implemented using a set of bases including least two pairs of conjugate photon polarization bases (e.g., a set of bases including, without limitation, a rectilinear photon basis [e.g., vertical 0° and horizontal 90° polarizations] and diagonal photon basis [e.g., 45° and 135° polarizations], or the circular basis of left- and right-handedness). The endpoints are connected by a quantum communication channel which allows quantum information (e.g., quantum states) to be transmitted. The quantum channel may be, for example, an optical fiber or optical free space. The sender device and receiver device also communicate over a non-quantum classical channel—for example, a fiber optic channel, telecommunications channel, radio channel, broadcast radio, or the internet [YAO24].
Component algorithm	Each cryptographic algorithm that forms part of a cryptographic scheme.
Composite PQ/T hybrid protocol	A protocol that incorporates one or more PQ/T hybrid schemes in such a way that the protocol fields and message flow are the same as those in a version of the protocol that uses single-algorithm schemes.

Term	Description
Continuous-variable (CV)-QKD protocols/ schemes	QKD protocols in which the sender encodes information using the position and momentum quadrature of a quantized electromagnetic field in a Hilbert space.
Cryptographic application	An application that receives a key (quantum encryption key or application key) stored in the key management system from the KM (Key Management) function and executes the application in a secure manner while performing cryptographic communication.
Cryptographically Relevant Quantum Computer (CRQC)	A QC that can run algorithms to crack or weaken existing ("classical") cryptography.
Discrete-variable (DV)-QKD protocols/ schemes	QKD protocols that detect signal pulses with single-photon detectors. Examples include the BB84 protocol and the E91 protocol.
E91 QKD protocol	The first QKD protocol using quantum entanglement; it detects information leakage by monitoring the violation of Bell inequality of data obtained by measuring the bipartite quantum states shared between the legitimate communication parties [EKE91].
Key control and management	A function that controls and manages the key management system.
Key relay	A function that transfers application keys to another site, encrypts the application keys using quantum cryptographic communication, and transfers the application keys to another site.
Key storage	A function that stores keys (quantum encryption keys or application keys).
Key supply	A function that supplies keys to the cryptographic application.
Non-composite PQ/T hybrid protocol	A protocol that incorporates multiple single-algorithm schemes of the same type, wherein at least one uses a post-quantum algorithm and at least one uses a traditional algorithm in such a way that the formats of the component cryptographic elements are the same as when they are used as part of single-algorithm schemes [DRI24].
One-time pad (OTP)	An encryption procedure that cannot be cracked but requires the use of a single-use pre-shared key that is larger than the size of the message being sent. In this procedure, a plaintext is paired with a random secret key—the one-time pad. An OTP quantum key exchange is uncrackable because a potential perpetrator eavesdropping on the transmission of a set of time-bin qubits representing a key introduces errors in the set of time-bin qubits due to quantum uncertainty, alerting the two parties to the attempted eavesdropping [ROU24].
Physical random number generators	A process that produces a random output and seek to obtain random numbers from fundamentally random physical phenomena.
Post-Quantum Cryptographic (PQE) Algorithm	An asymmetric cryptographic algorithm that is postulated to be secure against attacks using QCs as well as classical computers. In the NIST domain, the algorithms identified as post-quantum algorithms are defined in [NIS23]:

(continues)

Basic Glossary of Key Concepts in Chapter 7 *(cont.)*

Term	Description
Post-Quantum Cryptographic (PQE) Algorithm *(cont.)*	• FIPS 204 (Draft), Module-Lattice-Based Digital Signature Standard, specifies the "Module Learning with Errors Digital Signature Algorithm, or ML-DSA," which is based on the CRYSTALS603 Dilithium submission. • FIPS 205 (Draft), Stateless Hash-based Digital Signature Standard, specifies the "Stateless Hash-based Digital Signature Algorithm, or SLH-DSA," which is based on the SPHINCS+ submission. These initial quantum-resistant algorithms and stateless hash-based signature standards will augment the public-key cryptographic algorithms already contained in FIPS 186-5, Digital Signature Standard (DSS), as well as SP 800-56A Revision 3, Recommendation for Pair-Wise Key-Establishment Schemes Using Discrete Logarithm Cryptography, and SP 800-56B Revision 2, Recommendation for Pair-Wise Key Establishment Using Integer Factorization Cryptography. Additional quantum-resistant algorithms will be defined by NIST as the standardization effort progresses.
Post-Quantum Encryption (PQE)	Methods and techniques that aim to provide encryption methods that enjoy robust security without depending on the computational security of current mathematical algorithms. Encryption methods that can withstand attacks undertaken using QCs; for example, it is theoretically possible for QCs to factorize large primes using Shor's Algorithm and possibly solve other resistant one-way functions significantly more quickly than conventional computers.
Post-Quantum Traditional (PQ/T) Hybrid Scheme	A multi-algorithm scheme where at least one component algorithm is a post-quantum algorithm and at least one is a traditional algorithm.
PQ/PQ hybrid scheme	A multi-algorithm scheme where all components are post-quantum algorithms.
PQ/T hybrid combiner	A method that takes two or more component algorithms and combines them to form a PQ/T hybrid scheme.
PQ/T hybrid digital signature	A multi-algorithm digital signature scheme made up of two or more component digital signature algorithms where at least one is a post-quantum algorithm and at least one is a traditional algorithm. PQ/T hybrid KEMs, PQ/T hybrid PKE, and PQ/T hybrid digital signatures are all examples of PQ/T hybrid schemes [DRI24].
PQ/T Hybrid Key Encapsulation Mechanism (KEM)	A multi-algorithm KEM made up of two or more component KEM algorithms where at least one is a post-quantum algorithm and at least one is a traditional algorithm.
PQ/T hybrid protocol	A protocol that uses two or more component algorithms providing the same cryptographic functionality, where at least one is a post-quantum algorithm and at least one is a traditional algorithm.
PQ/T Hybrid Public Key Encryption (PKE)	A multi-algorithm Public Key Encryption (PKE) scheme comprises two or more component PKE algorithms where at least one is a post-quantum algorithm and at least one is a traditional algorithm.
Pseudo Random Functions (PRFs)	A type of ROWF function providing a long, disordered sequence of apparently random yet predictable numbers used to protect plaintext/input information and other types of information. PRFs are deterministic: by using a specific index into the sequence, called a key, the same random number is always returned.
Pseudorandom number generator	A process that produces random numbers from a deterministic algorithm.

Term	Description
QKD operation	A simplified operation entails the following [WOO23] (the protocol can use two basis, each with two orthogonal states: horizontal/vertical H/V, and diagonal/anti-diagonal D/A, 45/-45°):
	The sender in the protocol prepares states with one of H/V or D/A polarizations. The prepared states are selected from two orthogonal states (H and V or D and A) in one of two basis H/V and D/A. The pulses are attenuated so that they comprise, on average, just one photon. Thus, if a measurement is made on the pulse, the pulse is destroyed. In addition, it is not possible to split the pulse.
	The receiver uses a measurement basis for the polarization of a pulse selected from the H/V basis or the D/A basis. The selection of the measurement basis can be active or passive. In passive selection, the basis is selected using fixed components, such as a beam splitter. In active basis choice, the receiver makes a decision as to which basis to measure in (e.g., using a modulator with an electrical control signal). If the basis used to measure the pulse at the receiver is the same as the basis used to encode the pulse, then the receiver's measurement of the pulse is accurate; however, if the receiver selects the other basis to measure the pulse, then there will be a 50% error in the result measured by the receiver.
	To establish a key, the sender and receiver compare the basis used to encode and measure (decode). If they match, the results are kept; if they do not match, the results are discarded. The above method is secure in principle. If an eavesdropper intercepts the pulses and measures them, the eavesdropper must prepare another pulse to send to the receiver. However, the eavesdropper will not know the correct measurement basis and will therefore only have a 50% chance of correctly measuring a pulse. Any pulse recreated by the eavesdropper will cause a larger error rate to the receiver, which can be used to evidence the presence of an eavesdropper. The sender and receiver compare a small part of the key to determine the error rate and hence the presence of an eavesdropper.
QKDN controller	A device that controls devices related to quantum cryptographic communication.
QKDN mechanisms	Some of the mechanisms and/or technologies used in QKDNs include: • Key relay encryption methods (e.g., one-time pad [OTP], AES, and so on) • Key relay schemes (e.g., as defined in ITU-T Y.3800) • Key relay alternatives (i.e., XORs uniformly processed at destination node, which are specified in ITU-T Y.3803) Configurations of a QKDN controller (e.g., centralized QKDN or distributed QKDN as specified in ITU-T Y.3802 Protocols in the key management layer, the QKDN control layer, and the QKDN management layer.
Quantum Computer Resistant Encryption (QRE)	Encryption schemes aimed at resisting decryption using QCs. Resistance depends upon vulnerabilities within Resistant One-Way Functions (ROWFs), protecting publicly exchanged encryption keys, and/or within Pseudo-Random Functions (PRF) (a form of ROWF) producing very long cyclical, yet predictable, random number sequences.
Quantum Computer Safe Encryption (QSE)	Encryption schemes intended to resist decryption using QCs. Requires a Secure One-Way Function (SOWF) to exchange shared secrets/private keys and/or to encrypt plaintext/input information and decrypt ciphertext. No amount of computational power applied to either the exchange of private keys or the

(continues)

Basic Glossary of Key Concepts in Chapter 7 (cont.)

Term	Description
Quantum Computer Safe Encryption (QSE) (cont.)	encryption of plaintext/input information can decrypt the ciphertext. QSE is immune from advances in QCs and does not contain ROWFs or PRFs, both of which introduce potential insecurities [ROU24].
Quantum homomorphic encryption	An encryption method that allows quantum computation to be performed over encrypted quantum data.
Quantum Key Distribution (QKD)	A technique for sharing cryptographic keys between two parties: a transmitter (Alice) and a receiver (Bob). The value of this technique is that it provides a test of whether any part of the key can be retrieved by an unauthorized eavesdropper (Eve). In many forms of quantum key distribution, Alice and Bob use two or more non-orthogonal bases in which to encode the bit values. The laws of QM imply (dictate) that measurement of the photons by Eve without prior knowledge of the encoding basis of each causes an unavoidable change to the state of some of the photons; these changes to the states of the photons will cause errors in the bit values sent between Alice and Bob, and by comparing a part of their common bit string, Alice and Bob can thus determine if Eve has gained information [WOO23].
	QKD utilizes the no-cloning theorem and the Heisenberg uncertainty principle in combination with the one-time pad encryption algorithm, where messages are exchanged between two endpoints after being encrypted with a randomly generated shared secure quantum key. QKD networks can be deployed utilizing the underlying infrastructure of optical fibers in existing telecommunication networks, although the transfer rate is still relatively low.
Quantum Key Distribution Network (QKDN)	A QKDN is a cryptographic infrastructure aimed at providing secure symmetric keys to cryptographic applications in institutional networks. Realizing a large-scale QKDN that covers wide geographic areas may need multiple QKDNs that are interworked with each other.
	The ITU has defined a set of standards for QKDNs [ITU17]. The functional requirements and architecture of single QKDN are specified based on the functional requirements of QKDN in ITU-T Y.3801; the functional architecture and operational procedures of QKDN are defined in ITU-T Y.3802. This recommendation also specifies a framework for interworking QKDNs (including interworking between QKDNs supported by different QKDN providers and interworking between QKDNs with different technologies).
Quantum Key Distribution Network (QKDN) manager	A device that manages the entire quantum cryptographic communication network—the QKDN.
Quantum Random Number Generator (QRNG)	A device that uses quantum mechanical effects to produce random numbers
Quantum secret sharing	A protocol for sharing a secret among a number of participants such that only certain subsets of participants can collaboratively reconstruct it.
Random number generator	Physical device outputting unpredictable binary bit sequences.

Term	Description
Resistant One-Way Functions (ROWFs)	Computationally simple mathematical functions often requiring two or more parameters/values, whose reverse function produces (i) one correct solution and an impractically large number of incorrect solutions to check, and/or (ii) an impractically difficult polynomial time computation. Knowledge of the ROWF's result alone does not reveal the original key values. Thus, in this case, ROWFs require impractically large computational power to check all permutations of possible two key values to discover a valid result revealing an encryption key [ROU24].
Secure One-Way Function (SOWF)	A physical property or function requiring only one parameter/value, where no reverse function exists, such as entanglement within a quantum system or a private table of properties derived from a quantum effect or a QRNG. No amount of computational power with a SOWF is sufficient to create a reverse function [ROU24].
Single-algorithm cryptographic scheme versus Multi-algorithm cryptographic scheme	*Single-algorithm scheme:* A cryptographic scheme with one component algorithm. A single-algorithm scheme could use either a traditional algorithm or a post-quantum algorithm. *Multi-algorithm scheme:* A cryptographic scheme with more than one component algorithm. In a multi-algorithm scheme all component algorithms are of the same type—that is, all are signature algorithms or all are Public Key Encryption (PKE) algorithms [DRI24].
Traditional (classical) cryptographic algorithm	An asymmetric cryptographic algorithm based on integer factorization, finite field discrete logarithms, or elliptic curve discrete logarithms.

Chapter 8

Software Considerations

8.1 Introduction

A network is built to service—to connect—end nodes. In particular, it is built and deployed to connect servers and to provide access to users seeking to obtain some service on the end nodes, including applications processing and retrieval of useful information. Quantum Networks (Quanets) serve the same purpose. However, the nodes, the Quantum Computers (QCs), have to evolve first, and then the connectivity, as was the historical evolution with the (early, pre-) Internet. Eventually, direct access will be achievable over an institutional network, a multi-institution network, a Quantum Internet (Quaint), and/or a cloud platform.

A while back, QCs had only a handful of qubits and programs, as these devices were developed "by hand". At this juncture, QCs with hundreds of qubits are being built, and systems with thousands of qubits are rapidly emerging. As QCs become larger, the complexity of programming such QCs requires a more structured approach. QCs are best utilized when the overall architecture entails multiple functional layers. Preferably, application designers formulate quantum algorithms utilizing a high-level language; the high-level code can then undergo a compilation process to transform the algorithm description into a sequence of instructions that can be executed by the QC.

A *number of layered models* to facilitate application development were discussed in Chapter 6, including the European proposal of CEN-CENELEC (refer to Figure 6.17 on page 270). This stack is in some ways similar to the stack one would find in a classical setting, but with quantum-specific particularities. Table 8.1 and Figure 8.1, summarized from [OVA22], highlight some of the features of this QC layering model.

8.2 Quantum Execution Phases

Typically, in a QC execution process, it is desirable to utilize intermediate representations (IRs). An IR of a computation is some kind of coding that "falls" in between its source language description and the target machine instructions. Compilers may use several IRs during the process of translating and optimizing a program.

Table 8.1 CEN-CENELEC Layered Model with Emphasis on Application and Software Support (Summarized from [OVA22])

CEN-CENELEC Layer	Description
Layer 7, Applications/ services supporting use cases	Quantum applications need to be executed in a defined QC environment. Currently, quantum programs can be compiled locally on a classical computer, while considering the specific target QC, and then be submitted for processing to the underlying local or remote platform. The Quaint will enable networked quantum applications whose execution may well involve multiple QCs and may utilize interwoven digital (classical) and quantum message passing.
Layer 6, Programming layers	This layer encompasses the languages, libraries, and software development tools (e.g., software development kits, quantum compilers, debugging tools) utilized by developers for coding quantum algorithms or applications that (can) make use of predefined quantum algorithms as subroutines. Typically, programs are generated by a software library, from code written in a common programming language (e.g., Python). The executable specification of quantum algorithms can be achieved using Quantum Assembly Languages (QASMs, not an exact acronym). Quantum compilation is the process of translating an input quantum circuit into an efficient equivalent of itself, taking into consideration the characteristics of the device that will execute the computation while minimizing at the same time the number of required two-qubit gates.
Layer 5, Assembly/ register-level programming	This layer concerns using Quantum Assembly Languages that describe quantum computations according to a specific model (e.g., circuit model, measurement-based model, quantum annealing model), with an architecture-specific instruction set (for example, one can obtain an intermediate quantum circuit representation). QASM code is an intermediate code representation used in common quantum compilation and simulation tools to describe quantum circuits. The outcome of the compilation is preferably a hardware-independent quantum assembly code. Compilers can also perform optimizations, instruction scheduling, or qubits mapping. Given the variety of QC architectures, it is unlikely that a widely accepted universal QASM will emerge in the short term and become a standard. A QASM example is OpenQASM for IBM® Q devices, which enables experiments with small-depth quantum circuits; another example is NetQASM, a platform-independent and extendable instruction set with support for digital logic, local quantum gates, and quantum networking operations for distant entanglement generation. (Additional translation processes can be utilized to generate hardware-specific micro-code, which can then be executed on the target, providing the hardware-based control logic required to execute the instructions on the target QC).
Layer 4, Hardware Abstraction Layer (HAL)	This layer allows QC users (such as software engineers, application developers, and platform and system designers) to abstract the QC implementation details while retaining the performance; the underlying hardware may change, but the QASM-based programs are expected to still be able to work. The HAL may provide application programming interfaces (APIs) to the upper layer, decoupling from the different types of quantum hardware technologies.
Layer 3, Software drivers	This layer comprises components that are associated with the operating system (OS) and enable hardware-abstraction programs to invoke the control software of the underlying quantum hardware. If the hardware changes, the software drivers must also change.
Layer 2, Operating system (OS) resources	A QC must be provided with an OS, which acts as a resource manager for the underlying quantum hardware; the OS will include built-in networking functions to allow multiple users and applications to utilize the resources as remote clients. With an OS an application sees the QC as if it has its own (dedicated) *(continues)*

Table 8.1 CEN-CENELEC Layered Model with Emphasis on Application and Software Support (Summarized from [OVA22]) *(cont.)*

CEN-CENELEC Layer	Description
Layer 2, Operating system (OS) resources *(cont.)*	resource; the application is shielded from other applications. Applications can make use of facilities only if they are offered by the OS; for example, the OS supplies communication primitives to pass messages between client applications and the QC.
Layer 1, Control software	This layer may include (i) low-level code to translate instructions from higher software layers into commands for guiding the control electronics/optics, (ii) calibration means, (iii) techniques utilized to define error-robust physical operations, and (iv) supporting protocols designed to tune up and stabilize the hardware. Control software for a QC is typically stored on classical digital computers. In the long term, control software may work in conjunction with Quantum Error Correction (QEC) to reduce resource overheads required for encoding by homogenizing error rates and reducing error correlations.

In a general, there are several phases involved in the process of execution of a quantum program; typically, these phases include (i) specifying the quantum algorithm, (ii) transforming the algorithm into executable form, (iii) running the computation for an experiment or simulation, and (iv) analyzing the results. Figure 8.2—synthetized from [CRO17]— shows a high-level block diagram of processes and abstractions to transform and execute a quantum algorithm (post processing takes place on a classical computer after all real-time processing is complete [not shown]).

8.3 Quantum Assembly Languages (QASMs)

Quantum Assembly Languages (QASMs, not an exact acronym) are important tools (languages) that describe quantum computations according to a specific computing model (e.g., circuit model, quantum annealing model, as described in Chapter 5) with an architecture-specific instruction set. QASM code is an intermediate representation utilized in many quantum compilation and simulation tools to construe quantum circuits. QASMs are intended to be hardware-independent languages produced by quantum compilers as output of the translation process from a code formulated in a high-level language into a quantum-level assembly code [KHA18]. QASM originated as a markup language for defining a quantum circuit to generate images for visualization purposes; as quantum computing (Quacomp) evolved, the language was adapted and adopted as a mechanism to specify quantum circuits as input to QCs.

A QASM program defines the classical bits and qubits, describes the operations (gates) on those qubits, and specifies the measurements needed to obtain the classical result by probing the qubits [QUA24]. Many variants of QASM have emerged. Software architectures, compilers, and languages specifically for Quacomp have been researched by the academic community for more than two decades. (The initial version of QASM was first introduced in the mid-2000s [CRO05], [CRO17], [SEL04], [GAY06], [SVO06], [HAN16], [WEC14]).

The variants of QASM make the interworking between different tools challenging and time-consuming, given that this interworking typically requires translations between versions; furthermore, there is a risk of losing information in the translation process. Thus, a common

[Text continues on page 347]

Layer 7	Applications/Services/Use Cases	(similar to OSI/RM Layer 7): Commercial, industrial and research use cases developed for the environment in question, possibly also over cloud platforms (here quantum programs must be locally compiled for a specific device and submitted for batch processing to the remote platform)
Layer 6	Programming tools	Language tools, libraries, and software development facilities (e.g., software development kits, debugging tools, quantum compilers) used by developers for coding quantum algorithms. Quantum compilation addresses the translation of an input quantum circuit into code consistent with the characteristics of the device that will execute the computation
Layer 5	Assembly programming	Quantum assembly languages (QASM) that describe quantum computations according to a given model (e.g., circuit model, quantum annealing model, measurement-based model). Examples: OpenQASM (supports IBM Q devices with small-depth quantum circuits); NetQASM (platform-independent instruction set with support for local quantum gates, digital logic, and *Quacom* operations for remote entanglement generation)
Layer 4	Hardware Abstraction Layer	Provides APIs to the upper layer, decoupling from the different types of quantum hardware technologies. It allows QC users (e.g., application developers) to abstract away the QC implementation details while keeping the performance. The hardware may change, but the QASM-like programs (belonging to the upper assembly / register level programming layer) will still be able to work
Layer 3	Software drivers	Software components that allow hardware-abstraction programs to call the control software of the underlying quantum hardware
Layer 2	OS communication primitives	The Operating system (OS) is a resource manager for the underlying quantum hardware, provided with built-in networking functions to allow multiple users and applications to utilize the resources as remote clients. The OS provides communication primitives to pass messages between client applications and the QC
Layer 1	Control software	Calibration capabilities, low-level code to translate instructions from higher software layers into commands to control electronics/optics; includes techniques for error-robust physical operations and associated protocols to stabilize the hardware, including Quantum Error Correction (QEC). Control software for quantum hardware is typically stored on digital computers

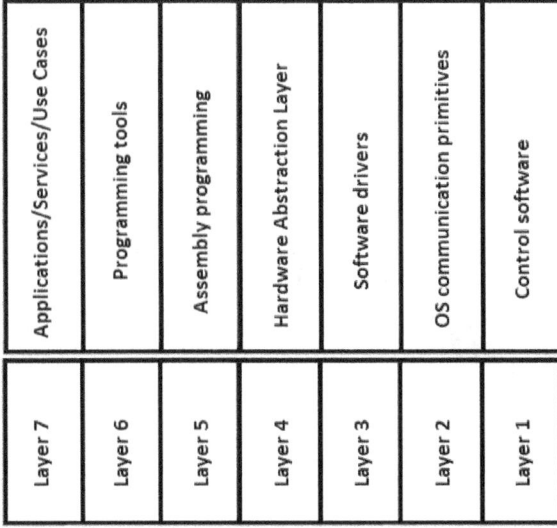

Figure 8.1 CEN-CENELEC layered model (summarized from [OVA22])

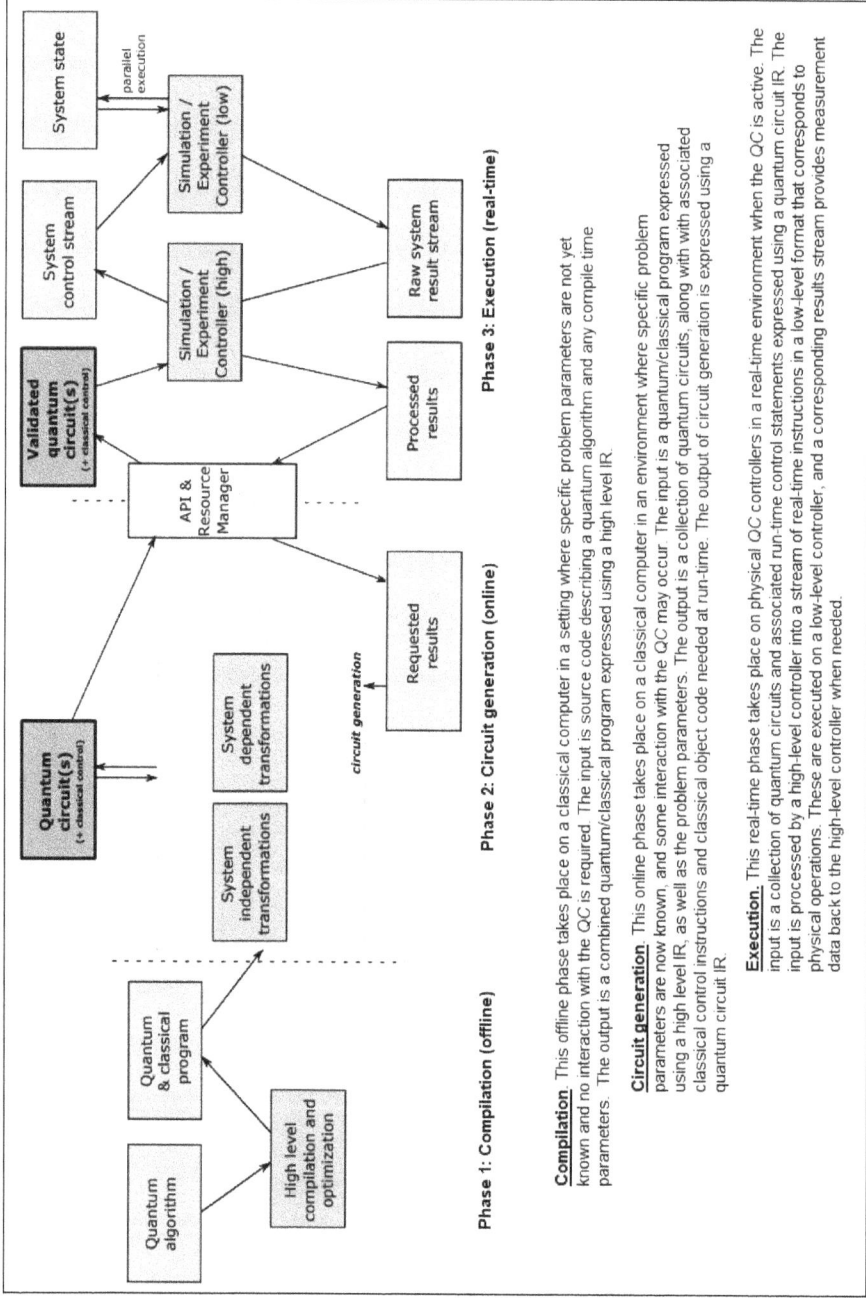

Phase 1: Compilation (offline)　　**Phase 2: Circuit generation (online)**　　**Phase 3: Execution (real-time)**

Compilation. This offline phase takes place on a classical computer in a setting where specific problem parameters are not yet known and no interaction with the QC is required. The input is source code describing a quantum algorithm and any compile time parameters. The output is a combined quantum/classical program expressed using a high level IR.

Circuit generation. This online phase takes place on a classical computer in an environment where specific problem parameters are now known, and some interaction with the QC may occur. The input is a quantum/classical program expressed using a high level IR, as well as the problem parameters. The output is a collection of quantum circuits, along with with associated classical control instructions and classical object code needed at run-time. The output of circuit generation is expressed using a quantum circuit IR.

Execution. This real-time phase takes place on physical QC controllers in a real-time environment when the QC is active. The input is a collection of quantum circuits and associated run-time control statements expressed using a quantum circuit IR. The input is processed by a high-level controller into a stream of real-time instructions in a low-level format that corresponds to physical operations. These are executed on a low-level controller, and a corresponding results stream provides measurement data back to the high-level controller when needed.

Figure 8.2 A high-level block diagram of processes and abstractions to transform and execute a quantum algorithm (synthetized from [CRO17])

QASM language (cQASM) has recently been proposed [KHA18]. See Figure 8.3 for a positioning of the proposed language in the broader context of compilation of QC algorithms. The cQASM instructions can subsequently be used as (i) input to a QC simulator, or (ii) as input to a lower-level compiler that generates hardware-specific instructions for the target QC. cQASM can be utilized to describe relatively simple quantum circuits, this capability being acceptable for the current generation of QCs (in the future, a higher level of abstraction may be required to address the billions of qubits that may make up a large-scale QC). Figures 8.4 and 8.5 depict simple examples of the use of two QC languages.

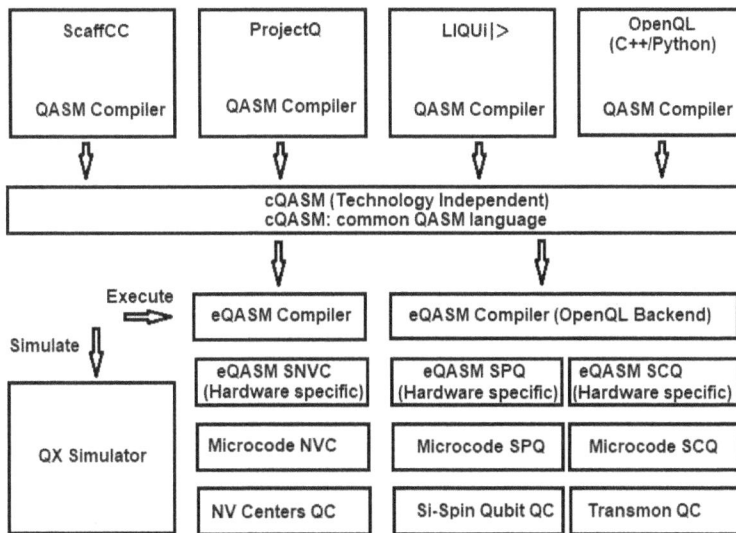

Figure 8.3 Compilation of quantum algorithms where an algorithm written in high-level language is optimized and compiled into a common technology-independent code and then into hardware-specific executable code (*Source:* [KHA18]).

8.4 Basic QASM Variants

Table 8.2 identifies some of the more popular in-use languages loosely based on references [FLI20] and [ION24]; other languages include, but are not limited to, Silq, PyQuil®, Quipper, and NetQASM.

At the time of this writing, the website https://github.com/qosf/awesome-quantum-software#quantum-full-stack-libraries provided an extensive, curated list of open-source developed quantum software projects, and it may continue to be available for consultation. Topics included:

- Quantum full-stack libraries
- Quantum annealing
- Quantum compilers
- Quantum assembly
- Quantum interoperability
- Quantum and post-quantum cryptography
- Quantum tools
- Quantum simulators
- Quantum algorithms
- Quantum converters
- Quantum control
- Quantum error correction
- Experimental quantum computing
- Quantum data

```
1   version 1 0
2
3   # a basic cQASM example
4   qubits 2
5
6   .prepare
7       prep_z q[0:1]
8
9   .entangle
10      H q[0]
11      CNOT q[0], q[1]
12
13   .measurement
14       measure_all
```

The file starts with the specification of the cQASM version on line 1. This line is always present.

Line 3 is a comment. Comments start with a hash (#), and the hash and everything after it until the end of current line is ignored by the cQASM parser.

Line 4 defines the size of the qubit register (and by extension, the size of the corresponding classical register). In this simple example, two qubits are defined. Each qubit in the register is identified by its index. The first qubit has index 0 and the second qubit has index 1. When displaying qubits or bits, the (qu)bit with index 0 is the right-most one.

Lines 6, 9 and 13 mark the beginning of a sub-circuit. A sub-circuit starts with a dot and the name of the sub-circuit and ends with the definition of another sub-circuit. The example defines three sub-circuits: prepare, entangle and measurement.

Line 7 contains the first quantum instruction, to prepare both qubits in their ground state. The Single-Gate Multiple-Qubits (SGMQ) syntax is used to address both qubits at once. Alternatively, we could have written line 7 as two lines, prep_z q[0] and prep_z q[1].

Line 10 and 11 describe the quantum gates that form the circuit. First, a Hadamard gate is applied to the qubit with index 0, followed by a CNOT where the qubit with index 0 is the control qubit and the qubit with index 1 is the target qubit.

At line 14 the state of all qubits is measured along the Z-axis to obtain the final result.

cQASM itself is case-insensitive.

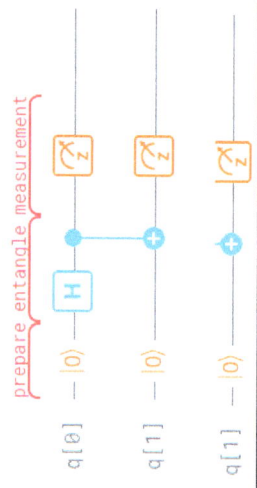

Bell states $|\Phi^{\pm}\rangle = \frac{|00\rangle = |11\rangle}{\sqrt{2}}$, $|\Psi^{\pm}\rangle = \frac{|01\rangle = |10\rangle}{\sqrt{2}}$ are specific quantum states of two qubits that represent the simplest (and maximal) examples of quantum entanglement. The measurement results for this program should give 0 for both qubits or 1 for both qubits, in equal amounts.

Figure 8.4 cQASM example to create a Bell state (synthetized from [QUA24])

```python
from qiskit import QuantumCircuit
# Create a basic Bell State.
qc = QuantumCircuit(2, 2) # 2 qubits, 2 bits
qc.h(0)
qc.cx(0, 1)

# Measure and put the result into the two bits.
qc.measure([0, 1], [0, 1])

# Display circuit.
print(qc)
```

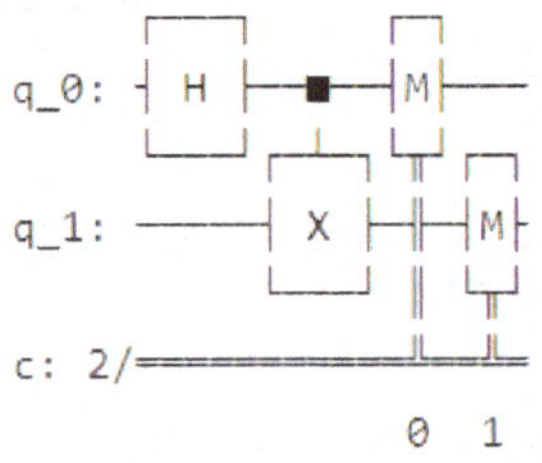

```
q_0: ┤ H ├──■──┤M├
           │  └╥┘
q_1: ──────┤ X ├─╫─┤M├
               ║  ║
c: 2/══════════╩══╩═
               0  1
```

Quantum circuit that creates an entangled Bell pair and measures it in Qiskit

Figure 8.5   Qiskit used to run quantum programs against IonQ's trapped ion QC [ION24]

Table 8.2 Popular In-Use QASM Languages[a]

QC Language	Description
Braket	Amazon's Quacomp service, which includes a Python®-based SDK for creating and submitting quantum programs. To run a job on Braket one needs an AWS account with the Amazon braket sdk.
common QASM (cQASM)	A recent language proposal [KHA18]. The cQASM instructions can be used as input to a lower-level compiler that generates hardware-specific instructions. cQASM can utilized to describe relatively simple circuits.
Open Cirq	This language and framework have been developed and supported by Google®. It is one of the newer languages. There are various adopters of Open Cirq, including but not limited to QC Ware®, Zapata®. Given that Cirq has been developed by Google, it is the most synergistic way to execute a quantum program over the Google Cloud Marketplace (GCM).
Open QASM	Open Quantum Assembly Language (Open QASM) originates from IBM and the IBM Q projects, embodying a low-level approach to manipulation of qubits. As the name implies, it is open source and quite universal. Open QASM is similar to a classical assembly language where instructions can be cobbled together to perform operations on registers and qubits, including, for example, manipulating qubits and logical bits, performing measurements, and aggregating those measurements.

Table 8.2 Popular In-Use QASM Languages *(cont.)*

QC Language	Description
ProjectQ	An open-source quantum programming framework developed by ETH Zurich (a public research university in Zürich, Switzerland). This framework incorporates several optimization techniques useful for emulation, such as mimicking the action of oracles in simulating quantum systems without having to compile these to low-level gates.
Pytket	A Python toolkit for quantum programming developed by Cambridge Quantum Computing.
Q#	This is a Microsoft-defined language, but it is open source. Q# adopts features from programming languages such as C#. Q# programs can run as a command-line application, or use a Python or .NET host program. The Microsoft QDK is a development kit for Q#; it can be used with tools and languages to develop quantum applications that can execute in various environments.
QISKit	Qiskit is an open-source language used by many practitioners that has been around since the late 2010s and is supported by IBM. It enjoys wide adoption and support from IBM and can work on a variety of hardware. It is a Python-based framework; it provides robust tools for writing quantum programs in Python and offers a variety of additional packages for working with quantum hardware and executing common algorithms. Qiskit has an extensive body of tutorial materials for learning the language
Strawberry Fields and PennyLane	The language has its origins from the photonic quantum computing company Xanadu. It has applications in Quantum Machine Learning.
XACC	XACC is a framework usable for hybrid quantum / classical architectures; it was developed by researchers at Oak Ridge National Lab, and has support for a large number of compilers, simulators, and quantum hardware.

a (loosely based on references [FLI20] and [ION24])

References

[CRO05] A. Cross, "QASM tools". 2005, https://www.media.mit.edu.

[CRO17] A. W. Cross, L. S. Bishop, et al., "Open Quantum Assembly Language". arXiv: 1707. 03429v2, 13 Jul 2017. https://doi.org/10.48550/arXiv.1707.03429.

[FLI20] R. Flint, "Which Quantum Programming Language Should You Learn?". July 12, 2020. Available online on June 3, 2024 at https://quantumzeitgeist.com/which-quantum-programming-language-should-you-learn/#google_vignette.

[GAY06] S. Gay, "Quantum Programming Languages: Survey and Bibliography". Math. Structures in Computer Science, 16:581–600, 2006.

[HAN16] T. Haner, D. Steiger, et al., "A Software Methodology for Compiling Quantum Programs". arxiv:1604.01401, 2016. https://doi.org/10.48550/arXiv.1604.01401. Also, Quantum Sci. Technol. 3 (2018) 020501. https://doi.org/10.1088/2058-9565/aaa5cc

[ION24] IonQ Staff, "Quantum Languages". March 14, 2024. Available online on June 3, 2024 at https://ionq.com/docs/hello-many-worlds-seven-quantum-languages.

[KHA18] N. Khammassi, G.G. Guerreschi, et al., "cQASM v1.0: Towards a Common Quantum Assembly Language". May 24, 2018. arXiv:1805.09607v1. https://doi.org/10.48550/arXiv.1805.09607.

[OVA22] O. van Deventer, N. Spethmann, et al., "Towards European Standards for Quantum Technologies". 3 Oct 2022, arXiv:2203.01622. https://doi.org/10.48550/arXiv.2203.01622. EPJ Quantum Technol. 9 (2022) 33. https://doi.org/10.1140/epjqt/s40507-022-00150-1.

[QUA24] Quantum Inspire staff, "cQASM: A Quantum Programming Language". Available online on June 1, 2024 at https://www.quantum-inspire.com/kbase/cqasm/.

[SEL04] P. Selinger, "A Brief Survey of Quantum Programming Languages". Proc. Seventh Int'l Symp. Functional and Logic Programming, pages 1–6, 2004.

[SVO06] K. Svore, A. Cross, et al., "A Layered Software Architecture for Quantum Computing Design Tools". IEEE Computer, (39(1)):74–83, 2006.

[WEC14] D. Wecker and K. Svore, "LIQUi |>: A Software Design Architecture and Domain-Specific Language for Quantum Computing". arXiv:1402.4467, 2014. https://doi.org/10.48550/arXiv.1402.4467.

Epilogue

This section captures *some* of the recent developments in the Quantum Computing (Quacomp) and Quantum Communications (Quacom) fields that emerged just prior to the publication of this text.[1] In 2024, developments were seen prevalently in Quacomp rather than in Quacom; the practical technical highlights of the year related to significant progress in Quantum Error Correction (QEC), and observers labeled accomplishments in that arena as the "breakthrough of the year". Further to that, the United Nations has designated 2025 as the International Year of Quantum Science and Technology [UNE25]. According to a 2024 McKinsey report, by the end of 2023 the global investment in Quacomp reached US$42 billion. By 2035 the commercial market was projected by the same research firm to be as high as US$90B: US$72B for Quacomp, $US15B for Quacom, and US$3B for quantum sensing [IDQ24]. At time of this writing Microsoft was urging businesses to get "quantum-ready" in 2025, stating that "We find ourselves in an exciting and pivotal time. We are at the advent of the reliable quantum computing era. And we are right on the cusp of seeing quantum computers solve meaningful problems and capture new business value. As we look toward the next 12 months, the pace of quantum research and development is only going to accelerate, making this a critical and catalyzing time for business leaders to act" [AZI25], [SUB25].

The industry goal is to create a physical million-qubit system with about 100 reliable logical qubits able to run circuit depths of 10,000. While a number of observers believe this is still a decade away, others prognosticate that the time-line may shrink [RUS25]. The key Quacomp's concern—the Achilles heel of every Quantum Computer (QC)—is whether, in a practical setting, a QC can perform long computations without aborting the process due to noise. Given that a QC is not able to undertake gate operations in a perfect manner, some pervasive residual error is inevitable; theoretically, this means that QCs with imperfect gates can only work through a number of gates before the computation is compromised by noise. Fortunately, the threshold theorem states that a QC with a physical error rate below a certain threshold can, with the application of QEC schemes, suppress the logical error rate to arbitrarily low levels [NIE11]. As noted, in 2024 there were significant advances in QEC, particularly led by Google®'s breaking of the QEC threshold barrier on Google's Willow chip (see Figure 5.27 on page 212). More efficient surface codes and improved alternative approaches (for example, Cat Qubits) emerged; also, MIT®, Harvard, and QuEra® demonstrated QEC on an atomic

---

[1] Vendors or products are mentioned only with the purpose of illustrating technology trends. No explicit or implicit endorsement of products or vendors by these authors should be inferred. The content of this section is based on public information.

processor with 48 logical qubits, and, Riverlane® and Rigetti® Computing demonstrated in small experiments that the decoding latency can be reduced.

In the Quacom context, and consistent with the premises of this text, of late some proponents (at Nokia) are heralding the "Era of Innovation Quantum 2.0 . . . reconceptualizing the way we communicate and the way we sense the world, as well as the way we compute" [BAT25], noting that "Quantum networks carry quantum information; as computing and communication have always been inextricably linked, these networks will play a vital role in creating a quantum internet that interconnects quantum computers and other quantum devices. Distributed quantum computing linked through a quantum network could scale to extremely complex problems no single quantum computer could tackle alone; this new internet will also connect quantum sensors and create secure communications systems that protect information with the immutable laws of physics".

E.1 Technical Advancements—Short List

Practical Quacomp requires error rates well below those achievable with physical qubits. QEC (as discussed in Chapter 5 and above) facilitates algorithmically managed error rates by encoding logical qubits within many physical qubits, for which increasing the number of physical qubits enhances protection against physical errors; however, traditionally, adding more qubits also increases the number of error sources, thus the density of errors must be sufficiently low for logical performance to improve with increasing code size. Google asserted that it had achieved the second of six steps in its strategy to build a QC with 1 million qubits (again, see Figure 5.27). Google's Willow was able to demonstrate that as additional qubits are added, the error rate actually dropped: the system of superconducting qubits has sufficient performance to overcome the additional errors from increasing qubit number [GOO23]. Keeping physical qubits from warming up is a practical operational challenge, given that extremely low temperatures (–460° F, absolute zero, 0 K) are required for particles to maintain an undisturbed quantum state. For example, improved temperature control enabled the Google Willow chip to perform at a manageable error correction threshold, utilizing cooling devices nicknamed "chandeliers". See Figure E.1.

Recall that common physical entities to construct qubits include:

Figure E.1   Cryostat refrigerator for cooling Quacomp chips such as Willow. Courtesy: Google/Handout

- Superconducting systems: Superconducting circuits are cooled to near absolute zero to create and manipulate quantum states (used by IBM®, Google and others).
- Trapped ions systems: Individual ions trapped by electromagnetic fields; quantum states are controlled by lasers—trapped ion qubits enjoy long coherence times (used by IonQ®, Honeywell®, and others).
- Photonics-based systems: Qubits based on photons have the potential to achieve long-distance Quacom (used by ORCA Computing, Xanadu®, and others); and,
- Neutral atoms systems: Individual atoms are held in place by laser fields; quantum states are manipulated using lasers (used by QuEra and others).

E.2 Novel Approaches for Error Control

As discussed extensively in previous chapters, state errors reduce the reliability—the fidelity—of a qubit. To perform complex computational tasks, a quantum chip must remain stable, enjoying high fidelity for a relatively long time—at least several minutes, if not longer. To address the issue, in recent years, researchers have developed systems that utilize a set of intrinsically lower fidelity qubits to encode an abstract "logical qubit", already mentioned in Chapter 1 and elsewhere; the logical qubit is designed to be protected from errors, achieving adequate overall fidelity. A useful QC will be based on many logical qubits. As noted above, at the time of writing, several firms made concrete progress on QEC capabilities, including, among others, Google, QuEra, IBM, and CSIRO® [USM25]. Quantum chips consisting of hundreds of physical qubits are already available, as noted in the text; at the time of writing, developers have created only single logical qubits, but chips with the ability to assemble multiple logical qubits together into a Quantum Processing Unit (QPU) are expected to appear in the near future.

One of the most widely used QEC codes (briefly cited by the references listed on page 198) is the *surface code,* which is a two-dimensional grid structure having squares of data qubits, with auxiliary qubits in the middle of each square checking whether the data qubits agree or not. Half of the auxiliary qubits check for phase-flip errors, while the other half check for bit-flip errors [RIV25] (see Figure E.2). The surface code is somewhat of an industry standard, but the

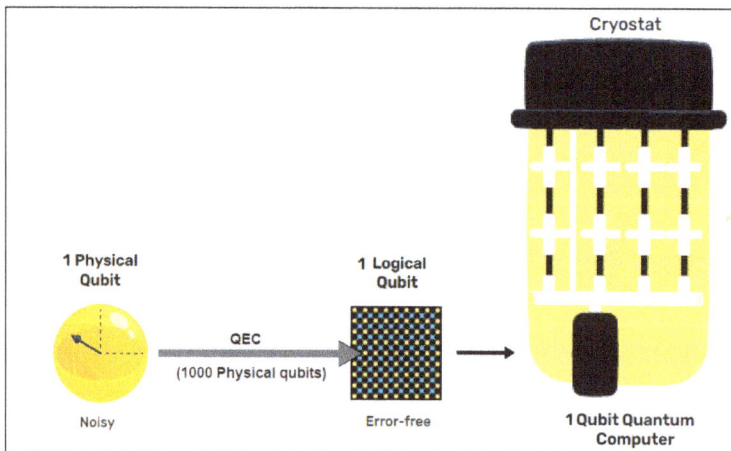

Figure E.2   Physical and logical qubits in support of QEC

required number of qubits to correct errors occurring simultaneously on multiple data qubits scales quadratically; for example, to protect against any two simultaneous errors requires 49 qubits, but to protect against any four simultaneous errors requires 161 qubits. For adequate protection for some practical algorithms, one needs between 500 and 1,250 physical qubits per logical qubit. These requirements were challenging at the time of writing, given that the largest superconducting QC was able to implement just over 1,000 qubits onto a chip (IBM's Condor chip supported 1,121 physical qubits).

Consider a system with transmon qubits using a surface code to factor a number with 2,048 binary digits. Shor's algorithm requires about 1,000 physical qubits per logical qubit; thus, the algorithm requires 20 million physical qubits [GID21]. The "overhead" of error correction and logical gates is substantial: only 0.06% of the qubits (a fraction of the total system qubits) are used to perform computation; the others are needed simply to correct errors. In summary, using the surface error correction code, one can create logical qubits, but this requires a substantial number of additional physical qubits and supporting hardware—depending on the fidelities one aims to achieve, thousands of physical qubits are necessary, thus scaling is problematic [COP24].

A quantum system interacts with its environment, no matter how much care is put into isolating it; consequently, qubits undergo corruption (degradation). One of the most common issues is bit-flip errors, which arise when a qubit's state changes unpredictably from 0 to 1 or from 1 to 0. As noted in earlier chapters, decoherence is a significant challenge in Quacomp. A proposed alternative to the current QEC approach is the use of "cat qubits"; cat states have been a theoretical concept for a while, but engineering the control hardware has proved challenging. New strategies are emerging to circumvent the corruption problem by encoding a quantum bit non-locally across the phase space of a physical system. Given that most sources of decoherence result from local fluctuations, the foundational promise is to exponentially suppress errors by increasing a measure of this non-locality. Prominent examples are topological qubits, which delocalize information over real space and where spatial extent measures non-locality. [LES20] proposes to encode a qubit in the field quadrature space of a superconducting resonator endowed with a special mechanism that dissipates photons in pairs; this process pins down two computational states to separate locations in phase space. By increasing this separation, one can achieve an exponential decrease of the bit-flip rate while only linearly increasing the phase-flip rate. Because bit-flips are autonomously corrected, only phase-flips remain to be corrected via a one-dimensional quantum error correction code. This exponential scaling demonstrates that resonators with nonlinear dissipation are promising building blocks for Quacomp, while enjoying significantly reduced hardware overhead.

Cat qubits make reference to Schrödinger's thought experiment which asserts that if one seals a cat in a box along with something that can potentially kill the cat, one would not know if the cat is alive or dead until one opens the box. So, until one opens the box and observes the cat, the cat is simultaneously dead and alive until observed by an outside observer. Thus, something can exist in two opposite states at the same time. Schrödinger's cat is a classical system in a superposition of two very different macroscopic states. Cat states are superpositions of quantum states that display almost-classical behavior. Cat qubits represent two opposite quantum states at the same time. Cat qubits offer several advantages over conventional qubits: while regular qubits demonstrate only a single superposition, cat qubits can exist in double superposition. Qubits with double superposition will not only improve fault tolerance but also lead to enhanced computing power and algorithmic efficiency [BRA25]. To create a cat qubit and protect its information from bit-flip errors, the proposal of [LES20] uses a coherent state of light (photons) in a superconducting chip. Cat qubits simplify quantum error correction by reducing the problem from two dimensions to one. Bit-flips are driven to the brink of extinction, leaving only phase-flips to contend with. Compared to surface codes, this approach is a more efficient path to fault tolerance.

Adding photons increases the "distance" between the two states; this distance suppresses unwanted jumps (bit-flip errors) between the two possible oscillations, making the states resilient to external noise. The protection against bit-flip errors grows exponentially with the number of photons. This protection becomes significant with just a few photons. Increasing the photon count to suppress bit-flip errors introduces a higher risk of phase-flip errors, because the more photons one has, the more likely we are to lose one. Fortunately, the increase is only linear with the number of photons, an acceptable trade-off considering the exponential decrease in bit-flip errors.

A single, well-controlled cat qubit is a strong building block for fault-tolerant quantum computers that are both powerful and efficient. Alice & Bob®, a Paris-based start-up company, has developed a single cat qubit chip, called *Boson 4,* that can average up to seven minutes between bit-flip errors, far beyond the tens of milliseconds of other superconducting quantum systems, with just 12 or more photons. Boson 4 was launched on Google Cloud [ALI25]. The ultimate goal is to develop a fault-less QC by 2030. The cat qubit QPU is a sapphire chip with a small apparatus of physical systems comprising mostly superconducting metals. The cat qubit encodes information in a quantum harmonic oscillator, which is set up in a superconducting resonator, a piece of tantalum and aluminum several hundred micrometers wide. This is the memory of the cat qubit, where quantum information is stored in the collective wave-particle behavior of the individual photons trapped within the metal.

E.3 Memory

Shifting to communications, Quacom systems require quantum storage in order to be able to scale. However, storing quantum information in solid-state devices is challenging because of the noise generated by the control pulses, which obscure the fragile single-photon signals needed for reliable storage and retrieval. Quantum memory is critical for enabling large-scale quantum networks (Quanets) by bridging short-distance entanglement into long-distance entanglement. Rare-earth ions doped in crystals have emerged as a candidate for implementing high-performance quantum memories, but existing integrated quantum memory systems rely on optically excited states, which limit storage times to the excited-state lifetime and do not allow on-demand retrieval. Researchers have recently developed an integrated spin-wave quantum memory that utilizes advanced noise-suppression techniques to achieve high-fidelity, long-duration, on-demand quantum storage. The spin-wave storage method encodes photons into ground-state spin-wave excitations, offering a solution by extending storage times to the spin coherence lifetime and supporting adjustable retrieval. Despite its promise, integrating spin-wave storage in solid-state devices has proven difficult to implement, but progress has been reported of late [SWA25a].

E.4 Business Applications

Advocates lobby for the concept of assessing QCs going forward as Fault-Tolerant Application Scale Quantum (FASQ) systems able to undertake specified levels of productive "work"

[RUS25]. Quacomp/Quacom applications now emerge on a routine basis. In addition to the applications noted in Chapter 1, the two applications listed below are indicative of the commercial scope of the evolving technology.

- Researchers have developed a multimodal quantum framework, called MultiQ-NLP, which integrates language and image data into a unified, structure-aware quantum model. By translating both text and images into quantum circuits, the approach leverages Quacom's ability to handle complex tensor structures, thus enhancing AI Natural Language Processing (NLP) systems. Tests demonstrated that the quantum-based model can perform on par with leading classical methods, suggesting that quantum-enhanced approaches may emerge as practical approaches in the near future. Words, sentences, and their grammatical roles can be represented as higher-order tensors, multi-dimensional arrays that capture how words relate to one another.

  On a QC, tensors can be encoded as states of qubits. The MultiQ-NLP framework extends an existing method for quantum NLP (QNLP) to incorporate images. By representing images as feature vectors extracted by a classical neural network and then mapping these vectors into quantum states, the method places language and visuals into the same mathematical framework in which each word and image feature corresponds to a set of qubits, and the meaning of sentences combined with images emerges from how these qubits interact. The operations that link words together are mapped onto quantum gates that entangle states, preserving the compositional structure of language in a quantum format. Similarly, image features undergo a dimensional reduction into a small vector that can be encoded as quantum rotations, capturing visual properties that the circuit can process alongside linguistic data. [HAW24], [SWA25].

- Existing predictive analytics works relatively well for relatively small problems but becomes impractical for large problems because of exponential complexity. Classical systems, even when endowed with artificial intelligence (AI), are designed to evaluate problems sequentially. Quacomp, by contrast, can process multiple possibilities simultaneously. Entanglement's interconnectedness enables quantum systems to collaborate in a distributed manner.

  One application entails optimizing campaign logistics in real time; marketing logistics pose a highly complex problem, involving thousands of variables, such as audience segmentation, channel preferences, budget allocation, and timing. AI systems can be Quacomp enhanced. These enhanced AI systems can (i) train faster by processing large datasets in parallel; (ii) Make predictions with greater precision by evaluating all potential outcomes simultaneously; (iii) solve previously intractable problems, such as hyper-personalizing real-time customer experiences.

  Annealing-based Quacomp technology can, and has, evaluated millions of potential configurations for a global marketing campaign, dynamically identifying the most effective strategies for resource allocation, ad placement, and campaign timing [SNY25]. In this instance, Quacomp reallocated budgets across multiple markets in real time; facilitated campaign launches inclusive of shifts in consumer sentiment, allowing quick response to changing conditions; and enabled activations that were optimized to align with audience demographics, location-specific engagement patterns, and real-time event dynamics. What would normally take weeks of manual analysis and iterative testing with classical systems and processes was reportedly completed in minutes using Quacomp resources.

E.5 Post-Quantum Cryptography

Stakeholders agree the risk posed by QCs to classical cryptographic systems is imminent. An RSA-based encryption algorithm can be cracked in 0.5 milliseconds on an HP® Frontier Supercomputer if it uses 512-bit keys, three months if it uses 1024-bit keys, and six trillion years if it uses 2048-keys, but the comparable times are 10 hours, one day, and one week with a QC [ALI25]. As noted in Chapter 7, Post-Quantum Cryptography (PQC) deals with the development and deployment of quantum-resistant algorithms that are usable across a spectrum of computing and communication systems, some such algorithms particularly optimized for embedded devices to support the security of Internet of Things (IoT) devices and critical infrastructure. PQC's cryptographic methods are designed to be secure against an attack by a QC. The PQC market is poised for significant growth as enterprises and governments urgently endeavor to apply solutions to protect critical infrastructure from upcoming quantum threats. PQC is expected to experience significant market growth well before Quacomp reaches maturity; PQC providers are already delivering actionable solutions that can be implemented on existing computing and networking infrastructure. Table E.1, developed by SEALSQ, depicts a six-year timeline (2025–2030) for developing a fully operational QC, alongside the SEALSQ advancements for developing PQC chips and post-quantum computing defense capabilities. Table E.2, also developed by SEALSQ, depicts a summary of possible timelines of PQC implementation by industry and government entities [SEA25].

E.6 Industry Commercial Snapshot

According to Hyperion Research® the worldwide QC market was around $1 billion in 2024 and is expected to reach about $1.6 billion in 2026. Just prior to this writing, as Alphabet/Google announced a chip breakthrough with Willow, Quacomp stocks for various providers, such as but not limited to, IonQ®, Rigetti Computing®, D-Wave® Quantum, and Quantum Computing "skyrocketed", rallying to new heights and drawing keen investor interest. However, some leading

Table E.1 SEALSQ Advancements for Developing Post-Quantum Chips and Post-Quantum Computing Defense Capabilities

Year	Quantum computing	Semiconductors	Defense Capabilities
2025	1,000+ qubits, error reduction with QEC	First PQC chips released	PQC standards pilot programs
2026	Small-scale quantum advantage	20M post-quantum chips produced	Hybrid encryption adoption
2027	10,000 qubits, mid-scale use cases	100M chips, consumer IoT devices	Quantum attack simulation tools
2028	Universal QC systems	PQC chips fully integrated	Quantum-ready networks expanded
2029	Hybrid QC-classical systems	Third-gen chips, global deployment	Quantum-resilient infrastructure
2030	100,000+ qubits, fault-tolerant QC	Quantum-ready semiconductors	Full global quantum security

Courtesy SEALSQ Corp., Geneva, Switzerland

**Table E.2 Possible Timelines of PQC Implementation by
Industry and Government Entities**

Sector	PQC activities
National Institute of Standards and Technology (NIST) Standards	Initial PQC standards finalized; additional standards expected by 2025.
CISA (US Critical Infrastructure)	PQC Roadmap recommends adoption of PQC by the late 2020s.
White House Federal Mandate	Federal systems to transition to quantum-safe cryptography by early 2030s.
European Union Critical Sectors	EU cybersecurity strategy encourages PQC adoption; mandates expected by 2025-2027.
IoT (Matter, Wi-SUN)	Quantum-safe integration encouraged; widespread adoption likely by 2026-2027.
Financial Services	SWIFT, PCI-DSS expected to incorporate PQC by 2026-2028.
Healthcare	Mandates for quantum-safe cryptography likely by the late 2020s

Courtesy SEALSQ Corp., Geneva, Switzerland

vendors of Graphics Processing Units (GPU) then asserted that the technology is not going to be "very useful" for 15–20–30 years; obviously, entrenched providers of legacy supercomputing chips do not appreciate the upcoming computing competition afforded by QC systems.

Nonetheless, some well-placed observers assert that, at the time of writing, Quacomp "looks like where AI was in the 2010s" [ALB25]. In addition, while a large-scale gate-based approach to Quacomp may be several years away, the annealing approach is currently deployed at this juncture (for example, D-Wave systems has already deployed at Mastercard® and NTT Docomo) [LES25]. Also, currently, cloud-based quantum platforms where users can access Quacomp as a Service (QaaS) without infrastructure development, include but are not limited to D-Wave's Leap and IBM's QISKit®. Table E.1 above, along with the roadmaps discussed in Chapter 5, provides a realistic view to the expected denouement.

A few recent vendor/commercial achievements are listed below.

- Quantum supremacy is achieved when a QC solves a problem that would take a classical computer an impractical amount of time to do. As of this writing, Google's Willow demonstrated quantum supremacy for a specific computational task.
- Vendors such as IBM, Google, Microsoft® and QuEra all have roadmaps to significantly scale up QPUs by 2030. Semiconductor firms include, but are not limited to, Intel®, Diraq®, SQC®, PsiQuantum® and Xanadu.
- Rigetti Computing is known to have supported full-stack cloud-based Quacomp since 2017 and for global enterprise, government, and research clients. The firm has developed a 9-qubit QPU called Novera, designed for local deployment, and an 84-qubit system called Ankaa®-3, which, according to the vendor, has enhanced precision. The firm also offers a QaaS platform. Their 2025 roadmap reportedly includes two new systems: a 36-qubit system and 100-qubit system, both designed with the goal of improved reliability, aiming at a 2x reduction in error rates.
- D-Wave® Quantum provides Quacomp systems, software, and services for quantum applications such as materials sciences, artificial intelligence, drug discovery, financial modeling, and cybersecurity.

- IonQ's current #AQ 36 Forte Enterprise systems already provide solutions for customers; the planned #AQ 64 Tempo systems, expected to be available in 2025, and the next-generation #AQ 256 systems will support increasingly complex business problems.
- SEALSQ Corp. announced around the time of writing that, in cooperation with WISeSat, they will launch six next-generation satellites in 2025; each satellite is expected to feature incremental advancements in PQC and secure communication technologies, with the goal of establishing a secure, scalable satellite constellation supporting IoT connectivity.
- More than a dozen localities around the world are endeavoring to become hotspots for quantum development and commerce, including, but limited to, the following: Illinois (Chicago Quantum Exchange®), Colorado (Elevate Quantum), Tennessee (Chattanooga Quantum Collaborative), Maryland (The Quantum Stater), Connecticut (QuantumCT), Massachusetts (Quantum Complex), and Germany (Munich Quantum Valley) [RUS25]. For example, Elevate Quantum (a multi-Western state effort—Colorado, New Mexico, and Wyoming) is a consortium of 120 organizations working to ensure that the region remains the global epicenter for Quacomp; it aims at creating a startup and scale-up ecosystem supporting a workforce of 15,000 workers, with $2 billion in funding for startups and $150 million in revenue by 2030.

As a closing observation, in early 2025 the U.S. Department of Energy announced a $625 million funding initiative to advance the development of Quacomp and Quacom; furthermore, Nvidia announced that on March 20, 2025, it was hosting a first-ever Quantum Day as part of its GTC 2025.

References

[ALB25] R. Albergotti, "Google CEO Talks About the Future of AI, Waymo and Quantum Computing". SEMAFOR, Dec 12, 2024. Available online on 1/8/2025 at https://www.semafor.com/article/12/12/2024/sundar-pichai-on-the-future-of-ai-waymo-and-quantum-computing

[ALI25] Alice & Bob, www.alice-bob.com, 49 Bd du Général Martial Valin, 75015 Paris.

[AZI25] M. Azizirad,"2025: The year to become Quantum-Ready". Jan. 15, 2025. Available online on Jan. 15, 2025 at https://azure.microsoft.com/en-us/blog/quantum/2025/01/14/2025-the-year-to-become-quantum-ready/

[BAT25] N. Batra, M. Eggleston, "Quantum Is More Than Just Computing". Forbes, Jan. 8, 2025. Available online on Jan. 10 at https://www.forbes.com/sites/nokia-industry-40/2025/01/08/quantum-is-more-than-just-computing/

[BRA25] R. Brahambhatt, "Meet cat qubit: The first step towards fault-tolerant quantum computing". interestingengineering.com, Jan. 5, 2025. Available online on Jan. 9, 2025, at https://interestingengineering.com/innovation/cat-qubits-for-fault-tolerant-quantum-computing

[COP24] N. Coppola, R. Lescanne, L. Prost, M. Žesko, "Think Inside the Box: Quantum Computing with Cat Qubits: An Introduction to Useful Quantum Computing by Alice & Bob". December 2024. The text is openly licensed via CC BY-SA 4.0, with credit given to Alice & Bob. Available online on Jan. 9, 2025 at https://alice-bob.com/wp-content/uploads/2024/12/Think-Inside-The-Box-Alice-Bob-Whitepaper.pdf

[GID21] C. Gidney, M. Ekerå. "How to Factor 2048 Bit RSA Integers in 8 Hours Using 20 Million Noisy Qubits". Quantum 5. April 2021, 433. https://doi.org/10.22331/q-2021-04-15-433

[GOO23] Google Quantum AI, "Suppressing Quantum Errors by Scaling a Surface Code Logical Qubit". Nature 614, 676–681 (2023). https://doi.org/10.1038/s41586-022-05434-1.

[HAW24] H. Hawashin, "Multimodal Structure-Aware Quantum Data Processing". 11 Nov 2024. arXiv:2411.04242v3

[IDQ24] IDQ staff, "McKinsey's Latest Quantum Technology Monitor Announces Progress Towards Quantum Advantage". IDQ. May 8, 2024. https://www.idquantique.com/mckinsey-quantum-technology-monitor-april-2024

[LES20] R. Lescanne, M. Villiers, *et al*, "Exponential Suppression of Bit-Flips in a Qubit Encoded in an Oscillator". Nature Physics 16 (2020): 509–513. https://doi.org/10.1038/s41567-020-0824-x

[LES25] K. Leswing, "Nvidia's Jensen Huang is 'Dead Wrong' About Quantum Computers, D-Wave CEO Says". CNBC, Jan 8, 2025. Available online on Jan. 9, 2025 at https://www.cnbc.com/2025/01/08/nvidia-ceo-jensen-huang-is-dead-wrong-about-quantum-d-wave-ceo.html

[NIE11] M. A. Nielsen, I. L. Chuang, *Quantum Computation and Quantum Information (10th anniversary ed.)*. Cambridge: Cambridge University Press. 2011. 9781107002173 and/or 9780511976667. https://doi.org/10.1017/CBO9780511976667

[RIV25] Riverlane Staff, "Quantum Error Correction Report 2024". Riverlane, 1 Broadway Ste 14, Cambridge, MA 02142. Available online on Jan. 9, 2025 at https://www.riverlane.com/quantum-error-correction-report-2024

[RUS25] J. Russell, "Quantum Computing 2025 — Is it Turning the Corner?". HPC Wire. Jan. 1, 2025. Available online on Jan. 11, 2025 at https://www.hpcwire.com/2025/01/01/quantum-computing-2025-is-it-turning-the-corner/

[SEA25] SEALQQ Staff, "Post-Quantum Cryptography Market Set for Accelerated Growth as SEALSQ and WISeKey Lead the Way in Securing Digital Infrastructure". Yahoo! Finance, Jan. 8, 2025. Available online on Jan. 11, 2025 at https://finance.yahoo.com/news/post-quantum-cryptography-market-set-142500489.html

[SNY25] J. Snyder, "Quantum Is 'Decades Away'? Not So Fast". Forbes, Jan 9, 2025. Available online on Jan 9, 2025, at https://www.forbes.com/sites/jasonsnyder/2025/01/09/quantum-is-decades-away-not-so-fast/

[SUB25] S. Subin, "The Crazy Ride In Quantum Computing Stocks Continues As Shares Rip Higher On Microsoft's 'Quantum-Ready' Directive", CNBC.com, Wed, Jan. 15, 2025. Available online on Jan. 15 at https://www.cnbc.com/2025/01/15/quantum-computing-stocks-rip-higher-on-microsofts-quantum-ready-directive.html.

[SWA25] M. Swayne, "Picture This: Quantum Approach Matches Classical AI in Text-Image Tasks". Quantum Insider, Jan. 5, 2025. Available online on Jan. 10, 2025 at https://thequantuminsider.com/2025/01/05/picture-this-quantum-approach-matches-classical-ai-in-text-image-tasks/

[SWA25a] M. Swayne, "Chinese Scientists' Quantum Memory Advance Could Help Scale Quantum Networks". Quantum Insider, Jan. 6, 2025. Available online on Jan. 10 at https://thequantuminsider.com/2025/01/06/chinese-scientists-quantum-memory-advance-could-help-scale-quantum-networks/

[UNE25] UNESCO, "100 Years of Quantum Is Just the Beginning". https://quantum2025.org/

[USM25] M. Usman, "2025 Will See Huge Advances in Quantum Computing. So What is a Quantum Chip and How Does it Work?". Quantum Insider. Jan. 8, 2025. Available online on Jan. 11, 2025 at https://thequantuminsider.com/2025/01/08/2025-will-see-huge-advances-in-quantum-computing-so-what-is-a-quantum-chip-and-how-does-it-work/

Super Glossary

Many of the concepts listed in this Super Glossary are covered in more detail in the text (in the appropriate chapter). These Super Glossary entries only refer to the formal definition of said concepts given in the relevant, chapter-level Glossary or tables. The reader should consult the text for enhanced understanding of the concepts (also in conjunction with the Index).

Term, concept	Where described
Acousto-Optic Modulators (AOMs)	Glos. Ch4
Actors in QKD	Glos. Ch7
Adiabatic algorithm time scaling factor	Glos. Ch5
Adiabatic approximation	Glos. Ch5
Adiabatic process	Glos. Ch5
Adiabatic Quantum Computation (AQC)	Glos. Ch5
Adiabatic quantum computer	Glos. Ch5
Adiabatic Quantum Optimization (AQO)	Glos. Ch5
Adiabatic Theorem (AT)	Glos. Ch5
Adversary-in-the-Middle Attacks	Glos. Ch7
Ancilla	Table 6.1
Automated Data Collection Attacks	Glos. Ch7
Basis	Glos. Ch1
	Table 2.A1
	Table 3.1
BB84 QKD protocol	Glos. Ch7
Bell inequality	Glos. Ch3
Bell State Measurement (BSM)	Glos. Ch1
	Table 3.1
Bell states	Glos. Ch1
	Glos. Ch3
Bell Theorem	Glos. Ch3

Term, concept	Where described
Blind Quantum Computation (BQC)	Table 1.2
Bra vector of $|\psi\rangle$, $\langle\psi|$	Table 2.A3
Bra (Dirac)	Table 2.A1
Bracket Language	Table 8.2
Category	Glos. Ch5
Circuit	Glos. Ch1
Clock synchronization	Table 1.2
Cluster state	Glos. Ch4
	Table 6.1
Coherence	Glos. Ch1
Cold atomic quantum computer	Glos. Ch5
common QASM (cQASM)	Table 8.2
Compatible/incompatible quantum observables/measurements	Glos. Ch3
Complete vector space	Table 2.A1
Complex conjugate c∗ of complex number $c = a \pm bi$	Table 2.A1
Complex vector space	Table 2.A1
Complexity theory	Glos. Ch5
Complexity theory, Boolean expression	Glos. Ch5
Complexity theory, Boolean satisfiability (SAT) problem	Glos. Ch5
Complexity theory, Maximum Satisfiability problem (MAX-SAT)	Glos. Ch5
Complexity theory, NP-complete	Glos. Ch5
Component algorithm	Glos. Ch7
Composite Post-Quantum/Traditional (PQ/T) Hybrid Protocol	Glos. Ch7
Composite systems/entanglement	Table 2.A2
Confidence interval of QC computation	Glos. Ch5
Continuous-Variable (CV)-QKD protocols/schemes	Glos. Ch7
Controlled NOT (CNOT) gate	Table 3.2
Cost of Quacomp, circuit model	Glos. Ch5
Cost of running an algorithm in AQC	Glos. Ch5
Creation operators and annihilation operators	Table 3.1
Cryogenic Solid State based QCs	Table 5.1
Cryptographic application	Glos. Ch7
Cryptographically Relevant Quantum Computer (CRQC)	Glos. Ch7
Decoherence	Glos. Ch1

Term, concept	Where described
Del	Table 2.A3
Demand for resources: "exponentially many resources"	Glos. Ch1
Density matrix	Table 2.A2
	Table 3.1
Density operator ρ	Table 2.A2
Depolarizing channel	Table 3.3
Depolarizing noise for a quantum gate	Table 3.3
Detection efficiency	Glos. Ch 4
Detector signal jitter	Glos. Ch 4
Diagonal matrix	Glos. Ch5
Dimension of a vector space	Table 3.1
Dimension of an inner product space	Table 2.A1
Dirac nomenclature, notation	Table 2.1
	Table 3.1
Discrete-Variable (DV)-QKD protocols/schemes	Glos. Ch7
Distributed quantum computing	Table 1.2
Distributed sensors	Table 1.2
E91 QKD protocol	Glos. Ch7
Eberhard bound	Table 6.1
Eigengap, Gap (of a system)	Glos. Ch5
Eigenstate	Glos. Ch1
	Table 2.A2
Eigenvalue	Glos. Ch1
	Table 2.A2
Eigenvector	Table 2.A2
Eigenvector of operator Op on $|v\rangle$	Table 3.1
Eigenvector of operator Op with eigenvalue a	Table 3.1
Electromagnetically induced Transparency (EIT)	Glos. Ch4
Entangled photon source	Table 6.1
Entanglement	Glos. Ch1
Entanglement distribution	Table 6.1
Entanglement swapping	Table 6.1
Entanglement/separable states	Table 2.A2
Entropy	Glos. Ch1
Etalons	Glos. Ch4

Term, concept	Where described
Expansion coefficients in a basis	Table 3.1
Fabrication of a time-dependent non-unitary approximate unitary transformation circuit	Glos. Ch5
Fidelity	Glos. Ch1
Fock state	Glos. Ch1
Gate	Glos. Ch1
	Table 3.2
General Phase Gate	Table 3.2
General Rotation Gate: U(3) (also represented as U_3)	Table 3.2
General terminology	Table 2.1
Graph state	Glos. Ch1
Ground state of a QM system	Glos. Ch1
Grover's algorithm	Glos. Ch1
Hadamard basis	Table 3.1
Hadamard Gate (H)	Table 3.2
Hamiltonian ($\hat{H}$) quantum operator	Table 2.A3
Hamiltonian $\hat{H}$	Glos. Ch1
Hamiltonian operator $\hat{H}$ (mathematics)	Table 2.A3
Heisenberg's Uncertainty Principle (HUP)	Glos. Ch1
Heralded sources	Glos. Ch4
Heralding	Glos. Ch4
Hermitian operator	Table 2.A3
	Table 3.1
Hermitian conjugate of the operator	Table 2.A3
Hermitian matrix	Table 2.A3
Hilbert space	Table 2.A1
Hybrid quantum applications (hQApps)	Glos. Ch5
Hybrid quantum computing	Glos. Ch5
Hybrid quantum machine learning	Table 1.2
Identity operator	Table 2.A1
Information Teleporting	Table 1.2
Inner product of tensor space	Table 2.A1
Inner product of two vectors—abstract formulation	Table 2.A1
Inner product of two vectors in complex spaces $\langle A \mid B \rangle$	Table 2.A1
Inner product of two vectors in complex spaces $\langle A \mid B \rangle$, comparison with traditional inner product	Table 2.A1

Term, concept	Where described
Inner product space	Table 2.A1
	Table 3.1
Intermediate representation (IR) of a computation	Glos. Ch5
Ions	Glos. Ch1
Ising Model	Glos. Ch5
Key control and management	Glos. Ch7
Key relay	Glos. Ch7
Key storage	Glos. Ch7
Key supply	Glos. Ch7
k-local Hamiltonian	Table 2.A3
k-local Hamiltonian problem	Glos. Ch5
Kraus operators	Table 3.1
Kronecker product	Table 2.A2
ℓ_2-norm	Table 2.A1
Linear Operator Op	Table 2.A1
Linear optical quantum computer	Glos. Ch5
Local realism	Glos. Ch3
Logical qubit	Glos. Ch5
Matrix multiplication	Table 2.A1
Maximally entangled state between two d-dimensional systems A and B	Table 3.1
Mean photon number	Glos. Ch 4
Measurement, Quantum Measurement	Glos. Ch1
Mixed states $|w\rangle$ in a $(A+B)$ aka $(A \otimes B)$ ensemble)	Table 3.1
Model	Table 2.A1
Modular quantum computing	Glos. Ch5
Multi-Algorithm Cryptographic Scheme	Glos. Ch7
Multi-photon signal	Glos. Ch 4
Neutral Atoms QCs	Table 5.1
Nitrogen-vacancy center	Glos. Ch4
No cloning theorem	Glos. Ch1
Nonclassical (quantum) light	Glos. Ch1
	Glos. Ch4
Non-composite Post-Quantum/Traditional Hybrid Protocol	Glos. Ch7
Non-unitary operators/matrices	Glos. Ch1

Term, concept	Where described
Norm	Table 2.A1
Normalized vectors	Table 3.1
NOT Gate, aka bit flip gate, or X (Pauli X)	Table 3.2
Observables	Table 3.1
	Glos. Ch3
Obtaining a ground state	Glos. Ch1
Obtaining a ground state of a quantum system	Glos. Ch5
One-time Pad (OTP)	Glos. Ch7
One-way quantum computer	Glos. Ch5
Open Cirq	Table 8.2
Open QASM	Table 8.2
Operator algebra	Table 3.1
Operator O, Op$|A$)	Table 2.A1
Operator O, Op$|A$), matrix representation	Table 2.A1
Orthogonal vector basis	Table 2.A1
Orthogonality	Table 2.A1
Outer product	Table 2.A2
Pauli Y	Table 3.2
Phase Flip, aka Z (Pauli Z)	Table 3.2
Photonic Quantum Computing QCs	Table 5.1
Physical random number generators	Glos. Ch7
Planck constant	Table 2.A3
Pockels effect	Glos. Ch4
Polarization	Glos. Ch1
Positive operator on a complex Hilbert space	Table 2.A3
Positive Operator Valued Measure (POVM)	Table 3.1
Post-Quantum Cryptographic (PQE) Algorithm	Glos. Ch7
Post-Quantum Cryptography (PQC)	Table 1.2
	Glos. Ch1
Post-Quantum Encryption (PQE)	Glos. Ch7
Post-Quantum/Traditional (PQ/T) Hybrid Scheme	Glos. Ch7
PPLN (Periodically Poled Lithium Niobate)	Glos. Ch4
PQ (Post-Quantum)/PQ Hybrid Scheme	Glos. Ch7
PQ/T Hybrid Combiner	Glos. Ch7

Term, concept	Where described
PQ/T Hybrid Digital Signature	Glos. Ch7
PQ/T Hybrid Key Encapsulation Mechanism (KEM)	Glos. Ch7
PQ/T Hybrid Protocol	Glos. Ch7
PQ/T Hybrid Public Key Encryption (PKE)	Glos. Ch7
Probabilistic outcomes and measurements	Glos. Ch3
Projection-Valued Measure (PVM)	Table 2.A1
Projective Hilbert space	Table 2.A3
ProjectQ	Table 8.2
Protocol	Table 6.1
Pseudo Random Functions (PRFs)	Glos. Ch7
Pseudorandom number generator	Glos. Ch7
Pure state	Table 2.A2 Table 3.1
Purification and error correction	Glos. Ch1 Table 6.1
Pytket	Table 8.2
Q factor (quality factor) of a resonator	Glos. Ch4
Q# Language	Table 8.2
Qiskit Language	Table 8.2
QKD operation	Glos. Ch7
QKDN controller	Glos. Ch7
QKDN mechanisms	Glos. Ch7
QM intrinsics	Table 3.1
Quantum advantage	Glos. Ch5
Quantum algorithm	Glos. Ch1
Quantum annealer	Glos. Ch5
Quantum applications	Table 1.2
Quantum blockchains	Table 1.2
Quantum Channel	Table 6.1
Quantum chip	Glos. Ch5
Quantum circuit	Glos. Ch1 Glos. Ch5
Quantum clock	Glos. Ch5
Quantum coding	Glos. Ch5

Term, concept	Where described
Quantum communications (Quacom)	Glos. Ch1 Table 6.1
Quantum computation	Glos. Ch1 Glos. Ch5
Quantum computer (QC)	Glos. Ch1
Quantum Computer Resistant Encryption (QRE)	Glos. Ch7
Quantum Computer Safe Encryption (QSE)	Glos. Ch7
Quantum computing (Quacomp)	Glos. Ch1
Quantum concurrence	Table 6.1
Quantum diagonal control	Glos. Ch5
Quantum dot (QD)	Glos. Ch4
Quantum dynamic network server load management	Table 1.2
Quantum entanglement	Table 6.1
Quantum Error Correction (QEC)	Glos. Ch1 Glos. Ch 5 Table 6.1
Quantum frequency transduction	Glos. Ch4
Quantum gate	Glos. Ch1 Table 2.A2 Glos. Ch5
Quantum gate array	Glos. Ch5
Quantum homomorphic encryption	Glos. Ch7
Quantum information	Glos. Ch1
Quantum Information Network (QIN)	Glos. Ch1
Quantum Information Technology (QIT)	Glos. Ch1
Quantum Interconnects (QuICs)	Glos. Ch1
Quantum Internet (Quaint)	Glos. Ch1
Quantum Key Distribution (QKD)	Table 1.2 Glos. Ch1 Glos. Ch7
Quantum Key Distribution Network (QKDN)	Glos. Ch7
Quantum Key Distribution Network (QKDN) manager	Glos. Ch7
Quantum measurement	Table 3.1
Quantum memory	Glos. Ch1 Table 6.1

Term, concept	Where described
Quantum Merlin Arthur (QMA)	Glos. Ch5
Quantum metrology	Table 1.2
Quantum network (Quanet)	Glos. Ch1
Quantum neural network	Glos. Ch1
Quantum node	Table 6.1
Quantum nondemolition measurements	Table 6.1
Quantum operations	Glos. Ch1
Quantum operator	Glos. Ch1
Quantum photon source	Table 6.1
Quantum processor	Glos. Ch5
Quantum program	Glos. Ch5
Quantum protocol	Table 6.1
Quantum Purification	Glos. Ch1
Quantum Random Number Generator (QRNG)	Glos. Ch7
Quantum repeater (Quarep, aka QR)	Glos. Ch1 Table 6.1
Quantum router (Quarou)	Table 6.1
Quantum secret sharing	Glos. Ch7
Quantum simulation	Glos. Ch1
Quantum simulator	Glos. Ch1
Quantum software	Glos. Ch1 Table 8.2
Quantum state	Glos. Ch1
Quantum storage	Table 6.1
Quantum Switch (QS)	Glos. Ch1 Table 6.1
Quantum teleportation	Table 6.1
Quantum transducer	Glos. Ch1 Table 6.1
Quantum tunneling	Glos. Ch1
Quantum-supported Natural Language Processing (NLP)	Table 1.2
Qubit (qbit, quantum bit)	Glos. Ch1
Qudit	Glos. Ch1
Qudit-based computation	Glos. Ch5
Random number generator	Glos. Ch7

Term, concept	Where described
Repeaterless bound	Table 6.1
Representation of vectors, matrix	Table 2.A1
Representation of vectors, wave function	Table 2.A1
Resistant One-Way Functions (ROWFs)	Glos. Ch7
Room-temperature Solid-state based QCs	Table 5.1
Rotation	Table 2.A1
Rotation Gate	Table 3.2
Rydberg atom	Glos. Ch4
Scalar multiplication of a vector	Table 2.A1
Schrödinger equation	Table 2.A3
Scientific and commercial applications	Table 1.2
Secure access to remote quantum computers	Table 1.2
Secure identification	Table 1.2
Secure One-Way Function (SOWF)	Glos. Ch7
Semiconductor quantum computer	Glos. Ch5
Separable states; entangled states	Table 3.1
Shor's algorithm	Glos. Ch1
Single photon	Glos. Ch1
Single photon source	Glos. Ch4
Single-algorithm cryptographic scheme versus Multi-algorithm cryptographic scheme	Glos. Ch7
Single-photon detector	Glos. Ch4
Source timing jitter	Glos. Ch 4
Spin operator gates	Table 3.2
Spin-photon entanglement	Glos. Ch 4
Spontaneous Parametric Down Conversion (SPDC)	Glos. Ch4
Squeezed light	Glos. Ch4
Squeezed state	Glos. Ch1
Stabilizer code	Glos. Ch5
States, superposition	Glos. Ch3
State-space of a composite system	Table 2.A1
State-space of a system	Table 2.A1 Table 3.1
Strawberry Fields and PennyLane	Table 8.2
Structure, general	Table 2.A1

Term, concept	Where described
Structure, mathematical	Table 2.A1
Superconducting quantum computer	Glos. Ch5
Superdense coding (SD)	Table 1.2
	Glos. Ch1
Superposition	Glos. Ch1
SWAP states gate	Table 3.2
Syndrome	Glos. Ch5
Syndrome measurements	Table 6.1
Systems interaction, composite system	Table 2.A2
Telecom heralding	Table 6.1
Teleportation	Table 1.2
	Glos. Ch1
Tensor	Table 2.A2
Tensor product $V \otimes U$ of two vector spaces V and U	Table 2.A2
Tensored gates (local gates)	Table 3.2
Three-dimensional (3D) networks	Glos. Ch1
Time-bandwidth product	Glos. Ch4
Time-dependent Schrödinger equation	Glos. Ch5
Time-evolution of quantum sate	Table 2.A3
Toffoli gate	Table 3.2
Trace of a matrix M	Table 2.A1
Traditional (Classical) Cryptographic Algorithm	Glos. Ch7
Transmon	Glos. Ch4
Transposed vector	Table 2.A1
Trapped Ion Quantum Computing (TIQC)	Glos. Ch4
Trapped Ions QCs	Table 5.1
Unit vector	Table 2.A1
Unitary matrix	Glos. Ch1
	Table 2.A1
Unitary operation gates	Table 3.2
Unitary operator U	Glos. Ch1
	Table 2.A1
Unitary operator, Time-evolution	Table 2.A3
Universal set of gates	Glos. Ch5
Value for observable	Table 2.A2

Term, concept	Where described
Variational Quantum Eigensolver (VQE)	Glos. Ch1
	Glos. Ch5
Vector	Table 2.A1
Vector addition	Table 2.A1
Vector basis	Table 2.A1
Vector length	Table 2.A1
Vector space	Table 2.A1
	Table 3.1
Voting using a Quanet	Table 1.2
Walltime	Glos. Ch1
Wave function	Glos. Ch1
	Table 3.1
Wave-particle duality	Glos. Ch1
Whispering Gallery Modes (WGMs) Microresonators (WGMRs)	Glos. Ch4
XACC	Table 8.2
Zero operator	Table 2.A1

Index

For Product Safety Concerns and Information please contact our EU
representative GPSR@taylorandfrancis.com
Taylor & Francis Verlag GmbH, Kaufingerstraße 24, 80331 München, Germany

www.ingramcontent.com/pod-product-compliance
Lightning Source LLC
Chambersburg PA
CBHW080702220326
41598CB00033B/5286